50 Years
K·G·Saur

History and Bibliography
1949 – 1999

Edited by Klaus G. Saur

History by Titus Arnu

Bibliography by
Andreas Brandmair and Konrad Kratzsch

K·G·Saur München 1999

Editorial Deadline: May 10, 1999

Die Deutsche Bibliothek - CIP-Einheitsaufnahme

K·G·Saur Verlag <München> :
50 years K·G·Saur : history and bibliography
1949 - 1999 / edit. by Klaus G. Saur.
Bibliography by Andreas Brandmair and Konrad Kratzsch.
History by Titus Arnu.
München : Saur, 1999
Dt.Ausg. u.d.T.: K·G·Saur Verlag <München>:
K·G·Saur Verlag - 50 Jahre

ISBN 3-598-10705-6

K·G·Saur Verlag GmbH & Co. KG, München 1999
Part of Reed Elsevier

Printed in the Federal Republic of Germany

Typesetting:
Buch & Grafik Design, Günther Herdin GmbH, München (History)
Michael Peschke, Berlin (Bibliography)

Printing:
Strauss Offsetdruck, Mörlenbach

Binding:
Buchbinderei Schaumann, Darmstadt

Expression of Thanks

We celebrate 50 years K.G. Saur – after a thorough check, this assertion only stands the test to a certain degree. In 1948 my father, Karl-Otto Saur founded an engineering office in Munich, on the site where the German Patent Authority is located today. By chance the office was commissioned to compile literature on the subject of "hydraulics" by a Lower Bavarian public authority. This commission was the incentive to deal with patent and literature information, and the first step towards the publishing of international reference works that have been able to spread the world over.

50 years are an occasion to take stock and to express thanks. In the first 20 years of its history the publishing house virtually only brought out publications edited by the editorial departments within the company itself or by individual freelance editors: Literature compilations, bibliographies, catalogs and indices which have been published in book and periodical form. To the present day publications such as *Publishers' International ISBN-Directory, World Guide to Libraries, International Directory of Arts, International Books in Print, Guide to Microforms in Print*, the *Allgemeines Künstlerlexikon* the *Deutsches Literaturlexikon* (Encyclopedia of German Literature) or the newly acquired *Kürschners Deutscher Literaturkalender* (Almanac of German Literature) and also the *Biographical Archives*, are projects which have been edited and managed by the editorial staff of our house. Therefore I would like to first give thanks to all of the colleagues, who have done such a good job over the past 50 years in order to create many thousands of publications.

In 1970 we were able to make a worldwide distribution agreement with the Börsenverein des Deutschen Buchhandels (Association of the German Booksellers and Publishers). One of the most important publishing decisions to this day is hidden behind these dry words. The Buchhändler-Vereinigung GmbH, the publishing arm of the Börsenverein, granted K.G. Saur, at the time called Verlag Dokumentation, the exclusive distribution rights, but also the distribution duties, for the *Verzeichnis lieferbarer Bücher (VLB)* (German Books in Print).

At first a print-run of 2,000 copies was foreseen – we were able to offer a guaranteed purchase of 6,000 copies which cut the fee per line for the entries by more than half and decisively contributed towards the success of this work. The *VLB* was distributed worldwide and the publishing house developed or acquired similar *books in print* such as *Guide to Microforms in Print, International Books in Print, Guide to Reprints*, etc.

Also in 1970 our contacts began with the IFLA International Federation of Library Associations and Institutions, which entrusted us with the edition and distribution of all of its publications starting in 1971. Agreements were to follow with numerous other international institutions and organizations such as the International Council on Archives, the International Council on Museums, etc. Our express thanks goes out to all of them.

In 1975 the Lower Saxon librarian Dr. Reinhard Oberschelp laid a novel idea on my desk: It was the concept for a *Gesamtverzeichnis des deutschsprachigen Schrifttums (GV 1700 to 1965)* (Bibliography of German-Language Publications), in approx. 300 volumes. The cue was "reprocumulation". In total 14,000 different alphabetical orders for two time periods, 1700–1910 and 1911–1965, had to be brought together into one system and one alphabet. The publishing company, still young and most importantly short of capital, was facing the largest financial challenge of its history. The result of my opinion polls among a few leading library directors lead to negative prognoses. The answer was that one was happy with the bibliographical situation as it was. One would not purchase a new bibliography at a retail price of c. 25,000 DM they said. In the end it was the trust of the President of Otto Harrassowitz the leading German Library supplier, Richard W. Dorn – who promised to purchase 100 copies and sell them – that lead to the decision to carry on with this risky project. Exactly 10 years after Richard W. Dorn's confirmation the bookseller Otto Harrassowitz ordered the 100th copy.

The cooperation with the British Library began in 1978 with the publication of the *British Library Catalogue of Printed Books to 1975*, which at first appeared in 360 volumes. In the meanwhile we have published numerous additional editions in cooperation with the British Library and are preparing new ones. Next year the number of volumes that we have published together with the British Library will surpass the mark of one thousand.

The next great moment came in 1980. In the middle of a crisis threatening the very existence of the company, which was caused by expanding too quickly, the idea arose to reproduce the biographical reference works from the 17th to the 19th centuries on microfiche. On the basis of, and arising out of this basic idea was the concept of building up a Biographical Archive, this being in one alphabetical order. Whilst working on this the experiences of the reprocumulation of specialist journals and books in the 50's and the experiences gained while working on the *GV* could be called upon. The *German Biographical Archive* was the start of an *International Biographical Information System* which is nearing its completion in the 50th year of the publishing house.

In 1981 we made an agreement with the International Union of Associations in Brussels for numerous publications, most of all for the *Yearbook of International Organizations*.

In 1983 the cooperation with the Library of Congress began as well the filming of the *Main Catalog of the Library of Congress* with 27 mil. library cards on microfiche.

From 1949 to the present day 5,035 titles have been published in a total of 9,291 volumes. 194 microfiche editions with 296,000 microfiches and 305 CD-ROM editions have come out. In total 2.5 million pictures and 31 mil. library cards have been recorded on microfiche, 24 mil. library cards on CD-ROM, 41 mil. pages of text on microfiche and 3.9 pages of text on paper.

Felix Dahn, one of most successful authors in history, determined already many years ago that it was relatively easy to write a book. It was also not unusually difficult to produce it. But the highest art of all laid in the corresponding distribution of this book as well. This goes in an especially large way for our publications. They can only be successful if they are known by every specialist in the areas we serve in the whole world, and when ways are found to sell them to libraries and institutions worldwide. That this really has become possible has to do with the unusually committed effort of numerous colleagues in the marketing and sales departments of the company, and naturally in a very enormous way in the cooperative work with our excellent representatives in the USA, in Great Britain, France, Italy, Spain, Australia, Japan and other parts of the world.

My thanks also go out to all booksellers, with whom we have worked closely and trustfully, for the most part for years and decades.

My very special thanks go to all of those in the history and bibliography of the publishing house, mentioned and non-mentioned authors, editors, compilers and institutions. Only by way of their trust and readiness to work in cooperation has the successful development been possible.

My thanks similarly go to the librarians who have purchased and put our publications to use. The nicest reward for a publishing house is fundamentally the distribution and use of its books. Our outstanding position is proven by the fact that it is impossible to visit any university or national library in the world without coming across the titles of K.G. Saur Publishing. The librarians have kindly forgiven the one or another mistake with our publications and have time and again shown patience during the delay of publication dates.

The worldwide changes in the information paths provide us with continuously new challenges and problems, and it is becoming ever more difficult to satisfy the classical task of the publishing company, namely providing scientific institutions of the world with information.

The connection with Reed Elsevier and the close cooperation with our sister companies, Butterworth and R.R. Bowker, offer the best starting basis for being able to fulfill this task in the future as well.

Klaus G. Saur

Table of Contents

History

Bibliography (1949 to May 1999)

Abbreviations

Laudatory speech

K.G. Saur Publishing: We thought that the 40th anniversary was yesterday, it is already 10 years ago. A half century of tradition and innovation at the same time, of sponsoring and of Cupertino with IFLA. A wide tradition of publishing not only classical bibliographies like the Gesamtverzeichnis des deutschsprachigen Schrifttums, but also for technical and professional publications.

Well known to the librarians throughout the world, K.G. Saur's books in a green or orange cover are to be found wherever a library needs technical information, and international approaches to library management of all kind. But Saur's CD-ROMs and microfiches show that the publishers of today know how to use new technologies for an old purpose, i.e. publishing. Bibliographies, catalogues and indexes of all kinds are available at Saur's. But also periodicals and serials: all the IFLA's publications are there, including IFLA Journal and IFLA's Annual Council Report.

But this in only one facet of Mr. Saur's activities: he also likes to help to the development of his profession, works with the International Publishers Association, is to be seen at the major Book Fairs, and specially in Frankfurt, and sponsors very generously IFLA for its Conference and its social events.

In the XVIth Century, this would have thought of as quite natural from such a "gentleman-publisher"; it is less so in the XXth Century, and indeed what will the XXIst Century feel regarding the development of the intricate relationship between the mere printing activity, all the other document-producing, or object-producing activities, and all the marketing and sponsoring activities?

IFLA believes that it can trust such people as Mr. Saur to see his profession as a global one, and one working together with all the associated professions. He has already taken the right turning, and is working in the good direction. Saur's people are albeit used to working in co-operation at a time where it is absolutely essential. They will therefore know how to work for a productive future under the guidance of Professor Dr. Saur. We at IFLA hope to accompany them along this road, for a better future of the book professions, working alongside in the same direction.

Christine Deschamps

President of the International Federation of Library Associations (IFLA)

Chronicle

1948

Karl-Otto Saur sr. and Margarethe Gringmuth found an engineering office for business and office organization. The office is opened on Erhardtstraße in Munich, across from the Deutsches Museum.

1949

The first commission comes from a Lower Bavarian public authority: Saur's office is to compile literature on the subject of hydraulics.

1950

Begin of the literature research service.

1951

Focus on technical literature information published monthly.
The *Internationaler Literaturdienst* (International Literature Service) appears in its first specialized editions.

1952

First *Jahrbuch der Technischen Dokumentation und Bibliographie* (Yearbook of Technical Documentation and Bibliography).

1954

Relocation to Rosenheimer Straße 2 in Munich. The company changes its name to "Dokumentationen der Technik". First edition of the *Indexbücher der Technik* (Index of Technical Books). The international literature service comprises up to 30,000 records monthly.

1957

The first two *Handbücher der Technischen Dokumentation und Bibliographie* (Handbooks of Technical Documentation and Bibliography) are published.

1958

Concentration of business focus on monographic records.

1959

Publishing start of the *Universalbibliographie Technik und Wirtschaft (UTW)* (Universal Bibliography of Technics and Business) in 37 specialized editions.

1960

First *Generalkatalog der Fachzeitschriften* (General Catalogue of Professional Journals) in 49 specialized editions.

First edition of the international bibliography *Technik und Wirtschaft in fremden Sprachen* (Technics and Business in Foreign Languages).

1963

Relocation from Munich to Pullach, Jaiserstraße 13. Klaus Gerhard Saur joins the publishing company.

1964

First edition of the *Internationales Verlagsadressbuch* (1979 renamed the *Publishers' International Directory with ISBN-Index*).

1966

Conversion of the company to Verlag Dokumentation Saur o.H.G. with the personally liable partners Klaus G. Saur and Karl-Otto Saur jr. First license edition from the VEB Bibliographisches Institut Leipzig for distribution in the Federal Republic of Germany, Austria and Switzerland.

Death of the company founder Karl-Otto Saur sr.

1969

Karl-Otto Saur jr. leaves the company. Dr. h.c. Thomas Karger, Basel enters as a silent partner. Transformation of the general partnership into a limited partnership.

1970

Contractual agreement with the Buchhändler-Vereinigung GmbH in Frankfurt/Main for worldwide distribution of the *Verzeichnis lieferbarer Bücher (VLB)* (German Books in Print) as of 1971.

1971

Contractual agreement with the International Federation of Library Associations (IFLA) to publish the book and journal program.

1972

Daniel Melcher of Charlottesville, VA, founder of the first Books in Print worlwide in 1946, joins the publishing house as a silent partner.

1975

Relocation from Pullach to Munich-Solln on Pössenbacherstraße 2b.

1976

The first volumes of the *Gesamtverzeichnis des deutschsprachigen Schrifttums 1911–1965 (GV)* (Bibliography of German-Language Publications 1911–1965) are published.

1977

First installment of the *Marburger Index*, art documentation on microfiche. Founding of K.G. Saur Inc., New York.

Founding of Minerva Publikation Saur GmbH.

1978

Change of the company's name to K.G. Saur Verlag. Acquisition of the publishing house Clive Bingley and founding of K.G. Saur Publishing Ltd., London. Founding of K.G. Saur Editeur S.A.R.L., Paris.

1979

Start of the *British Library General Catalogue of Printed Books*.

1980

Financial crisis in Munich, New York and Paris. Sale of the publishing house Clive Bingley to the Library Association in London. The office in New York is closed. Acquisition of Hans Zell Publishers Oxford.

Thomas Karger quits silent partnership.

1981

Die Akten der Partei-Kanzlei der NSDAP (Records of the Party Chancellery of the National Socialist Party) are published.
Completion of the *Gesamtverzeichnis des deutschsprachigen Schrifttums 1911 bis 1965* (Bibliography of German-Language Publications 1911 to 1965) in 150 volumes. Agreement with the Union of International Associations in Brussels on the publication of the *Yearbook of International Organizations*.

1982

Start of the *German Biographical Archive*.

1983

Refounding of K.G. Saur Inc., New York.

1984

Start of publication of the *Main Catalog of the Library of Congress* with a total of more than 26 million cards. Begin of the *British Biographical Archive*, the *Italian Biographical Archive* and the *American Biographical Archive*.

1985

Klaus G. Saur is granted the honorary doctorate from Philipps-Universität Marburg.

1986

Contractual agreement with the Vatican library for publication of the *Bibliotheca Palatina* on microfiche. Opening of an office in the Vatican library.

Relocation to Heilmannstraße 17 in Munich-Solln.

1987

Sale of the publishing house to Reed International P.L.C., London.

Start of a close cooperative working relationship with Butterworth, London and R.R. Bowker, New York.

1988

The *VLB* (German Books in Print) comes out for the first time in CD-ROM form.

1989

Acquisition of Francke AG Bern with its *Deutsches Literatur-Lexikon* (Encyclopedia of German Literature), *Großes Sängerlexikon* (Biographical Dictionary of Singers) and further reference works.

1990

Acquisition of Art Address, Frankfurt with its *International Directory of Arts*.

1991

Relocation to Ortlerstraße 8 in Munich-Sendling.

Takeover of Meckler Publishing, New Haven (*Microforms in Print, Microform Review*, political science and sociology publications).

1996

Klaus G. Saur is made a Visiting Honorary Professor of the University of Glasgow.

1995

The first volume of the *Deutsche Biographische Enzyklopädie* (Dictionary of German National Biography) is published.

1997

Acquisition of the journals *Libri* and *Restaurator* from Munksgaard Publishing, Copenhagen.

1998

The 61st edition of *Kürschners Deutscher Literaturkalender* (Kürschner's German Literature Almanac) is published for the first time under K.G. Saur's aegide.

The *World Biographical Index* is offered on the Internet.

1999

Completion of the *Deutsche Biographische Enzyklopädie* (Dictionary of German National Biography) with the publication of volume 10.

Start of the *Japanese* and the *Korean Biographical Archive* to complete the *International Biographical Information System*.

History

1949 – 1999

Introduction

"Facts, Data, Information" – this could be the main motto of the K.G. Saur Publishing house. This being due to the company's focus always remaining in the sector of information and reference: millions of alphabetically listed names, hundreds of thousands of book titles, year numbers, page numbers, addresses. As a documentation and information publisher K.G. Saur is an important transport-vehicle for facts. Information is the important raw material, from which the publisher forms books, catalogues, microfiche, CD-ROMs or on-line publications.

In order to make clear how meticulous and extensive K.G. Saur Publishing deals with information, here are just a few chosen figures: the *Allgemeines Künstlerlexikon* (World Biographical Dictionary of Artists) stretches out over 500,000 entries. The *Deutsche Biographische Enzyklopädie* (Dictionary of German National Biography) lists articles on over 60,000 important people from the German language area. The *American Biographical Archive I* is made up of 1,842 microfiches and contains information on over 300,000 people, and the *Main Catalog of the Library of Congress*, filmed by K.G. Saur Publishing, is made up of over 26 million cards.

The fact is, however as well, that the facts-conscious K.G. Saur Publishing company cannot prove its exact date of foundation. The reason is not due to forgetfulness (that would hardly fit in with the K.G. Saur reputation), but rather because of the fact that Karl-Otto Saur sr. essentially never intended to found a publishing house. He opened an engineering office in Munich in 1948 which in the beginning was committed not to books but rather to "business- and office organization".

One could judge the following note from Karl-Otto Saur sr. as the begin of K.G. Saur Publishing: "11th of March,

1949: Closing of the contract with the Amt für Landeskunde, Landshut, (a Lower Bavarian public authority) for the compilation of literature research". Saur was commissioned more or less by chance by the Lower Bavarian authority to compile literature on the subject of hydraulics. This led the engineer to concern himself intensively with the scientific literature from the fields of technics and business.

The publishing house is, in other words, built on water and paper, a positively solid foundation, as later years showed. That from this mini-company, which found its main theme rather unintentionally 50 years ago – the documentation of scientific literature – would one day become an institution in scholarly life, was however hardly foreseeable at the beginning of the 1950's. The edition of the first publication, exactly that documentation on the subject of hydraulics, had a print run of one copy. In the meanwhile the publisher has 2,000 titles available. K.G. Saur Publishing has today attained a central function in libraries, universities and bookshops worldwide. "Research would break down without handbooks, archive collections, on the cover of which is written: K.G. Saur Verlag, München London, New York, Paris", as was stated by Rolf Michaelis in the German newsweekly *Die Zeit*. There is perhaps virtually not a single librarian in the world who hasn't heard of the name K.G. Saur. And when one enters "K.G. Saur" in an Internet search engine, one receives 1,790 entries.

The publisher is internationally successful and has a good reputation, not only in scientific circles. K.G. Saur Publishing is "known by way of its glowing merits surrounding the area of library science in total", praised Hans Wollschläger in the German daily newspaper *Süddeutsche Zeitung*. K.G. Saur Publishing achieved international renown especially due to its publications on politics and con-

Title pages show the ten-year development of an idea: In 1954 the first Taschenbuch der Dokumentationen der Technik *(Handbook of Documentations of Technics) (left) came out. In 1964 the third newly revised edition of the* Handbuch der technischen Dokumentation *(Handbook of Technical Documentation) (right) was published.*

temporary German history. One example is the publication of the NSDAP-Akten (Records of the Party Chancellery of the National Socialist Party of German), of the *International Biographical Dictionary of Central Euopean Emigrés 1933–1945* or the diaries of Josef Goebbels. Especially the publications on the Nazi period of a specialized nature found great resonance in the media; among others *Stern, Der Spiegel, Die Zeit* and the *Frankfurter Allgemeine Zeitung*, as well as the *New York Times* and the *Times* in London have repeatedly reported extensively on K.G. Saur Publishing.

Mr. Saur himself does an important part of the public relations work. Through his classes and guest lectures held regularly at library schools and universities Klaus G. Saur has played a decisive role in the view of the publishing house in the public eye. "A publishing company is like a

February, 1992: The President of the Ludwig-Maximilians-Universität in Munich, Prof. Dr. Wulf Steinmann (left) presents to Klaus G. Saur (right) a diploma and medal during the bestowal of an honorary Senator title.

tree,that continuously develops new branches", says Klaus G. Saur.

To count the annual rings of K.G. Saur Publishing is a relatively simple thing. There are 50. It becomes more difficult when the task involves sketching the tree with all of its branches true to scale. This documentation of the history of K.G. Saur Publishing cannot and is not meant to be as watertight and extensive as its works. A complete documentation of the 50-year-old firm would slightly overwhelm the reader (and the author), in terms of time but also physically. Who would drag home a commemorative anniversary publication several thousand pages thick? And then read it as well?

The selected main points here are only supposed to give a general overview of the work of the publishing house. In order to stay within the 'tree'-metaphor we are limiting ourselves to the trunk, the thickest branches and some singular leaves. This history may serve the employees and friends of the publishing house to remember the most important activities of the firm. And the tree is turning a proud 50 years old, it is finally high time to describe, to laud and praise the growth of the trunk, branches and fruits.

The Early History of the Publishing House from 1949 to 1963

Karl-Otto Saur sr., from 1940 to 1945 director of the main authority for technics in the Nazi-German defense ministry, was imprisoned for three years following the war. After having testified as a witness for the American prosecutors at a trial against the company Krupp, Karl-Otto Saur was released from confinement in 1948. Because Saur was sentenced to a ban from his profession he founded a consultancy firm together with his secretary of many years, Margarethe Gringmuth, which was called "Engineering Office for Business and Office Organization". He opened the office in the rebuilt and patched up "Isarpassage" on Erhardtstraße in Munich, across from the Deutsches Museum.

When Karl-Otto Saur sr. was commissioned to produce a compilation of literature on the subject of hydraulics in 1949, he found out that there was no comprehensive documentation for technical literature. He subsequently developed a system that classified and utilized more than 10,000 journals. Such technical-specialized magazines were available, for example, via the library of the Deutsches Museum and the German Patent Office. The system of classification was made up of 50 main groups and 1,400 sub-groups, which were then used to be able to sort the essays, dissertations and book publications according to subjects such as "Energy", "Railroads", "Shipping", "Waste Water" or "Storage Dam". In total around two million titles were compiled in the time period from 1950 to 1958. The 50's were marked by the extreme difficulties during rebuilding after World War II. Both in 1953 and 1958 it looked as if the young publishing company was to already face the end of its possibilities. At the end of the 50's the firm, which still had 32 employees up until the middle of the 1950's, had to reduce its personnel to five employees due to economic difficulties.

On the 1st of January, 1959 Karl-Otto Saur sr. published the first edition of the *Universalbibliographie Technik und Wirtschaft (UTW)* (Universal Bibliography of Technics and Business). All book titles and dissertations relevant to the fields were listed in 37 specialized editions. For the universal bibliography Saur turned to the process of reprocumula-

Christmas, 1958: Annual work outing to the Residenztheater (at far left: Karl-Otto Saur sr.)

tion. The titles were photographed from bibliographies, cut out, pasted and subsequently converted into a photoprint copy, from which reproduction copies were produced.

However, none of these first publications were commercially successful. Klaus G. Saur, who completed a trainee program for the book trade and publishing in Northrhine-Westfalia from 1959 to 1963, entered the publishing house in 1963 as authorized representative and started the first edition of the *Internationales Verlagsadressbuch* in 1964. It became the company's first successful publication: the starting print-run of 750 copies was sold out after only a few weeks. From the proceeds the firm bought its first electric typewriter, used of course. With this typewriter work on the second edition was begun immediately. In 1999 the 26th edition of the *Publishers' International ISBN Directory / Internationales Verlagsadressbuch* will be published in 3 volumes, as well as on CD-ROM and online.

The Development of the Publishing House from 1963 to 1999

The success of the *Internationales Verlagsadressbuch* influenced the further development of the company quite decidedly. Bibliographies, reference works and documentations, which filled a market niche and were designed for an international audience, rose to become the most important area.

Following the *Internationales Verlagsadressbuch* came the *Internationales Bibliotheksadressbuch / World Guide to Libraries* in 1966. In this field as well there was no comparable index on the market. The *World Guide to Libraries* came to be a respectable international success. From then on the term *World Guide to ...* would be copied more than one hundred times by other publishing companies.

Both handbooks were such a success that Klaus G. Saur determinately followed the path of the tried and true. In 1970 the publishing agency in Germany of the UNESCO in Paris was conferred upon him. As the official publisher K.G. Saur Publishing has been bringing out the publications of the International Federation of Library Associations (IFLA) since 1971. Saur closed an agreement of similar strategic importance in 1975 with the International Council on Archives (ICA).

With the *Verzeichnis lieferbarer Bücher* (VLB) (German Books in Print) Klaus G. Saur started a project with the Buchhändler-Vereinigung in 1970 which has become an absolutely essential working tool for booksellers who simply couldn't imagine doing without it. The distribution of the VLB was the initial spark that set off a series of further large projects. In 1975 Saur developed the plan to bring out a *Gesamtverzeichnis des deutschsprachigen Schrifttums* (GV) (Bibliography of German-Language Publications) from 1700 to 1965. Despite early skepticism from numerous library directors, the mammoth undertaking in 311 volume was a success. Today the GV is considered a standard work for scientific operations.

Projects were developed of similarly gigantic dimensions on international levels as well. In 1979 the company began publishing the *British Library General Catalogue of Printed Books to 1975* in a total of 360 volumes. In 1984 the publishing house dared to take on the catalog of the most important public library in the world: the Library of Congress in Washington. Over a period of three years K.G. Saur Publishing filmed the main catalog of the library with 7,5 million book titles from the years 1898 to 1980. The result was a microfiche-edition with 9,000 fiches and 2.6 million cards.

K.G. Saur Publishing works intensively together with virtually all of the most important university and state libraries, both foreign and domestic. This does not just have to do with the sales of publications, but rather in terms of contents and editorial cooperative work. Among others K.G. Saur Publishers cooperates with Harvard University, Bibliothèque Nationale in Paris and the Bodleian Library in Oxford.

Time and again it occurs that further branches develop from the main trunks of the company, and they usually bear fruit.

Klaus G. Saur formed new projects out of the successful fundamental principles which seemed even more extensive and time consuming than the previous ones, but have proven to be no less successful. Here is one example: from the project *Die Deutschen Literaturzeitschriften 1815 bis 1850* (The German Literature Journals 1815 to 1850). In cooperation with the Deutsches Literaturarchiv in Marbach, it was possible to develop further publications *Die deutschen Literaturzeitschriften 1850 bis 1945* (The German Literature Journals 1850 to 1945) and the *Inhaltserschliessung deutscher Literaturzeitschriften 1815 bis 1880* (Index of the Contents of the German Literature Journals 1815 to 1880) in twelve volumes.

From 1964 on the *Internationales Verlagsadressbuch* turned into the *Publishers' International ISBN-Directory*. Shortly thereafter the first editions of the handbooks *World Guide to Libraries, World Guide to Scientific Associations, World Guide to Trade Associations* and the *Music Publishers' International ISMN-Directory* came out. The technical and organizational experiences developed by K.G. Saur Publishing during its large-scale projects were beneficial factors for subsequent projects. The *Marburger Index*, a photographic documentation on German art history appearing with K.G. Saur Publishing since 1976, provides another example. The company was able to use the *Marburger Index* to develop a series of European art documentation with the Bildarchiv – Foto Marburg. Some examples are the *Index photographique de l'art en France*, the *Index der Antiken Kunst und Architektur / Index of Ancient Art and Architecture*, the *Italien-Index* (Italian Index), the *Ägypten-Index* (Egyptian Index), the *Spanien- und Portugal –Index*.

The business development was similar for the most economically important product of the company to date. In the spring of 1980 the idea arose to create a *German Biographical Archive*, a cumulation of all biographical reference works on German personalities ever published. It contains

480,000 biographies on 250,000 individuals in one single alphabetical order by name. According to the same principle Biographical Archives were completed for North America, France, Great Britain and all other major countries. Over the years this project, already gigantic in dimension, turned into a unique plethora of information: the *International Biographical Information System*. It enables access to eight millions articles. The most large-scale project also promises to bring in the richest yields. The "World Archives" are the foliage of the publishing tree so to speak.

The Economic Development

The beginning of the economic history of the house, insofar as this was documented at all, was not exactly promising. On the 21st of February, 1949 Karl-Otto Saur sr. wrote a letter to a friend of the family in which he expressed his thanks for a loan of 500 marks. On the 16th of September, 1949 the daily proceeds from the Electronics Fair are proudly recorded as being DM 61,30. When the 22-year-old Klaus G. Saur joined his fathers publishing company after his booktrade apprenticeship in 1963, the firms turnover was just DM 127,000 per year. In 1966 as well, when he took over the reigns of the company, the future of the business did not look especially rosy at first.

Nevertheless, Klaus G. Saur discovered a market niche: the library, booktrade and archive businesses. Other German publishers had neglected these areas at the time. In 1964 the first *Internationales Verlagsadressbuch* came out in a small operation in Pullach, in 1966 the *World Guide to Libraries* joined the program. Both books, under Klaus G. Saur's personal editorial auspices, were so successful that he decided to continue to take that same road.

Professor Klaus-Dieter Lehmann, at that time General

Frankfurt Book Fair, 1959: Karl-Otto Saur sr. (middle) and Klaus G. Saur (at right) in a talk with Mr. Nahm of the bookseller Herder, Barcelona.

Director of the Deutsche Bibliothek, acknowledged Klaus G. Saur on the occasion of the 40 year anniversary of his publishing house for his "publishing strategy, which combined a continuously innovative approach and a sense for market developments with a readiness to take risks and a personal engagement." This also includes showing the courage to invest when the publishing house was experiencing difficulties. In 1980 a financial crisis arose after the rapid expansion of the company had collided with the budget cuts in the acquisition funds of many libraries. With a large degree of personal engagement Klaus G. Saur was able to increase the turnover of the publishing house in the years 1980 to 1986 from 12 to 25 million DM.

Klaus G. Saur caused a commotion in the publishing business when he sold his company to the British publisher Butterworth, a member of the Reed group, effective as of the 1st of July, 1987. This decision enabled the company to continue its program without any changes and to build up its international activities. In the end analysis the balance is impressive: from 1963 to 1998 Klaus G. Saur caused the turnover of his publishing house to multiply by a factor of 300. Out of its humble circumstances at the beginning the firm was able to achieve a turnover of 40 million marks in 1998.

From Paste to the Internet:

Changing work techniques

K.G. Saur Publishing resolutely put its methodological and technical knowledge into practice from the very beginning. Karl Otto Saur sr. used the process of reprocumulation for his technical documentations. The titles were photographed from bibliographies, cut out, pasted and subsequently converted into a photoprint copy. The process of reprocumulation, first carried out in 1959 for the *Universalbibliographie Technik und Wirtschaft* (Universal Bibliography of Technics and Business), was the technical basis of many successful projects for a long time, for example for the *Gesamtverzeichnis des deutschsprachigen Schrifttums 1700–1965* (Bibliography of German-Language Publications 1700–1965). The experiences made during this project also formed the foundation for work on the *British Library General Catalogue of Printed Books*, appearing since 1979, and for the microfiche edition of the *Main Catalog of the Library of Congress* begun in 1984.

K.G. Saur Publishing made use of the good experiences with the microfiche medium not just for bibliographical projects, but rather also for text and object collections like the *Marburger Index* or the *Flugschriftensammlung Gustav Freytag* (Pamphlet collection of Gustav Freytag) of the Public and University Library of Frankfurt.

Just how important the technical development for a publishing house specializing in documentation can be, is shown by the example of the *British Library General Catalogue of Printed Books*. The 360 volume large basic work was done by hand. Twenty editors worked on it for many years. They pasted, sorted and collated the entire edition by

Klaus G. Saur (right) presents the Bibliothek der Deutschen Literatur *(Library of German Literature) to the Prime Minister of North Rhine-Westphalia at that time, Johannes Rau.*

hand. The 50-volume supplement of the *British Library General Catalogue of Printed Books* for the time period from 1976 to 1982 was compiled electronically and sent on magnetic tape. This product achieved such a high profit that a large part of the losses incurred by the basic work was balanced out.

CD-ROMs and databases in the Internet have proven to be very useful media. These new methods of information transfer are best suited for documentations and reference works due to the fact that they require little space, are quick, and easily edited. In spite of this K.G. Saur Publishing continues to also build upon the printed edition of most titles. Publisher Klaus G. Saur is convinced that indices and lexicons in book form have a future in any case: "Despite all of the new forms of carrying information the book form will not just be accepted by the user, but rather appreciated as well."

With the International Market in View

From the beginning the strength of K.G. Saur Publishing has been that it has thought and acted internationally. As of 1950 the founder of the firm Karl-Otto Saur sr. tried to bring out international publications, in the beginning with very limited success. An international literature research service for the patent and technical businesses was offered starting in 1950. As of 1951 it published the *Internationaler Literaturdienst* (International Literature Service). The *Universalbibliographie Technik und Wirtschaft* (Universal Biblio-

graphy of Technics and Business) included the most important foreign literature as well starting in 1959.

The first successful work Klaus G. Saur published for the company came in 1964, and was created for the international market. The *Internationales Verlagsadressbuch / Publishers' International ISBN-Directory* concentrated not only on company addresses in the German-speaking area, but brought a record of all book and journal publishers worldwide.

Klaus G. Saur knew how to make strength out of the weakness of most other publishers. At the time there was not one noteworthy publisher with the courage to open branches in France, England and the USA and publish books in English. By way of the agreement closed in 1971 with the International Federation of Library Associations (IFLA) to bring out all IFLA publications at K.G. Saur Publishing, the percentage of English-language publications rose to over 50 percent at the beginning of the 70's. In 1977 and 1978 Saur founded offices in New York and London.

At the IFLA Conference 1990 in Stockholm: W R H Koops, Director of the University Library of Groningen (left), Klaus G. Saur (middle) and Dr. Franz Georg Kaltwasser, Director of the Bavarian State Library (right) in a discussion.

Within the course of the 70's the French-language library program of K.G. Saur Publishing was also developed. Among other things handbooks and textbooks appeared for the French library market, to a degree in cooperative work with the national research center CNRS or with the Bibliothèque Nationale in Paris. When K.G. Saur Publishing took up the task of publishing the French national bibliography for the time period 1926 to 1933 the opening of an office in Paris became necessary.

After the publishing house had fallen into a serious financial crisis in 1980, the international contacts and knowledge of K.G. Saur proved to be of essential importance for the company. The turnaround in the years after 1980 was made possible through a series of international agreements. In 1981 K.G. Saur Publishing closed an agreement with the Union of International Organizations in Brussels for the publishing of the *Yearbook of International Organizations*. Previously the work appeared every two years in one volume; Saur expanded it to three, as of 1996 to four volumes and published the work annually – which greatly in-

Gordon Graham, Chairman of the Board of Butterworth and Chairman of the Board of R.R. Bowker (left), and Klaus G. Saur (right) during the signing of the agreement to sell the publishing company to Reed International P.L.C., London in 1987.

creased the level of information, the turnover and the profits. The *Yearbook of International Organizations* became just as important to K.G. Saur Publishing as the classical international reference works of the company, *Publishers' International ISBN-Directory* and *World Guide to Libraries*. In the meanwhile the 35th edition of the *Yearbook* has appeared; it presents 30,000 international organizations from 300 countries.

Also in terms of the most important break in the publishing company's history – the sale of the firm to the Reed group in the year 1987 – international aspects were a deciding factor. Thoughts about the international competitiveness provided the final push for this important decision. Klaus G. Saur took it as his starting point that the internationally adapted products of his publishing house would be far easier to market through an international group than this could otherwise ever be the case for a relatively small publisher in Munich.

His decision would turn out to be right. In the years after the sale K.G. Saur Publishing worked ever more closely together with the publishers R.R. Bowker, New York and Butterworth, London. Through the integration of K.G. Saur Publishing in the international Reed group a whole series of international projects and cooperation were developed.

In 1990 K.G. Saur took over the reference works *Microforms in Print, Subject Guide to Microforms in Print* and the journal *Microform Review* from the American publishing house Meckler. K.G. Saur Publishing also closed an agreement with Meckler in 1991 for an extensive social sciences program with large series such as *History of Women in the United States* or *Japanese-American World War II* as well as a row of projects on the subject of the Holocaust. These new acquisitions strengthened the position on the international market to an even further degree. Since 1992 K.G. Saur Publishing, together with the American publishing house R.R. Bowker and the British company Whitaker,

publishes an index of all English-language titles in the world: *Global Books in Print*. It is only available on CD-ROM.

With the International Musicological Society K.G. Saur Publishing developed a CD-ROM with manuscripts of famous compositions. A further international project came about together with the Bildarchiv – Foto Marburg and the Institut Computer & Letteren, Utrecht: the so-called DISKUS-series, a series of museum collections on CD-ROM.

Klaus G. Saur received several awards for his international activities. Among others he has honorary doctorates from the Universities of Marburg, Boston, Ishewsk and Pisa. The University of Glasgow proclaimed Klaus G. Saur a Visiting Honorary Professor without teaching duties, the Humboldt-Universität in Berlin made him an Honorary Professor with regular teaching duties.

The VLB as a Milestone

The *Verzeichnis lieferbarer Bücher* (*VLB*) (German Books in Print) has belonged to the everyday life of booksellers and libraries since 1971. The thick volumes, in which all books in print are alphabetically ordered by authors, titles and key words, have proven themselves from the very beginning as perhaps the most important and irreplaceable tools for daily research. The *VLB* has made possible the process of cooperative work between publisher, intermediate book trade and retail bookseller, and has greatly improved the process of placing orders.

The model for the *VLB* was the American index *Books in Print*, brought out by the US publisher R.R. Bowker. Up until 1970 nobody was able to decide on bringing out a similar reference work for the German language area. Indeed, this subject had already been discussed for many years in the different committees of the Börsenverein des

Professor Kunst, Dean of the department of German literature and art sciences at Philipps-Universität, Marburg (left), gives Klaus G. Saur (right) a diploma on the occasion of the bestowal of his honorary doctorate in 1985.

Deutschen Buchhandels, but the expense of bringing out a sort of German *Books in Print* still seemed too high. The book trade found remedy in the stock catalogs of the book wholesalers, even though only less than a third of all books in print were listed. The complete index was thus a market niche – though the realization of this mammoth work was very difficult to shape.

The costs – the publishers were to pay fees by line of page for the listing of their titles – spoke at first against the project. When the Börsenverein des deutschen Buchhandels conducted a poll of the booksellers as to how many copies of the *VLB* they would purchase the result was very disappointing. Only 138 booksellers showed an interest. Due to this the Börsenverein planned a print-run of 2,000 copies, which would have driven the price to astronomic heights. But Klaus G. Saur was convinced of the success of the *VLB* and courageously guaranteed a purchase of 6,000 copies, which thus paved the way for a better calculation basis for the fee per line. And his feeling proved him right: only 800 copies of the first print-run (6,000 copies) remained unsold. The *VLB* has been appearing in the publishing company of the Buchhändler-Vereinigung since 1971, and is distributed by K.G. Saur Publishing worldwide. In contradiction to the entire skepticism at the beginning the project has turned into the most important distribution titles of K.G. Saur Publishing.

In the meanwhile the *VLB*, along with the ISBN-index, the subject index and two supplement volumes, is comprised of 18 volumes. Additionally, a CD-ROM version of *VLB* has been available since 1988.

Who`s Who?

In 1970 K.G. Saur Publishing brought out a small handbook with the title *Who's Who at the Frankfurt Bookfair* for the first time. And in 1977 the company published the first edition of the *Who's Who in the Socialist Countries*, edited by Borys Lewytzky and Julius Stroynowski. This reference work was so unique and so well researched that the American Library Association chose it to be the book of the year.

The first biographical index under the name *Who's Who* appeared in 1848. *Who's Who*-books offer a service which the general lexicons and biographies do not: namely information on very special circles of individuals.

Normally the purchasers of the *Who's Who*-books are identical with the listed individuals of these reference works. However, in this case it was different because information on individuals from socialist countries was not yet, or only partially, available.

A series of specialized *Who's Who*-indices appeared in the K.G. Saur Publishing house which appealed to a small, but thus all-the-more interesting circle of customers, as with the *Who's Who in Mass Communication*, the *Who's Who in International Organizations* or the *Who's Who in the People's Republic of China*. That the *Who's Who*-volumes have a unifying effect among peoples over and above their practical use has, by the way, been confirmed from the highest of places. After reading it Pope John Paul II judged of the three volume biographical encyclopaedia *Who's Who in the Socialist Countries of Europe:* "The value of this work lies in the advancement of mutual understanding between individuals from different peoples of different cultures who live under different circumstances."

IFLA – A Model of Cooperation

In 1971 Klaus G. Saur agreed with the International Federation of Library Associations (IFLA) that the company Verlag Dokumentation should publish all book publications and the *IFLA Journal*; K.G. Saur Publishing was still known at the time as Verlag Dokumentation.

This agreement with the international federation of library associations and institutions was a milestone for K.G. Saur Publishing. With this agreement to take charge of IFLA-publications, K.G. Saur Publishing thus became internationally renowned as an important specialized publisher for library science.

K.G. Saur Publishing brings out the official journal of the International Federation (*IFLA Journal*), the *IFLA Publications* as well as the *IFLA Directory*. K.G. Saur also publishes reference works, handbooks and catalogs edited by the IFLA.

The *IFLA Journal* is an international forum in which specialists discuss different questions regarding library science. This is where a librarian can receive information about all news relevant to the field. The articles chosen reflect the bandwidth of the business: Management, Library

Services for the Blind, Conservation, Transfer of Documents, and Copyrights. Naturally also of importance are the reports on IFLA-meetings as well as of the international calendar of events.

Its function as the "court publisher" of the international federation of all library associations and institutions brought K.G. Saur Publishing worldwide renown and a whole series of agreements to follow with similar institutions. In 1974 K.G. Saur Publishing also became the publisher of the International Council on Archives, the International Federation of Film Archives (Fédération Internationale des Archives du Film – FIAF) and the International Council of Museums – ICOM.

GV – An Idea gets its Way

In January 1975 a letter from librarian Reinhard Oberschelp in Hanover reached the publisher. Oberschelp had written an exposé, in which he described how he envisioned a "Gesamtverzeichnis des deutschsprachigen Schrifttums 1700 bis 1965" (Bibliography of German-Language Publications 1700 to 1965). According to this idea all available bibliographical entries on German-language publications should be brought into one continuous alphabetical order.

When Klaus G. Saur then developed the plan to actually bring out a *Gesamtverzeichnis des deutschsprachigen Schrifttums (GV)* (Bibliography of German-Language Publications), at first for the time period from 1911 to 1965, scholars shook their heads. Four of five directors of state and university libraries answered the question, of whether they would purchase a copy of the *GV*, with "no, thanks".

The dimensions and the price may have acted forbidding at first: 155,000 pages, 150 volumes, total price 22,000 DM. In the meantime all major libraries have the index in their collections. The *GV* has long become a standard work in scholarly operations.

Especially helpful was the documentation of literature from 1933 to 1945, a period which was neglected by scholars in the two decades following the end of World War II. The 'Bibliography' has closed "a noticeable hole in the national biographical records", wrote Professor Klaus-Dieter Lehmann in 1989, at that time General Director of the Deutsche Bibliothek.

In the meanwhile the continuation followed: The *Gesamtverzeichnis des deutschsprachigen Schrifttums 1700–1910* (Bibliography of German-Language Publications 1700–1910). It consists of 160 volumes, which together cost DM 26,880. The *Gesamtverzeichnis des deutschsprachigen Schrifttums außerhalb des Buchhandels* (Bibliography of German-Language Publications outside of the Booktrade) and the *Gesamtverzeichnis Deutschsprachiger Hochschulschriften* (Bibliography of German-Language Academic Publications) appeared as supplements for the time period from 1966 to 1980. In the meantime there are also affordable microfiche editions of the 'Bibliographies' available.

Dictionaries, Lexicons and Handbooks

What is the correct German technical term for "simplified cataloguing"? What are the goals of the Canadian Coady International Institute? And what does "aleatory" mean? Whoever is tortured by such tricky questions should reach for a reference work from K.G. Saur Publishing. The *Dictionary of Librarianship / Wörterbuch des Bibliothekswesens*, for example, is a German-English/English-German dictionary which devotes itself extensively to the terminology of library science and serves as a basis for international understanding. Whoever is looking for orientation in the world of religion can turn to the *World Guide to Religious and Spiritual Organizations*. It contains addresses and descriptions of 3,495 federations, orders, brotherhoods and institutes – from the "Aethenius Society" to the "World Mind Group".

In 1997 K.G. Saur Publishing brought out the tenth volume of the renowned *Allensbacher Jahrbuch der Demoskopie* (Allensbach Yearbook of Opinion Research), edited by Elisabeth Noelle-Neumann and Renate Köcher. It contains opinion poll results and statistics on 1,300 pages – among other things the not very surprising statement that 46 percent of the population trust neither politicians, nor journalists round the corner.

K.G. Saur Publishing acquired the Verlag Francke AG Bern in 1989. This acquisition meant the takeover of a series of renowned and quality lexicons fitting in perfectly into the existing publishing program. Some of these were the *Deutsches Literatur-Lexikon* (Encyclopedia of German Literature), the *Großes Sängerlexikon* (Biographical Dictionary of Singers), the *Deutsches Theater-Lexikon* (Encyclopedia of German Theater), *Die Geschichte der Griechischen Literatur* (History of Greek Literature) and *Die Geschichte der Römischen Literatur* (History of Roman Literature).

PRVD? GDDM? Pr Vyzk Ustavu CS Naft Dolu? No worries, the *International Encyclopedia of Abbreviations and Acronyms of Science and Technology* provides further assistance. These have to do namely not with typing errors,

At the presentation of the Allensbacher Jahrbuch 1997: Klaus G. Saur (left), Prof. Elisabeth Noelle-Neumann (middle) and Chancellor Dr. Helmut Kohl (right).

but rather with technical abbreviations, which are however virtually non-understandable for the layman even when written out. Nevertheless, this encyclopaedia is often the last hope for technicians and scientists.

However, one cannot find a direct answer to this question about the meaning of life even in the reference works of K.G. Saur Publishing, but still at least an official reference to the *omnipresence of God* as well as the confirmation that a *human reason* exists – can be read in Elmar Waibl's *Dictionary of Philosophical Terms*.

Catalogs of Great Libraries: MCLC, BLC and Harvard University

Every student knows the frustrating card system in the catalogs of university libraries. Such catalogs are creatures that have grown through history. Parts of the catalog can be removed, other cards inserted. In the end the confusion is absolute. On many cards there is virtually no remaining bibliographical information, but rather only cross references to other libraries and other institutes.

The K.G. Saur Publishing company can provide further help in such cases. The positive experiences with the *Gesamtverzeichnis des deutschsprachigen Schrifttums* (Bibliography of German-Language Publications) lead to similar projects on an international level. K.G. Saur Publishing brought out the main catalogs of many university and state libraries. Among these were such honorable institutions as the Library of Congress in Washington and the British Library.

During a talk with a librarian of the British Library Klaus G. Saur once mentioned that it would surely make sense to publish a complete cumulation of all catalogs available in the British Library. The British Library was favorable to this idea and invited him to a working talk. On the very first day an agreement was made. The first six volumes of the *British Library General Catalogue of Printed Books (BLC)* appeared in 1979. This general catalog is the most comprehensive scholarly complete bibliography in existence. In the meanwhile more than 700 volumes have appeared.

K.G. Saur Publishing was commissioned with a similarly large-scale and also honorable project from the USA: the filming of the *Main Catalog of the Library of Congress 1898-1980 (MCLC)*. The Library of Congress in Washington is the largest library in the world. It is in possession of the so-called Dictionary-Catalog, the largest bibliographical complete catalog, containing all books and writings organized by author, title, subject field and organs. The catalog is made up of more than 26 millions cards.

In 1983 K.G. Saur Publishing was entrusted with filming the catalog, and in 1989 the project was successfully completed. The *Main Catalog of the Library of Congress* is the largest subject bibliography in the world.

K.G. Saur Publishing closed an agreement with Harvard University in 1983 to publish a whole series of bibliographies. Among these are publications of international meaning, for example the *Card Catalogs of the Harvard University Fine Arts Library 1895-1981*, the *Card Catalog of the Harvard University Music Library to 1985* and the *Card Catalogs of the Harvard University Law School Library 1817-1981*. The library of the law school at Harvard University is the largest and oldest law library in the USA. K.G. Saur Publishing brings out the microfiche-edition of

Klaus G. Saur in front of the volumes of the BLC in the library at Simmons College in Boston, MA.

the card catalog of this law and jurisprudence collection, which is, with 1.5 million volumes, the second largest in the world.

The important *Fine Arts Library* came into existence in 1962/63 when the art book collections of the *Widener Library* and the *Fogg Museum Library* were combined. They comprise more than 200,000 volumes and 25,000 presentation- and auction catalogs, as well as the collections of the *Rübel Asiatic Research Collection*, one of the most comprehensive and best collections of books and magazines on art of the Far East, Southeast Asia and the Indian subcontinent.

The Marburger Index

The *Marburger Index*, published since 1977, is an inventory of art in Germany. The picture section of the edition contains 1,250,000 photographs, supplemented by an additional 55,000 pictures per year. The text and index section came out from 1983 to 1991 on microfiche and in the meanwhile, in an annually expanded form, on CD-ROM as well.

The microfiche version of the *Marburger Index* was the first ever microfiche edition of K.G. Saur Publishing. Upon completion of the edition the *Marburger Index* will offer the most comprehensive photo documentation on German art and architecture history. K.G. Saur is trying, together with the Bildarchiv – Foto Marburg, to fulfill bit by bit the dream of all those interested in culture of being able to possess a collection of art history photos. Numerous photo documentations on art from different countries and eras on microfiche have been published since 1979: the *Index of Ancient Art and Architecture* (edited by the Deutsches Archäologisches Institut, Abteilung Rom), the *Index photographique de l'art en France*, the *Italien-Index*, the *Spanien- und Portugal-Index*, the *Schweiz-Index*, the *Österreich-Index*, the *Ägypten-Index* and soon the *Benelux-Index*.

The CD-ROM database of the *Marburger Index* offers many search functions. Photos can be searched according to type of object, subjects, artist names, and place and time. The database is expanded upon by way of ca. 60,000 new object descriptions for each edition and a large part of the museums' collections are illustrated with digital pictures.

The Herzog August Bibliothek in Wolfenbüttel

The Herzog August Bibliothek is counted among the most famous German libraries. Gotthold Ephraim Lessing accepted a nomination to become a librarian there in April, 1770. The Herzog August Bibliothek is internationally known mostly due to its rich collections of German baroque literature.

Inventory of the precious baroque prints began more than two decades ago. When the publishing house Kraus,

which had published the first four volumes of the bibliography, withdrew from the project in 1984, Prof. Dr. Dr. h.c. mult. Paul Raabe, at that time Director of the Herzog August Bibliothek, invited Klaus G. Saur to a talk in Wolfenbüttel. Following this K.G. Saur Publishing took over the stock of the volumes already published and continued the work, completing it in 1996.

With 43 catalog volumes and 3 index volumes this was thus the most extensive bibliography of the entire German baroque literature. As a supplement to this physical inventory of baroque literature there appeared the catalogs *Ungarische Drucke und Hungarica 1480 bis 1720* (Hungarian Prints and Hungarica 1480 to 1720) as well as the *Polnische Drucke und Polonica 1501 bis 1700* (Polish Prints and Polonica 1501 to 1700), and the *Verzeichnis medizinischer und naturwissenschaftlicher Drucke 1472–1830* (Index of Medical and Natural Sciences Prints 1472–1830).

In 1996 the catalogs of the *German Prints of the Baroque* were cumulated together with the *Polonica* and *Hungarica* in one microfiche edition.

The cooperative work of K.G. Saur Publishing with the Herzog August Bibliothek was, however, to bear even more fruit: the *Wolfenbütteler Bibliographie zur Geschichte des Buchwesens im deutschen Sprachgebiet 1840–1980* (Wolfenbütteler Bibliography on the History of the Book Trade in the German-Speaking Countries 1840–1980) in twelve volumes, the *Katalog der Graphischen Portraits in der Herzog August Bibliothek Wolfenbüttel: 1500–1850* (Catalog of Graphical Portraits in the Herzog August Bibliothek Wolfenbüttel: 1500–1850) in ca. 38 volumes, as well as the *Leser und Lektüre vom 17. zum 19. Jahrhundert* (Reader and Reading from the 17th to 19th Century), an index of the books on loan from 1664–1806 in eight volumes.

The Bibliotheca Palatina

The Vatican Library houses many treasures, but one of the best known is surely the *Bibliotheca Palatina*. This precious collection comes from Heidelberg University and was deemed in the 17th century as being the most important library north of the Alps. In the commotion of the Thirty-Year War the *Bibliotheca Palatina* was taken to Rome. Duke Maximilian of Bavaria sent the books to the Pope as partial payment for his share of the war costs.

That was 1623. A part of the *Bibliotheca Palatina* was given back in 1816. However, almost no work from the *Bibliotheca Palatina* has ever been available in a German library since then. For a long time the material – 3,500 documents and 5,000 prints – remained closed to the public.

After lengthy negotiations between the Vatican, the University of Heidelberg and K.G. Saur Publishing the company received permission to bring out the *Bibliotheca Palatina* on microfiche. For the scholarly world this was a sensation because large sections of literature from the 15th to the 17th century became accessible then again. The edition of the printed books of the *Bibliotheca Palatina* comprises approximately three million pages on 21,100 fiches.

Among these are treasures such as the Grammatica Palatina Heidelbergensis, the original 'Falkenbuch' manuscript of King Friedrich II, or the first printed book with Arabic letters in Germany. Also world famous is the Ottheinrich's encyclopaedic structured book collection. The exhibition *Bibliotheca Palatina* on the occasion of the 600th year of foundation of the University of Heidelberg made clear in 1986 what kind of treasure had been hidden in the Vatican for such a long time. The microfiche edition of the Palatina prints made this treasure available to the public once again.

As with all books of the Vatican Library the *Bibliotheca Palatina* collections were naturally not allowed to be removed from the Vatican. K.G. Saur Publishing was in the end therefore able to arrange to film the books on location. For this purpose it was necessary to set up a recording studio in the Vatican where three photographers worked for five years. A specially designed book-protecting cradle was used for this purpose which ensured that the difficult work with the precious collections would not damage the books. The books could only be opened at a 90 degree angle on this machine so that the binding would be preserved.

Exile Literature

During a short stay in Washington, D.C. Klaus G. Saur wanted to visit the Holocaust Memorial Museum. When he saw the long line at the entrance he almost changed his mind – he would have had to wait several hours, and he did not have so much time before continuing his travels. He went to the back entrance and said what he wanted and his name. The doors were immediately opened wide for him. The reason is clearly shown below.

The name K.G. Saur is closely connected with research of the historical period between 1933 and 1945. Titles on the history of the Third Reich, the anti-fascist resistance, exile literature and Judaica are very important to Klaus G. Saur. The reason for the employees at the Holocaust Memorial Museum being so friendly also has to do with the fact that scientists whose field of interests are National Socialism and persecution of the Jews could hardly do without the reference works of the Munich publishing house. At any rate the most important handbooks of K. G. Saur Publishing on the subject were lying out on the table of the Holocaust Memorial Museum library. They were being used at the time – which especially pleased the publisher of course.

The main focus on contemporary history began in the years 1980–83, when K.G. Saur Publishing brought out the *International Biographical Dictionary of Central European Émigrés 1933–1945* in a cooperative effort with the Institut für Zeitgeschichte and the Research Foundation for Jewish Immigration. After the National Socialist Party seized power in 1933 a half million persecuted people emigrated, the majority of them Jews, from Germany, Austria and the Czechoslovakian. Among them were many artists, scientists and writers. The *International Biographical Dictionary of Central European Émigrés 1933–1945* documents the cultural losses resulting from this exodus. The handbook presents nearly 9,000 biographies of emigrants substantiated in a systematic and specific fashion.

Presentation of the Bibliotheca Palatina *in the Vatican: (from right) Prof. Elmar Mittler, Director of the State and University Library of Lower Saxony in Göttingen, Klaus G. Saur and Father Leonard Boyle, Prefect of the Vatican Library, (far left) Cardinal Stickler, Chairman of the Librarian Archive Museums and Libraries.*

At the presentation (in 1983) of the International Biographical Dictionary of Central European Émigrés 1933-1945: (from left) Klaus G. Saur, the editors Prof. Dr. Herbert A. Strauss and Dr. Werner Röder, Dr. Manfred Briegel from the DFG, Prof. Dr. Walter Fabian, official speaker and listed in the dictionary himself, and Prof. Dr. Martin Broszat, Director of the Institut für Zeitgeschichte in Munich.

The *International Biographical Dictionary of Central European Émigrés 1933–1945* found international renown everywhere. Since then numerous titles in the main program area of contemporary history belong to the subject areas Emigration, National Socialism, History and Culture of Judaism. Among these is, for example, a series of *Bibliographien zur deutsch-jüdischen Geschichte* (Bibliographies on German-Jewish History) published on behalf of the Salomon Ludwig Steinheim Institute for German-Jewish history. As part of this the *Bibliographie zur Geschichte der Juden in Bayern* (Bibliography on the History of Jews in Bavaria) appeared in 1989; thereafter followed bibliographical records on Jewish literature in Germany 1933–1943, on the history of Jews in Schlesia, in Hamburg and in Switzerland.

The series *Deutschsprachige Exilliteratur seit 1933* (German-Language Exile Literature since 1933), ed. by John M. Spalek a.o., gives a precise overview on exile literature in the United States. The first volume concerns itself with the authors who went to Hollywood, the second volume deals with those who emigrated to New York. Volume 3 will contain further essays on authors in exile in the USA who were not taken into account in the first two volumes. Volume 4 will bring another 225 bibliographies on writers in exile such as Oskar Maria Graf or Julius Berstl.

The documentations of K.G. Saur Publishing have served as a definitive source of help for research of the exiles because the literary bequest of the emigrants is mostly spread out over the whole of America, Israel and France. Viewing and evaluation of the scattered material was partially made possible for the first time by way of a series of literary bequest indices. The *Guide to the Archival Material of the German-Speaking Emigration to the United States after 1933*, edited by John M. Spalek and Sandra H. Hawrylchak, contains for example in ca. 1,000 entries the bequests of artists, scientists and writers driven out by the NS regime. Among other things the following are listed and described: diaries, letters, manuscripts, interviews, photos, radio, cassette and LP recordings.

Another record of the literary bequests of emigrants is offered by the *Inventory of the Archival Materials of German-Speaking émigré Scientists in Archives and Libraries of the Federal Republic of Germany*, edited by the Deutsche Bibliothek. In the volume *Schicksale deutscher Emigranten* (The Fate of German Emigrants), which supplements the index *Quellen zur deutschen politischen Emigration 1933–1945* (Sources on German Political Emigration 1933–1945) , the Herbert and Elsbeth Weichmann Foundation showed the results of their work. Herbert Weichmann returned to Germany after 15 years in exile in 1948 and became the Mayor of Hamburg in 1965. His wife started a foundation devoted to the researching of political exiles.

K.G. Saur Publishing offers a whole series of additional publications on the subject 'Exiles': the *Lexikon deutsch-jüdischer Autoren* (Lexicon of German-Jewish Authors), ed. by the Archiv Bibliographia Judaica, documenting the 200-year development of German-Jewish Culture, which was brutally interrupted in 1933.

The edition *Expatriation Lists as Published in the "Reichsanzeiger" 1933–45*, edited by Michael Hepp became an essential research instrument for scientists. More than 39,000 individuals had their citizenship officially revoked in the Third Reich. The three-volume documentation makes this infamous chapter of the German past accessible with the help of a name and place index.

Documents of National Socialism

Everything had to be done orderly in Auschwitz. Indices, lists, daily orders, personnel forms and reports were filled out on a daily basis in the writing rooms of the camp. The

Klaus G. Saur presenting the Auschwitz Memorial Books *to the German Federal President Prof. Dr. Roman Herzog.*

largest part of these documents was destroyed during the evacuation of the concentration camp in January of 1945 in order to get rid of evidence for the crimes committed by the SS.

A part of the so-called 'Death Books' have remained. From the 27th of July, 1941 to the 31st of December 1943 approx. 69,000 entries were listed in 46 volumes with the death dates of prisoners in Auschwitz. In total more than twice as many prisoners died in Auschwitz.

When K.G. Saur Publishing brought out the *Death Books from Auschwitz* in 1995 in three volumes, supplemented by a selection of 2,482 photos and reports of the experiences of various Auschwitz survivors, Jeshajahu Weinberg, Director of the Holocaust Memorial Museum in Washington, D.C. recommended the publication to Holocaust survivors and researchers. This work made it possible for survivors and their children and grandchildren to receive reliable information on relations in the camps – even if the cause of death was falsified.

K.G. Saur Publishing sought a close cooperative working relationship with leading historians in the area of history of the Third Reich from the very beginning. Together with the Munich Institut für Zeitgeschichte the K.G. Saur Publishing was able to realize the large edition *Akten der Partei-Kanzlei der NSDAP* (Records of the Party Chancellery of the National Socialist Party). From 1983 to 1992 the reconstituted records of the Party Chancellery of the National Socialist Party were published on a total of 200,000 pages in a microfiche edition cumulated on 491 silver-fiches and completed by four descriptive volumes (in co-production with R. Oldenbourg Publishing) as well as two index volumes.

The Party Chancellery of the NSDAP took a special place in the NS power apparatus. It was the central leading organ of the party and at the same time the most important

place of state organization that was involved in all essential decisions. Every researcher whose focal interest is the Third Reich can thus not get around the Party Chancellery.

The Munich Institut für Zeitgeschichte had started reconstituting the records of the Party Chancellery on the basis of the papers of the recipients at the beginning of the 70's, for which a collection of copies from letters, review documents and file reports was created. Together with K.G. Saur Publishing the Institute edited the microfiche edition of the NS Party files starting in 1983. This project had to do with one of the most ambitious and large-scale plans ever undertaken for the research of the National Socialist power apparatus.

The recording and compilation of sources on the National Socialist history of Germany is an important task, as most of the sources were destroyed or scattered and lost by the war. The division of Germany also hindered the free researching of the Third Reich up until ten years ago.

In 1991 and 1995 K.G. Saur Publishing brought out the *Inventar archivalischer Quellen des NS-Staates* (Inventory of Archive Sources of the NS-State) in two volumes by Heinz Boberach in the series *Texte und Materialien zur Zeitgeschichte* (Texts and Materials on Contemporary History), edited by the Institut für Zeitgeschichte. This work presents the archive records of higher and middle authorities of the Third Reich, SS and police, of the Justice Department, culture and science authorities, of the 'Wehrmacht', directorates of the Reich and undersecretaries of the German National Socialist Workers' Party and much more. The inventory is an essential work tool that helps to find the most varying of documents from the NS period, for example for political scientists, lawyers, historians as well as for scholars of many other fields of study.

A further helping tool for NS research appeared in 1997 in the same series: the handbook *Ämter, Abkürzungen, Aktionen des NS-Staates* (Authorities, Abbreviations, Actions of the Nazi Regime). This research work explains descriptions of official authorities, ranks and the administrative chain of command and general setup, however also non-military cover terms.

K.G. Saur Publishing also documented resistance against the NS regime: the microfiche edition *Widerstand als Hochverrat 1933–1945* (Resistance as High Treason 1933–1945) edited by the Institut für Zeitgeschichte, makes court briefs, pamphlets, names and code names on the subject resistance available to research. And in cooperative work with the Federal Archive the documentation *Lageberichte (1920–1929) und Meldungen (1930–1933)* (Status Reports 1920–1929 and Entries 1930–1933) of the Reich's commissioner for the surveillance of public order was published.

When the historian Elke Fröhlich discovered the glass plates, on which Joseph Goebbels had copied his diaries towards the end of World War II, in the former Special Archives in Moscow in 1992, the Munich Institut für Zeitgeschichte and K.G. Saur Publishing jointly decided to bring out the Goebbels-Diaries as a completely new edition. Fragments of the diaries were published at K.G. Saur already in 1987. The complete diaries, published since 1994, open new perspectives to research of the period, for example on the seizure of power in 1933 or on the "Röhmputsch" 1934. The press was unanimous that the Goebbels records, which span the time period from 1923 to 1945, had to do with a first rate documentation. "The diaries, as an unprecedented contemporary witness account of a social background enables a look into a person full of self-deception", judged the German daily *Süddeutsche Zeitung*. Goebbels' legacy is proof "of a brain flooded with insanity and a dried out soul", wrote the German daily *Frankfurter Allgemeine Zeitung*, "This offers contemporary historians expanded grounds for research."

Of similar importance for research of the NS period is the publication *Hitler. Reden, Schriften, Anordnungen* (Hitler. Speeches, Papers, Orders) has been published since 1991 in twelve part volumes with a four volume supplement section on the Hitler trial in 1924. The edition of the documents for this immense project, which K.G. Saur Publishing produced together with the Munich Institut für Zeitgeschichte, was completed by the beginning of 1999.

At the center point of the series are Hitler's speeches, which were the most important method of propaganda for the Nazis until 1933. They throw light upon Hitler's worldview, but also upon the change of NS propaganda over the years. The Hitler edition makes available for research all of the statements made by Adolf Hitler from February, 1925 up until the seizure of power on the 30th of January, 1933. Besides the speeches and papers, inner-party orders are also contained.

Specialists do not doubt to the fundamental importance of the Hitler-Edition. It offers "a welcome possibility to follow the development of Hitler's personal 'Führer'-role during the rise of National Socialism in an authentic and virtually unbroken manner", remarked the magazine *Das historisch-politische Buch*.

The four volumes on the Hitler trial of 1924 constitute an important supplement to the Hitler-Edition. The wording of the main hearing before the people's court Munich 1 is documented on 1,600 pages for the first time. From today's point of view this trial must be judged to be the historically most momentous court trial of the century.

Judaica

Whereas historical Jewish books in Germany were for the most part completely lost to destruction by the Nazis, there are important Jewish libraries in the USA. One of the largest collections of Judaica worldwide is at the Harvard College Library. It comprises works from the antiquity, the Middle Ages to the present.

The Judaica Department in Harvard was established in 1962. It was the result of the acquisition of the libraries of three important collectors – Ephraim Deinard, Felix Friedmann and Lee M. Friedman. The Judaica collection in Harvard is made up of 150,000 volumes in the Yiddish and Hebrew languages.

The catalog of the entire collection was compiled electronically in the past years, by which the largest database of Hebrew and Yiddish bibliographical data resulted. Together with K.G. Saur Publishing the Harvard University also planned the filming of the collection. The selection of the books to be filmed was determined by three factors: research importance, condition and rarity. The edition *Hebrew Books from the Harvard University* is made up of two parts, which comprise eight subject areas each. Rabbi texts and bible commentaries are contained, as are history, fiction and biographies. The complete edition is made up of 4,934 volumes on 11,453 microfiches.

K.G. Saur Publishing has brought out a whole series of additional titles on the subject area Judaica, such as the *Jewish Biographical Archive*, and, together with the Archiv Bibliographia Judaica, the *Lexikon deutsch-jüdischer Autoren* (Lexicon of German-Jewish Authors) as well as a *Dokumentation zur jüdischen Kultur in Deutschland* (Documentation of Jewish Culture in Germany). This documentation offers a collection of newspaper clippings, in which Jewish culture in Germany is discussed – from music critics to reports on premieres to anniversary articles.

Thousands of clippings from the years 1840 to 1940 document the cultural works of Jews in Germany. Among the musicians are, to name a few, Otto Klemperer, Lotte Schöne and Frida Goldstein. The department 'Music' contains material on 2,400 Jewish composers, conductors, singers, musicians and music scholars. The documentation, which arose out of the newspaper clipping collection of Carl Steininger, shows in the most unique way how very much Jewish artists influenced German culture prior to 1939.

Allgemeines Künstlerlexikon (World Biographical Dictionary of Artists)

After extremely complicated negotiations with the E.A. Seemann-Verlag in Leipzig and the trust office in Berlin, K.G. Saur took over the *Allgemeines Künstlerlexikon (AKL)* (World Biographical Dictionary of Artists) in 1991. The *Allgemeines Künstlerlexikon* is the new edition of the *Thieme-Becker*, published between 1907 and 1950 in 37 volumes. The new edition is projected in 80 volumes and covers the artists of the world throughout all ages.

It had become clear even before the reunification of Western and Eastern Germany that a new publisher was urgently needed for the *Allgemeines Künstlerlexikon*. Continuation of the work under the conditions, as they had prevailed in German Democratic Republic, seemed unthinkable. From 1969 to 1989 only three volumes were published in Leipzig – an absolutely unprofitable business.

The editorial team in Leipzig was quickly brought up to a modern, technically highly developed standard.

Since 1996 three volumes have been published per year. When the *Allgemeines Künstlerlexikon* is completed by volume 80, it will contain around 500,000 articles – thus the most comprehensive documentation on the Plastic Arts worldwide. The encyclopaedia is now considered to be an important work tool by specialists in the field. "The first volume of the *Allgemeines Künstlerlexikon* already belongs in the essential part of the collection of every library offering even only a very little amount of art literature, and it will be able to be found in every museum, every art history institute, and in many places in the world", determined the professional journal *Bildende Kunst* (Plastic Arts).

In 1993 K.G. Saur published the first CD-ROM edition of the *World Biographical Dictionary of Artists*. The seventh edition from 1998 already contains biographical information on over 420,000 artists. In the meanwhile the information is partially also available online.

The electronic distribution of the artists' biographies is a consequential further development of the *World Biographical Dictionary of Artists*. Today it is the worldwide largest database with artists' biographies. Exact biographical and bibliographical information on painters, graphic artists, sculptors, architects and representatives of the applied arts of the world from the antiquity up to present are available at the touch of a key.

Deutsche Biographische Enzyklopädie (The Dictionary of German National Biography)

"One will never tire of reading biographies", wrote Johann Wolfgang von Goethe, "all that is truly biographical, whether they count among the leftover letters, the diaries, the memoirs and so many others, bring past life back to the forefront."

Whether Goethe would have perhaps at some point become tired of reading biographies after all, had he have had the *Dictionary of German National Biography* from K.G. Saur at his disposal? This biographical reference work contains 60,000 names. Not only the big names, as they are contained in general encyclopaedias, but rather also many lesser known personalities who were however representative for their period nevertheless, be it in public, political, economic, theological or cultural life.

The *Dictionary of German National Biography* has been published since 1995. The tenth and final volume came out in the Spring of 1999.

At first the *Dictionary of German National Biography* was edited by Walther Killy, and after his death in 1995 by Rudolf Vierhaus. The undertaking was only possible through the support of an advisory board, the members of which were Dietrich von Engelhardt, Wolfram Fischer, Franz Georg Kaltwasser and Bernd Moeller.

The *Dictionary of German National Biography* provides articles on 60,000 personalities from the time of

Talks surrounding the DBE *(Dictionary of German National Biography): (from left) Klaus G. Saur, Dr. Hanna-Renate Laurien, President of the Berlin Parliament, Prof. Dr. Walther Killy, editor, and the German Federal President Dr. Richard von Weizsäcker during the presentation of the first volume in Berlin in 1995.*

Charlemagne to the present day. In terms of area the encyclopaedia covers the German territories within the borders of the Holy Roman Empire of the German Nation including Austria, and the German-speaking area of Switzerland.

Out of the masses of 60,000 names the *Dictionary of German National Biography* especially highlights around 1,000 individuals because they "were of great influence for their own period as well as for the further development of Germany." These people, whose lives and accomplishments had an especially large impact in German-speaking areas and also far beyond, are presented in articles by known authors on one to two pages – among them Johann Wolfgang von Goethe as well.

In the beginning there was the German Biographical Archive...

The youngest and most successful part of the history of K.G. Saur is marked by the publication of the first installment of the *German Biographical Archive*. The resonance in response to the first part, which appeared in 1982 on microfiche, was however still quite catastrophic. A grand total of only nine new orders came in for the first edition. Yet, libraries started ordering more and more copies as time went by. There was a time lag before the librarians first realized what sort of treasure chest of names and sources the *German Biographical Archive* was.

For the *German Biographical Archive*, the publishing company combined its experiences with reprocumulation, a technique which was used for the *Gesamtverzeichnis des deutschsprachigen Schrifttums* (Bibliography of German-Language Publications), with the high quality of its microfilms, known by way of the editorial work on the art history collection *Marburger Index*.

The *Biographical Archive* brings together texts, biographical and bibliographical data. Hundreds of thousands of data from thousands of individual works are reproduced and transferred onto a data carrier, such as microfiches, and the index onto a CD-ROM. The result is a comprehensive documentation, easy to use and as complete as possible.

The idea for a *German Biographical Archive* arose in 1980. The starting point of considerations was the fact that biographical handbooks, lexicons and indices were not even close to being fully available in the vast majority of libraries. And where they are indeed available, the physical condition of the books leaves a lot to be desired.

Inventory taking of the materials which would come into question for the *German Biographical Archive* showed that there were around 750 volumes for the time period between 1650 and 1900 in existence. They contained 450,000 articles on 230,000 individuals. In comparison: the *Brockhaus-Enzyklopädie* contains 12,000 name articles on German individuals, the *Allgemeine Deutsche Biographie* 27,000 entries.

From the beginning the undertaking had thus taken on editorial dimensions never seen before in this area. Under the editorial direction of Willi Gorzny the material was completely photographed, photocopied and enlarged. Then the articles were individually pasted onto cards. Each entry was noted as to its original source work. In the end the complete material was put into one alphabetical order. A suborder by chronological order of appearance was maintained for those persons, on whom more than one entry was in existence.

Whoever looks up Johann Wolfgang von Goethe for example, for whom the *German Biographical Archive* contains 17 articles, can determine when and how Goethe was mentioned in a lexicon for the first time ever, how the view of Goethe has changed – and who has copied from whom.

Such a huge project as the *German Biographical Archive* would never have been able to be realized in book form. The immense dimensions first really become apparent when converted into printed pages: 1.3 million pages, a mass of 2,000 volumes, which would require 80 meters of shelf space. Microfiches are able to reduce up to 465 pages to the size of a postcard. By this means the entire archive only takes up one and a half meters of space.

The reproductions of old lexicon articles on microfiches also additionally have an aesthetic appeal say the fans of the *Biographical Archives*. This is because the typeface of the original usually remains the same. This is an advantage over the electronic recording process.

Asides from the ideal availability of the data scholars point out the care with which the work was produced. The *German Biographical Archive* is "surely the largest act ever undertaken against – here especially the Germans – forgetting history", lauded Hans Wollschläger in the German daily newspaper *Süddeutsche Zeitung*.

The World Biographical Archive

The success of the *German Biographical Archive* had immense consequences. From the national project a *World Biographical Archive* was developed, the largest part of which is now complete. When it is fully completed a dream of all librarians and scholars will have become true: all biographical entries from all lexicons available in the world will then be compiled and ordered in one universal index, in total approx. eight million articles.

Both in terms of its basic concept as well as its scope the Biographical Archives are unique. The success of the *German Biographical Archive* motivated K.G. Saur to develop similar projects for other countries.

The first to be taken up was the *British Biographical Archive* because a team of editors in London had just become free who had worked on the *British Library Catalogue*. Following that Victor Herrero, who had also worked on the *German Biographical Archive*, proposed the publication of a *Spanish, Portuguese and Latin-American Archive*.

Then followed the *Italian Biographical Archive*, the *Biographical Archive of the Benelux-Countries*, the *American, French* and *Scandinavian Biographical Archives*, as well as the *Australasian Biographical Archive*. In work and partially complete are Poland, Czeck/Slovak, all of Africa, India, China, Korea, Japan, in addition a *Jewish Biographical Archive* as well as an *Arab-Islamic Biographical Archive*.

The *Biographical Archives*, which each took into account all biographical lexicons up to ca. 1910, found such a great resonance in the libraries of the world that the idea to also include the 20th century arose quite soon. The second series covering the time period of ca. 1910 to ca. 1960 is already complete for a whole series of *Biographical Archives*. Besides this K.G. Saur has made it its goal to cover all countries geographically by the year 2000.

Each *Biographical Archive* is completed by an index. This index in book form lists alphabetically all names, birth- and death dates, information on occupations and sources. The Biographical Indices are joined together in a *World Biographical Index*, first published in 1994. With the help of the *World Biographical Index* it is very easy to determine, for example, that there are 11 articles on Georg Friedrich Händel contained in the *German Biographical Archive*, 20 in the *British*, and in the *Italian* on the other hand just one.

The *Biographical Indices* have proven themselves to be extremely helpful work materials. Mostly small libraries often only order a *Biographical Index* and pass up the main works, the *Biographical Archives* – which naturally also has to do with the costs. After all, the archives are so extensive that they have to cost as much as a mid-sized sedan.

In Germany, Austria, Switzerland, Italy, Spain, the USA and in Japan the *Biographical Archives* have already been purchased by the majority of libraries, and demand is also rising in the other countries. In the area of historical biographical information K.G. Saur has a worldwide monopoly position. In the meanwhile the *Biographical Archives* have thus become one of the most important product lines of K.G. Saur. In 1995 the business with the *Biographical Archives* made up more than 30 percent of the annual profits.

One of the few private purchasers of the *German Biographical Archive* is Hans Wollschläger, who was so impressed with the work that he had to admit in the German daily newspaper *Süddeutsche Zeitung*: "With such a truly overwhelming undertaking the reviewer has nothing critical to say." It would also be quite difficult to find a name forgotten by K.G. Saur. Rather, Wollschläger wanted to act as "town crier: people, run to your libraries and demand to see the great thing!"

The *Biographical Archives* have long since become a staple part of the information area of libraries. The national editions stand next to each other in different colored boxes and document to an ever-greater degree the biography of the world. It is now just a question of time until the entire work is on hand- a complete biographical archive of the world. "The World Biographical Archive System is coming into existence", determined Gert Hagelweide three years ago, "and whoever knows the publisher Saur knows that it will be completed as well."

The Future: CD-ROM and Internet

The *Biographical Indices* are appropriate for CD-ROM usage. The silver discs enable the user quick and complex search- and query possibilities that a book cannot provide – most of all not on such little space. Other editions planned are, among others, a worldwide *Who's Who*-CD-ROM, part editions of the *Biographical Indices*, updates for the *Marburger Index* and much more.

The amount of data makes practically no difference for electronic publications. It would be possible to bring out all of the data on individuals in the *Who's Who* works on one CD-ROM.

The first biographical big object to be published on CD-ROM was the *World Biographical Dictionary of Artists*.

Each *Biographical Archive* is followed by a *Biographical Index*. Since 1994 K.G. Saur has been publishing a collection of these indices on CD-ROM, the *World Biographical Index*.

K.G. Saur together with the Research Libraries Group in California developed a series with world bibliographies. The first to appear was a Spanish bibliography, the *Bibliografía General Española Siglo XV-1998*, which lists on CD-ROM all Spanish-language titles published between 1400 and the present day. Following the success of this edition similar editions were brought out for Italian, French and English literature. Russian, Portuguese and Latin world bibliographies are presently being worked on.

In the future data are to be offered to an ever-greater extent in the Internet as well. Online-databases have long since become a work tool for professionals, especially in the area of librarianship. In the meanwhile the electronic publications make up a large part of K.G. Saur Publishing's turnover.

For this reason it is clear that K.G. Saur Publishing has to take multiple paths in the future. Each information can be published in electronic form or in book form. The two areas do not necessary take away market share from one another, to an extent they can certainly complement each other well. When, for example, the *World Biographical Dictionary of Artists* is completed, special compilations of these data can be edited which can then appear in book form again – like a compilation of all Munich artists or an index of all garden architects.

Further future perspectives arise from the electronic combination of different encyclopaedias and databases. Thus it is quite possible to imagine the collection of all data of the *Biographical Archives* in one index. Even more: one could combine and cumulate this biographical world alphabetical system with all of the lexicons in the world, with the *Verzeichnis lieferbarer Bücher* (German Books in Print) or with *Gesamtverzeichnis des deutschsprachigen Schrifttums* (Bibliography of German-Language Publications), such that a gigantic database arises which in principle 'knows it all'.

Bibliographie

(1949 bis Mai 1999)

Abkürzungen

Bibliography

(1949 to May 1999)

Abbreviations

Abhandlungen und Sitzungsberichte der Deutschen Akademie der Wissenschaften (Königlich-Preußische Akademie) zu Berlin 1900 - 1960. Bibliographie. Zsgst. von Pál Vezényi / 1968. 264 S. (3-7940-3208-X)

Abrechnungstabelle für das Gastgewerbe
• 10 % Mehrwertsteuer / 1968. 48 S. / Kt
• 11 % Mehrwertsteuer / 1968. 48 S. / Kt

Actes de la conférence internationale de la table ronde des archives. Ed.: Conseil International des Archives Paris / International Council on Archives Paris.
• (10e) Copenhague 1967. Ed.: Direction des Archives de France / 1969. 95 S. / Br (3-598-11034-0)
• (11e et 12e) Bucarest 1969 / Jérusalem 1970 / 1970. 171 S. / Br (3-598-11036-7)
• (15e) Ottawa 1974 / 1977. 138 S. / Br (3-598-11037-5)
• (20e) Oslo 1981 / 1982. 111 S. / Br (3-598-11038-3)
• (21e) Kuala Lumpur 1982 / 1983. 111 S. / Br (3-598-11039-1)
• (22e) L'Archiviste et l'information des archives contemporaines. Bratislava 1983 / The archivist and the inflation of contemporary records. Proceedings of the twenty-second international archival round table conference Bratislava 1983 / 1984. 204 S. / Br (3-598-11040-5)
• (23e) Accès aux archives et vie privée Actes de la vingt-troisième conférence internationale de la table ronde des archives Austin 1985 / Access to archives and privacy. Proceedings of the twenty-third international archival round table conference Austin 1985 / 1987. 182 S.. / Br (3-598-11031-6)
• (24e) Centralisation, Décentralisation et Archives / Centralization, Decentralization and Archives. Actes de la vingt-quatrième conférence internationale de la table ronde des archives Helsinki 1986 / Proceedings of the twenty-fourth international archival round table conference Helsinki 1986 / 1987. 172 p. / Br (3-598-11032-4)

Adam, Paul: Der Bucheinband. Seine Technik und seine Geschichte. - Nachdr. d. Ausg. Leipzig 1890. - Nachw. v. Walter Bergner / 1993. X, 288 S., 194 Abb. / Ln (3-598-07270-8)

Ägypten-Index. Bilddokumentation zur Kunst in Ägypten. Mikrofiche-Edition. Hrsg.: Bildarchiv Foto Marburg - Deutsches Dokumentationszentrum f. Kunstgeschichte Philipps-Universität Marburg / 1997. ca. 11.000 Fotos auf 111 Silberfiches. Lesefaktor 24x (3-598-33634-9)

Ämter, Abkürzungen, Aktionen des NS-Staates. Handbuch für die Benutzung von Quellen der nationalsozialistischen Zeit. Amtsbezeichnungen, Ränge und Verwaltungsgliederungen, Abkürzungen und nichtmilitärische Tarnbezeichnungen. Im Auftr. d. Instituts f. Zeitgeschichte, München. Bearb. v. Heinz Boberach, Rolf Thommes und Hermann Weiss (Texte und Materialien zur Zeitgeschichte Bd. 5. Red.: Werner Röder u. Christoph Weisz) / 1997. 416 S. / Gb (3-598-11271-8)

Africa Index to Continental Periodical Literature. Ed. by Colin Darch, Mascarenhas, D.C. Published on behalf of the Africa Bibliographic Centre (ISSN 0378-4797) [Hans Zell Publishers]
• Vol. 2 (1977) / 1978. 53 p. / Soft (3-598-21821-4)
• Vol. 3 (1978). Ed. by Colin Darch, Mascarenhas D. C. Published on behalf of the Africa Bibliographic Centre Dar es Salaam / 1981. 191 p. (3-598-21820-6)
• Vol. 4/5 (1979/80) / 1983. XXI, 375 p. / Hard (3-598-21822-2)
• Vol. 6 (1981) / 1985. XIV, 215 p. / Hard (3-598-21823-0)
• Vol. 7 (1982). Ed. by Colin Darch and Alice Nkhoma-Wamunza / 1989. 220 p. / Hard (3-598-21824-9)

African Biographical Archive / Archives Biographiques Africaines / Afrikanisches Biographisches Archiv (AfBA). A single-alphabet cumulation of 231 original biographical reference works published up to 1990 covering approx. 87,000 individuals with a total of 121,747 articles. Microfiche edition. Comp. by Victor Herrero Mediavilla. Source list comp. by Ulrike Kramme and Zelmira Urra Muena / 1994-1997. 457 fiches. Reader factor 24x / Silverfiche (3-598-33100-2) / Diazofiche (3-598-33101-0)
• African Biographical Index / Index Biographique Africain / Afrikanischer Biographischer Index (AfBI). Ed. by Victor Herrero Mediavilla / 3 vols. 1998. Cplt. LIV, 1400 p. / Hard (3-598-33128-2)

The **African** Book Publishing Record. Ed. by Hans M. Zell (Vol. 14-20), ed. by Hans M. Zell, Mary Jay (Vol. 21) / Vol. 14/1988 - Vol. 21/1995 (4 issues annually) (ISSN 0306-0332) [Hans Zell Publishers]

African Book Trade Directory 1971. Ed. by Sigfred Taubert / 1971. 319 S. / Lin (3-7940-7014-3)

The **African** Book World & Press: A Directory / Répertoire du livre et de la presse en Afrique. Comp. by The African Book Publishing Record. Ed. by Hans M. Zell / 1977. XXVI, 296 S. / Hard (3-7940-7014-3) [Hans Zell Publishers]
• 2. ed. Ed. by Hans M. Zell / 1980. XXIV, 244 S. / Hard (3-598-10134-1)
• 3rd rev. and exp. ed. Ed. by Hans M. Zell and Carol Bundy / 1983. XX, 285 p. / Hard (3-598-10439-1)

African Population Census Reports. A Bibliography and Checklist. Ed.: Standing Conference on Library Materials on Africa SCOLMA. Comp. and ed. by John R. Pinfold / 1984. XII, 100 p. / Hard (3-598-10571-1) [Hans Zell Publishers]

Afrika 69/70 / 1969. 448 S., 46 Landkarten, Abb., graph. Darst., Tafeln u. Fotos / Ln (3-7940-6507-7)

Afrika Literatur-Zeitschriften. Aufsätze - Rezensionen - Werke. Bearb. v. Ellen Rüggeberg.
• Bd. 1: (Stand: Dez. 1983) / 1985. 610 S. / Br (3-598-07512-X)

Akten der Britischen Militärregierung in Deutschland / Control Commission for Germany British Element Sachinventar 1945-1955 / Inventory 1945-1955. Hrsg. v. Adolf M. Birke; Hans Booms; Otto Merker, unter Mitw. d. Deutschen Historischen Instituts, London u. d. Niedersächsischen Hauptstaatsarchivs, Hannover / 11 Bde. 1993. Zus. DXCIV, 4.535 S. / Gb (3-598-22900-3)

Akten der Partei-Kanzlei der NSDAP. Rekonstruktion eines verlorengegangenen Bestandes. Sammlung der in anderen Provenienzen überlieferten Korrespondenzen, Niederschriften von Besprechungen usw. mit dem Stellvertreter des Führers und seinem Stab bzw. der Parteikanzlei, ihren Ämtern, Referaten und Unterabteilungen sowie mit Hess und Bormann persönlich – durch Helmut Heiber in Zusammenarbeit mit Hildegard von Kotze und Ino Arndt. Mikrofiche-Edition mit Erschließungsbänden. Hrsg.: Institut f. Zeitgeschichte / 1983/1984. ca. 200.000 S. auf 491 Silberfiches, Lesefaktor 48x (3-598-30260-6)
• Teil I. Bearb. v. Helmut Heiber.
  - Regestenbd. 1. Unter Mitw. von Hildegard von Kotze / 1983. XXXII, 1042 S. / Gb (3-598-30261-4)
  - Regestenbd. 2. Unter Mitw. von Gerhard Weiher / 1983. XII, 1095 S. / Gb (3-598-30262-2)
  - Registerbd. 1/2. Unter Mitw. von Volker Dahm / 1983. 852 S. / Gb (3-598-30263)
• Teil II. Bearb. u. m. e. Einf. „Hitlers Stellvertreter" v. Peter Longerich
  - Regestenbd. 3 / 1992. VI, 287, 383 S. / Gb (3-598-30276-2)
  - Regestenbd. 4 / 1992. VI, 659 S. / Gb (3-598-30277-0)
  - Registerbd. 3/4 / 1992. VI, 412 S. (3-598-30278-9)

Akten der Prinzipalkommission des Immerwährenden Reichstages zu Regensburg 1663-1806. Haus-, Hof- und Staatsarchiv, Wien, Reichskanzlei. Mikrofiche-Edition. Geleitw. v. Hans Booms. / 1990/1991. 493.653 S. auf 5.118 Silberfiches, 1 Begleitbd. Lesefaktor 24x (3-598-33080-4)
• Begleitband mit chronologischem Inhaltsverzeichnis / 1993. 145 S. / Br (3-598-33091-X)

Aktenstücke zur Zensur, Philosophie und Publizistik aus dem Jahre 1842. Hrsg. v. Arnold Ruge. - Reprint d. Orig.-Ausg. 1847 nach d. Exemplar d. Universitätsbibliothek Berlin / 1988. 208 S. / Ln (3-598-10320-4)

Aktuelle Probleme des EDV-Einsatzes in Erwerbung und Katalogisierung. Bericht eines Symposiums, veranst. von der Arbeitsstelle für Bibliothekstechnik am 18. / 19. Juni 1975. Arbeitsstelle für Bibliothekstechnik bei der Staatsbibliothek Preuß. Kulturbesitz / 1976. 150 S. (3-7940-3052-4)

Akzeptanz neuer Kommunikationsformen. Forschung als Begleitung, Programm oder Folgenabschätzung?. Hrsg. v. Petra Schuck-Wersig und Gernot Wersig / 1985. II, 239 S. / Br (3-598-10599-1)

Albrecht, Michael von: Geschichte der römischen Literatur / 2 Bde. 1992. Zus. XXXII, 1.469 S. / Ln (3-907820-89-4) [Vorher im Francke Vlg., Bern]
• 2. verb. u. erw. Aufl / 1994. Zus. XXXII, 1.466 S. / Ln (3-598-11198-3)

Alisch, Alfred: Richtlinien für den Satz fremder Sprachen / 3. Aufl. 1971. 145 S. (3-7940-5112-2)

Allensbacher Jahrbuch der Demoskopie. Hrsg. vom Institut für Demoskopie Allensbach. Bis Bd. V u. d. T.: Jahrbuch der öffentlichen Meinung [1983 übernommen. Koprod. m. d. Vlg. für Demoskopie (VfD), Allensbach. Bis Bd. V im Vlg. für Demoskopie (VfD), Allensbach; bis Bd. VII im Vlg. Fritz Molden, Wien]
• Bd. I: Jahrbuch der öffentlichen Meinung 1947 - 1955. Hrsg. v. Elisabeth Noelle-Neumann und Erich P. Neumann / 3. durchges. Aufl. 1975. XLVIII, 388 S., zahlr. Tab. u. Abb. / Ln (3-598-20711-5)

- Bd. II: 1957. Hrsg. v. Elisabeth Noelle-Neumann und Erich P. Neumann / 1957. XLII, 412 S., zahlr. Tab. u. Abb. / Ln (3-598-20712-3)
- Bd. III: 1958 - 1964. Hrsg. v. Elisabeth Noelle-Neumann und Erich P. Neumann / 1965. XXIV, 642 S., zahlr. Tab. u. Abb. / Ln (3-598-20713-1)
- Bd. IV: 1965 - 1967. Hrsg. v. Elisabeth Noelle-Neumann und Erich P. Neumann / 1967. LVI, 545 S., zahlr. Tab. u. Abb. / Ln (3-598-20714-X)
- Bd. V: 1968 - 1973. Hrsg. v. Elisabeth Noelle-Neumann und Erich P. Neumann / 1974. LVI, 666 S., zahlr. Abb. u. Tab. / Ln (3-598-20715-8)
- Bd. VI: Allensbacher Jahrbuch der Demoskopie 1974 - 1976. Hrsg. v. Elisabeth Noelle-Neumann / 1976. XXXIV, 355 S., zahlr. Abb. u. Tab. / Lin (3-598-20716-6)
- Bd. VII: 1976 - 1977. Hrsg. v. Elisabeth Noelle-Neumann / 1977. XLX, 349 S., zahlr. Abb. u. Tab. / Lin (3-598-20717-4)
- Bd. VIII: 1978 - 1983. Hrsg. v. Elisabeth Noelle-Neumann / 1983. LI, 716 S. / Lin (3-598-20710-7)
- Bd. IX: 1984 - 1992. Hrsg. v. Elisabeth Noelle-Neumann / 1993. 33, 1207, 25 S. / Lin (3-598-20718-2)
- Bd. X: 1993 - 1997. Demoskopische Entdeckungen. Hrsg. v. Elisabeth Noelle-Neumann und Renate Köcher / 1997. 1300 S. / Lin (3-598-20719-0)
- Bd. I - X: 1947-97. 50 Jahre Demoskopie Allensbach. Hrsg. v. Elisabeth Noelle-Neumann; Renate Köcher / 10 Bde i. Kass. 1997. Zus. 7052 S. (3-598-20700-X)

Allgemeine deutsche Musik-Zeitung. (Wochenschrift für das gesammte musikalische Leben der Gegenwart) (ab 9. 1882, Nr. 15: Wochenschrift für die Reform des gesammten Musiklebens der Gegenwart); (ab 9. 1882, Nr. 16: Wochenschrift für die Reform Leipzig und Cassel); (ab 2. 1875): Cassel, (ab 3. 1876): Berlin, (ab 8. 1881): Charlottenburg, (ab 34. 1907): Berlin. Jahrgang 1. 1874-70. 1943, Nr. 6. Gegründet von Otto Reinsdorf. Verantwortlicher Redacteur (ab 5. 1878): Wilhelm Tapert, (ab 8. 1881): Otto Lessmann, (ab 34. 1907, Nr. 26): Paul Schweers, (ab 66. 1939, Nr. 3): Richard Petzold. Microfiche-Edition / 1988. DXXIII (7139) / 53.436 S. auf 670 Fiches. Lesefaktor 24x / Silberfiche (3-598-32530-4) / Diazofiche (3-598-32531-2)

Allgemeine Grundlagen der marxistischen Pädagogik. Hrsg.: Akademie der Pädagogischen Wissenschaften der UdSSR. Mit einem Nachw. f. d. dt. Ausg. v. F. F. Koroljow; W. J. Gmurman / 1973. 399 S. / Kst. (UTB 237) (3-7940-2616-0)

Allgemeine Systematik der öffentlichen Büchereien (ASB). Erarbeitet vom Ausschuß für Systematik beim Verband der Bibliotheken des Landes Nordrhein-Westfalen / 2., überarb. Aufl. 1977. 191 S. / Lin (3-7940-5139-4) [1. Aufl. u. d. T.: Allgemeine Systematik für Büchereien]

Allgemeines Künstlerlexikon. Die Bildenden Künstler aller Zeiten und Völker. Begr. u. mit-hrsg. v. Günther Meissner / ca. 78 Bde. 1992 ff. (3-598-22740-X) [Früher im Vlg. E. A. Seemann, Leipzig]
- Bd. 1 A-Alanson / 1992. LII, 744 S. / Hld (3-598-22741-8)
- Bd. 2 Alanson-Alvarez / 1992. XXXVI, 743 S. / Hld (3-598-22742-6)
- Bd. 3 Alvarez-Angelin / 1992. XXXVI, 755 S. / Hld (3-598-22743-4)

- Bd. 4 Angelin-Ardon / 1992. XXXVI, 736 S. / Hld (3-598-22744-2)
- Bd. 5 Ardos-Avogaro / 1992. XXXVI, 748 S. / Hld (3-598-22745-0)
- Bd. 6 Avogaro-Barbieri / 1992. LVI, 680 S. / Hld (3-598-22746-9)
- Bd. 7 Barbieri-Bayona / 1993. LIX, 688 S. / Hld (3-598-22747-7)
- Bd. 8 Bayonne-Benech / 1993. LX, 640 S. / Hld (3-598-22748-5)
- Bd. 9 Benecke-Berrettini / 1994. LIX, 640 S. / Hld (3-598-22749-3)
- Bd. 10 Berrettini-Bikkers / 1994. LII, 640 S. / Hld (3-598-22750-7)
- Register zu den Bänden 1 - 10 Teil 1: Länder / 1995. VIII, 682 S. / Hld (3-598-22801-5)
- Register zu den Bänden 1 - 10 Teil 2: Künstlerische Berufe / 1995. X, 675 S. / Hld (3-598-22802-3)
- Bd. 11 Biklar-Bobrov / 1995. XIV, 680 S. / Hld (3-598-22751-5)
- Bd. 12 Bobrov-Bordacev / 1995. XIV, 680 S. / Hld (3-598-22752-3)
- Bd. 13 Bordaligo-Braun / 1996. XIV, 690 S. / Hld (3-598-22753-1)
- Bd. 14 Braun-Buckett / 1996. XXIV, 696 S. / Hld (3-598-22754-X)
- Bd. 15 Bucki-Campagnari / 1996. XV, 695 S. / Hld (3-598-22755-8)
- Bd. 16 Campagne-Cartellier / 1997. LI, 636 S. / Hld (3-598-22756-6)
- Bd. 17 Carter-Cesaretti / 1997. XIV, 637 S. / Hld (3-598-22757-4)
- Bd. 18 Cesari-Choupay / 1997. XIV, 642 S. / Hld (3-598-22758-2)
- Bd. 19 Chouppe-Clovio / 1998. LII, 604 S. / Hld (3-598-22759-0)
- Bd. 20 1998. LII, 604 S. / Hld (3-598-22760-4)
- Reg. zu d. Bdn 11-20 Teil 1: Länder / 1998. IX, 622 S. / Hld (3-598-22803-1)
- Reg. zu d. Bdn 11-20 Teil 2: Künstlerische Berufe / 1998. XVIII, 624 S. / Hld (3-598-22804-X)
- Bd. 21 / 1998. LII, 604 S. / Hld (3-598-22761-2)

Allgemeines Künstlerlexikon - Internationale Künstlerdatenbank / Allgemeines Künstlerlexikon - World Biographical Dictionary of Artists (IKD). CD-ROM-Edition / 1993 (3-598-40246-5)
- 2. Ausg. 1995 (3-598-40272-4)
- 3. Ausg. 1996 (3-598-40281-3)
- 4. Ausg. 1997 (3-598-40356-9)
- 5. Ausg. 1997 (3-598-40357-7)
- 6. Ausg. 1998 (3-598-40387-9)
- 7. Ausg. 1998 (3-598-40388-7)

Almanach für Freunde der Schauspielkunst. Jahrgang (1)- (6), Jahrgang 7-10. Fortgesetzt unter den Titeln: Wolffs Almanach für Freunde der Schauspielkunst, Jahrgang 11. Almanach für Freunde der Schauspielkunst, Jahrgang 12-17. Deutscher Bühnen-Almanach, Jahrgang 18-57. Berlin 1837-1893. Mikrofiche-Edition. Hrsg. v. Ludwig Wolff (Jg. 11 ff. v. Alois Heinrich, Jg. 24 v. Ludwig Schneider, Jg. 26 ff. v. Albert Entsch, Jg. 48 ff. v. Th. Entsch) / 1988. DCVC / 40. 890 / (151) S. auf 463 Fiches, Lesefaktor 24x / Silberfiche (3-598-32323-9) / Diazofiche (3-598-32324-7)

Alphabetischer Musikalienkatalog der Pfälzischen Landesbibliothek Speyer. Mikrofiche-Edition / 1991. 74.591 Karteiktn. auf 55 Fiches. Lesefaktor 42x, Beih. / Diazofiche (3-598-33195-9)

Altmann, Günter: Musikalische Formenlehre. Mit Beispielen und Analysen / 1981. 485 S. / Br (UTB 1109) (3-598-02667-6)

- 2. Aufl. 1984. 485 S., davon 80 S. Anh. „Musikbeispiele" / Br (UTB 1109) (3-598-02674-9)
- 3. Aufl.: neue Broschurausg. der unveränd. Aufl. 1984/1989. 405, 80 S. / Br (3-598-10873-7)

American Biographical Archive / Amerikanisches Biographisches Archiv (ABA I). A single-alphabet cumulation of 367 original biographical reference works in approx. 600 volumes covering 300.000 individuals from the earliest period of North American history through to the early 20th century. Microfiche edition. Ed. by Nanette Gibbs and Garance Worters / 1986-1991. 1,842 fiches. Reader factor 24x / Silverfiche (3-598-30951-1) / Diazofiche (3-598-30950-3)
- American Biographical Index / Amerikanischer Biographischer Index (ABI I). Ed. by Laureen Baillie / 6 vols. 1993. Cplt. 3,600 p. / Hard (0-86291-382-9)

American Biographical Archive. Series II / Amerikanisches Biographisches Archiv. Neue Folge (ABA II). A single-alphabet cumulation of 135 original biographical reference works published to 1980. Microfiche edition. Ed. by Laureen Baillie / 1993-1996. 734 Fiches. Reader factor 24x / Silverfiche (3-598-33534-2) / Diazofiche (3-598-33520-2)
- American Biographical Index / Amerikanischer Biographischer Index (ABI II). Comp. by Laureen Baillie / 10 vols. 2nd cumulative and enl. ed. 1998. Cplt. CCXL, 4,200 p. / Hard (3-598-33548-2)

American Libraries 1986. U.S. and Canadian Libraries. Ed. by Barbara Verrel / 1986. 500 p. / Soft (3-598-10613-0)
- 1988 U.S. and Canadian Libraries / 2nd ed. 1987. IX, 715 p. / Soft

American Museums. United States and Canada. Ed. by Barbara Verrel; Assistant: Angelika Riedl / 1987. 550 p. (3-598-10612-2)

American publishers
- 1. Aufl. u.d.T.: American Publisher's Directory. A guide to publishers of books, journals, magazines, directories, reprints, maps, microeditions, braille books, and book clubs. Comp. by Michael Zils / 1. ed. 1978. 390 S. / Soft (3-7940-8000-9)
- 2.ed. 1986. Manag. Ed.: Barbara Verrel. Ed. by Marianne Albertshauser, Astrid Kramuschka. Vol. 1 and 2: U. S. publishers A-Z. ISBN index. Vol. 3: Canadian Publishers, ISBN Index / 3 vols. 1986. XV, 1690 p. / Soft (3-598-10314-9)
- 3rd ed. 1987 / 3 vols. 1987. Cplt. 1720 p. / Soft (0-89664-376-X)

Amir, Dov: Leben und Werk der deutschsprachigen Schriftsteller in Israel. Eine Bio-Bibliographie / 1980. 95 S. / Br (3-598-10070-1)

Aminde, Hans J. / Heinrich W. Wichmann: Der städtebauliche Entwurf von Baugruppen. Entwurfsmethode und Grundlagen für Baumassenkonzepte dargestellt am Beispiel einer Hohschulnutzung. Mit einer Untersuchung zur Festpunktanordnung in Bebauungsarten von Stefan Eichler, Ulrich Gothe und Gerd Peter von der Planungsgruppe für den Institutsbau des Landes Baden-Württemberg / Bearb. von Barbara Aminde, Jochen Jaich und Hans W. Liebert / 1975. 220 S. (3-598-10018-3)

Analytische Bibliographien deutschsprachiger literarischer Zeitschriften. Begr. v. Gerhard Seidel. Redakt.: Volker Riedel, Barbara Voigt

- 8: Internationale Literatur Moskau 1931-45. Bibliographie einer Zeitschrift. Veröffentlichung der Akademie der Künste der Deutschen Demokratischen Republik. Vorw. v. Heinz Willmann. Bearb. v. Christa Streller, Volker Riedel / 2 Bde. 1985. Zus. 535 S. / Gb (3-598-07254-6)
- 11: Heute und Morgen Schwerin 1947-1954. Bibliographie einer Zeitschrift. Vorw. v. Rolf Richter. Veröffentlichung d. Akademie d. Deutschen Demokratischen Republik. Bearb. v. Herbert Riedel / 2. Aufl. 1987. 342 S. / Gb (3-598-07248-1)
- 12: Die Wiener Weltbühne. Wien 1932-1933. Die neue Weltbühne. Prag / Paris 1933-1939. Bibliographie einer Zeitschrift. Veröffentlichung d. Akademie der Künste zu Berlin. Bearb. v. Jörg Armer / 2 Halbbde. 1992. Zus. 797 S. / Gb (3-598-11087-1)
- 13: Neue Deutsche Literatur Berlin 1953-62 (Jahrgang 1-10). Bibliographie einer Zeitschrift. Titelverzeichnis und Register. Veröffentlichung d. Akademie der Künste d. Deutschen Demokratischen Republik. Vorw. v. Günther Deicke. Bearb. v. Siegfried Scheibe / 2 Bde. 1989. Zus. 792 S. / Gb (3-598-07595-2)
- 14: Ost und West. Berlin 1947-49 Bibliographie einer Zeitschrift Veröffentlichung d. Akademie der Künste zu Berlin. Bearb. v. Ewald Birr. Vorbem. u. Redakt.: Barbara Voigt / 1993. 125 S. / Gb (3-598-11141-X)

L'Anarchisme / Anarchism. Catalogue de Livres et Brochures des XIXe et XXe Siècles / Catalogue of XIXth and XXth Centuries Books and Pamphlets from Different Countries Ed. par.: Institut Français d'Histoire Sociale, Paris. Redigé sous la direction de Denise Fauvel-Rouif par Janine Gaillemin; Marie A. Sowerwine-Mareschal / Diana Richet / 1982. 170 S. / Br (3-598-10442-1)
- Vol. II. Sous la direction de Denise Fauvel-Rouif. Réalisé par Hélène Strub / 1993. VII, 303 S. / Gb (3-598-11134-7)

Ancient Roman Architecture. A Photographic Index of Art and Architecture on Microfiche. Ed. by Fototeca Unione, American Academy in Rome, International Union of Institutes of Archaeology, History, and History of Art in Rome / 1979. 141 Microfiches (3-598-06180-3)

Anderson, Dorothy: Universal Bibliographic Control. A long term policy. A plan for action / 1974. 87 S. / Br (3-7940-4420-7)

Anderson, Robert: Wagner. A Bibliography with a Survey of Books, Editions and Recordings / 1980. 154 p. (3-598-10327-1)

Anekdoten-Lexikon. - Nachdr. der Originalausg. von 1843 / 44 / 1969. 426 S. (3-7940-4282-4)

An **Annotated** Bibliography of English fiction representing Africa. Ed by Gordon Killam / 1987. 280 p. (3-598-10658-0)

Antisemitism. An Annotated Bibliography. Ed. by Vidal Sassoon International Center for the Study of Antisemitism; The Hebrew University of Jerusalem; Susan Sarah Cohen
- Vol. 1 Antisemitism (1984-1985) / 1987. XXIX, 392 p. / Hard (3-598-23697-2)
- Vol. 2 Antisemitism (1986-1987) / 1991. XXXIV, 559 p. / Hard (3-598-23698-0)

- Vol. 3 Antisemitism (1987-1988) / 1994. XXXIV, 544 p. / Hard (3-598-23699-9)
- Vol. 4-6 Antisemitism (1988-1990) / 3 vols. with index to vols. 4-6 1997. Cplt. LXXV, 1450 p. / Hard (3-598-23703-0)
- Vol. 7-9 Antisemitism (1991-1993) / 3 vols. with index to vols. 7-9 1997. Cplt. LXXXV, 1449 p. / Hard (3-598-23710-3)
- Vol. 10-11 Antisemitism (1994-1995) / 2 vols. with index to vols. 10-11 1999. Cplt. XV, 1001 p. / Hard (3-598-23707-3)

Apianus, Peter (Petrus): Abbreviationes Vetustorum Monumentorum in Ordinem Alphabeticum Digestae. Verzeichnis der im Mittelalter gebräuchlichen Abkürzungen. - Nachdr. d. Ausg. Ingolstadt 1554. - Nachwort von Johannes Müller / 1968. 30 S. / Br (Vlg.-Nr. 02108)

Appell, Johann Wilhelm: Die Ritter-, Räuber- und Schauerromantik. Zur Geschichte d. dt. Unterhaltungsliteratur. - Nachdruck d. Ausg. Leipzig: Engelmann 1859 / 1968. V, 92 S. / Br (Vlg.-Nr. 04097) [Lizenzausg. m. Genehm. v. Zentralantiquariat d. DDR, Leipzig]

Arab-Islamic Biographical Archive / Arabisch-Islamisches Biographisches Archiv (AIBA). Microfiche edition. Comp. by Ulrike Kramme and Zelmira Urra Muena / 1995-1999. Approx. 450 fiches in 12 instalments, Reader factor 24x / Silverfiche (3-598-33881-3) / Diazofiche (3-598-33880-5)

Arbeiten zur sozialwissenschaftlich orientierten Freiraumplanung. Hrsg. v. Ulfert Herlyn; Gert Gröning (ISSN 0941-0783) [Minerva Publikation]
- 1: Seyfang, Volkmar: Freiraumnutzung im Geschosswohnungsbau Theoretische Überlegungen und empirische Grundlagen zu einer nutzerorientierten Planung und Gestaltung der Freiräume im Geschosswohnungsbau. Hrsg. v. Ulfert Herlyn; Gert Gröning / 1980. XII, 420 S., 120 S. Anh / Br (3-597-10265-4)
- 2: Gröning, Gert: Dauercamping Analyse und planerische Einschätzung einer modernen Freizeitwohnform / 2. verb. Aufl. 1984. X, 192 S., 12 schw.-w. Abb. / Br (3-597-10266-2)
- 3: Spitthöver, Maria: Freiraumansprüche und Freiraumbedarf Zum Einfluss von Freiraumversorgung und Schichtzugehörigkeit auf die Anspruchshaltungen an innerstädtischen Freiraum / 1982. 346 S., 5 Ktn, 14 Abb., 40 Tab. / Br (3-597-10267-0)
- 4: Buchholz, Rüdiger / Gert Gröning; Maria Spitthöver: Grün in alten Stadtvierteln Eine empirische Untersuchung zur Nutzung und Beurteilung innerstädtischer Freiraumqualität / 1984. X, 181 S. / Br (3-597-10268-9)
- 5: Milchert, Jürgen: Tendenzen der städtischen Freiraumentwicklung in Politik und Verwaltung / 1984. X, 324 S. / Br (3-597-10269-7)
- 6: Amar, Laure: Parks und Plätze in Paris Eine sozialpsychologische Analyse städtischer Freiraumqualitäten / 1986. II, 225 S. / Br (3-597-10534-3)
- 7-9: Gröning, Gert / Joachim Wolschke-Bulmahn: Die Liebe zur Landschaft
  - 7: Tl. I. Natur in Bewegung Zur Bedeutung natur- und freiraumorientierter Bewegungen der ersten Hälftedes 20. Jahrhunderts für die Entwicklung der Freiraumplanung / 1986. XII, 289 S. / Br (3-597-10537-8)
  - 8: Tl. II. Vom Gartenarchitekten zum Landschaftsplaner / ca. 300 S. / Br (3-597-10603-X)

- 9: Tl. III. Der Drang nach Osten Zur Entwicklung der Landespflege im Nationalsozialismus und während des Zweiten Weltkrieges in den „eingegliederten Ostgebieten" / 1987. X, 279 S. / Br (3-597-10535-1)
- 10: Landschaftswahrnehmung und Landschaftserfahrung Texte zur Konstitution und Rezeption von Natur als Landschaft. Hrsg. v. Gröning, Gert; Herlyn, Ulfert / 1990. 174 S. / Br (3-597-10536-X)
- 11: Wolschke-Bulmahn, Joachim: Auf der Suche nach Arkadien Zu Landschaftsidealen und Formen der Naturaneignung in der Jugendbewegung und ihrer Bedeutung für die Landespflege / 1989. X, 285 S., 85 Abb. / Br (3-597-10604-8)
- 12: Von grossen Plätzen und kleinen Gärten Beiträge zur Nutzungsgeschichte von Freiräumen in Hannover Vorw. u. Hrsg. v. Herlyn, Ulfert; Ursula Poblotzki / 1992. 209 S., 130 schw.-w. Abb. / Br (3-597-10701-X)
- 13: Poblotzki, Ursula: Menschenbilder in der Landespflege 1945-1970 / 1992. XII, 470 S. / Br (3-597-10702-8)
- 14: Breckwoldt, Michael: Das „Landleben" als Grundlage für eine Gartentheorie. Eine literaturhistorische Analyse der Schriften von Christan Cay Lorenz Hirschfeld / 1995. 153 S. / Br (3-597-10703-6)

Arbeitsbuch Geschichte. Hrsg. v. Eberhard Büssem; Michael Neher.
- Mittelalter. 3. bis 16. Jh. Repetitorium. Bearb. v. Karl Brunner. (UTB 411) / 4., überarb. Aufl. 1975. 320 S., 1 Faltbl / Br (3-7940-2629-2)
  - 5., verb. Aufl. 1977. 320 S., Abb. / Br (3-7940-2629-2)
  - 6., verb. Aufl. 1980. 327 S., 1 Falttaf / Br (3-598-02629-3)
  - 7., verb. Aufl. 1983. 329 S., mehrere Abb. u. Tab., 1 Faltpl. / Br (3-598-02630-7)
  - 8. Aufl. 1987. 329 S., mehrere Abb. u. Tab., 1 Faltpl., Reg / Br (3-598-02631-5)
  - 9. durchges. Aufl. 1990. 331 S. / Br (3-598-10890-7)
- Neuzeit 1. 16.-18. Jahrhundert. Repetitorium. Bearb. v. Eberhard Büssem; A. Faust; H. F. Kopp; Michael Neher. Bibliograph. Mitarb.: Ulrike Schiller-Peters. (UTB 569) / 1976. 391 S. (3-7940-2630-6)
  - 3., überarb. Aufl. 1976. 387 S., Ktn, Tab. / Br (3-7940-2630-6)
  - 4., erg. Aufl. 1979. 394 S., Ktn. u. Tab. / Br (3-598-02630-7)
  - 5., verb. u. erg. Aufl. 1983. 417 S., Ktn, Tab., Reg / Br (3-598-02634-X)
  - 6. durchges. Aufl. 1991. 424 S. / Br (3-598-10969-5)
- Neuzeit 1. 16.-18. Jahrhundert. Quellen. Mit einer Einführung in die hilfswissenschaftlichen Disziplinen Bearb. v. Leopold Auer. (UTB 625) / 1977. 358 S. / Br (3-7940-2631-4)
- Neuzeit 3 / 1. 1871-1914. Die imperialistische Expansion. Repetitorium 1.Teil. Bearb. v. Gerd Höhler. (UTB 1143) / 2. völlig überarb. Aufl. 1982. 386 S., 3 Ktn, mehrere Tab. / Br (3-598-02632-3)
- Neuzeit 3 / 2. 1871-1914. Die imperialistische Expansion. Repetitorium 2.Teil. Bearb. v. Gerd Höhler, unter Mitarb. v. Eberhard Büssem; Anselm Faust. (UTB 1144) / 2., völlig überarb. Aufl. 1982. XIV, S. 387-681, 1 herausnehmb. Schaubl. / Br (3-598-02633-1)

Archäographie. Archäologie u. elektronische Datenverarbeitung. Hrsg.: Cornelius Ankel und Rolf Gundlach. Jg. 3/1974 - Jg.6/1977 (1 Heft jährlich)

(ISSN 0587-3460) [Erscheinen eingestellt. Vorher im Vlg. Bruno Hessling, Berlin]

Archives Biographiques Françaises / Französisches Biographisches Archiv / French Biographical Archive (ABF I). A single-alphabet cumulation of 180 biographical reference works covering approx. 140,000 individuals. Microfiche edition. Ed. by Susan Bradley / 1989-1991. 1,065 fiches. Reader factor 24x / Silverfiche (3-598-32579-7) / Diazofiche (3-598-32564-9)
• Index Biographique Français / Französischer Biographischer Index / French Biographical Index (IBF). Ed. by Helen Dwyer / 4 vols. 1993. Cplt. 2155 p. / Hard (0-86291-801-4)

Archives Biographiques Françaises. Deuxième Série / Französisches Biographisches Archiv. Neue Folge / French Biographical Archive. Series II (ABF II). A single-alphabet cumulation of 122 original biographical reference works published to 1986 covering approx. 175,000 individuals. Microfiche edition. Comp. by Tommaso Nappo / 1993-1996. 664 fiches. Reader factor 24x / Silverfiche (3-598-33568-7) / Diazofiche (3-598-33555-5)
• Index Biographique Français / Französischer Biographischer Index / French Biographical Index (IBF II). Comp. by Tommaso Nappo / 7 vols. 2ème édition cumulée et augmentée / 1997. Cplt. CLXI, 3310 p. / Hard (3-598-33581-4)

Archivio Biografico Italiano / Italienisches Biographisches Archiv / Italian Biographical Archive (ABI I). Eine Kumulation der 321 wichtigsten biographischen Nachschlagewerke für den italienischen Bereich bis zum Anfang des 20. Jahrhunderts. Insgesamt ca. 200.000 biographische Einträge. Mikrofiche-Edition Redakt. Leitung: Tommaso Nappo. Beratender Hrsg.: Silvio Furlani / 1987-1990. 1.046 Fiches. Lesefaktor 24x / Silberfiche (3-598-31520-1) / Diazofiche (3-598-31540-6)
• Indice Biografico Italiano / Italienischer Biographischer Index / Italian Biographical Index (IBI I). Bearb. v. Tommaso Nappo und Paolo Noto / 4 Bde. 1993. Zus. XXXIV, 1440 S. / Ln (3-598-31555-4)

Archivio Biografico Italiano. Nuova Serie / Italienisches Biographisches Archiv. Neue Folge / Italian Biographical Archive. Series II (ABI II). Mikrofiche-Edition Redakt. Leitung: Tommaso Nappo. Beratender Hrsg.: Silvio Furlani / 1991-1994. 710 Fiches. Lesefaktor 24x / Silberfiche (3-598-33154-1) / Diazofiche (3-598-33140-0)
• Supplemento / 1997. 95 Fiches / Silberfiche (3-598-33331-5) / Diazofiche (3-598-33330-7)
• Indice Biografico Italiano / Italienischer Biographischer Index / Italian Biographical Index (IBI II). Der Index kumuliert in einem Alphabet die biographischen Kurzinformationen zu den ca. 350198 Artikeln im ABI I und II. Vorw. v. Silvio Furlani. Bearb. v. Tommaso Nappo / 7 Bde. 2. kumulierte u. erw. Ausgabe 1996. Zus. CCXXIV, 2.650 S. / Ln (3-598-33168-1)
• Indice Biografico Italiano. Edizione CD-ROM / Italienischer Biographischer Index. CD-ROM-Ausgabe / Italian Biographical Index. CD-ROM Editon / 1998 (3-598-40379-8)

Archivio Biografico Italiano sino al 1996 / Italienisches Biographisches Archiv bis 1996 / Italian Biographical Archive to 1996 (ABI III). Mikrofiche-Edition Bearb. v. Tommaso Nappo / 1998-2001. Ca. 470 Fiches in 12 Lfgn, Lesefaktor 24x / Silberfiche (3-598-34301-9) / Diazofiche (3-598-34300-0)

Archivo Biográfico de España, Portugal e Iberoamérica / Spanisches, Portugiesisches und Iberoamerikanisches Biographisches Archiv / Spanish, Portuguese and Latin-American Biographical Archive (ABEPI I). Mikrofiche-Edition. Bearb. v. Victor Herrero Mediavilla, Lolita R. Aguayo Nayle / 1986-1989. 1.144 Fiches. Lesefaktor 24x / Silberfiche (3-598-32045-0) / Diazofiche (3-598-32030-2)
• Indice Biográfico de España, Portugal e Iberoamérica / Spanischer, Portugiesischer und Iberoamerikanischer Biographischer Index / Spanish, Portuguese and Latin-American Index (IBEPI I). Bearb. v. Victor Herrero Mediavilla, Lolita R. Aguayo Nayle / 4 Bde. 1990. Zus. XXXVII, 1429 S. / Ln (3-598-32060-4)

Archivo Biográfico de España, Portugal e Iberoamérica. Nueva Serie / Spanisches, Portugiesisches und Iberoamerikanisches Biographisches Archiv. Neue Folge / Spanish, Portuguese and Latin-American Biographical Archive. Series II (ABEPI II). Die neue Folge des Spanischen, Portugiesischen und Iberoamerikanischen Archivs umfaßt mit der Auswertung von 328 biographischen Lexika bis 1959 ca. 200.000 biographische Artikel. Mikrofiche-Edition Bearb. v. Victor Herrero Mediavilla / 1991-1993. 1018 Fiches, Lesefaktor 24x / Silberfiche (3-598-32964-4) / Diazofiche (3-598-32977-6)
• Indice Biográfico de España, Portugal e Iberoamérica / Indice Biográfico deEspanha, Portugal e Ibero-America / Spanish, Portuguese and Latin-American Biographical Index / Spanischer, Portugiesischer und Iberoamerikanischer Biographischer Index (IBEPI II). Der Index kumuliert in einem Alphabet die biographischen Kurzinformationen zu den 410.209 Artikeln im ABEPI I und II Bearb. v. Victor Herrero Mediavilla / 7 Bde. 2. kumulierte u. erw. Ausg. 1995. Zus. LXVIII, 3315 S. / Ln (3-598-32990-3)
• Indice Biográfico de España, Portugal e Iberoamérica. Edición en CD-ROM / Indice Biográfico de Espanha, Portugal e Ibero-América. Ediçao em CD-ROM / Spanish, Portuguese and Latin American Biographical Index. CD-ROM-Edition / Spanischer, Portugiesischer und Iberoamerikanischer Biographischer Index. CD-ROM-Edition / 1997 (3-598-40285-6)

Archivo Biográfico de España, Portugal e Iberoamérica 1960-1995 / Spanisches, Portugiesisches und Iberoamerikanisches Biographisches Archiv 1960-1995 / Spanish, Portuguese and Latin-American Biographical Archive 1960-1995 (ABEPI III). Eine alphabetische Kumulation von biographischen Einträgen aus 227 Nachschlagewerken von 1960 bis 1995. Microfiche-Edition. Bearb. v. Victor Herrero Mediavilla / 1996-1998. 471 Fiches in 12 Lfgn, Lesefaktor 24x / Silberfiche (3-598-34031-1) / Diazofiche (3-598-34030-3)

Archivum. International review on archives / Revue Internationale des Archives. Ed. by the International Council on Archives (ISSN 0066- 6793)
• Vol. I: Proceedings of the International Congress on Archives. 1st Congress, Paris 1950 / 1951. 144 p. (3-7940-3751-0)
• Vol. II: The Archives of International Organisations. Archiv Repair. The Information of the Public in Archives. International Archival Bibliography 1945-1952 / 1953. 230 p. (3-7940-3752-9)
• Vol. III: Proceedings of the International Congress on Archives. 2nd Congress, The Hague / 1953 (3-7940-3753-7)

• Vol. IV: Professional Training of Archivists. Ecclesiastical Archives. Internaional Archival Bibliography 1953 / 1955. 289 p. (3-7940-3754-5)
• Vol. V: International Directory of Archives / 1956. 253 p. (3-7940-3755-3)
• Vol. VI: Proceedings of the International Congress on Archives. 3rd Congress, Florence 1956 / 1958. (3-7940-3756-1)
• Vol. VII: Archive Buildings. Part 2. International Archival Bibliography 1954-1956 / 1958. 287 p. (3-7940-3757-X)
• Vol. VIII: Birth, Marriage and Death Registers, Part I. Archive Repair. Private Archives. International Archival Bibliography 1957, Part I / 1959. 197 p. (3-7940-3758-8)
• Vol. IX: Birth, Marriage and Death Registers, Part II. International Archival Bibliography 1957, Part II / 1961. 279 p. (3-7940-3759-6)
• Vol. X: Proceedings of the International Congress on Archives. 4th Congress, Stockholm 1960 / 1962. (3-7940)
• Vol. XI: Activities of Archives in the World / 1963. 280 p. (3-7940-3761-8)
• Vol. XII: Notarial Archives / 1965. 128 p. (3-7940-3762-6)
• Vol. XIII: Municipal Archives / 1965. 144 p. (3-7940-3763-4)
• Vol. XIV: Proceedings of the International Congress on Archives. 5th Congress, Brussels 1964 / 1968. (3-7940-3764-2)
• Vol. XV: The Great Archive Repositories of the World. Notices on the most important repositories with holdings of international interest / 1969. 374 p. (3-7940-3765-0)
• Vol. XVI: Proceedings of he Extraordinary International Congress on Archives, Washington 1966 / 1969. 237 p. (3-7940-3766-9)
• Vol. XVII: Archive Legislation. Europe. Part 1 / 1971. 270 p. (37-7940-3767-7)
• Vol. XVIII: Proceedings of the International Congress on Archives. 6th Congress, Madrid 1968 / 1970. (3-7940-3768-5)
• Vol. XIX: Archive Legislation. Europe. Part 2 / 1972. 260 p. (3-7940-3769-3)
• Vol. XX: Archive Legislation. Africa-Asia. Part 3 / 1972. 246 p. (3-7940-3770-7)
• Vol. XXI: Archive Legislation. America-Oceania. Part 4 / 1973. 239 p. (3-7940-3771-5)
• Vol. XXII / XXIII: International Directory of Archives / Annuaire international des archives / (1972-73) 1975. 480 p. / Soft (3-7940-3772-3)
• Vol. XXIV: Proceedings of 7th International Congress on Archives, Moscow 1972 / Actes du 7e Congrés international des Archives, Moscov 1972 / (1974) 1976. 388 p. / Soft (3-7940-3774-X)
• Vol. XXV: Basic International Bibliography of Archive Administration / Bibliographie internationale fondamentale a 'archivistique / 1978. 250 p. / Soft (3-7940-3775-8)
• Vol. XXVI: Proceedings of the 8th International Congress on Archives, Washington 1967 / Actes du 8e Congrés international des Archives / 1979. 207 p. / Soft (3-598-03776-7)
• Vol. XXVII: Labour and Trade Union Archives / Les Archives des Syndicats et Mouvements Ouvriers / 1980. 190 p. / Soft (3-598-21227-5)
• Vol. XXVIII: Archival legislation / Législation Archivistique. 1970-80 / 1981. 447 p. / Soft (3-598-21228-3)
• Vol. XXIX: Proceedings of the 9th International Congress on Archives / Actes du 9e Congrès international des Archives London 1980 / 1982. 204 p. / Soft (3-598-21229-1)
• Vol. XXX: Archives, Libraries, Museums and Documentation Centres. General Index of

Archivum Vol. I to XXX / Archives, Bibliothèques, Musées et Centres de Documentation. Index général d'Archivum Vol. I à XXX / 1984. 118 p. / Soft (3-598-21230-5)
- Vol. XXXI: Modern Buildings of National Archives / Bâtiments Modernes d'Archives Nationales / 1986. 142 p., 60 illus. / Soft (3-598-21231-3)
- Vol. XXXII: Proceedings of the 10th International Congress on Archives. Bonn 17.-21. September 1984 / Actes du 10e Congrès International des Archives / 1986. 332 p. / Soft (3-598-21232-1)
- Vol. XXXIII: International Directory of Archives / Annuaire International des Archives / 1988. XLV, 351 p. / Soft (3-598-21233-X)
- Vol. XXXIV: Professional Training of Archivists / Formation Professionnelle des Archivistes Frwd by Vanrie, André / 1988. XII, 236 p. / Soft (3-598-21234-8)
- Vol. XXXV: Proceedings of the 11th International Congress on Archives (Paris, 22.-26. August 1988) / Actes du 11e Congrès International des Archives (Paris, 22.-26. ao-t 1988) / 1989. 284 p. / Soft (3-598-21235-6)
- Vol. XXXVI: International Bibliography of Directories and Guides to Archival Repositories / Bibliographie Internationale des Guides et Annuaires Relatifs aux Dépôts d'Archives / 1990. XXIX, 195 p. / Soft (3-598-21236-4)
- Vol. XXXVII: Archives and Genealogical Sciences / Les Archives et les Sciences Généalogiques / 1992. XV, 254 p. / Soft (3-598-21237-2)
- Vol. XXXVIII: International Directory of Archives / Annuaire International des Archives / 1992. XLIII, 427 p. / Soft (3-598-21238-0)
- Vol. XXXIX: Proceedings of the 12th International Congress on Archives / Actes du 12e Congrès International des Archives. Montreal, 6-11 September 1992 / Montreal, 6-11 Septembre 1992 / 1993. XXII, 534 p. / Soft (3-598-21239-9)
- Vol. XL: Archival Legislation 1981-94 / Législation Archivistique 1981-94 Albania-Kenya / 1995. XX, 343 p. / Soft (3-598-21240-2)
- Vol. XLI: Archival Legislation 1981-94 / Législation Archivistique 1981-94 Latvia-Zimbabwe / 1995. XXIII, 338 p. / Soft (3-598-21241-0)
- Vol. XLII: Memory of the World at Risk: Archives Destroyed, Archives Reconstituted / 1996. XI, 359 p. / Hard (3-598-21243-7)
- Vol. XLIII: Proceedings of the 13th International Congress on Archives. Beijing 2-7 September 1996 / Actes du 13e Congrès International des Archives. Pékin 2-7 Septembre 1996 / 1997. XIV, 381 p. / Soft (3-598-21244-5)
- Vol. XLIV: Basic Archival Problems. Strategies for Development / 1999. XIV, 264 p. / Soft (3-598-21245-3)
- Special Vol. 1: Proceedings of the General Conference on the Planing of Archival Developments in the Third World, Dakar 1975 / Cahier hors 1976. 117 p. / Soft (3-7940-3750-2)
- Special Vol. 2: Actes de la Seconde Conference des Archives antillaises / Proceedings of the Second Caribbean Archives Conference. Guadeloupe, Martinique 1975 / Cahier hors 1980. 160 p. / Soft (3-598-21002-7)

Archivwesen in der DDR. Theorie und Praxis. Von einem Autorenkollektiv unter Leitung v. Botho Brachmann / 1984. 480 S., 1 Kte, 13 Schemata / Ln (3-598-07212-0) [Lizenzausg. m. Genehm. v. VEB Deutscher Verlag der Wissenschaften, Leipzig]

Arndt, Alfred: Kleines Formelexikon. Geleitw. v. Manfred von Ardenne / 1973. 560 S., 370 Abb. / Kst. (UTB 193) (3-7940-2615-2)

Arndt, Karl J. / May E. Olson: The German Language Press of the Americas / Die deutschsprachige Presse der Amerikas [1973 v. Vlg. Quelle & Meyer, Wiesbaden übernommen]
- Vol. 1: History and Bibliography 1732-1968: U.S.A / Geschichte und Bibliographie 1732-1968: Vereinigte Staaten von Amerika / 3rd, rev. Ed. (and enlarged by an Appendix) 1976. IV, 845 p. / Hard (3-7940-3422-8)
- Vol. 2: History and Bibliography 1732-1968: Argentinia, Bolivia, Brazil, Canada, Chile, Columbia, Costa Rica, Cuba, Dominican Republic, Ecuador, Guatemala, Guayana, Mexico, Paraguay, Peru, Uruguay, USA (Addenda), Venezuela / Geschichte und Bibliographie 1732-1968: Argentinien, Bolivien, Brasilien, Chile, Costa Rica, Dominikanische Republik, Ecuador, Guatemala, Guayana, Kanada, Kolumbien, Kuba, Mexiko, Paraguay, Peru, Uruguay, Venezuela, Vereinigte Staaten von Amerika (Nachtrag) / 1973. 709 p., 24 ill / Lin (3-7940-3421-X)
- Vol. 3: German American Press Research from the American Revolution to the Bicentennial / Deutsch-amerikanische Presseforschung von der amerikanischen Revolution bis 1976 / 1980. XIII, 848 p. / Lin (3-598-10152-X)

Arntz, Helmut: Die Kognakbrenner. Der Geschichte vom Geist des Weines anderer Teil. Ed.: Gesellschaft für Geschichte des Weines e.V., Wiesbaden / 1990. 360 S. / Ln (3-598-10936-9)

Ars Impressoria. Entstehung und Entwicklung des Buchdrucks. Eine internationale Festgabe für Severin Corsten zum 65. Geburtstag. Hrsg. v. Hans Limburg / 1985. 354 S. / Lin (3-598-10587-8)

L'**Art** et ses adresses en France 1996/97. French Art Directory 1996/97 / 1996. 304 S. / Gb (3-598-23077-X)

Art Nouveau / Jugendstil. Architecture in Europe. Ed. by Hans D. Dyroff, Deutsche UNESCO-Kommission. Scientific Coordination: Manfred Speidel / 1988. 244 p. / Hard (3-598-10818-4)

Atom - Kartei. Literaturnachweise von 1 / 55 - 6 / 59 mit insgesamt 14.000 Literaturtiteln / 9 Ringordner 1959

August Bebel. Ausgewählte Reden und Schriften / 10 Bde. in 14 Tlbdn. 1995-97. / Gb (3-598-11372-2)
- Bd. 1: Reden und Schriften, Briefe 1863 bis 1878. Anmerkungen, Bibliographie und Register. Hrsg. v. Horst Bartel; Rolf Dlubek; Heinrich Gemkow,. Bearb. v. Rolf Dlubek; Ursula Herrmann, unter Mitarb. v. Dieter Malik / 1995. 765 S. / Ln (3-598-11265-3)
- Bd. 2: Reden und Schriften, Briefe 1878 bis 1890. Anmerkungen, Bibliographie und Register. Hrsg. v. Horst Bartel u.a. Bearb. v. Ursula Herrmann; Heinrich Gemkow, unter Mitarb. v. Anneliese Beske; Marga Beyer; Wilfried Henze; Gudrun Hofmann; Ruth Rüdiger; Gerhard Winkler / 2 Tlbde. 1995. Zus. 1.491 S. / Ln (3-598-11266-1)
- Bde. 3-5: Reden und Schriften, Briefe 1890 bis 1899. Band 3: Reden und Schriften. Oktober 1890 bis Dezember 1895 / Band 4: Reden und Schriften. Januar 1896 bis Oktober 1899 / Band 5: Briefe 1890 bis 1899. Anmerkungen, Bibliographie und Register zu den Bänden 3 bis 5. Hrsg. v. Gustav Seeber. Bearb. v. Bärbel Bäuerle;

Anneliese Beske; Gustav Seeber; Walter Wittwer. Endredakt.: Anneliese Beske; Eckhard Müller / 3 Bde. 1995. Zus. 1.564 S. / Ln (3-598-11235-1)
- Bd. 6: Aus meinem Leben. Hrsg. v. Horst Bartel u.a. Bearb. v. Ursula Herrmann, unter Mitarb. v. Wilfried Henze; Ruth Rüdiger / 1995. 811 S. / Ln (3-598-11267-X)
- Bd. 7: Reden und Schriften 1899 bis 1905. Hrsg.: Internationales Institut f. Sozialgeschichte, Amsterdam. Bearb. v. Anneliese Beske; Eckhard Müller / 2 Tlbde. 1997. Zus. 22, 919 S. / Ln (3-598-11236-X)
- Bd. 8: Reden und Schriften 1906 bis 1913. Hrsg.: Internationales Institut f. Sozialgeschichte, Amsterdam. Bearb. v. Anneliese Beske; Eckhard Müller / 2 Tlbde. 1997. Zus. 649 S. / Ln (3-598-11277-7)
- Bd. 9: Briefe 1899 bis 1913 August Bebel Anmerkungen, Bibliographie und Register zu den Bänden 7 bis 9. Hrsg.: Internationales Institut f. Sozialgeschichte, Amsterdam. Bearb. v. Anneliese Beske; Eckhard Müller / 1997. 595 S. / Ln (3-598-11278-5)
- Bd. 10: Die Frau und der Sozialismus. Erste und 50. Auflage. Beilagen, Anmerkungen, Bibliographie und Register. Hrsg.: Internationales Institut f. Sozialgeschichte, Amsterdam. Bearb. v. Anneliese Beske; Eckhard Müller. Vorw. v. Susanne Miller / 2 Tlbde. 1996. Zus. 834 S. / Ln (3-598-11237-8)

August Bebel. Zum siebzigsten Geburtstag, 22. Februar 1910. - Reprint des Unikats aus dem Zentralen Parteiarchiv der SED, Dietz Verlag, Berlin. - Nachw. v. Ursula Herrmann / 1989. 343 S. / Ln (3-598-07256-2)

Aus alten Börsenblättern. Ein Anzeigen-Querschnitt durch das Börsenblatt für den Deutschen Buchhandel 1834 - 1945. Auswahl u. hrsg. v. Klaus Gerhard Saur. Einführung v. Ehrhardt Heinold / 2. Aufl. 1968. XVI, 303 S. / Ln (3-7940-2088-X) / Br (3-7940-2098-7)

Die **Ausbürgerung** deutscher Staatsangehöriger 1933 - 45 nach den im Reichsanzeiger veröffentlichten Listen / Expatriation Lists as Published in the „Reichsanzeiger" 1933-45. Einl. v. Hans G. Lehmann; Michael Hepp. Hrsg. v. Michael Hepp / 3 Bde. 1985-1988 Zus. LXXII, 1376 S. / Lin (3-598-10537-1)
- Bd. 1: Listen in chronologischer Reihenfolge / Lists in chronological order / 1985. LVIII, 724 S. / Lin (3-598-10538-X)
- Bd. 2: Namensregister / Name Index / 1985. VII, 356 S. / Lin (3-598-10539-8)
- Bd. 3: Register der Geburtsorte und der letzten Wohnorte / Index to Place of Birth. Index to Place of last-known Residence / 1988. VII, 296 S. / Lin (3-598-10540-1)

Australasian Biographical Archive / Australasiatisches Biographisches Archiv (ANZO-BA). A single-alphabet cumulation of 181 biographical reference works covering approx. 175,000 biographical entries. Mikrofiche edition. Ed. by Victor Herrero Mediavilla. 1990-1993. 423 fiches. Reader factor 24x / Silverfiche (3-598-32930-X) / Diazofiche (3-598-32944-X)
- Australasian Biographical Index / Australasiatischer Biographischer Index (ANZO-BI). Ed. by Victor Herrero Mediavilla / 3 vols. 1996. Cplt. VL, 1328 p. / Hard (3-598-32958-X)
- Supplement Ed. by Victor Herrero Mediavilla / 1995. 119 fiches / Silverfiche (3-598-32925-3) / Diazofiche (3-598-32924-5)

Automatisierung bei der Zeitschriftenbearbeitung. Berichte e. Kolloquiums veranst. v. d. Arbeitsstelle f. Bibliothekstechnik am 25. u. 26. Jan. 1972 Arbeitsstelle f. Bibliothekstechnik bei d. Staatsbibliothek Preuß. Kulturbesitz / 1972. 220 S. (3-7940-320003-9)

Die **Auswirkungen** neuer Technologien auf das Verlagswesen. Bericht über ein von der Kommission der Europäischen Gemeinschaften, Generaldirektion Wissenschaftliche und Technische Information und Informationsmanagement, veranstaltetes Symposium, Luxemburg, 6.-7. November 1979. Hrsg.: Kommission d. Europ. Gemeinschaften, Generaldirektion Wissenschaftl. u. Techn. Information u. Informationsmanagement / 1981. 209 S. / Br (3-598-10127-9)

Bärwinkel, Roland / Natalija I. Lopatina / Günther Mühlpfordt: Schiller-Bibliographie 1975-85 / 1989. XIII, 527 S. / Ln (3-598-07273-2)

Baeyer, Alexander von / Bernhard Buck: Wörterbuch Kommunikation und Medienpraxis für Erziehung und Ausbildung. Begriffe u. Erl. aus Praxis, Wiss. u. Technik / 1979. 158 S. / Br (3-598-10001-9)

Bakewell, K. G.: Classification and indexing Practice / 1978. 216 p. (0-85157-247-2) [Clive Bingley]

Baltisches Biographisches Archiv / Baltic Biographical Archive (BaBA). Eine alphabetische Kumulation von biographischen Einträgen aus 218 Nachschlagewerken von ca. 1650 bis 1993. Mikrofiche-Edition. Bearb. v. Axel Frey. Wiss. Berater: Paul Kaegbein / 1995-1998. 436 Fiches, Lesefaktor 24x / Silberfiche (3-598-33821-X) / Diazofiche (3-598-33820-1)

Das **Bamberger** Blockbuch. Inc. typ. Ic I 44 der Staatsbibliothek Bamberg. Ein xylogryphisches Rechenbuch aus dem 15. Jahrhundert. Hrsg. u. erl. von Kurt Vogel. Mit e. buchkundl. Beschreibung von Bernhard Schemmel (Veröffentlichungen des Forschungsinstituts des Deutschen Museums für die Geschichte der Naturwissenschaften und Technik) / 1980. 105 S. / Gb (3-598-10150-3)

Bartel, Hans / Erhard Anthes: Mengenlehre. Theoretische und didaktische Grundlagen / 1975. 212 S., zahlr. Tab. / Kst. (UTB 466) (3-7940-2642-X)

Bartke, Wolfgang: Biographical Dictionary and Analysis of China's Party Leadership 1922-88 / 1990. X, 482 p. / Hard (3-598-10876-1)

Bartke, Wolfgang: The Agreements of the People's Republic of China with Foreign Countries 1949-1990 / 2nd rev. and enl. ed. 1992. IX, 231 p. / Hard (3-598-10840-0)

Bartke, Wolfgang: The Economic Aid of the PR China to Developing and Socialist Countries. A Publ. of the Institute of Asian Affairs, Hamburg / 2nd rev. and enl. ed. 1989. 160 p. / Hard (3-598-10839-7)

Bartke, Wolfgang: Who's who in the People's Republic of China
• 2nd ed. With more than 1000 portraits / 1987. IX, 786 p. / Hard (3-598-10610-6)
• 3rd ed. With more than 2400 portraits / 1990. 909 p. / Hard (3-598-10771-4)

Bartke, Wolfgang: Who was who in the People's Republic of China. With more than 3100 Portraits / 2 vols. 1997. Cplt. XVIII, 700 p. / Hard (3-598-11331-5)

Bartsch, Eberhard: Die Bibliographie. Einführung in Benutzung, Herstellung, Geschichte / 1979. 280 S., Reg / Br (UTB 948) (3-598-02665-X)
• 2. durchges. Aufl. 1989. 280 S. / Br (3-598-10878-8)

Baudenkmäler in Bayern. Hrsg. von Michael Petzet, Bayerisches Landesamt für Denkmalpflege, München
• Bd. 12: Landkreis Fürstenfeldbruck. Bearb. v. Klaus Kraft und Florian Hufnagel: Fotos: Werner Neumeister / 1978. 102 S., 3 Farbtaf. / Br (3-7940-5812-7)
• Bd. 63: Stadt Schwabach. Bearb. v. Klaus Kratzsch und Alexander Rauch. Fotos: Joachim Sowieja / 1978. 104 S., 5 Farbtaf., 1 Kte. / Br (3-7940-5863-1)

Bautechnische Flächenarten. Ergebnisbericht Naturwissenschaften. Von Artur Bartke u.a. Tl. 1: Allgemeiner Teil. (3-7940-5409-1); Tl. 2: Biologie. (3-7940-5410-5); Tl. 3: Chemie / Pharmazie. (3-7940-5411-3); Tl. 4: Physik. (3-7940-5412-1) / 4 Bde. 1974. 4.132 S.; 49 S.; 47 S.; 42 S. (3-7940-5409 / 5412-1; 5; 3;1)

Bayerische Staatsbibliothek. Alphabetischer Katalog 1501-1840 / Bavarian State Library. Alphabetical Catalogue 1501-1840 (BSB-AK 1501-1840).
• Voraus-Ausgabe / Preliminary Edition: 60 Bde. 1987-1990 Zus. 30.720 S. / Ln (3-598-30800-0)

Bayerische Staatsbibliothek. Katalog 1501-1840 / Bavarian State Library. Catalogue 1501-1840. CD-ROM-Edition / 1996 (3-598-40332-1)

Bayerische Staatsbibliothek. Katalog der Geschichtszeitschriften (BSB-GeZ) / Bavarian State Library. Catalogue of Historic Periodicals (BSB-GeZ). Hrsg.: Bayerische Staatsbibliothek (Zeitschriften der Bayerischen Staatsbibliothek 2 / Periodicals of the Bavarian State Library 2) / 2 Bde. 1991. Zus. 640 S. / Ln (3-598-22242-4)

Bayerische Staatsbibliothek. Katalog der Musikdrucke / Bavarian State Library. Catalogue of Printed Music (BSB-Musik / BSB-Music) / 17 Bde. 1988 ff. Zus. 7.690 S. / Ln (3-598-30560-5)

Bayerische Staatsbibliothek. Katalog der Musikzeitschriften (BSB-MuZ) / Bavarian State Library. Catalogue of Music Periodicals (BSB-MuZ). Hrsg.: Bayerische Staatsbibliothek (Zeitschriften der Bayerischen Staatsbibliothek 1 / Periodicals of the Bavarian State Library 1) / 1990. X, 242 S. / Ln (3-598-22241-6)

Bayerische Staatsbibliothek. Katalog der Notendrucke, Musikbücher und Musikzeitschriften / Bavarian State Library. Catalogue of Printed Music, Books and Periodicals on Music. CD-ROM-Edition / 1999 (3-598-40381-X)

Bayerische Staatsbibliothek. Katalog der Osteuropazeitschriften (BSB-OeZ) / Bavarian State Library. Catalogue of Eastern Europe Periodicals. Hrsg.: Bayerische Staatsbibliothek (Zeitschriften der

Bayerischen Staatsbibliothek 3 / Periodicals of the Bavarian State Library 3) / 4 Bde. 1993. Zus. XLIV, 1744 S. / Ln (3-598-22243-2)

Die **Bayerische** Staatsbibliothek in historischen Beschreibungen Ausw. u. Kommentierung d. Texte: Klaus Haller. Ausw. d. Abb. u. Liste d. Zimelien: Karl Dachs. Übers. u. Reg.: Claudia Fabian / 1992. 227 S. / Ln (3-598-11149-5) [Herrn Dr. phil. Franz Georg Kaltwasser. Direktor der Bayerischen Staatsbibliothek von 1972 bis 1992]
• 2. durchges. Aufl. 1998. 227 S. / Ln (3-598-11395-1)

Bayerisches Jahrbuch. Auskunfts- und Adressenwerk über Behörden, Ministerien, Verbände und Gemeinden [Vorher im Gerber Vlg., München]
• 75. Jg. 1996 / 1995. 563 S. / Br (3-598-23644-1)
• 76. Jg. 1997 / 1996. 454 S. / Br (3-598-23646-8)
• 77. Jg. 1998 / 1997. X, 534 S. / Br (3-598-23648-4)
• 78. Jg. 1999 / 1998. X,539 S. / Br (3-598-23650-6)

Beaudiquez, Marcelle: Guide de bibliographie générale. Méthodologie et pratique (Bibliothèques et organismes documentaires) / 1983. 280 S. / Br (3-598-20454-X)
• Nouvelle édition revue et mise à jour 1989. 277 S. / Br (3-598-10828-1)

Bebel, *August* → *August* Bebel

Der **befragte** Leser. Buch und Demoskopie. Vorw. v. Elisabeth Noelle-Neumann. Hrsg. v. Ludwig Muth. Beitr. v. Renate Köcher, Ludwig Muth, Elisabeth Noelle-Neumann, Gerhard Schmidtchen, Rüdiger Schulz / 1993. XVIII, 220 S. / Gb (3-598-11131-2)

Behrendt, Ethel Leonore: Recht auf Gehör. Grundrecht und Grundwert / 1978. XXXIV, 484 S. (3-597-10028-7) [Minerva Publikation]

Beiträge des Instituts für Zukunftsforschung [Minerva Publikation]
• 1: Zweiweg-Kabelfernsehen und Datenschutz. Hrsg. von Klaus Dette; Rolf Kreibich; Wilhelm Steinmüller / 1979. VI, 150 S. (3-597-10083-X)
• 2: Kabelfernsehen und gesellschaftlicher Dialog. Vorstudien deer interdisziplinärer Arbeitsgruppe Kabelkommunikation Berlin (IKB) zur wissenschaftlichen Vorbereitung und Begleitung von Pilotprojekten zum Zweiweg-Kabelfernsehen. Hrsg. von Klaus Dette; Rolf Kreibich; Heidrun Kunert-Schroth / 1979. XVI, 688 S. (3-597-10091-0)
• 3: Zweiweg-Kabelfernsehen und bürgernahe Verwaltung. Hrsg. von Helgomar Pichlmayer; Manfred Birreck; Dieter Kolb; Johanna Kratzsch. Vorwort von Hans Buchholz / 1979. XVI, 265 S. (3-597-10092-9)
• 4: Neue technische Kommunikationssysteme und Bürgerdialog. Hrsg. von Fred Grätz; Maksut Kleemann; Rolf Kreibicj; Helgomar Pichlmayer / 1979. VI, 281 S. (3-597-10151-8)
• 5: Anforderungen an die Ausbildung von Führungsnachwuchs. Fragen zur künftigen Bildungspolitik und zur Entwicklung der Qualifikation von Führungskräften. Expertengespräche zwischen Managern der Deutschen Unilever GmbH und Bildungsökonomen. Hrsg. von Hans Buchholz; Ralf Schröter / 1980. VI, 98 S. (3-597-10152-6)
• 6: Futurum. Hrsg.von Ossip K. Flechtheim / 1980. XVIII, 378 S. (3-597-10153-4)

- 7: Wenzel, Catrin: Organisationsstruktur und Behandlungsauftrag im Strafvol lzug. Darstellung und Analyse am Beispiel der Teilanstalt IV (Sozialtherapie) der Justizvollzugsanstalt Berlin-Tegel / 1979. VI, 246 S. (3-597-10154-2)
- 8: Lutz, Wulf R.: Soft Systems Design. Images and Impacts of Societal Transitions. / 1979. VIII, 128 S. (3-597-10155-0)
- 9: Computergestützte Planungsmodelle in der Verwaltung. Hrsg. von Christian E. Riethmüller / 1980. 232 S. (3-597-10208-5)
- 10: Die Situation ausländischer Kinder. Fachtagung im Internationalen Jahr des Kindes am 14. Juli 1979 in Berlin (West). Hrsg. von Ilse Reichel / 1980. VI, 171 S. (3-597-10209-3)
- 11: Arbeitslosigkeit - Beschäftigung - Beruf. Systembegrenzung und Lebensneugestaltung. Preisträgerarbeiten des GZ-Wettbewerbs. / Von Peter Perutz; Walter R. Stahel; Rolf Roschmann; Joseph Huber / 1980. 104 S. (3-597-10210-7)
- 12: Haas, Peter: Kritik der Weltmodelle. Philosophische Aspekte globaler Modellierungen. Vorwort von Helmut Arnaszus / 1980. XVI, 249 S. (3-597-10211-5)
- 13: Herrmann, Egbert: Umweltökonomische Grundlagen zur Planung von Fremdenverkehrs- und Erholungsgebieten / 1980. XXX, 583 S. (3-597-10212-3)
- 14: Vief, Bernhard: Ein Modell zur Lehrerbedarfsberechnung. Lehrerbedarf in Berlin als bildungs- und arbeitsmarktpolitisches Problem. 14 / 1981. IV, 284 S. (3-597-10213-1)
- 15: Lutz, Wulf R. / Dieter Kolb: Mittlere Technologie für Industrieländer. Prüfung der Anwendungsmöglichkeiten Mittlerer Technologien in der Abfallbeseitigung und Abfallverwertung / 1980. XIV, 619 S. (3-597-10214-X)
- 16: Plutoniumgesellschaft. Folgen der Kernenergienutzung für das Leben in der Zukunft. Hrsg. von Lutz Mez; Manfred Richter / 1981. 146 S. (3-597-10296-4)
- 17: Situation der Ausländerinnen. Fachtagung am 19., 20. und 21. September 1980 in Berlin (West). Hrsg. v. Ute Welzel / 1981. VI, 137 S. / Br (3-597-10297-2)
- 18: Oel, Hans U.: Sozialräumliche Beziehungen und Bürgerbeteiligung in der Stadtteilentwicklungsplanung. Eine vergleichende Untersuchung partizipatorischer Planungsansätze / 1982. IV, 208 S. (3-597-10298-0)
- 19: Bückmann, Walter: Aspekte der Sozialverträglichkeitsprüfung. Theoretische Erörterungen und erste Handlungsanleitungen für die Verbindung der Sozialverträglichkeitsprüfung mit der Unverträglichkeitsprüfung / 1982. XX, 184 S. / Br (3-597-10299-9)
- 20: Bückmann, Walter / Hans U. Oel: Bürgerbeteiligungen bei kommunalen Planungen / 1981. XVIII, 200 S. (3-597-10300-6)

Beiträge zur Bedarfsmessung an wissenschaftlichen Hochschulen
- 4: Schoser, Gustav / Anton Lang: Bedarfsplanung der Fachgruppe Biologie. Ein Vorschlag zur Planung der Fachgruppe Biologie an wissenschaftlichen Hochschulen. Stand November 1966 / Hrsg.: Arbeitsgruppe Bedarfsbemessung wissenschaftlicher Hochschulen im Finanzministerium Baden-Württemberg; Zentralarchiv für Hochschulbau, Stuttgart / 1967. 24 S. (3-598-20300-4)
- 12: Bedarfsrichtwerte und Kostenlimits im britischen Universitätsbau. Auszugsweise Übersetzung der Verfahrensvorschriften zur Bewilligung einmaliger Ausgaben für Bau und Ausstattung britischer Univeritäten / 1979. XIV, 56 S. (3-598-20301-2)
- 13: Hub, Ernst: Ergebnisbericht über die Bedarfsbemessung von Hochschul-Sportanlagen. Stand: 1967/68. Einl. von O. Grupe / 1969. 97 S. (3-598-20302-0)
- 14: Quantitativer Datenvergleich von Raumprogrammen in den Fachbereichen Physik und Chemie. Ergebnisbericht über Auswertungsmethoden als Planungshilfen zur Bemessung des Flächenbedarfs. / 1970. 136 S. (3-598-20303-9)
- 15: Christes, Hans / J. G. Helm; Ulrich Hempel: Bemessung des Flächenbedarfs geisteswissenschaftlicher Fachrichtungen. Band 1: Ergebnisse. 223 S. Band 2: Dokumentarischer Anhang. 116 S. / 2 Bände 1968. (3-598-20304-7)
- 16: Bemessung des Flächenbedarfs im Fachbereich Physik. Ergebnisbericht. Diskussionsfassung. Von Günter Busch; Gerhard Gramm; Josef Hidasi u.a / 1968. VII, 142 S. (3-598-20305-5)
- 18: Bayer, Werner / Alfred Bingert: Lineare Programmierung und mathematische Statistik in der Bedarfsbemessung wissenschaftlicher Hochschulen / 1969. 67 S. (3-598-20311-X)

Beiträge zur Bibliothekstheorie und Bibliotheksgeschichte / Contributions to Library Theory and Library History. Ed. by Paul Kaegbein; Peter Vodosek; Peter Zahn
- 1: Happel, Hans G.: Das wissenschaftliche Bibliothekswesen im Nationalsozialismus. Unter besonderer Berücksichtigung der Universitätsbibliotheken / 1989. 190 S., 16 Abb. / Lin (3-598-22170-3)
- 2: Die Universitätsbibliotheken Heidelberg, Jena und Köln unter dem Nationalsozialismus. Hrsg. v. Ingo Toussaint. Beitr. v. Lothar Bohmüller; Hans G. Happel; Konrad Marwinski; Hildegard Müller; Ingo Toussaint / 1989. 407 S., 8 schw.-w. Fotos / Lin (3-598-10858-3)
- 3: Studies on Research in Reading and Libraries. Approaches and Results from Several Countries Ed. on behalf of the IFLA Round Table on Research in Reading by Paul Kaegbein; Bryan Luckham; Valeria Stelmach / 1991. VIII, 284 p. / Hard (3-598-22171-1)
- 4: Wang, Jingjing: Das Strukturkonzept einschichtiger Bibliothekssysteme Idee und Entwicklung neuerer wissenschaftlicher Hochschulbibliotheken in der Bundesrepublik Deutschland / 1990. XII, 378 S., 8 Abb. / Lin (3-598-22172-X)
- 5: Die Neugründung wissenschaftlicher Bibliotheken in der Bundesrepublik Deutschland. Symposium, veranstaltet vom Institut für Buchwesen der Johannes-Gutenberg-Universität Mainz vom 23. bis 25. Februar 1988 mit Unterstützung der Fritz-Thyssen-Stiftung. Hrsg. v. Hans J. Koppitz / 1990. 290 S., 7 Abb. / Lin (3-598-10906-7)
- 6: Deutsche Bibliothekskataloge im 19. Jahrhundert. Analytisches Repertorium. Bearb. v. Martin Schenkel; Jens Ahlers; Waltraud Richartz-Malmede; Birgit Seipt / 2 Teilbde. 1992. Zus. LXVI, 590 S. / Lin (3-598-22173-8)
- 7: Seela, Torsten: Bücher und Bibliotheken in nationalsozialistischen Konzentrationslagern. Das gedruckte Wort im antifaschistischen Widerstand der Häftlinge / 1992. XII, 252 S. / Lin (3-598-22174-6)
- 8: Bibliothek - Kultur - Information. Beiträge eines internationalen Kongresses anlässlich des 50jährigen Bestehens der Fachhochschule für Bibliothekswesen, Stuttgart vom 20. bis 22. Oktober 1992. Hrsg. v. Peter Vodosek, in Zus.-Arb. mit Askan Blum; Wolfram Henning; Hellmut Vogeler / 1993. 362 S. / Lin (3-598-22175-4)
- 9: Marwinski, Felicitas: Sozialdemokratie und Volksbildung Leben und Wirken Gustav Hennigs als Bibliothekar / 1994. 130 S., Abb. / Lin (3-598-22176-2)
- 10: Bibliotheca Baltica Symposium vom 15. bis 17. Juni 1992 in der Bibliothek der Hansestadt Lübeck im Rahmen der Initiative ARS Baltica. Hrsg. v. Jörg Fligge; Robert Schweitzer. Bearb. v. Frauke Büter / 1994. 186 S. / Lin (3-598-22177-0)

Beiträge zur empirischen Kriminologie. Hrsg. v. Hans Göppinger [Minerva Publikation]
- 4: Dolde, Gabriele: Sozialisation und kriminelle Karriere. Eine empirische Analyse der sozio-ökonomischen und familialen Sozialisationsbedingungen männlicher Strafgefangener im Vergleich zur „Normal-Bevölkerung" / 1978. (3-597-10022-8)
- 5: Wulf, Bernd R.: Kriminelle Karrieren von Lebenslänglichen. Eine empirische Analyse ihrer Verlaufsformen und Strukturen anhand von 141 Straf- und Vollzugsakten / 1979. (3-597-10067-8)
- 6: Schmehl, Hans H.: Jugendliche und heranwachsende Straftäter während ihrer Ausbildung. Eine Untersuchung über die Bedeutung schulischer und beruflicher Ausbildung für die Legalbewährung. Beschreibung einer Gruppe von Straffälligen und einer Vergleichsgruppe / 1980. XII, 250 S. 20, 9 x 14, 7 cm / Br (3-597-10069-4)
- 7: Kofler, Rolf: Beruf und Kriminalität. Eine empirische Untersuchung der Zusammenhänge zwischen Beruf und Straffälligkeit bei den Probanden der Tübinger Jungtäter-Vergleichsuntersuchung / 1980. VIII, 203 S. 20, 9 x 14, 7 cm / Br (3-597-10070-8)
- 8: Diesing, Ulrich: Psychische Folgen von Sexualdelikten bei Kindern. Eine katamnestische Untersuchung Vorw. v. Hermann Witter / 1980. X, 139 S. / Br (3-597-10235-2)
- 9: Keske, Monika B.: Die Kriminalität der „Kriminellen". Eine empirische Untersuchung von Struktur und Verlauf der Kriminalität bei Strafgefangenen sowie ihrer Sanktionierung / 1983. VIII, 311 S. / Br (3-597-10469-X)
- 10: Jehle, Jörg M.: Untersuchungshaft zwischen Unschuldsvermutung und Wiedereingliederung. Ein empirischer Beitrag zur Ausgestaltung des Untersuchungshaftvol lzugs unter besonderer Berücksichtigung kriminalpolitischer Reformvorstellungen / 1985. XII, 302 S. 20, 9 x 14, 7 cm / Br (3-597-10539-4)
- 11: Maschke, Werner: Das Umfeld der Straftat Ein erfahrungswissenschaftlicher Beitrag zum kriminologischen Tatbild / 1987. VIII, 161 S. / Br (3-597-10590-4)
- 12: Friedrichsmeier, Hans: Das Sozialverhalten von Prostituierten. Eine empirische Untersuchung auch zur Vergleichbarkeit der Lebensläufe von weiblichen Prostituierten und männlichen Straftätern / 1991. X, 152 S. / Br (3-597-10694-3)

Beiträge zur Informations- und Dokumentationswissenschaft. Hrsg. von Hans-Werner Schober,
- Folge 1: Sprache und Begriff / 1970. 94 S.
- Folge 2: Dokumentenanalyse als sprachlich-informationstechnisches Problem / 1970. 45 S
- Folge 3: Kschenka, Wilfried: Benutzeranalyse und pädagogische Literaturinformation. Studie zur Verbesserung d. Informationsmittel / 1970. 75 S. (3-7940-3403-1)
- Folge 4: Vergleichende Analyse von Referatediensten. Ein Bericht über e. empir. Unters. Unter

Mitarb. von Ilona Gaszak / 1971. 78 S. (3-940-3404-X)
- Folge 5: Wersig, Gernot: Information - Kommunikation - Dokumentation. Ein Beitrag zur Orientierung d. Informations- und Dokumentationswissenschaften / 1971. 351 S. (3-7940-3405-8) - 2. Aufl. 1974. 362 S. (3-7940-3405-8)
- Folge 6: Beling, Gerd / Gernot Wersig: Zur Typologie von Daten und Informationssystemen Terminologie, Begriffe u. Systematik Gutachten. Verfertigt im Auftr. d. Bundesministeriums d. Innern / 1973. 144 S. (3-7940-3406-9)
- Folge 7: Ockenfeld, Marlies: Das Informationsverhalten von Chemikern E. Pilotstudie / 1975. 219 S. (3-7940-3407-4)
- Folge 8: Strauch, Dietmar: Wissenschaftliche Kommunikation und Industrialisierung. Einheit u. gesellschaftl. Bedeutung d. Wissenschaft als Kommunikationsprobleme / 1976. 81 S. (3-7940-3408-2)
- Folge 9: Seeger, Thomas: Ausbildungsgänge im Tätigkeitsbereich Information und Dokumentation. Eine Untersuchung / 1977. 183 S. (3-7940-3409-0)
- Folge 10: Zum Verhältnis von Staat, Wissenschaft zu Information und Dokumentation Beitr. zur Entwicklung in Ost u. West. Hrsg. von Marianne Buder u. Gunther Windel / 1978. 316 S. (3-7940-3720-0)
- Folge 11: Supper, Reinhard: Neuere Methoden der intellektuellen Indexierung. Brit. Systeme unter bes. Berücks. von PRECIS / 1978. 255 S. (3-7940-3721-9)
- Folge 12: Delphi-Prognose in Information und Dokumentation. Untersuchungen über zukünftige Entwicklungen des Bibliotheks-, Informations- und Dokumentationswesens in der Bundesrepublik Deutschland uns Österreich. Hrsg. von Wolf Rauch und Gernot Wersig / 1978. 296 S. (3-7940-3722-7)
- Folge 13: Informations- und Dokumentationswissenschaft an der Freien Universität Berlin. Entwicklung, Stand und Perspektiven nach 10 Jahren Red. Günther Lehmann; Ernst Michael Oberdieck / 1979. 148 S. (3-598-20463)

Beiträge zur Kommunalwissenschaft / Studies in Local Government and Politics. Interdisziplinäre Reihe. Hrsg. v. Rüdiger Voigt, Hans F. Illy, Helmut Köser, Hans J. Siewert, Oscar W. Gabriel, Heinrich Mäding [Minerva Publikation]
- 1: Vorholz, Fritz: Interkommunale Einrichtungen. Ein Beitrag zur Stadt-Umland-Problematik / 1981. 188 S. / Br (3-597-10353-7)
- 2: Demetriou, Georgios: Kommunalpolitik in Griechenland. Handlungsspielräume der Kommunalverwaltung und lokale Einfluss- und Entscheidungsstrukturen am Beispiel der Stadt Thessaloniki Vorw. v. Köser / 1981. VIII, 218 S. / Br (3-597-10354-5)
- 3: Kommunale Macht- und Entscheidungsstrukturen. Hrsg. v. Klaus M. Schmals; Hans J. Siewert / 1982. VIII, 335 S. / Br (3-597-10355-3)
- 4: Ammann, Wolfgang: Die Bezirksplanung als Bindeglied staatlicher und kommunaler Planung? Organisationssoziologische und planungstheoretische Aspekte / 1982. II, 90 S. / Br (3-597-10356-1)
- 5: Fälker, Margot: Integration bei kommunalen Entscheidungsprozessen / 1982. II, 186 S. / Br (3-597-10357-X)
- 6: Savelsberg, Joachim J.: Ausländische Jugendliche Assimilative Integration, Kriminalität und Kriminalisierung und die Rolle der Jugendhilfe / 1982. XVIII, 438 S. / Br (3-597-10358-8)

- 7: Kommunikationstechnologien und kommunale Entscheidungsprozesse. Hrsg. v. Dietrich Thränhardt; Herbert Uppendahl / 1982. II, 291 S. / Br (3-597-10359-6)
- 8: Platzer, Renate: Bürgerinitiativen in Salzburg Eine vergleichende Untersuchung der Bürgerinitiative „Schützt Salzburgs Landschaft" mit der „Initiative für mehr Lebensqualität in Lehen" / 1983. XIV, 293 S. / Br (3-597-10360-X)
- 9: Quentmeier, Regine: Regionalplanung im Planungsverbund Zur Funktion des Bezirksplanungsrates im Rahmen der Politikverflechtung / 1983. II, 243 S. / Br (3-597-10437-1)
- 10: Krischmann, Helmut: Der Einfluss staatlicher Raumplanung auf die kommunale Planungshoheit Die Regionalplanung Nordhessen im Verhältnis zu den Bauleitplanungen der Städte Kassel und Baunatal / 1983. II, 103 S. / Br (3-597-10438-X)
- 11: Eichenauer, Hartmut: Das zentralörtliche System nach der Gebietsreform. Geographisch-empirische Wirkungsanalyse raumwirksamer Staatstätigkeit im Umland des Verdichtungsgebietes Siegen / 1983. 368 S. / Br (3-597-10439-8)
- 12: Gau, Doris: Politische Führungsgruppen auf kommunaler Ebene. Eine empirische Untersuchung zum Sozialprofil und den politischen Karrieren der Mitglieder des Rates der Stadt Köln / 1983. 146 S. / Br (3-597-10440-1)
- 13: Bürgerbeteiligung und kommunale Demokratie. Hrsg. v. Oscar W. Gabriel / 1983. 393 S. / Br (3-597-10441-X)
- 14: Jarren, Otfried: Kommunale Kommunikation. Eine theoretische und empirische Untersuchung kommunaler Kommunikationsstrukturen unter besonderer Berücksichtigung lokaler und sublokaler Medien / 1984. XII, 390 S. / Br (3-597-10442-8)
- 15: Rapior, Roland: Interkommunale Zusammenarbeit. Eine empirische Untersuchung über die Stadt-Umland-Beziehungen in der Stadtregion Freiburg / 1984. VI, 304 S. / Br (3-597-10443-6)
- 16: Casasus, Gilbert: Kommunalpolitik in Penzberg und Saint-Fons. Ein Vergleich der Gemeindepolitik der deutschen Sozialdemokraten und der französischen Sozialisten / 1985. II, 393 S. / Br (3-597-10566-1)
- 17: Pehle, Heinrich: Kommunale Entscheidungsstrukturen in Schweden und Deutschland. 4 Fallstudien zum Stellenwert Kommunaler Aussenpolitik bei Verkehrsinvestitionen / 1985. II, 427 S. / Br (3-597-10444-4)
- 18: Keim, Karl D.: Macht, Gewalt und Verstädterung. Vorstudien zur Theoriebildung / 1985. IV, 171 S. / Br (3-597-10445-2)
- 19: Richter, Rudolf: Soziokulturelle Dimensionen freiwilliger Vereinigungen. USA, Bundesrepublik Deutschland und Österreich im soziologischen Vergleich / 1985. VI, 354 S. / Br (3-597-10446-0)
- 20: Schneider, Herbert: Kreispolitik im ländlichen Raum. Eine vergleichende Untersuchung über Landkreispolitik / 1985. 291 S. / Br (3-597-10447-9)
- 21: Jurczek, Peter: Städtetourismus in Oberfranken. Stand und Entwicklungsmöglichkeiten des Fremdenverkehrs in Bamberg, Bayreuth, Coburg und Hof / 1986. 232 S. / Br (3-597-10448-7)
- 22: Heye, Werner: Entstehung und Akzeptanz von Bürgerhäusern. Analysen zu Bedingungen und Formen eines speziellen Typs sozialer Infrastruktur / 1986. II, 405 S. / Br (3-597-10583-1)
- 23: Haasis, Hans A.: Bodenpreise, Bodenmarkt und Stadtentwicklung. Eine Studie zur sozial-

räumlichen Differenzierung städtischer Gebiete am Beispiel von Freiburg i. Br. / 1987. 403 S. / Br (3-597-10584-X)
- 24: Entwicklung durch Dezentralisierung? Studien zur Kommunal- und Regionalverwaltung in der Dritten Welt. Hrsg. v. Hans F. Illy; Klaus Schimitzek / 1986. 441 S. / Br (3-597-10585-8)
- 25: Tessin, Wulf: Stadtwachstum und Stadt-Umland-Politik. Die politisch-administrative Subsumtion des Landes unter die Herrschaft der Stadt im Prozess der Stadt-Umland-Entwicklung im Raum Wolfsburg / 1986. II, 270 S. / Br (3-597-10586-6)
- 26: Baumheier, Ralph: Altlasten als aktuelle Herausforderung der Kommunalpolitik. Zu den Schwierigkeiten politisch-administrativer Problemverarbeitung in der Kommune / 1988. XII, 279 S. / Br (3-597-10587-4)
- 27: Fruth, Hanno: Sind unsere ehrenamtlichen Stadträte überfordert? Kommentar zu einer in Ansbach, Bamberg, Erlangen, Fürth, München und Nürnberg durchgeführten Untersuchung Vorw. v. Rüdiger Voigt / 1989. 155 S. / Br (3-597-10588-2)
- 28: Schwiderowski, Peter: Entscheidungsprozesse und Öffentlichkeit auf der kommunalen Ebene. Erweiterte Bürgerbeteiligung durch die Nutzung neuer lokaler Massenmedien? / 1989. 297 S. / Br (3-597-10589-0)
- 29: Kommunale Demokratie zwischen Politik und Verwaltung. Hrsg. v. Oscar W. Gabriel / 1989. 362 S. / Br (3-597-10650-1)
- 30: Efrat, Elisha: The New Towns of Israel (1948-1988). A Reappraisal / 1989. 184 p., 27 illus. / Br (3-597-10651-X)
- 31: Mayer, Alexander: Die Sperrung des Rathausplatzes in Nürnberg. Eine Fallstudie zur Machtverteilung und zu Einflussstrukturen in einer deutschen Grossstadt / 1989. XII, 203 S. / Br (3-597-10652-8)
- 32: Gau, Doris: Kultur als Politik. Eine Analyse der Entscheidungsprämissen und des Entscheidungsverhaltens inder kommunalen Kulturpolitik / 1989. 260 S. / Br (3-597-10653-6)
- 33: Gabriel, Oscar W. / Volker Kunz; Thomas Zapf-Schramm: Bestimmungsfaktoren des kommunalen Investitionsverhaltens. Eine empirische Untersuchung der Investitionsausgaben rheinland-pfälzischer Städte auf den Gebieten Kultur, Sport, Erholung und Verkehr in den Jahren 1978 bis 1985 / 1990. 218 S. / Br (3-597-10654-4)
- 34: Frank, Rainer: Kultur auf dem Prüfstand. Ein Streifzug durch 40 Jahre kommunaler Kulturpolitik / 1990. 374 S. / Br (3-597-10655-2)
- 35: Schneider, Herbert: Kommunalpolitik auf dem Lande / 1991. 252 S. / Br (3-597-10656-0)
- 36: Czarnecki, Thomas: Kommunales Wahlverhalten. Die Existenz und Bedeutsamkeit kommunaler Determinanten für das Wahlverhalten. Eine empirische Untersuchung am Beispiel Rheinland-Pfalz. Hrsg. v. Voigt, Rüdiger / 1992. XVI, 111 S. / Br (3-597-10657-9)
- 37: Haller, Michael / Thomas Mirbach: Medienvielfalt und kommunale Öffentlichkeit / 1994. 218 S. / Br (3-597-10658-7)
- 38: Frevel, Bernhard: Funktion und Wirkung von Laienmusikvereinen im kommunalen System. Zur sozialen, kulturellen und politischen Bedeutung einer Sparte lokaler Freizeitvereine / 1993. II, 257 S. / Br (3-597-10659-5)
- 39: Pohlmann, Markus: Kulturpolitik in Deutschland. Städtisch organisierte Kultur und Kuturadministrationen / 1993. II, 335 S. / Br (3-597-10660-9)

Beiträge zur Sozialökonomik der Arbeit. Hrsg. v. Michael Bolle; Burkhard Strümpel [Minerva Publikation]
- 1: Volz, Joachim: Arbeitslosenversicherung und Arbeitslosigkeit in integrierten Wirtschaftsräumen. Dargestellt am Beispiel der Europäischen Gemeinschaft und der USA Vorw. v. Burkhard Strümpel / 1980. XII, 406 S. / Br (3-597-10240-9)
- 2: Feuerstein, Stefan: Aufgabenfelder und Informationsbedarf kommunaler Wirtschaftsförderungspolitik / 1981. XXIV, 446 S. / Br (3-597-10241-7)
- 3: Analytische und empirische Aspekte der Arbeitszeitgestaltung auf betrieblicher und überbetrieblicher Ebene. Hrsg. v. Michael Bolle; Ulrike Fischer; Burkhard Strümpel / 1982. VIII, 395 S. / Br (3-597-10242-5)
- 4: Die Dynamik der Arbeitsmärkte aus der Sicht internationaler Forschung. Hrsg. v. Michael Bolle; Jürgen Gabriel / 1983. VIII, 335 S. / Br (3-597-10243-3)
- 5: Conradi, Hartmut: Teilzeitarbeit Theorie, Realität, Realisierbarkeit / 1982. VIII, 178 S. / Br (3-597-10244-1)
- 6: Olle, Werner: Strukturveränderungen der internationalen Direktinvestitionen und inländischer Arbeitsmarkt. Empirisch-quantitative Probleme der makroökonomischen Relevanzanalyse der deutschen Direktinvestitionen im Ausland / 1983. 418 S. / Br (3-597-10245-X)
- 7: Hoff, Andreas: Betriebliche Arbeitszeitpolitik zwischen Arbeitszeitverkürzung und Arbeitszeitflexibilisierung / 1983. XII, 409 S. / Br (3-597-10246-8)
- 8: Pawlowsky, Peter: Arbeitseinstellungen im Wandel. Zur theoretischen Grundlage und empirischen Analyse subjektiver Indikatoren der Arbeitswelt / 1986. XVI, 262 S. / Br (3-597-10247-6)
- 10: Gabriel, Jürgen: Flexibilisierung der Arbeit und wirtschaftliche Instabilität. Überlegungen zur Theorie und Empirie des Beschäftigungsverhaltens von Unternehmen bei Unsicherheit / 1985. X, 316 S. / Br (3-597-10540-8)
- 11: May, Karl / Eveline Mohr: Probleme und Realisierungschancen individueller Arbeitszeitmodelle / 1985. XVIII, 281 S. / Br (3-597-10541-6)
- 12: Scholz, Joachim: Wertwandel und Wirtschaftskultur / 1987. XII, 309 S. / Br (3-597-10542-4)
- 13: Stach, Hannelore: Gleichstellung im Erwerbsleben? Situation der Frauen zwischen Forderungen und Realität / 1987. X, 129 S. / Br (3-597-10543-2)
- 14: Bolle, Michael / Ellen R. Schneider: Neue Technologien und neue Qualifikationsanforderungen für Frauen In Zus.-Arb. mit Angela Ehrmann / 1988. XVI, 252 S. / Br (3-597-10544-0)

Beiträge zur Universitätsplanung
- 5: Harnest, A. / E. Sparas: Der Komplexraum als wissenschaftliche Arbeitsfläche. Überlegungen zu einer neuen Bauform im Hochschulbereich. Hrsg.: Zentralarchiv für Hochschulbau, Stuttgart / 1969. 93 S. (3-598-20330-6)
- 6: Dunkl, Walther / F. Geyer: Rationale Lösungen von Laborbau- und Laboreinrichtungsproblemen in Hochschulen, Industrieunternehmungen und im Gesundheitswesen der USA. Bericht über eine Studienreise / 1970. 79 S. (3-598-20331-4)
- 7: Kemmerich, C. / H. Küsgen u.a.: Bauliche Planungsalternativen im Gesamthochschulbereich. Bedarfsmessung, Bestandsnutzung, Kostenverläufe von Planungsalternativen. Dargestellt am Gesamthochschulbereich Stuttgart / 1970. 210 S. (3-598-20332-2)

Bemann, Rudolf / Jakob Jatzwauk: Bibliographie der sächsischen Geschichte. 3 Bde. in 5 Tlbdn. - Nachdruck der Ausg. Leipzig 1918 - 32 / 1971. XVI, XII, 521 S.; XVIII, 614 S.; XI, 199 S. (3-7940-5024-X)

Bender, Wolfgang F. / Siegfried Bushuven / Michael Huesmann: Theaterperiodika des 18. Jahrhunderts Bibliographie und inhaltliche Erschliessung deutschsprachiger Theaterzeitschriften, Theaterkalender und Theatertaschenbücher / 3 Tle. in ca. 8 Bdn. 1994 ff. Zus. ca. 4000 S. / Gb (3-598-23181-4)
- Tl. 1: 1750-80 / 2 Bde. 1994. Zus. XC, 1.028 S. / Gb (3-598-23182-2)
- Tl. 2: 1781-90 / 3 Bde. 1997. Zus. LXXVIII, 1.327 S. / Gb (3-598-23183-0)
  Tl. 3: 1791-99 [i.Vb.]

Benutzerverhalten an deutschen Hochschulbibliotheken. Ergebnisse e. mit Unterstützung d. Dt. Forschungsgemeinschaft durchgeführten vergleichenden Untersuchung. Mit Beitr. von Gunter Bock. Hrsg.von Karl Wilhelm Neubauer, Mitarb. Heinrich Meister / 1979. 656 S. (3-598-10077-9)

Bergk, Johann Adam: Die Kunst, Bücher zu lesen. Nebst Bemerkungen über Schriften und Schriftsteller. - Nachdruck der Ausg. Jena: Hempel 1799. - Mit einem Nachwort von Horst Kunze / 1971. XVI, 416, 8 S. / Gb (3-7940-3012-5)

Bergk, Johann Adam: Die Kunst zu denken. Ein Seitenstück zur Kunst, Bücher zu lesen. - Nachdruck der Ausg. Leipzig: Hempel 1802 / 1973. XXXII, 447 S. / Gb (3-7940-3011-7)

Bergmann, Joachim: Die Schaubühne - Die Weltbühne 1905-1933 Bibliographie mit Annotationen. Teil I: Bibliographie mit biographischen Annotationen. Alphabetisches Titelregister / 1991. XXIII, 382 S. / Gb (3-598-10831-1)

Bericht der Tonmeistertagung. Hrsg.: Bildungswerk des Verbands Deutscher Tonmeister (VDT). Redakt.: Michael Dickreiter
- (13.) München 1984. Internationaler Kongress mit Fachausstellung im Kongressbau des Deutschen Museums vom 21.-24. November 1984 / 1984. 635 S. / Br (3-598-20354-3)
- (14.) München 1986 Internationaler Kongress mit Fachausstellung im Kongressbau des Deutschen Museums vom 19.-22. November 1986 / 1987. 528 S. / Br (3-598-20355-1)
- (15.) Mainz 1988. Internationaler Kongress mit Fachaustellung in der Rheingoldhalle und im Hotel Hilton vom 16. bis 19. November 1988 / 1989. 642 S. / Br (3-598-20356-X)
- (16.) Karlsruhe 1990. Internationaler Kongress mit Fachausstellung vom 20. bis 23. November 1990 in der Stadthalle / 1991. 836 S. / Br (3-598-20357-8)
- (17.) Karlsruhe 1992. Internationaler Kongress mit Fachausstellung vom 17. bis 20. November 1992 / 1993. 893 S. / Br (3-598-20358-6)
- (18.) Karlsruhe 1994. International Covention on Sound Design vom 15. bis 18. November 1994 / 1995. 1053 S. / Br (3-598-20359-4)
- (19.) Karlsruhe 1996. International Convention on Sound Design vom 15. bis 18. November 1996 / 1997. 940 S. / Br (3-598-20360-8)
- (20.) Karlsruhe 1998. International Convention on Sound Design vom 20. bis 23. November 1998 / 1999. 1.276 S. / Soft (3-598-20361-6)

Bericht über Projekte und Studien der Studiengruppe für Systemforschung in den Jahren 1968 und 1969. Hrsg. von der Studiengruppe für Systemforschung e.V., Heidelberg / 1970. 120 S. / Br (3-7940-3501-1)

Berichte und Arbeiten aus der Universitätsbibliothek Gießen
- 26: Schüling, Hermann: Die Dissertationen und Habilitationsschriften der Universität Gießen im 18. Jahrhundert / 1976. XX, 317 S. (3-598-10076-0)

Berichte und Materialien des Zentralinstituts für sozialwissenschaftliche Forschung (ZI 6) der Freien Universität Berlin. Bd 1 bei der Leitstelle Politische Dokumentation, Berlin erschienen
- Bd. 2: Schmollinger, Horst W.; Stöss, Richard: Die Parteien und die Presse der Parteien und Gewerkschaften in der Bundesrepublik Deutschland 1945-74. Materialien zur Parteien- u. Gewerkschaftsforschung / 1975. XXIX, 480 S. / Lin (3-7940-3223-3)
- Bd. 3: Sozialwissenschaftliche Forschungen. Arbeitsbericht des Zentralinstituts 6 der Freien Universität Berlin 1972-1975 / 1975. XX, 272 S. / Lin (3-7940-3225-X)
- Bd. 4. Wahlstatistik in Deutschland. Bibliographie der deutschen Wahlstatistik von 1848-1975. Bearb. von Nils Diederich / 1976. IX, 206 S. / Lin (3-7940-3220-9)

Berichte zur Erziehungstherapie und Eingliederungshilfe. Schriftenreihe des EREW Institutes, Viersen [Minerva Publikation]
- 1: Hirschfeld, Ursula / Karl J. Kluge: Spielen und Spielverhalten. Basisbefunde zur Entwicklung einer kindzentrierten Spielpädagogik für Behinderte und Nichtbehinderte / 1980. 346 S. / Br (3-597-10192-5)
- 2: Wir wollen hier raus Tl. I: Obdachlose Jugendliche - Eingliederungshilfen für Familie, Schule, Beruf und Freizeit. Ergebnisse eines bundesweiten Modellversuches. Hrsg. v. Karl J. Kluge / 1980. XVI, 403 S. / Br (3-597-10193-3)
- 3: Wir wollen hier raus Tl. II / 1980. XIV, 482 S. / Br (3-597-10194-1)
- 4: Abram, Ada / Beate Berkemeier; Karl J. Kluge: Suicid im Jugendalter Tl. I: Es tut weh, zu leben. Darstellung des Phänomens aus pädagogischer Sicht / 1980. XIV, 289 S. / Br (3-597-10195-X)
- 5: Abram, Ada / Beate Berkemeier; Karl J. Kluge: Suicid im Jugendalter Tl. II: Wir könnten weiterhin zusammenleben. Ursachenforschung - Pädagogik - Therapie - Prophylaxe / 1980. XIV, 396 S. / Br (3-597-10196-8)
- 6: Flissikowski, Renate / Karl J. Kluge; Klaus Schauerhammer: Vom Prügelstock zur Erziehungsklasse für „schwierige" Kinder Zur Sozialgeschichte abweichenden Verhaltens in der Schule / 1980. XII, 382 S. / Br (3-597-10197-6)
- 7: Kluge, Karl J. / Irmtraud Schnell; Helga Plum: Eine kindgerechte Umwelt schaffen. Das pädagogische System von Janusz Korczak und seine Bedeutung für Sondererziehung und Rehabilitation / 1981. XIV, 285 S. / Br (3-597-10198-4)
- 8: Fitting, Klaus / Karl J. Kluge; Dorothee Steinberg: Sich auf seine Schüler einlassen. Zur Konfliktregelung und Kommunikationsverbesserung im erziehungstherapeutischen Unterricht / 1981. XVIII, 450 S. / Br (3-597-10199-2)
- 9: Kluge, Karl J. / Raimund Patt: Kreativierender Unterricht - Eine Lernchance für Problemschüler in allen Schulstufen. Basisbefunde für erziehungs-

therapeutisches Lernen / 1982 .XII, 348 S. / Br (3-597-10200-X)
- 10: Kluge, Karl J. / K. Suermondt-Schlembach: Hochintelligente Schüler verhaltensauffällig gemacht? / 1981. VI, 145 S. / Br (3-597-10201-8)
- 11: Seht her, was ich kann! Erlebnisse, Lernerfahrungen und Beobachtungen in Europa. Jugendseminaren für Behinderte und Nichtbehinderte. Hrsg. v. Karl J. Kluge / 1981. X, 166 S. / Br (3-597-10316-2)
- 12: Gafni, Dov: Zur Berufswahl und zum Berufsbild von Sonderschullehrern. Eine bilaterale Studie in Köln und Tel-Aviv / 1981. XXVIII, 341 S. / Br (3-597-10317-0)
- 13: Anstelle von studentischer Revolution, Resignation und Vereinsamung - Engagement und „Lebendiges Lernen" Zur psycho-sozialen Situation von Studenten und „neue" Kooperationsformen in der Hochschulausbildung Von Karl J. Kluge; Peter Roeben; Jörg Ohmland; Anna von Kessel / 1982. 375 S. / Br (3-597-10318-9)
- 14: Kluge, Karl J. / Rolf Schramm; Joachim Stopp: Mit ein bisschen Hilfe meiner Freunde. Angewandte Ferienpädagogik als integraler Bestandteil von Rehabilitation / 1982. XIV, 245 S. / Br (3-597-10319-7)
- 15: Der Heilpädagoge im Rehabilitationsteam Tl. I: Aktuelle Stellungnahmen und zukunftsweisende Perspektiven eines interdisziplinären Gutachterteams zugunsten eines benachteiligten Berufsstandes in der BRD. Hrsg. v. Karl J. Kluge / 1982. XXX, 364 S. / Br (3-597-10320-0)
- 16: Der Heilpädagoge im Rehabilitationsteam Tl. II / 1982. XXX, S. 365-718 / Br (3-597-10321-9)
- 17: Fitting, Klaus: Sonderschullehrer zwischen Pädagogik und Therapie? Eine Studie zur humanistisch orientierten Hochschulausbildung der Verhaltensauffälligenpädagogen / 1983. X, 322 S. / Br (3-597-10322-7)
- 18-22: Kluge, Karl J. / Sibylle Hemmert-Halswick: Familie als Erziehungsinstanz
- 18: Tl. I: Eltern in Not. Probleme in der Familienerziehung. Zur Notwendigkeit von Elternberatung und Elternarbeit / 1982. IV, 440 S. / Br (3-597-10323-5)
- 19: Tl. II: Wie Eltern in Not geholfen werden kann. Aspekte zur Elternförderung im Bereich von Familienerziehung / 1982. II, S. 441-716 / Br (3-597-10324-3)
- 20: Tl. III: Endlich weiss ich weiter! Angewandte Elternförderung im Bereich von Familienerziehung. Multiplikatorentrainings und ihre Ergebnisse / 1982. IV, S. 717-1169 / Br (3-597-10325-1)
- 21: Tl. IV: Eltern helfen sich selbst Für Eltern praktizierbare Möglichkeiten der Prävention und Bewältigung von Problemen in Eltern-Kind-Beziehungen / 1984. X, 417 S. / Br (3-597-10381-2)
- 22: Tl. V: Elternmitarbeit in Schulen - Umwege und pädagogisch-therapeutische Wege / 1984. XXI, 662 S. / Br (3-597-10382-0)
- 23: Schildt, Jörg: Pädagoge sein heisst, selbst wieder Kind werden. Janusz Korczak und seine Waisenhauserziehung / 1982. VIII, 97 S. / Br (3-597-10383-9)
- 24: Ursprünge und Anfänge der Erziehungstherapie in Deutschland - Menschenbild, Hypothesen und Erfahrungen in der Anwendung der Humanistischen Psychologie. Hrsg. v. Bodo Januszewski; Karl J. Kluge / 1984. XII, 762 S. / Br (3-597-10384-7)
- 25: Elternförderung und Familientherapie als angewandte Erziehungstherapie. Hrsg. v. Karl J. Kluge / 1984. XX, 615 S. / Br (3-597-10385-5)

- 26-31: Entwicklung im Heim
- 26: Tl. I: Kluge, Karl J. / Hans J. Kornblum: Einmal aus der Bahn, immer aus der Bahn? Gedanken und Ideen zur Weiterentwicklung moderner Heimerziehung / 1984. XXVIII, 575 S., 12 S. Anh / Br (3-597-10386-3)
- 27: Tl. II / A u. B: Kluge, Karl J. / Hans J. Kornblum: Ob Unterricht noch weiterhilft? Wege und Chancen für Unterricht mit Schülern in besonderen Problemlagen / 1984. A: XXVIII, 306 S., B: IV, S. 307-1035 / Br (3-597-10387-1)
- 28: Tl. III: Kluge, Karl J. / Hans J. Kornblum: Am liebsten mache ich Sport und tanze. Was Heimbewohner denken, hoffen, fühlen / 1984. XXVIII, 318 S., 12 S. Anh / Br (3-597-10388-X)
- 29: Tl. IV: Karl J. / Hans J. Kornblum: Freund oder Feind? Über Erzieher und Jugendliche im Viersener Heim / 1984. XXVIII, 356 S., 12 S. Anh / Br (3-597-10389-8)
- 30: Tl. V: Karl J. / Hans J. Kornblum: Was Heimerzieher zu leisten vermögen Berichte aus dem Heimalltag / 1984. XXVIII, 568 S., 12 S. Anh / Br (3-597-10390-1)
- 31: Tl. VI: Fürderer-Schoenmackers, Heidi / Karl J. Kluge: Berufsprobleme und -chancen von Heimerziehern / 1984. XXXII, 234 S., 12 S. Anh / Br (3-597-10391-X)
- 32: Mertens, Beatrix / Karl J. Kluge: Die Veränderung der Sonja. Eine Frau erarbeitet sich, was ihr die Eltern verweigerten - und erlebt über eine andere Beziehung die Welt weniger schmerzhaft und zerrissen / 1984. XVI, 328 S. / Br (3-597-10502-5)
- 33: Kornblum, Hans J.: Therapie in der Berufsschule? Das „Krefelder Modell" erziehungstherapeutischen Unterrichts mit Jungarbeitern / 1984. X, 458 S. / Br (3-597-10503-3)
- 34: Deckers, Petra / Karl J. Kluge: Erfolgreich durch die Familienkrise? Eine „handfeste" Materialsammlung und Hilfestellung zur subjektzentrierten Elternförderung und zur Analyse von Elternkonflikten / 1984. 214 S. / Br (3-597-10504-1)
- 35: Kluge, Karl J. / Thomas M. Scheuer: Elternarbeit in der Sonderschule / 1984. 80 S. / Br (3-597-10505-X)
- 36: Conrads, Matthieu / Karl J. Kluge: Die spieltherapeutische Grundausstattung. Eine erste Zusammenstellung von Spielmaterialien, Spielgeräten und Spielprogrammen zur Diagnostik und für die Spieltherapie / 1985. X, 284 S. / Br (3-597-10506-8)
- 37: Hangert, Michael: Muss ich mich ändern, wenn sich Kinder ändern sollen? Hilfreiche Beziehungen zwischen Erzieherinnen und Kindern im Kindergarten / 1985. VIII, 139 S. 20, 9 x 14, 7 cm / Br (3-597-10507-6)
- 2. veränd. Aufl. 1988. IV, 122 S. / Br (3-597-10517-3)
- 38-39: Die verborgene Kraft Hochbegabung - Talentierung - Kreativität. Ansätze - Erfahrungen. Eine Übersicht der Begabtenhilfe sowie zur Vermeidung von benachteiligten Begabten in demokratischen Gesellschaften / 2 Bde. / Br (3-597-10510-6)
- 38: Tl. I. Hrsg. v. Klaus Bongartz; Ulrich Kaisser; Karl J. Kluge / 1985. XII, 285 S. / Br (3-597-10508-4)
- 39: Tl. II. Hrsg. v. Dov Gafni; Karl J. Kluge; Klaus Weinschenk / 1985. 305 S. / Br (3-597-10509-2)
- 40-41: Lernen als Dialog. „Werkstatt"-Berichte über Beziehungen zwischen Schülern, Lehrern und Erziehern unter erschwerten Bedingungen. Hrsg. v. Karl J. Kluge; Uta Sievert
- 40: Bd. I, 2 Bde. (Tl. A u. B) 1990. Tl. A: XXII, 389 S., Tl. B: VIII, 318 S. / Br (3-597-10592-0)
- 41: Bd. II, 2 Bde. (Tl. A u. B) 1990. Tl. A: 382 S., Tl. B: VIII, 259 S. / Br (3-597-10593-9)

- 42: Erziehungstherapie - eine kommentierte Bibliographie. Teil III. Hrsg. v. Klaus Fitting; Karl J. Kluge / 1993. VI, 364 S. / Br (3-597-10594-7)
- 43: Kollmar-Masuch, Rita: Hat der Lehrer in der stationären Kinder- und Jugendpsychiatrie eine Chance? Eine empirische Studie zur Situation von Lehrern an stationären Kinder- und jugendpsychiatrischen Einrichtungen der BRD und sonderpädagogische Konsequenzen / 1987. 263 S. / Br (3-597-10595-5)
- 44: Erziehungstherapie - eine kommentierte Bibliographie. Teil I / II. Hrsg. v. Klaus Fitting; Karl J. Kluge / 1988. 378 S. / Br (3-597-10596-3)
- 47: Kluge, Karl J. / Elisabeth Sander: Körperbehindert - und deswegen soll ich 'anders' sein als du? Eine vergleichende empirische Untersuchung zum Sexualerleben und -verhalten von körperbehinderten und nichtbehinderten Jugendlichen / 1987. XXIV, 442 S. / Br (3-597-10608-0)
- 48: Mit Vergnügen forschen und lernen - 1985. Aussergewöhnliches Lernen in universitären Sommercamps. Hrsg. v. Achim Bröcher; Nicola Griffel; Karl J. Kluge / 1987. 176 S. / Br (3-597-10609-9)
- 49: Mit Vergnügen Forschen und Lernen - 1986. Aussergewöhnliches Lernen in universitären Sommercamps. Hrsg. v. Achim Bröcher; Nicola Griffel; Karl J. Kluge / 1987. 320 S., 20 Abb. / Br (3-597-10610-2)
- 50: Oversberg, Michael P.: Sonderpädagogische Diagnostik an Schulen für Erziehungshilfe in Nordrhein-Westfalen und in der universitären Sonderschullehrerausbildung in Köln / 1988. IV, 305 S. / Br (3-597-10611-0)
- 51: Persönlichkeit oder Fachlichkeit? Eine Einführung in die aktuelle Verhaltensauffälligkeit und Erziehungstherapie. Von Achim Bröcher; Klaus Fitting; Bodo Januszewski; Helga Karsch; Karl J. Kluge; Eva M. Sassenrath / 1989. XIV, 203 S. / Br (3-597-10640-4)
- 52: Pinnow, Ulrich: Schüler-Uni. Ein Enrichmentprogramm für Kinder und Jugendliche mit besonderen Bedürfnissen, Fähigkeiten und hoher Motivation / 1989. 219 S. / Br (3-597-10641-2)
- 53: Bröcher, Achim: Kreative Intelligenz und Lernen. Eine Untersuchung zur Förderung schöpferischen Denkens und Handelns unter anderem in einem Universitären Sommercamp / 1989. 397 S., 18 Abb. / Br (3-597-10642-0)
- 54: Grobel, Anna: Hochbegabung in Familien. Eine Untersuchung über Beziehungen zwischen Eltern und ihren hochbegabten Kindern / 1990. XI, 366 S. / Br (3-597-10643-9)
- 55: Sassenrath, Eva M.: Intelligenz und Elternhaus. Erwartungen von besonders befähigten Kindern an ihre Eltern: „Mentoring" als pädagogische Konsequenz. Eine Studie in Verbindung mit einem Elterntrainingsprogramm / 1990. 371 S. / Br (3-597-10644-7)
- 56: Kluge, Annett / Peter V. Zysno: Teamkreativität. Eine Untersuchung zum Training der Ideenfindung mit klassischen Kreativitätsmethoden / 1993. 153 S. / Br (3-597-10645-5)

Berlin-ABC. Geschichte - Politik - Wirtschaft - Kultur. Hrsg. v. Walter Krumholz unter Mitarb. v. Wilhelm Lutze, Oskar Kruss, Richard Höpfner u.a. Im Auftrag des Presse- und Informationsamtes des Landes Berlin / 1969. 798 S / Lin (Vlg.-Nr. 03309)

Berlin-Bibliographie
- 1985 bis 1989. Hrsg. v. Senatsbibliothek Berlin. Bearb. v. Gabriele Feldmann-Kruse und Bruno

Gomez de Ortega / 2 Bde. 1995. Zus. 1.166 S. / Ln (3-598-23430-9)
- 1990. Hrsg.: Berliner Stadtbibliothek, in Zus.-Arb. mit d. Senatsbibliothek Berlin. Bearb. v. Horst Dedlow und Roswitha Arndt / 1995. XVII, 247 S. / Ln (3-598-23431-7)
- 1991. Hrsg.: Berliner Stadtbibliothek, in Zus.-Arb. mit d. Senatsbibliothek Berlin. Bearb. v. Roswitha Arndt, Horst Dedlow und Uta Scheibner / 1996. XVI, 248 S. / Ln (3-598-23432-5)
- 1992. Hrsg.: Zentral- u. Landesbibliothek Berlin, in Zus.-Arb. mit d. Senatsbibliothek Berlin. Bearb. v. Roswitha Arndt, Horst Dedlow und Uta Scheibner / 1997. XV, 244 S. / Ln (3-598-23433-3)

Berliner Adressbuch / Berlin Directory. Adressbuch für Berlin und seine Vororte 1919-1932. Vollständige Mikrofiche-Edition / Directory for Berlin and its Suburbs 1919-1932. Complete Microfiche edition. Vorw. v. Peter K. Liebenow. Hrsg. v. Konrad Umlauf. Veranlaßt v. d. Amerika Gedenkbibliothek. Berliner Zentralbibliothek / 1984. 104.850 S. auf 981 Fiches. Lesefaktor 24x / Silberfiche (3-598-30284-3)

Berliner allgemeine musikalische Zeitung 1824-1830. Mikrofiche-Edition / 1994. Ca. 3.500 S. auf 37 Fiches mit Begleitband. Lesefaktor 24x / Silberfiche (3-598-33805-8) / Diazofiche (3-598-33804-X)
- Berliner allgemeine musikalische Zeitung 1824-1830. Bibliographie und Indizes zur Mikrofiche-Edition / 1994. X, 72 S. / Ln (3-598-33806-6)

Berliner China-Studien. Hrsg. v. Kuo Heng-yü [Minerva Publikation]
- 1: Lao She: Blick westwärts nach Changan. Aus d. Chin. v. Ursula Adam, Thomas Kampen, Kuo Heng-yü und Eva Sternfeld / 1983. 166 S., 13 schw.-w. Fotos / Br (3-597-10457-6)
- 2: Gransow, Bettina: Soziale Klassen und Schichten in VR China. Theoretische Transformationskonzepte und reale Entwicklungsformen von 1949-1979 unter besonderer Berücksichtigung der städtischen Arbeiterklasse / 1983. XVI, 380 S. / Br (3-597-10458-4)
- 3: Keen, Ruth: Autobiographie und Literatur. Drei Werke der chinesischen Schriftstellerin Xiao Hong / 1984. II, 145 S., 1 schw.-w. Abb. / Br (3-597-10459-2)
- 4: Junkers, Elke: Leben und Werk der chinesischen Schriftstellerin Lu Yin (ca. 1899-1934). Anhand ihrer Autobiographie / 1984. II, 191 S., 1 schw.-w. Abb. / Br (3-597-10460-6)
- 5: Groeling-Che, Hui-wen von: Leben und Werk des chinesischen Sprachforschers Wang Li / 1984. II, 119 S., 1 schw.-w. Abb. / Br (3-597-10461-4)
- 6: Esser, Alfons: Bibliographie zu den deutsch-chinesischen Beziehungen 1860-1945 / 1984. XVIII, 120 S. / Br (3-597-10462-2)
- 7: Li, Erna: Deutsche Frauen in China. 12 Plaudereien über deutsch-chinesische Ehen / 1984. XVI, 328 S. / Br (3-597-10463-0)
- 8: Zhao Baoxu: Vorlesungen zur aktuellen Politik der Volksrepublik China. Aus d. Chines. v. Kuo Heng-yü, Mechthild Leutner, Erling von Mende und Klaus Stermann / 1985. 257 S. / Br (3-597-10464-9)
- 9: Der ewige Fluss. Chinesische Erzählungen aus Taiwan. Hrsg. v. Kuo Heng-yü / 1986. 300 S. / Ln (3-597-10465-7)
- 10: Rudolph, Jörg M.: Die Kommunistische Partei Chinas und Taiwan (1921-1981) / 1986. 325 S. / Br (3-597-10466-5)

- 11: Klaschka, Siegfried: Die Rehabilitierung Liu Shaoqis in der chinesischen Presse / 1987. 164 S. / Br (3-597-10598-X)
- 12: Beiträge zu den deutsch-chinesischen Beziehungen. Hrsg. v. Kuo Heng-yü; Mechthild Leutner / 1986. 176 S. / Br (3-597-10599-8)
- 13: Von der Kolonialpolitik zur Kooperation. Studien zur Geschichte der deutsch-chinesischen Beziehungen. Hrsg. v. Kuo Heng-yü / 1986. 518 S. / Ln (3-597-10600-5)
- 14: Meissner, Werner: Das rote Haifeng Peng Pais. Bericht über die Bauernbewegung in Südchina / 1987. 250 S. / Br (3-597-10601-3)
- 15: Kolonko, Petra: Im fremden Spiegel. Weltgeschichte und deutsche Geschichte in der VR China / 1988. VIII, 260 S. / Br (3-597-10602-1)
- 16: Jamann, Wolfgang / Thomas Menkhoff: Make big profits with a small capital. Die Rolle der Privatwirtschaft und des „Informellen Sektors" für die urbane Entwicklung der VR China / 1988. 246 p. / Soft (3-597-10612-9)
- 17: Deutsch-chinesische Beziehungen 1928-38. Eine Auswertung deutscher diplomatischer Akten. Hrsg. v. Kuo Heng-yü. Bearb. v. Frank Suffa-Friedel unter Mitw. v. Cornelia Anderer und Werner Meissner / 1988. 186 S. / Br (3-597-10613-7)
- 18: Exotik und Wirklichkeit. China in Reisebeschreibungen vom 17. Jahrhundert bis zur Gegenwart. Ed. by Mechthild Leutner und Dagmar Yü-Dembski / 1990. 120 S. / Br (3-597-10614-5)
- 19: Deutsch-chinesische Beziehungen vom 19. Jahrhundert bis zur Gegenwart Beiträge des Internationalen Symposiums in Berlin. Hrsg. v. Kuo Heng-yü und Mechthild Leutner / 1991. 445 S. / Br (3-597-10615-3)
- 20: Frauenstudien Beiträge der Berliner China-Tagung 1991. Hrsg. v. Cheng Ying, Bettina Gransow und Mechthild Leutner / 1993. 317 S., 6 schw.-w. Abb. / Br (3-597-10616-1)
- 21: Deutschland und China Beiträge des Zweiten Internationalen Symposiums zur Geschichte der deutsch-chinesischen Beziehungen, Berlin 1991. Hrsg. v. Kuo Heng-yü und Mechthild Leutner / 1994. 423 S. / Br (3-597-10617-X)
- 22: Hettler, Joachim: Shandong. Ein chinesischer Wirtschaftsraum und seine Entwicklung seit 1979 / 1993. 224 S. / Br (3-597-10618-8)
- 23: Mühlhahn, Klaus: Geschichte, Frauenbild und kulturelles Gedächtnis. Der mingzeitliche Roman Shuihu zhuan / 1994. 218 S. / Br (3-597-10619-6)
- 24: Spakowski, Nicola: Die Autorität der Vergangenheit. Funktionen der chinesischen Geschichtsschreibung am Beispiel der Rezeption Li Dazhaos / 1993. 205 S. / Br (3-597-10620-X)
- 25: Brandtstädter, Susanne: Früher hat uns das Meer ernährt, heute ernähren uns unsere Töchter. Frauen und wirtschaftlicher Wandel in einem taiwanesischen Fischerdorf / 1994. 178 S. / Br (3-597-10621-8)
- 26: Gransow, Bettina / Li Hanlin: Chinas neue Werte. Einstellungen zu Modernisierung und Reformpolitik / 1994. 117 S. / Br (3-597-10646-3)
- 27: Frick, Heike: Kunst und Politische Artikulation. Die chinesische Karikatur 1934-1937 / 1994. 231 S. / Br (3-597-10647-1)
- 28: Frauenforschung in China. Analysen, Texte, Bibliographie. Hrsg. v. Heike Frick, Mechthild Leutner und Nicola Spakowski / 1995. 312 S. / Br (3-597-10648-X)

Berthold, Werner / Brita Eckert; Frank Wende: Deutsche Intellektuelle im Exil. Ihre Akademie und die „American Guild for German Cultural Freedom". Eine Ausstellung des Deutschen Exilarchivs 1933-

1945 der Deutschen Bibliothek, Frankfurt / 1993. XI, 584 S. / Ln (3-598-11153-3) / Br (3-598-11152-5)

Bertrand Russell. A Bibliography of his Writings 1895-1976 / Eine Bibliographie seiner Schriften 1895-1976. Comp. by Werner Martin / 1981. XLV, 332 p. / Soft (3-598-10348-4)

Der **Bestand** Preussische Akademie der Künste - Kaiserreich, Weimarer Republik, Nationalsozialismus, Nachkriegszeit (1871-1955) Mikrofiche-Edition. Hrsg. v. d. Stiftung Archiv d. Akademie d. Künste Berlin, unter d. Leitung v. Norbert Kampe
- Tl. 1: Die Sektion für die bildenden Künste, für Musik und Dichtkunst / 1994. 1218 Fiches, Lesefaktor 24x, 1 Findbuch / Silberfiche (3-598-33705-1) / Diazofiche (3-598-33704-3)
  - Findbuch. Bearb. v. Silvia Diekmann, Kerstin Diether, Jörg Feßmann, Norbert Kampe, Kerstin Köhntopp, Annemarie Menke-Schwinghammer und Gudrun Schneider / 1994. XLII, 486 S. / Br (3-598-33712-4)
- Tl. 2: Präsident, Mitglieder, Ständige Sekretäre, Statuten und Senatsprotokolle / 1995. 969 Fiches, Lesefaktor 24x, 1 Findbuch / Silberfiche (3-598-33715-9) / Diazofiche (3-598-33714-0)
  - Findbuch. Bearb. v. Silvia Diekmann, Kerstin Diether, Jörg Feßmann, Norbert Kampe, Kerstin Köhntopp, Annemarie Menke-Schwinghammer und Gudrun Schneider / 1996. VI, XXXII, 538 S. / Br (3-598-33722-1)
- Tl. 3: Ausstellungen und Kunstpreise / 1997. 1150 Fiches, Lesefaktor 24x, 1 Findbuch / Silberfiche (3-598-33725-6) / Diazofiche (3-598-33724-8)
  - Findbuch. Bearb. v. Silvia Diekmann, Norbert Kampe, Kerstin Köhntopp, Annemarie Menke-Schwinghammer und Gudrun Schneider / 1998. XXXII, 344 S. / Br (3-598-33732-9)

Bestandskatalog Archiv- und internationale Dokumentationsstelle für das Blinden- und Sehbehindertenwesen (AiDOS). Hrsg.: Deutsche Blindenstudienanstalt, Marburg a. d. Lahn / 1987. XII, 759 S. / Lin (3-598-10641-6)

Bestandverzeichnis der deutschen Heimatvertriebenenpresse. Hrsg.: Stiftung Ostdeutscher Kulturrat OKR. Bearb. v. Horst von Chmielewski und Gert Hagelweide / 1982. XXIX, 284 S. / Lin (3-598-10386-7)

Bewahren und Ausstellen - die Forderung des kulturellen Erbes in Museen. Bericht über ein internationales Symposium veranstaltet von den ICOM-Nationalkomitees der Bundesrepublik Deutschland, Österreichs und der Schweiz Mai 1982 am Bodensee. Hrsg. v. Hermann Auer / 1983. 253 S. / Br (3-598-10481-2)

Bibliografía General Española Siglo XV - 1995 / Spanische Bibliographie 15. Jahrhundert - 1995 / Spanish Bibliography 15th Century - 1995 / CD-ROM-Edition 1995. (3-598-40283-X)
- Bibliografía General Española Siglo XV - 1997 / Spanische Bibliographie 15. Jahrhundert - 1997 / Spanish Bibliography 15th Century to 1997 / Bibliographie espagnole du XVe siècle à 1997 / Bibliografia Generale Spagnola dal XV secolo al 1997 / CD-ROM-Edition 2. Ausg. 1998. (3-598-40401-8)

Bibliografia Generale Italiana dal XV secolo al 1997 / Italienische Bibliographie 15. Jahrhundert - 1997 / Italian Bibliography 15th Century to 1997 /

Bibliographie italienne du XVe siècle à 1997 / Bibliografi General Italiana Siglo XV - 1997 / CD-ROM-Edition 1998. (3-598-40376-3)

Bibliographia Academica Germaniae. Abhandlungen u. Sitzungsberichte. Reports and proceedings. Hrsg. von Pál Vezényi
• Bd. 1: Königliche Gesellschaft d. Wissenschaften (Akademie d. Wissenschaften), Göttingen; Bayerische Akademie der Wissenschaften München:. Abhandlungen, Sitzungsberichte, Forschungen zur deutschen Geschichte / 1971. XVI, 715 S. / Lin (3-7940-3349-3)

Bibliographia Cartographica. Internationale Dokumentation des kartographischen Schrifttums / International documentation of cartographical literature / Documentation internationale de la litterature cartographique. Redaktion: Lothar Zögner. Hrsg.: Staatsbibliothek Preussischer Kulturbesitz in Verb. mit der Deutschen Gesellschaft für Kartographie
• Nr. 1. 1974 / 1975. 184 S. / Br (3-7940-3471-6)
• Nr. 2. 1975 / 1976. 195 S. / Br (3-7940-3472-4)
• Nr. 3. 1976 / 1977. 209 S. / Br (3-7940-3473-2)
• Nr. 4. 1977 / 1978. 206 S. / Br (3-7940-3474-0)
• Nr. 5. 1978 / . 1979. XII, 212 S. / Br (3-598-20619-4)
• Nr. 6. 1979 / 1980. XII, 255 S. / Br (3-598-20620-8)
• Nr. 7. 1980 / 1981. 244 S. / Br (3-598-20622-4)
• Nr. 8. 1981 / 1982. XII, 223 S. / Br (3-598-20624-0)
• Nr. 9. 1982 / 1983. XII, 232 S. / Br (3-598-20625-9)
• Nr. 10. 1983 / 1984. XII, 173 S. / Br (3-598-20626-7)
• Nr. 11. 1984 / 1985. XII, 213 S. / Br (3-598-20627-5)
• Nr. 12. 1985 / 1986. XII, 255 S. / Br (3-598-20628-3)
• Nr. 13. 1986 / 1987. XI, 247 S. / Br (3-598-20629-1)
• Nr. 14. 1987 / 1988. XI, 289 S. / Br (3-598-20630-5)
• Nr. 15. 1988 / 1989. XII, 260 S. / Br (3-598-20631-3)
• Nr. 16. 1989 / 1990. 260 S. / Br (3-598-20632-1)
• Nr. 17. 1990 / 1991. XVIII, 336 S. / Br (3-598-20633-X)
• Nr. 18. 1991 / 1992. XVIII, 360 S. / Br (3-598-20634-8)
• Nr. 19. 1992 / 1993. XVII, 396 S. / Br (3-598-20635-6)
• Nr. 20. 1993 / 1994. XVIII, 359 S. / Br (3-598-20636-4)
• Nr. 21. 1994 / 1995. XVIII, 422 S. / Br (3-598-20637-2)
• Nr. 22. 1995 / 1996. VIII, 424 S. / Br (3-598-20638-0)
• Nr. 23. 1996 / 1997. XVIII, 401 S. / Br (3-598-20639-9)
• Nr. 24. 1997 / 1998. XVIII, 372 S. / Br (3-598-20640-2)
• Nr. 25. 1998 / 1999. XVIII, 384 S. / Br (3-598-20641-0)
• Sonderh. 1. Schulatlanten in Deutschland und benachbarten Ländern vom 18. Jahrhundert bis 1950 Bibliographie und Standortverzeichnis. Hrsg. v. Lothar Zögner. Bearb. v. Astrid Badziag; Petra Mohs, unter Mitarb. v. Wolfgang Meinecke / 1982. VIII, 200 S. / Lin (3-598-20621-6)
• Sonderh. 2.. Zögner, Lothar: Bibliographie zur Geschichte der deutschen Kartographie von den Anfängen bis 1982. Unter Mitarb. von Evelyn Schulte / 1984. 267 S. (3-598-20623-2)

Bibliographie Bauwesen - Architektur - Städtebau. Veröffentlichungen der Bauakademie Berlin 1951 bis 1991. Hrsg.: BQG Bauakademie. Bearb. v. Wolfgang Tripmacker / 1993. 284 S. / Gb (3-598-11142-8)

Bibliographie Bildende Kunst / Bibliography of the Fine Arts Deutschsprachige Hochschulschriften und Veröffentlichungen ausserhalb des Buchhandels 1966-1980 / Bibliography of German-Language

University Dissertations and Publications outside the Book Trade 1966-1980. Bearb. v. Irene Butt; Monika Eichler / 6 Bde. 1992. Zus. XXVI, 2390 S. / Gb (3-598-22380-3)

Bibliographie Bildung, Erziehung, Unterricht / Bibliography of Education Deutschsprachige Hochschulschriften und Veröffentlichungen ausserhalb des Buchhandels 1966-1980 / Bibliography of German-Language University Dissertations and Publications outside the Book Trade 1966-1980. Bearb. v. Irene Butt; Monika Eichler / 8 Bde. 1994. Zus. XXXIV, 3358 S. / Gb (3-598-22950-X)

Bibliographie Buch- und Bibliothekswesen, Medienkunde, Hochschulwesen / Bibliography of Book Trade, Library Science, Communication Science and Higher Education Deutschsprachige Hochschulschriften und Veröffentlichungen ausserhalb des Buchhandels 1966-1980 / Bibliography of German-Language University Dissertations and Publications outside the Book Trade 1966-1980. Bearb. v. Irene Butt; Monika Eichler / 6 Bde. 1994. Zus. XXVIII, 2433 S. / Gb (3-598-21380-8)

Bibliographie Darstellende Kunst und Musik. Deutschsprachige Hochschulschriften und Veröffentlichungen ausserhalb des Buchhandels 1966-1980. Bearb. v. Margarethe Wolf, / 3 Bde. 1992. Zus. XIV, 813 S. / Gb (3-598-21950-4)

Bibliographie der Bibliotheksadressbücher. Bearb. von Swolke von Uslar und Manhard Schütze / 1967. 25 Bl.

Bibliographie der Buchherstellung. (Sonderdr. aus: Handbuch der technischen Dokumentation und Bibliographie Bd. 2) / 1968. S. 305-426
• 2. Ausg. 1968. 122 S. / Br (Vlg.-Nr. 02228)

Bibliographie der deutschsprachigen Lyrikanthologien 1840-1914. Hrsg. v. Günter Häntzschel unter Mitarb. v. Sylvia Kucher und Andreas Schumann / 2 Bde. 1991. Zus. XIV, 695 S. / Ln (3-598-10838-9)

Bibliographie der Filmbibliographien / Bibliography of Film Bibliographies. Comp. and ed. by Hans J. Wulff; Karl D. Möller; Jan Ch. Horak. Bibliographie d. slawischen Filmbibliographien zsgst. v. Andrzej Gwozdz und Anna Wastkowska / 1987. XXIX, 326 p. / Hard (3-598-10630-0)

Bibliographie der geheimen DDR-Dissertationen / Bibliography of Secret Dissertations in the German Democratic Republic Band 1: Bibliographie / Band 2: Register / Volume 1: Bibliography / Volume 2: Index. Hrsg. und eingel. v. Wilhelm Bleek und Lothar Mertens / 2 Bde. 1994. Zus. XLIII, 945 S. / Ln (3-598-11209-2)

Bibliographie der Sozialforschung in der Sowjetunion 1960 - 1970 / Bibliography of Social Research in the Soviet Union 1960 - 1970. Zsgst. von Sergej Woronitzin / 1973. 215 S. / Lin (3-7940-3650-5)

Bibliographie der versteckten Bibliographien. Aus deutschsprachigen Büchern und Zeitschriften der Jahre 1930-1953. Bearb.: Deutsche Bücherei. - Nachdr. d. Ausg. Leipzig 1956 / 1983. 372 S. / Ln (3-598-07210-4) [Lizenzausg. m. Genehm. v. Zentralantiquariat der DDR, Leipzig]

Bibliographie der Zeitschriftenliteratur zum Stand der Technik. Bibliography of periodical literature on the state-of-the-art for the areas of technology. Utz-Friedebert Taube / Jg. 1. 1976 ff. 1978 ff.
• Bibliographie der Zeitschriftenliteratur zum Stand der Technik. Jg. 2 (1977). Bd1: Sektion A:Täglicher Lebensbedarf. Sektion B: Arbeitsverfahren / Section A: Human Necessities. Section B: Performing Operations. 1979. XIV, 155 S. (3-598-20701-8);. Bd. 2: Sektion C: Chemie und Hüttenwesen / Section C: Chemistry and Metallurgy. 1979. VI, 248 S. (3-598-20702-6) Bd. 3: Sektion D: Textil und Papier. Sektion E: Bauwesen, Bergbau. Sektion F: Maschinenbau, Beleuchtung. Heizung, Waffen, Sprengen. Sektion G: Physik / Section D: Textiles and Paper. Section E: Fixed Constructions. Section F: Mechanical Engineering, Lighting, Heating, Weapons, Blasting, Section g: Physiks 1979. X, 235 S. / Bd4: Sektion H: Elektrotechnik / Section H: Electricity / 1979. 211 S. (3-598-20704-2) / (3-598-20703-4)
• Bibliographie der Zeitschriftenliteratur zum Stand der Technik. Jg. 3 (1978). Bd1: Sektion A:Täglicher Lebensbedarf. Sektion B: Arbeitsverfahren / Section A: Human Necessities. Section B: Performing Operations. 1980. 126 S. (3-598-20705-0). Bd. 2: Sektion C: Chemie und Hüttenwesen / Section C: Chemistry and Metallurgy. 1980. 249 S. (3-598-20706-9) Bd. 3: Sektion D: Textil und Papier. Sektion E: Bauwesen, Bergbau. Sektion F: Maschinenbau, Beleuchtung. Heizung, Waffen, Sprengen. Sektion G: Physik / Section D: Textiles and Paper. Section E: Fixed Constructions. Section F: Mechanical Engineering, Lighting, Heating, Weapons, Blasting, Section g: Physiks 1980. 205 S. / Bd4: Sektion H: Elektrotechnik / Section H: Electricity / 1980. 205 S. (3-598-20708-5) / (3-598-20707-7)

Bibliographie des Bibliothekswesens. (Sonderdr. aus: Handbuch der technischen Dokumentation und Bibliographie. Bd. 2) / 1968. S. 91-304
• 2. Ausg. 1968. 214 S. / Br (Vlg.-Nr. 02068)
• 3. Ausg. 1970. 240 S. (3-7940-2068-5)

Bibliographie des Buchhandels. (Sonderdruck aus: Handbuch der technischen Dokumentation und Bibliographie. Bd 2: Die Fachliteratur zum Buch- und Bibliothekswesen) / 1968. 90 S.
• 2. Ausgabe 1968. 90 S. (Vlg. –Nr. 02058)
• 3. Ausgabe 1970, 100 S. / Br (Vlg. –Nr. 02058)
Neuausgabe → Werkstatt des Buches. Bibliographie des Buchhandels

Bibliographie française du XVe siècle à 1997 / Französische Bibliographie 15. Jahrhundert - 1997 / French Bibliography 15th Century to 1997 / Bibliografia Generale Francese dal XV secolo al 1997 / Bibliografi General Francesa Siglo XV 1997 / CD-ROM-Edition 1998. 2 discs (3-598-40373-9)

Bibliographie Geographie, Kartographie, Reisen. Deutschsprachige Hochschulschriften und Veröffentlichungen ausserhalb des Buchhandels 1966-1980. Bearb. v. Irene Butt; Monika Eichler / 4 Bde. 1994. Zus. XVIII, 1420 S. / Gb (3-598-22640-3)

Bibliographie Geschichte, Volkskunde, Völkerkunde / Bibliography of History, Folklore and Ethnology Deutschsprachige Hochschulschriften und Veröffentlichungen ausserhalb des Buchhandels 1966-1980 / Bibliography of German-Language University

Dissertations and Publications outside the Book Trade 1966-1980 Bearb. v. Irene Butt; Monika Eichler / 7 Bde. 1994. Zus. XXX, 2844 S. / Gb (3-598-22920-8)

Bibliographie Hochschulplanung 1977. Hrsg.: Zentralarchiv für Hochschulbau, Stuttgart / 1978. 371 S. / Br (3-598-20007-2)

Bibliographie Land-, Forst- und Ernährungswirtschaft / Bibliography of Agriculture, Forestry and the food Sciences Deutschsprachige Hochschulschriften und Veröffentlichungen ausserhalb des Buchhandels 1966-1980. Bibliography of German-Language University Dissertations and Publications outside the Book Trade 1966-1980. Bearb. v. Irene Butt und Monika Eichler / 8 Bde. 1994. Zus. XXXIV, 3298 S. / Gb (3-598-23340-X)

Bibliographie Medizin / Bibliography of Medicine. Deutschsprachige Hochschulschriften und Veröffentlichungen ausserhalb des Buchhandels 1966-1980 / Bibliography of German-Language University Dissertations and Publications outside the Book Trade 1966-1980. Bearb. v. Irene Butt und Monika Eichler / 34 Bde. 1993. Zus. CXXXVIII, 14950 S. / Gb (3-598-22499-0)

Bibliographie Militärwesen / Bibliography of Military Affairs. Deutschsprachige Hochschulschriften und Veröffentlichungen ausserhalb des Buchhandels 1966-1980 / Bibliography of German-Language University Dissertations and Publications outside the Book Trade 1966-1980. Bearb. v. Irene Butt; Monika Eichler / 1993. VI, 306 S. / Gb (3-598-22390-0)

Bibliographie Pädagogik / Educational Bibliography. Hrsg.: Dokumentationsring Pädagogik (DOPAED), Berlin / Jg. 6/1971 - Jg. 22/1987 [Erscheinen eingestellt. Vorher im Beltz Verlag, Weinheim]
- Jg. 6/1971 - Jg. 11/1976 / Br (6 Ausg. jährl.) (ISSN 0523-2678)
- Jg. 12/1977 / 1980. XXII, 520 S. / Br
- Jg. 13/1978 / 1981. XXXIV, 889 S. / Br (3-598-00066-9)
- Jg. 14/1979 / 1982. VII, 886 S. / Br (3-598-00068-5)
- Jg. 15/1980 / 1983. 1011 S. / Br (3-598-00069-3)
- Jg. 16/1981 / 1984. 931 S. / Br (3-598-00071-5)
- Jg. 17/1982. Reihe / Series A: Zeitschriftenaufsätze / Articles. Hrsg.: Dokumentationsring Pädagogik (DOPAED) (ISSN 0176-2567). Reihe / Series B: Bücher. Sondersammelgebiet Bildungsforschung in Erlangen / Books. Special Collection Educational Research in Erlangen. Hrsg.: Dokumentationsring Pädagogik (DOPAED); Universitätsbibliothek Erlangen-Nürnberg; Gesellschaft für sozial- und erziehungswissenschaftliche Forschung und Dokumentation e.V. (GFD) (ISSN 0176-2575). Reihe / Series C: Projekte, Bildungsforschung / Educational Research. Hrsg.: Dokumentationsring Pädagogik (DOPAED) und Gesellschaft für sozial- und erziehungswissenschaftliche Forschung und Dokumentation e.V. (GFD) in Zus-Arb. m. Informationszentrum Sozialwis-senschaften (IZ) (ISSN 0176-2583) / 3 Tle. 1985
  - Reihe / Series A 1982 / 1985. VIII, 911 S. / Br (3-598-21729-3)
  - Reihe / Series B 1982/83 / 1985. X, 93 S. / Br (3-598-21730-7)
  - Reihe / Series C 1982/83 / 1985. IX, 248 S. / Br (3-598-21731-5)
- Jg. 18/1983 / 2 Tle. 1986 (3-598-21732-3)
  - Reihe / Series A 1983 / 1986. VII, 650 S. / Br (3-598-21733-1)
  - Reihe / Series B 1983 / 1986. VII, 218 S. / Br (3-598-21734-X)
- Jg. 19/1984 / 3 Tle.1987 (3-598-21738-2)
  - Reihe / Series A 1984 / 1987. VII, 786 S. / Br (3-598-21735-8)
  - Reihe / Series B 1984 / 1987. VII, 242 S. / Br (3-598-21736-6)
  - Reihe / Series C 1984 / 1987. VII, 153 S. / Br (3-598-21737-4)
- Jg. 20/1985 / 3 Tle. 1989 (3-598-21739-0) [nur Reihe A u. B erschienen]
  - Reihe / Series A 1985 / 1989. XIII, 656 S. / Br (3-598-21740-4)
  - Reihe / Series B 1985 / 1989. XIII, 242 S. / Br (3-598-21741-2)
- Jg. 21/1986 / 3 Tle. 1990 (3-598-21743-9) [nur Reihe A u. B erschienen]
  - Reihe / Series A ,1986 / 1990. XIII, 31, 783 S. / Br (3-598-21744-7)
  - Reihe / Series B 1986 / 1990. XIII, 228 S. / Br (3-598-21745-5)

Bibliographie Paul Raabe. Zu seinem 60. Geburtstag Geleitw. u. zsgst. v. Barbara Strutz / 1987. 99 S. / Gb (3-598-10679-3)

Bibliographie Politik und Zeitgeschichte / Bibliography of Politics and Contemporary History. Deutschsprachige Hochschulschriften und Veröffentlichungen ausserhalb des Buchhandels 1966-1980 / Bibliography of German-Language University Dissertations and Publications outside the Book Trade 1966-1980. Bearb. v. Irene Butt und Monika Eichler / 6 Bde. 1993. Zus. XXVI, 2298 S. / Gb (3-598-22934-8)

Bibliographie Programmierter Unterricht. Hrsg.: Pädagogisches Zentrum, Berlin / Jg. 7/1971 - Jg. 9/1973 / Gh (ISSN 0067-7027). Ab Auswertungs-Jg. 1974 vereinigt mit *Bibliographie Pädagogik* [Vorher im Beltz-Verlag, Weinheim]

Bibliographie Psychologie. Deutschsprachige Hochschulschriften und Veröffentlichungen ausserhalb des Buchhandels 1966-1980. Bearb. v. Monika Eichler; Margarethe Wolf / 2 Bde. 1992. Zus. 625 S. / Gb (3-598-21927-X)

Bibliographie Religion und Philosophie / Bibliography of Religion and Philosophy. Deutschsprachige Hochschulschriften und Veröffentlichungen ausserhalb des Buchhandels 1966-1980 / Bibliography of German-Language University Dissertations and Publications outside the Book Trade 1966-1980. Bearb. v. Irene Butt; Monika Eichler / 5 Bde. 1993. Zus. 2.091 S. / Gb (3-598-22485-0)

Bibliographie Sport und Freizeit. Deutschsprachige Hochschulschriften und Veröffentlichungen ausserhalb des Buchhandels 1966-1980. Bearb. v. Monika Eichler; Margarethe Wolf / 2 Bde. 1991. Zus. 639 S. / Gb (3-598-21955-5)

Bibliographie Sprache und Literatur. Deutschsprachige Hochschulschriften und Veröffentlichungen ausserhalb des Buchhandels 1966-80. Bearb. v. Irene Butt; Monika Eichler / 8 Bde. 1991 / 1992. Zus. 3228 S. / Gb (3-598-22470-2)

Bibliographie und Berichte. Festschrift für Werner Schochow, den langjährigen Redakteur der Bibliographischen Berichte. Geleitw. v. Richard Landwehrmeyer. Hrsg. v. Hartmut Walravens / 1990. 299 S. / Gb (3-598-10907-5)

Bibliographie Veterinärmedizin / Bibliography of Veterinary Medicine. Deutschsprachige Hochschulschriften und Veröffentlichungen ausserhalb des Buchhandels 1966-1980 / Bibliography of German-Language. University Dissertations and Publications outside the Book Trade 1966-1980. Bearb. v. Irene Butt; Monika Eichler / 3 Bde. 1992 XIV, 1036 S. / Gb (3-598-22494-X)

Bibliographie 'Widerstand'. Einl. v. Karl O. von Aretin. Hrsg.: Forschungsgemeinschaft 20. Juli e.V. Bearb. v. Ulrich Cartarius / 1984. 326 S. / Lin (3-598-10468-5)

Bibliographie zu den Biographischen Archiven / Biographical Archives Bibliography / 1994. 185 S. / Br (3-598-33752-3)
- 2. korr. u. erw. Aufl. 1998. 238 S. / Br (3-598-33753-1)

Bibliographie zum Antisemitismus / A Bibliography on Antisemitism. Die Bestände der Bibliothek des Zentrums für Antisemitismusforschung der Technischen Universität Berlin / The Library of the Zentrum für Antisemitismusforschung at the Technical University of Berlin. Hrsg. v. Herbert A. Strauss. Bearb. v. Lydia Bressem. Sachreg. bearb. v. Uta Lohmann; Lucie Renner; Martina Strehlen / 4 Bde. 1990 / 1993. Zus. XXX, 1377 S. / Lin (3-598-10868-0)

Bibliographie zur deutschen Literaturgeschichte des Barockzeitalters. Begr. v. Hans Pyritz. Fortgef. v. Ilse Pyritz [1991 vom Francke Vlg., Bern übernommen]
- Tl. 1. Kultur- und Geistesgeschichte - Poetik - Gattungen - Traditionen - Bezeichnungen - Stoffe. Bearb. v. Reiner Bölhoff / 1991. Zus. XXVII, 738 S. / Gb (3-907820-58-4)
- Tl. 2. Dichter und Schriftsteller. Anonymes. Textsammlungen Bearb. v. Ilse Pyritz / 1985. XXI, 810 S. / Gb (3-907820-62-2)
- Gesamtregister. Bearb. v. Reiner Bölhoff / 1994. 220 S. / Gb (3-907820-63-0)

Bibliographie zur deutschen Literaturgeschichtsschreibung 1827-1945. Hrsg. v. Andreas Schumann / 1994. XXXII, 278 S. / Gb (3-598-11229-7)

Bibliographie zur Dokumentation. Bibiography of documentation. Ausgabe 3 / 1970. 77 S. (3-7940-2500-8)

Bibliographie zur Frühpädagogik. Bearb. von Helga Kochan-Döderlein; Hella Maria Erler; Erika Kolb; Helga Riedel; Hartmut Schirm. Institut für Frühpädagogik des Bayerischen Staatsministeriums für Unterricht und Kultus / 1972. 328 S. / Br (3-7940-3169-5)

Bibliographie zur Geschichte des Weines. Zsgst. v. Renate Schoene. Hrsg.: Gesellschaft für Geschichte des Weines e. V. [1982 übernommen. Vorher bei d. Südwestdeutschen Verlagsanstalt, Mannheim u. im Eigenverlag (Suppl. 1)]

- Im Auftr. d. Gesellschaft für Geschichte des Weines hrsg. v. Karl Fill. Vorworte v. Helmuth Arntz und Hartwig Lohse / 1976. XXXV, 543 S. / Lin (3-598-10388-3)
- Supplement 1 / 1978. XXXI, 175 S. / Lin (3-598-10387-5)
- Supplement 2 / 1982. XXXV, 237 S. / Ln (3-598-10389-1)
- Supplement 3 / 1984. XLII, 206 S. / Lin (3-598-10507-X)
- 2. mit allen Supplementen kumulierte und aktualisierte Aufl. 1988. XXVI, 480 S. / Ln (3-598-10748-X)

Bibliographie zur Geschichte von Stadt und Hochstift Bamberg 1945-1975. Mit Bamberger Zeitschriftenbeiträgen 1919-1964. Hrsg.: Historischer Verein Bamberg für die Pflege der Geschichte des ehemaligen Fürstbistums Bamberg e.V. / 1980. XIV, 576 S. / Lin (3-598-10075-2)

Bibliographie zur lateinischen Wortforschung. Hrsg. v. Otto Hiltbrunner (3-907820-71-1) [1991 vom Francke Vlg., Bern übernommen]
- Bd. 1. A-acutus / 1981. XXII, 298 S. / Gb (3-907820-72-X)
- Bd. 2. Adeo-atrocitas / 1984. 324 S. / Gb (3-907820-73-8)
- Bd. 3. Atrox-causa / 1988. 310 S. / Gb (3-907820-74-6)
- Bd. 4. Censeo-cura / 1992. 348 S. / Gb (3-907820-75-4)

Bibliographie zur Zeitgeschichte 1953-1995. Kumulation der Bibliographie zu den Vierteljahresheften für Zeitgeschichte. Im Auftr. d. Instituts f. Zeitgeschichte München begr. v. Thilo Vogelsang. Bearb. v. Hellmuth Auerbach, Christoph Weisz, Ursula von Laak und Hedwig Straub-Waller / 5 Bde. 1982-1996 / Lin (3-598-10420-0)
- Bd. I. Allgemeiner Teil. Hilfsmittel - Geschichtswissenschaft - Gesellschaft und Politik - Biographien / 1982. XVI, 445 S. / Ln (3-598-10421-9)
- Bd. II. Geschichte des 20. Jahrhunderts bis 1945. Allgemeine Geschichte - Europäische Geschichte - Geschichte des I. Weltkriegs - Deutsche Geschichte - Geschichte einzelner Staaten - Geschichte des II. Weltkriegs / 1982. VIII, 559 S. / Lin (3-598-10422-7)
- Bd. III. Geschichte des 20. Jahrhunderts seit 1945. Allgemeine Geschichte - Europäische Geschichte - Deutsche Geschichte - Geschichte einzelner Staaten / 1983. 501 S. / Lin (3-598-10423-5)
- Bd. IV. Supplement 1981-1989 / 1991. XV, 585 S. / Lin (3-598-10924-5)
- Bd. V. Supplement 1990-1995 / 1996. XIV, 374 S. (3-598-11124-X)

Bibliographien zur deutsch-jüdischen Geschichte. Hrsg. im Auftr. d. Salomon Ludwig Steinheim-Instituts für deutsch-jüdische Geschichte v. Michael Brocke; Julius Schoeps; Falk Wiesemann,
- 1: Wiesemann, Falk: Bibliographie zur Geschichte der Juden in Bayern / 1989. XV, 263 S. / Lin (3-598-10832-X)
- 2: Wassermann, Henry: Bibliographie des Jüdischen Schrifttums in Deutschland 1933-1943 Bearb. f. d. Leo Baeck Institut, Jerusalem, unter Mitw. v. Joel Golb; Lydia Katzenberger; Ada Walk / 1989. XXVII, 153 S. / Gb (3-598-10750-1)
- 3: Komorowski, Manfred: Bio-bibliographisches Verzeichnis jüdischer Doktoren im 17. und 18. Jahrhundert Inauguraldissertationen und biographische Fundstellen / 1992. 128 S. / Lin (3-598-10980-6)

- 4: Studemund-Halévy, Michael: Bibliographie zur Geschichte der Juden in Hamburg / 1994. XIV, 256 S. / Gb (3-598-11178-9)
- 5: Bibliographie zur Geschichte der Juden in der Schweiz Auf der Basis des Werkes von Annie Fraenkel. Hrsg. u. aktualis. v. Uri R. Kaufmann / 1993. XX, 151 S. / Lin (3-598-11139-8)
- 6: Heitmann, Margret / Andreas Reinke: Bibliographie zur Geschichte der Juden in Schlesien In Zus.-Arb. mit Harald Lordick; Andreas Reinke; Heike Teckenbrock / 1995. 254 S. / Gb (3-598-11230-0)

Bibliographien zur deutschen Barockliteratur. Hrsg. v. John D. Lindberg [1991 vom Francke Vlg., Bern übernommen]
- 1: Otto, Karl F. jr.: Philipp von Zesen. A Bibliographical Catalogue / 1972. 310 S., 8 Abb. / Gb (3-907820-79-7)
- 2: Hardin, James: Johann Beer. Eine beschreibende Bibliographie / 1983. 181 S., 12 Abb., 1 Taf / Gb (3-907820-80-0)
- 3: Hardin, James: Johann Christoph Ettner. Eine beschreibende Bibliographie / 1988. 182 S. / Gb (3-907820-81-9)

Bibliographien zur Regionalen Geographie und Landeskunde / Bibliographies on Regional Geography and Area Studies. Hrsg. v. Sperling, Walter; Zögner, Lothar
- 1: Sperling, Walter: Landeskunde DDR. Eine annotierte Bibliographie / 1978. XXII, 456 S. / Lin (3-7940-7038-0)
- 2: Ehlers, Eckart: Iran. Ein bibliographischer Forschungsbericht / A Bibliographic Research Survey. Mit Kommentaren und Annotationen / With Comments and Annotations / 1980. XII, 441 S. / Lin (3-598-21132-5)
- 3: Hetzer, Armin / Viorel S. Roman: Albania. Ein bibliographischer Forschungsbericht. Mit Titelübersetzungen und Standortnachweisen / A Bibliographic Research Survey. With Location Codes / 1983. 653 S. / Lin (3-598-21133-3)
- 4: Philipp, Hans J.: Saudi Arabia. Volume I / Saudi-Arabien. Band I. Bibliography on Society, Politics, Economics / Bibliographie zur Gesellschaft, Politik, Wirtschaft / Literatur seit dem 18. Jahrhundert in westeuropäischen Sprachen mit Standortnachweisen / 1984. XCI, 405 p. / Hard (3-598-21134-1)
- 5: Sperling, Walter: Landeskunde DDR. Eine kommentierte Auswahlbibliographie. Ergänzungsband 1978-1983 / 1984. 623 S. / Lin (3-598-21135-X)
- 6: Tauschinsky, Aja / Ute Krauss-Leichert: Italien. Bibliographie der deutschsprachigen sozialwissenschaftlichen Monographien und Zeitschriftenartikel 1945-1980 / 1985. XXXI, 460 S. / Lin (3-598-21136-8)
- 7: Philipp, Hans J.: Saudi Arabia. Volume II / Saudi Arabien. Band II. Bibliography on Society, Politics, Economics. Literature in West European Languages with Location References / Bibliographie zu Gesellschaft, Politik, Wirtschaft. Literatur in westeuropäischen Sprachen mit Standortnachweisen / 1989. XLV, 634 p. / Hard (3-598-21137-6)
- 8: Sperling, Walter: DDR-Bibliographie. Ergänzungsband 1984-1986. Unter Mitarb. v. Heinz P. Brogiato / 1991. 625 S. / Gb (3-598-21138-4)
- 9: Ostafrika-Bibliographie. Kenia-Tansania-Uganda 1945-1993. Bearb. v. Hans Hecklau, unter Mitarb. weiterer Mitarbeiterinnen und Mitarbeiter / 1996. XVI, 278 S. / Gb (3-598-21140-6)

Bibliographischer Alt-Japan-Katalog 1542-1853. Bearb. u. Hrsg.: Japaninstitut in Berlin u. Deutsches

Forschungsinstitut in Kyoto. -Nachdr. d. Ausg. Kyoto 1940 / 1977. XXXVIII, 416 S. / Lin (3-7940-3173-3)

Bibliographischer Dienst. Hrsg.: Internationales Zentralinstitut für das Jugend- und Bildungsfernsehen (IZI), München [Bis Bd. 2 beim IZI erschienen]
- 3: Lernen durch Fernsehen in Schule und Ausbildung. Eine Bibliographie ausgewählter Forschungsliteratur der Jahre 1967 bis 1979. Bearb. v. Manfred Meyer, Frieder von Krusenstjern, Ursula Nissen und Sylvia Huth / 1980. 121 S. / Br (3-598-20683-6)
- 4: Kabelfernsehen: Eine Bibliographie. Konzepte, Projekte, Erfahrungen. Bearb. v. Silvia Huth und Paul Löhr / 1982. 205 S. / Br (3-598-20684-4)
- 5: Kind und Familie vor dem Bildschirm. Eine Bibliographie ausgewählter Forschungsergebnisse zur Fernsehrezeption. Bearb. v. Werner Müller; Manfred Meyer / 1985. 150 S. / Br (3-598-20685-2)
- 6: Medienpädagogik: Fernsehen Theorie und Praxis. Eine Bibliographie internationaler Fachliteratur. Hrsg. v. Werner Müller; Paul Löhr / 1988. 172 S. / Br (3-598-20686-0)
- 7: Fernsehen und Jugend Eine Bibliographie internationaler Fachliteratur 1969-1989. Bearb. v. Kurt Aimiller; Paul Löhr; Manfred Meyer / 1989. 204 S. / Br (3-598-20687-9)

Bibliography of American Imprints to 1901. Comp.: Databases of the Research Libraries Group (RLG); American Antiquarian Society / 92 vols. 1992 / 1993. Cplt. 4,0711 p. / Hard (3-598-33340-4)
- Vol. 43-56: Author-Index / 14 vols. 1993 / Hard (3-598-33498-2)
- Vol. 57-71: Subject-Index / 15 vols. 1993 / Hard (3-598-33499-0)
- Vol. 72-82: Place-Index / 11 vols. 1993 / Hard (3-598-33500-8)
- Vol. 83-92: Date-Index / 10 vols. 1993 / Hard (3-598-33501-6)

Bibliography of Publications issued by UNESCO or under its auspices in the first twenty-five years: 1946 – 1971 / 1973. XVIII, 365 S. / Ln (3-7940-5124-6) [Koproduktion mit UNESCO, Paris]

Bibliography of the International Congresses of Philosophy / Bibliographie der Internationalen Philosophie-Kongresse Proceedings 1900-78 / Beiträge 1900-1978. Ed. by Lutz Geldsetzer / 1981. 208 p. / Hard (3-598-10331-X)

Bibliography on German Unification / Bibliographie zur deutschen Einigung. Economic and Social Developments in Eastern Germany November 1989 to June 1992 / Wirtschaftliche und soziale Entwicklung in den neuen Bundesländern November 1989 bis Juni 1992. Comp. by Horst Thomsen; Frauke Siefkes / 1992. XVII, 345 p. / Gb (3-598-11147-9)

Bibliography on Peace Research and Peaceful International Relations / Bibliographie Friedensforschung und Friedenspolitik. The Contributions of Psychology 1900-1991 / Der Beitrag der Psychologie 1900 bis 1991. Ed. by Marianne Müller-Brettel. Comp. by Gerhild Richter / 1993. 384 p. / Hard (3-598-11072-3)

Bibliography on the History of Chemistry and Chemical Technology. 17th to the 19th Century / Bibliographie zur Geschichte der Chemie und Chemischen Technologie. 17. bis 19. Jahrhundert.

Ed. by Valentin Wehefritz, University of Dortmund Library, with the assist. of Zoltán Kováts, Budapest / 3 Vols. 1994. Cplt. 1749 p. / Hard (3-598-11200-9)

Bibliotheca Palatina. Druckschriften - Stampati Patini - Printed Books. Mikrofiche-Edition. Hrsg. u. Einf. v. Leonard Boyle; Elmar Mittler / 1989-1995. 21.103 Silberfiches u. Index. Lesefaktor 24x / Silberfiche (3-598-32880-X)
• Katalog und Register zur Mikrofiche-Ausgabe. Druckschriften der Bibliotheca Palatina. Hrsg.: Elmar Mittler / 4 Bde. 1998. LX, 1946 S. / Gb (3-598-32886-9)
• Bd. 1: Bibliographie A - K / 1998. XXVI, 508 S. / Gb (3-598-32887-7)
• Bd. 2: Bibliographie L - Z / 1998. XII, 533 S. / Gb (3-598-32888-5)
• Bd. 3: Register der Signaturen, Mikrofiches, Erscheinungsjahre / 1998. XI, 576 S. / Gb (3-598-32889-3)
• Bd. 4: Register der Personen, Verleger bzw. Drucker, Verlags- bzw. Druckorte / 1998. XI, 429 S. / Gb (3-598-32882-6)

Bibliotheca Philologica. - Nachdr. d. Ausg. 1870-1878 / 2 Bde. 1978
• Bd. 1. 229, 341 S. / Ln (3-7940-5064-9)
• Bd. 2. 182, 221, XXII, 181 S. / Ln (3-7940-5065-7)

Bibliotheca Scatologica ou catalogue raisonné des livres traitant des vertus faits et gestes de très noble et très ingénieux Messire Luc (a rebours) seigneur de la chaise et autres lieux. Zus.gest. u. d. Reg. erschlossen v. Pierre Jannet; Jean François Payen; Alexandre Auguste Veinant. - Nachdr. d. Ausg. Paris 1849 / 1970. XXXI, 143 S. / Ln (3-7940-5059-2) [Koprod. m. d. Zentralantiquariat d. DDR, Leipzig]

Bibliotheca Trinitariorum. Internationale Bibliographie trinitarischer Literatur / International Bibliography of Trinitarian Literature. Hrsg. v. Erwin Schadel / 2 Bde. / Lin (3-598-10568-1)
• Bd. I. Autorenverzeichnis / Author Index Unter Mitarb. v. Dieter Brünn; Peter Müller / 1984. CXII, 624 S. / Lin (3-598-10572-X)
• Bd. II. Registerband / Index Volume Zus.-Arb. mit Leonore Bazinek; Peter Müller / 1988. XXXVI, 594 S. / Lin (3-598-10573-8)

Bibliothek der Deutschen Literatur. Mikrofiche-Gesamtausgabe nach Angaben des Taschengoedeke. Hrsg.: Kulturstiftung der Länder / 1990-1994. 9,7 Millionen S. auf 19.963 Fiches. Inkl. Bibliographie und Register. Lesefaktor 42x / Silberfiche (3-598-50001-7) / Diazofiche (3-598-50772-9)
• Bibliographie und Register. Bearb. unter d. Leitung v. Axel Frey / 1994. 581 S. / Gb (3-598-50100-5) / Br (3-598-53763-8)
• Supplement. Bearb. v. Axel Frey / 1997/99. 320.736 S. auf 718 Fiches. Lesefaktor 42x, 800 S. / Silberfiche (3-598-53301-2) / Diazofiche (3-598-53300-4)

Bibliothek des Deutschen Museums. Alphabetischer Katalog und Schlagwortkatalog. Erscheinungszeitraum: 15. Jahrhundert - 1976. Hrsg.: Bibliothek des Deutschen Museums München / Mikrofiche-Edition / 2 Lfg. 1981 / 1982. 612 Fiches. Lesefaktor 42x / Silberfiche (3-598-30397-1) / Diazofiche (3-598-30400-5)
• Lfg. 1: Alphabetischer Katalog. Erscheinungszeitraum: 15. Jahrhundert-1976. Erwerbungsstand: 31. 3. 1981 / 1981. 426.000 Ktn. auf 313 Fiches. Lesefaktor 42x / Diazofiche (3-598-30401-3)

• Lfg. 2: Schlagwortkatalog. Erscheinungszeitraum: 15. Jahrhundert - 1976. Erwerbsstand: 31. 3. 1982 / 1983. ca. 382.000 Ktn. u. Reg. auf 299 Fiches. Lesefaktor 42x / Diazofiche (3-598-30402-1)

Bibliothek des Deutschen Patentamtes München. Kreuzkatalog. Erscheinungszeitraum 1945-1974. Stand 31. 12. 1982. Mikrofiche-Edition / 1983. ca. 20.743 Ktn. auf 167 Fiches. Lesefaktor 42x / Silberfiche (3-598-30453-6) / Diazofiche (3-598-30454-4)

Bibliothek - Dokument - Information. Symposium, Wien 27. u. 28. Nov. 1980. Tagungsbericht mit Wortlaut der Referate. Hrsg.: Österreichisches Institut für Bibliotheksforschung, Dokumentations- und Informationswesen / 1981. 206 S. / Br (3-598-10362-X)

Bibliothek Forschung und Praxis. Hrsg. v. Paul Kaegbein; Hans J. Kuhlmann; Elmar Mittler / seit Jg. 1/1977. (3 Hefte jährlich) (ISSN 0341-4183)

Bibliothek und Buch in Geschichte und Gegenwart. Festgabe für Friedrich Adolf Schmidt-Künsemüller zum 65. Geburtstag am 30. Dez. 1975. Hrsg. von Otfried Weber / 1975. 336 S. / Ln (3-7940-3311-6)

Bibliotheken bauen und führen. Eine internationale Festgabe für Franz Kroller zum 60. Geburtstag. Hrsg. v. Sigrid Reinitzer / 1983. 412 S. / Lin (3-598-10499-5)

Bibliotheken im Netz. Funktionswandel Wissenschaftlicher Bibliotheken durch Informationsverarbeitungsnetze. Konstanzer Kolloquium (19.-21. 2. 1986). Vorträge. Joachim Stoltzenburg zu Ehren. Von Richard Landwehrmeyer; Klaus Franken; Ulrich Ott; Günther Wiegand / 1986. 148 S. / Ln (3-598-10644-0)

Bibliotheken in München. Hrsg.: Generaldirektion der Bayerischen Staatlichen Bibliotheken / 1983. 96 S. / Br (3-598-10488-X)

Bibliotheken wirtschaftlich planen und bauen. Tendenzen - Ausblicke - Empfehlungen. Ergebnisse des IFLA-Bibliotheksbau-Seminars, Bremen 1977. Hrsg. v. Horst Meyer / 1981. 188 S., 67 Abb. / Br (3-598-10334-4)

Bibliotheksarbeit für Parlamente und Behörden. Festschrift zum 25jährigen Bestehen der Arbeitsgemeinschaft der Parlaments- und Behördenbibliotheken. Hrsg. v. Wolfgang Dietz, Hildebert Kirchner und Kurt G. Wernicke / 1980. 385 S. / Ln (3-598-10125-2)

Bibliotheksautomatisierung, Benutzererwartungen und Serviceleistungen. Bericht e. Symposiums, veranstaltet vom Deutschen Bibliotheksinstitut u. d. Gesamthochschulbibliothek Essen. Redaktion: Detlef Schwarz / 1980. 146 S. (3-598-10030-2)

Bibliotheksforum Bayern (BFB). Hrsg.: Generaldirektion der Bayerischen Staatlichen Bibliotheken, München. Redakt.: Michael Mücke (bis Jg. 22); Klaus Kempf (seit Jg. 23) / seit Jg. 1/1973 (3 Hefte jährlich) (ISSN 0340-000X)

Bibliothekspraxis
• 1: Niewalda, Paul: Die elektronische Datenverarbeitung im Bibliothekswesen / 1971. 136 S. (3-7940-4001-5)
  - 2. Aufl. 1977. 120 S. / Lin (3-7940-4000-7)
• 2: Schoch, Gisela: Die Informationsmittel einer Universitätsbibliothek. Ihre Nutzung durch d. student. Leser Dargest. am Beispiel d. Staats- u. Universitätsbibliothek Hamburg / 1971. 163 S. (3-7940-4002-3)
• 3: Heidtmann, Frank: Materialien zur Benutzerforschung. Aus e. Pilotstudie ausgew. Benutzer d. Universitätsbibliothek d. Techn. Univ. Berlin / 1971. 191 S. (3-7940-4003-1)
• 4: Kissel, Gerhard: Betriebswirtschaftliche Probleme wissenschaftlicher Bibliotheken / 1971. 127 S. (3-7940-4004-X)
• 5: Ascher, Werner: Organisationsprobleme eines Hochschulbibliothekssystems / 1972. 107 S. (3-7940-4005-8)
• 6: The Exchange of Bibliographic Data and the MARC Format. Austausch bibliographischer Daten und das MARC-Format. Proceedings of an international Seminar on the MARC Format and the Exchange of Bibliographic Data in Machine Readable Form, sponsored by the Volkswagen Foundation, Berlin, June 14-16, 1971. Arbeitsstelle für Bibliothekstechnik bei der Staatsbibliothek Preußischer Kulturbesitz, Berlin / Vorwort von W. Lingenberg / 1972. 196 S. (3-7940-4006-6)
• 7: Audio-visuelle Medien in Hochschulbibliotheken. Seminar in d. Universitätsbibliothek Bochum vom 6.-8. März 1972. Hrsg. von Eckhard Franzen u. Günther Pflug / 1972. 103 S. (3-7940-4007-4)
• 8: Poggendorf, Dietrich: Anleitung für die Katalogisierung in Institutsbibliotheken. / 1974. 133 S. / Lin (3-7940-4008-2)
• 9: EDV in englischen und irischen Bibliotheken und Dokumentationseinrichtungen. Hrsg. von Rudolf Frankenberger; Paul Niewalda / 1973. 88 S. (3-7940-4009-0)
• 10: Bibliotheksautomation in den USA und in Kanada. Bericht über eine Studienreise von 5 dt. Bibliiothekaren im Jahre 1972. Hrsg.von Walter Lingenberg / 1973. 149 S. (3-7940-4010-4)
• 11: Bibliotheksbau und Bibliothekstechnik. Ein Kompendium für Bibliothekare. Hrsg. v. Franz H. Philipp / 1973. 144 S. / Lin (3-7940-4011-2)
• 12: Informationen zum Bibliotheksbau. In Zusammenarbeit mit der Kommission für Baufragen im Verein Deutscher Bibliothekare. Hrsg. v. Franz Künzl / 1974. 176 S. / Lin (3-7940-4012-0)
• 13: Klar, Rainer H. / Werner Sämann; Gabriele Daume: Personalbedarfsermittlung in dezentralen Bibliotheken / 1974. 136 S. (3-7940-4013-9)
• 14: Arnold, Ekkehard: Approval Plans als Instrument der Literaturwerbung / 1975. 102 S. (3-7940-4014-7)
• 15: Selbmann, Sibylle: Zur Öffentlichkeitsarbeit wissenschaftlicher Bibliotheken. E. theoret. u. empir. Beitr / 1975. 180 S. (3-7940-4115-1)
• 16: Rogalla von Bieberstein, Johannes: Archiv, Bibliothek und Museum als Dokumentationsbereiche. Einheit und gegenseitige Abgrenzung / 1975. 116 S. (3-7940-4116-X)
• 17: Funk, Robert: Kostenanalyse in wissenschaftlichen Bibliotheken. E. Modelluntersuchung an d. Universitätsbibiothek d. Techn. Univ. Berlin / 1975. 180 S. (3-7940-4117-8)
• 18: Stoltzenburg, Joachim / Günther Wiegand: Die Bibliothek der Universität Konstanz 1965-74. Erfahrungen und Probleme / 1975. 119 S. (37940-4114-3)

- 19: Bibliotheksverbund in Nordrhein-Westfalen. Planung u. Aufbau d. Gesamthochschulbibliotheken u. d. Hochschulbibliothekszentrums 1972 - 1975. Hrsg. von Klaus Barckow / 1976. 398 S. (3-7940-4119-4)
- 20: Funk, Robert: Arbeitsablaufuntersuchungen und Personalbedarfsermittlung für die Buchbearbeitung an zentralen Hochschulbibliotheken. Ergebnisse einer mit Unterstützung der Deutschen Forschungsgemeinschaft durchgeführten Untersuchung. Unter Mitarb. v. Renate Dopheide; Werner Sämann; Ulla Usemann-Keller / 1977. 351 S. / Lin (3-7940-4120-8)
- 21: Kohl, Ernst: Zeichensatz und Zeichenverschlüsselung für die elektronische Datenverarbeitung in Bibliotheken. Verarb., Ausg., Austausch unter bes. Berücksichtigung d. BSBEBCDIC-Zeichensatzes d. maschinellen Austauschformats für Bibliotheken (MAB 1) / 2., völlig neugestalt. Aufl. 1977. 246 S. (3-7940-4121-6)
- 22: Hauser, Hans-Jörg: Literaturproduktion und -preise im Publikationswesen außerhalb der BRD E. statist. Sekundäranalyse. Unter Mitarb. von Inge Stumpf. Erarb. im Auftr.d. Dt. Forschungsgemeinschaft an d. Bad. Landesbibliothek Karlsruhe / 1978. 220 S. (3-7940-4122-4)
- 23: Dissertationen in Wissenschaft und Bibliotheken. Hrsg. von Rudolf Jung und Paul Kaegbein / 1979. 175 S. (3-598-21123-6)
- 24: Sinogowitz, Bernhard: Leihverkehrs-Fibel. Kurzgefaßte Anleitung zur Praxis des Leihverkehrs der deutschen Bibliotheken. Mit Text und Kommentar der Leihverkehrsordnung 1979 und einer Auswahlbibliographie zum Leihverkehr 1967-1979 / 1980. 144 S. (3-598-21124-4)
  - 2. durchges. u. erg. Aufl. Mit Text und Kommentar der Leihverkehrsordnung 1979 und einer Auswahlbibliographie zum Leihverkehr 1971-1982. Unter Mitarb. v. Dieter Karasek / 1983. 146 S. / Lin (3-598-21126-0)
- 25: Nagelsmeier-Linke, Marlene: Automatisierte juristische Informationssysteme. Gegenwärtiger Stand ihrer Entwicklung u. ihre Bedeutung für d. bibliohekar. Praxis / 1980. 208 S. (3-598-21125-2)
- 26: Lux, Claudia: Das Bibliothekswesen der Volksrepublik China / 1986. 119 S. / Lin (3-598-21127-9)
- 27: Schmüser, Eike: Zeitschriftenerwerbung und Lieferantenwahl in wissenschaftlichen Bibliotheken der Bundesrepublik Deutschland / 1986. 125 S. / Lin (3-598-21128-7)
- 28: Barton, Walter: Die Gesamthochschulbibliothek Erfahrungen im Bibliotheksverbund Nordrhein-Westfalen / 1990. IX, 133 S. / Lin (3-598-21129-5)
- 29: Krauss-Leichert, Ute: Einsatz neuer Technologien im Bibliothekswesen. Eine Expertenbefragung / 1990. 227 S. / Br (3-598-10970-9)
- 30: Möller, André: CD-ROM-Einsatz in Bibliotheken / 1991. 120 S. / Lin (3-598-21130-9)
- 31: Retrospective Cataloguing in Europe: 15th to 19th Century Printed Materials Proceedings of the International Conference, Munich 28th - 30th November 1990. Ed. by Franz G. Kaltwasser; John M. Smethurst / 1992. 194 p. / Hard (3-598-21131-7)
- 32: Strzolka, Rainer: PC-Software-Lösungen für Bibliotheken / 1994. X, 247 S. / Gb (3-598-21122-8)
- 33: Pörzgen, Rainer / Martin Schreiber: Die Informationsvermittlungsstelle. Planung - Einrichtung - Betrieb / 1993. IV, 124 S. / Lin (3-598-21164-3)
- 34: Braun, Traute: Regionale Verbundsysteme in der Bundesrepublik Deutschland. Ihre Portabilität für wissenschaftliche Bibliotheken in den neuen Bundesländern / 1993. 293 S. / Lin (3-598-21165-1)
- 35: Zeitschriften in deutschen Bibliotheken Bestand - Erwerbung - Erschliessung - Benutzung. Hrsg. v. d. Staatsbibliothek zu Berlin - Preussischer Kulturbesitz, unter Leitung v. Hartmut Walravens / 1995. VIII, 371 S. / Gb (3-598-21166-X)

Bibliotheksstudien. Hrsg. v. Harro Heim.
- 1 A: Datenerfassung und Datenverarbeitung in der Universitätsbibliothek Bielefeld. Eine Materialsammlung. Hrsg. v. Elke Bonness; Harro Heim / 1972. 413 S., 10 Abb. / Lin (3-7940-3201-2)
- 1 B: Bonneß, Elke / Harro Heim: Oneline Übernahme von Fremddaten in der Universitätsbibliothek Bielefeld. E. neues Verfahren d. Katalogisierung mit Hilfe von Datensichtgeräten. Unter Mitarb. v. Carola Glienke / 1974. 205 S. (3-7940-3214-4)
- 1 C: Briesenick, Christa / Hartmut Felsch:. Automatisierte Ausleihe in der Universitätsbibliothek Bielefeld Ein neues Verfahren mit Hilfe von Stichcodeetiketten Ein Lehrbuch für Bibliotheken u. e. Anwendung für Rechenzentren Unter Mitarb. von Harro Heim / 1976. 225 S. (3-7940-3216-0)
- 2: Stock, Karl F.: Grundlagen und Praxis der Bibliotheksstatistik / 1974. XV, 397 S. (3-7940-3205)
- 3: Das Datenbanksystem IBAS in der Universitätsbibliothek Bielefeld. E. Ergebnis aus 10 Jahren Praxis mit automatisierter Datenverarbeitung. Hrsg. von Elke Bonneß u. Harro Heim, unter Mitarb. von Gottfried Fellgiebel / 1979. 646 S. (3-598-20593-7)
- 4: Heim, Harro: Die Universitätsbibliothek Bielefeld 1968-1984. Aufbau und Entwicklung / 1984. 160 S. / Lin (3-598-20595-3)
- 5: Aufbau, Organisation und Funktion eines neuen Informationszentrums am Beispiel der Vorarlberger Landesbibliothek. Ed. by Eberhard Tiefenthaler / 1990. 193 S. / Gb (3-598-11002-2)
- 6: Offene Systeme in offene Bibliotheken! Propagierung, Bedeutung, Auswirkungen, Probleme. Hrsg. v. Elke Bonness; Harro Heim / 1993. 134 S. / Gb (3-598-11188-6)

Bibliothekswelt und Kulturgeschichte. Eine internationale Festgabe für Joachim Wieder zum 65. Geburtstag. Hrsg. von Peter Schweigler / 1977. 311 S., 23 Abb. / Ln (3-7940-7018-6)

Bild- und Tonträger-Verzeichnisse. Hrsg.: Deutsches Rundfunkarchiv, Frankfurt a. M. [Bis Bd. 18/1987 beim Deutschen Rundfunkarchiv erschienen]
- 19: Lexikon des Musiktheaters im Fernsehen / Encyclopedia of Music Theatre on TV in German-Speaking Europe 1973-1987. Zus.gest. u. bearb. v. Achim Klünder; Christina Voigt / 1991. XX, 439 S. / Ln (3-598-10874-5)
- 20: Lexikon der Fernsehspiele / Encyclopedia of Television Plays in German-Speaking Europe 1978-1987. Zsgst. u. bearb. v. Achim Klünder / 3 Bde. 1991. Zus. XXXXII, 2410 S. / Gb (3-598-10836-2)
- 21: Lexikon der Fernsehspiele / Encyclopedia of Television Plays in German-Speaking Europe 1988 / 1991. XXIII, 460 S. / Ln (3-598-10856-7)
- 22: Lexikon der Fernsehspiele / Encyclopedia of Television Plays in German-Speaking Europe 1989 / 1992. XV, 416 S. / Ln (3-598-11088-X)
- 23: Lexikon der Fernsehspiele / Encyclopedia of Television Plays in German-Speaking Europe 1990 / 1992. XIX, 450 S. / Ln (3-598-11089-8)
- 24: Lexikon der Fernsehspiele / Encyclopedia of Television Plays in German-Speaking Europe 1991 / 1993. XXIII, 404 S. / Ln (3-598-11151-7)
- 25: Bild- und Tonträger-Verzeichnisse 1992 / 1994. XIX, 460 S. / Ln (3-598-11189-4)

Bildung und Beschäftigung. Probleme, Konzepte, Forschungsperspektiven. Mitarb.: Dirk Hartung; Reinhard Nuthmann, Ulrich Teichler / 1981. 271 S. (3-598-10332-8)

Bio-Bibliographisches Verzeichnis von Universitäts- und Hochschuldrucken (Dissertationen) vom Ausgang des 16. bis Ende des 19. Jahrhunderts / 4 Bde. 1977/80 / Ln (3-598-02815-6)
- Bd. 1: Aalst Schouten - Kühn. Hrsg. v. Hermann Mundt. - Nachdr. d. Ausg. Leipzig 1936 / 1965. 720 S. / Ln (3-7940-2811-2)
- Bd. 2: Kühn - Ritter. Hrsg. v. Hermann Mundt. - Nachdr. d. Ausg. Leipzig 1942 / 1965. 320 S. / Ln (3-7940-2812-0)
- Bd. 3: Ritter - ZZ. Begr. v. Hermann Mundt. Hrsg. v. Konrad Wickert / 1977. X, 569 S. / Ln (3-7940-2813-9)
- Bd. 4: Personenregister. Begr. v. Hermann Mundt. Hrsg. v. Konrad Wickert / 1980. V, 176 S. / Ln (3-598-02814-8)

Biografisch Archief van de Benelux / Archives Biographiques des Pays du Benelux / Biographical Archive of the Benelux Countries / Biographisches Archiv der Benelux-Länder (BAB). Eine Kumulation von Einträgen aus ca. 120 der wichtigsten biographischen Nachschlagewerke der Niederlande, Belgiens und Luxemburgs vom Ende des 16. bis zum Beginn des 20. Jahrhunderts. Mikrofiche-Edition. Bearb. v. Willi Gorzny, Willemina van der Meer / 1992-1994. 762 Fiches, Lesefaktor 24x / Silberfiche (3-598-32630-0) / Diazofiche (3-598-32610-6)
- Biografische Index van de Benelux / Index Biographique des pays du Bénélux / Biographischer Index der Benelux-Länder (BIB) / 4 Bde. 1996. Zus. LXXVI, 1604 S. / Gb (3-598-32645-9)

Biographische *Archive / Biographical Archive* →
   African *Biographical Archive (AfBA)*
   American *Biographical Archive (ABA I)*
   American *Biographical Archive. Series II (ABA II)*
   Arab-Islamic *Biographical Archive (AIBA)*
   Archives *Biographiques Françaises (ABF I)*
   Archives *Biographiques Françaises. Deuxième Série (ABF II)*
   Archivio *Biografico Italiano (ABI I)*
   Archivio *Biografico Italiano. Nuova Serie (ABI II)*
   Archivio *Biografico Italiano sino al 1996 (ABI III)*
   Archivo *Biográfico de España, Portugal e Iberoamérica (ABEPI I)*
   Archivo *Biográfico de España, Portugal e Iberoamérica. Nueva Serie (ABEPI II)*
   Archivo *Biográfico de España, Portugal e Iberoamérica 1960-1995 (ABEPI III)*
   Australasian *Biographical Archive (ANZO-BA)*
   Baltisches *Biographisches Archiv (BaBA)*
   Biografisch *Archief van de Benelux (BAB)*
   Biografisch Archief van de Benelux. Deel II (BAB II) [i.Vb.]
   Biographisches *Archiv der Antike (BAA)*
   British *Biographical Archive (BBA I)*
   British *Biographical Archive. Series II (BBA II)*
   Český *biografický archiv a Slovenský biografický archiv (CSBA)*

Chinese Biographical Archive (CBA)
Deutsches Biographisches Archiv (DBA I)
Deutsches Biographisches Archiv, Neue Folge bis zur Mitte des 20. Jahrhunderts (DBA II)
Deutsches Biographisches Archiv 1960-1999 (DBA III) [i.Vb.]
Griechisches Biographisches Archiv (GBA)
Indian Biographical Archive (IBA)
Japanese Biographical Archive (JaBA) [i.Vb.]
Jüdisches Biographisches Archiv (JBA)
Korean Biographical Archive (KBA) [i.Vb.]
Polskie Archiwum Biograficzne (PAB)
Polskie Archiwum Biograficzne. Seria Nowa (PAB II)
Russisches Biographisches Archiv (RBA)
Scandinavian Biographical Archive (SBA)
South-East Asian Biographical Archive (SEABA)
Südosteuropäisches Biographisches Archiv (SOBA)
Türkisches Biographisches Archiv (TBA) [i.Vb.]
Ungarisches Biographisches Archiv (UBA)

Biographisches Archiv der Antike / Biographical Archive of the Classical World (BAA). Ca. 120.829 Personen in ca. 170.830 Artikeln. Mikrofiche-Edition Bearb. v. Hilmar Schmuck / 1996-1999. Ca. 700 Fiches in 12 Lfgn, Lesefaktor 24x / Silberfiche (3-598-33971-2) / Diazofiche (3-598-33970-4)

Biographisches Handbuch der deutschsprachigen Emigration nach 1933 / International Biographical Dictionary of Central European Émigrés 1933-1945
• 3 Bde. / Ln (3-598-10087-6)
 - Bd. I: Biographisches Handbuch der deutschsprachigen Emigration nach 1933. Politik, Wirtschaft, Öffentliches Leben. Hrsg.: Institut f. Zeitgeschichte, München / Research Foundation of Jewish Immigration, Inc., New York. Gesamtltg.: Werner Röder und Herbert A. Strauss. Bearb. v. Dieter M. Schneider und Louise Forsyth / 1980. LVIII, 875 S. / Ln (3-598-10088-4)
 - Vol. II: International Biographical Dictionary of Central European Émigrés 1933-1945. The Arts, Science and Literature. Sponsored by Research Foundation of Jewish Immigration, Inc., New York and Institut f. Zeitgeschichte, München. Ed. by Herbert A. Strauss and Werner Röder. Comp. by Hannah Caplan, Egon Radvany, Horst Möller and Dieter M. Schneider / 2 vols. 1983. Cplt. XCIV, 1304 p. / Hard (3-598-10089-2)
 - Bd. III: Biographisches Handbuch der deutschsprachigen Emigration nach 1933. Gesamtregister / Index. Hrsg.: Institut f. Zeitgeschichte, München / Research Foundation of Jewish Immigration, Inc., New York. Gesamtltg.: Werner Röder und Herbert A. Strauss. Zsgst. unter d. Leitung v. Werner Röder v. Sybille Claus, Daniel Niederland und Beatrix Schmidt / 1983. XX, 281 S. / Ln (3-598-10090-6)
• Unveränd. Broschurausg. / 3 Bde. 1999. CLX, 2472 S. / Br (3--598-11420-6)

Biographisches Handbuch der SBZ / DDR 1945-1990. Hrsg. v. Gabriele Baumgartner und Dieter Hebig / 2 Bde. 1995/96. Zus. XXV, 1057 S. / Ln (3-598-11130-4)

Biographisches Wörterbuch zur deutschen Geschichte. Band 1: A - H, Band 2: I - R, Band 3: S - Z, Namenregister für das Gesamtwerk. Begr. v. Hellmuth Rössler und Günther Franz. Bearb. v. Karl Bosl, Günther Franz und Hanns H. Hofmann / 3 Bde. 2. völlig neubearb. u. stark erw. Aufl. 1973-1975. Bd. 1: XII S., 1.266 Spalten, Bd. 2: X S., 1.150

Spalten, Bd. 3: X S., 914 Spalten / Gb (3-907820-83-5) [1991 vom Francke Vlg., Bern übernommen]

Biologie-Dokumentation. Bibliographie der deutschen biologischen Zeitschriftenliteratur 1796-1965. Hrsg. v. Martin Scheele und Gerhardt Natalis / 24 Bde. 1981-82. 13.280 S. / Ln (3-598-30295-9)
• 1-19: Bibliographie
• 20-22: Stichwortregister
• 23-24: Systematisches Register

The **Black** Women Oral History Project. Ed. by Ruth E. Hill, The Arthur and Elizabeth Schlesinger Library on the History of Women in America, Radcliffe College / 11 vols. 1990 / 1991. Cplt. 5,149 p. / Hard (3-598-41350-5) [1991 von Meckler Publishing, Westport, CT übernommen]
• Vol. 1. Jessie Abbott, Christia Adair, Frankie V. Adams, Kathleen Redding Adams, Frances Mary Albrier / 1990. XXII, 547 p. / Hard (3-598-41351-3)
• Vol. 2. Margaret Walker Alexander, Sadie Alexander, Elizabeth Barker, Etta Moten Barnnett, Norma Boyd, Melnea A. Cass, May Edward Chinn / 1991. XXII, 528 p. / Hard (3-598-41352-1)
• Vol. 3. Juanita Jewel Craft, Clara Dickson, Alice Dunnigan, Alfreda Duster, Eva B. Dykes, Mae Massie Eberhardt, Florence Jacobs Edmonds, Lena Edwards, Dorothy Boulding Ferebee / 1991. XXII, 493 p. / Hard (3-598-41353-X)
• Vol. 4. Minnie L. Fisher, Katherine Stewart Flippin, Virginia Clark Gayton, Zelma George, Frances O. Grant / 1991. XXII, 438 p. / Hard (3-598-41354-8)
• Vol. 5. Ardie Clark Halyard, Dorothy I. Height, Beulah S. Hester / 1991. XXII, 384 p. / Hard (3-598-41355-6)
• Vol. 6. May Edwards Hill, Margaret Holmes, Clementine Hunter, Ellen Jackson, Fidelia Johnson, Lois Mailou Jones, Susie Jones / 1991. XXII, 379 p. / Hard (3-598-41356-4)
• Vol. 7. Virginia Lacy Jones, Maida Springer Kemp, Abna Aggrey Lancaster, Eunice Laurie, Catherine Cardozo Lewis, Inabel Burns Lindsay, Eliza Champ McCabe, Miriam Matthews / 1991. XXII, 467 p. / Hard (3-598-41357-2)
• Vol. 8. Lucy Miller Mitchell, Audley Moore, Annie M. Nipson, Rosa Parks, Lucy Rucker Aiken, Neddie Rucker Harper, Hazel Rucker, Esther Mae Scott, Julia Hamilton Smith / 1991. XXII, 501 p. / Hard (3-598-41358-0)
• Vol. 9. Muriel S. Snowden, Olivia Pearl Stokes, Ann Tanneyhill, Ruth Janetta Temple, Constance Allen Thomas, Era Bell Thompson, Mary C. Thompson D. S. / 1991. XXII, 546 p. / Hard (3-598-41359-9)
• Vol. 10. Bazoline Usher, Charleszetta Waddles, Dorothy West, Addie Luck Williams, Frances H. Williams, Ozeline Wise, Deborah Wolfe, Arline J. Yarbrough / 1991. XXII, 479 p. / Hard (3-598-41360-2)
• Vol. 11. Guide to the Transcripts of the Black Women Oral History Project / 1991. XV, 152 p. / Hard (3-598-41361-0)

Blana, Hubert: Die Herstellung ➤ *Grundwissen Buchhandel - Verlage Bd. 5*

Blana, Hubert / Hermann Kusterer; Peter Fliegel: Partner im Satz. Ein Handbuch für Autoren, Hersteller, Produktioner, Setzer / 1988. XI, 275 S. / Lin (3-598-10633-5)

Bleuler, Eugen: Dementia Praecox oder Gruppe der Schizophrenien. - Autoris. Reprint / 1978. XII, 420 S. / Ln (3-597-20001-X) [Minerva Publikation]

Bobeth, Johannes: Die Zeitschriften der Romantik. - Nachdr.der Ausg. Leipzig 1911 / 1970. 442 S. m. 17 Faks.-Beilagen (Vlg.-Nr. 05116)

Böhm, Wolfgang: Biographisches Handbuch zur Geschichte des Pflanzenbaus / 1997. IX, 398 S. / Ln (3-598-11324-2)

Bönsch, Manfred: Beiträge zu einer kritischen und instrumentellen Didaktik / 1975. 166 S. / Kst. (UTB 516) (3-7940-2646-2)

Börsenblatt für den Deutschen Buchhandel 1834-1945. Mikrofiche-Edition / 1979-1981. Zus. ca. 875.000 S. auf 3.057 Fiches. Lesefaktor 42x. Diazofiche (3-598-10177-5)

Bonner Katalog. Verzeichnis reversgebundener musikalischer Aufführungsmaterialen. Bearb. u. Hrsg.: Deutsches Musikarchiv, in Verb. mit dem Deutschen Musikverlegerverband e.V. / 2. neubearb. Aufl. 1982. XIII, 530 S. / Ln (3-598-10327-1)
• 3. neubearb. Aufl. 2 Bde. 1997. Zus. XVII, 618 S. / Ln (3-598-11356-0)

Bonner Katalog. CD-ROM-Edition / . Hrsg.: Deutsches Musikarchiv, in Verb. mit d. Deutschen Musikverlegerverband e.V. / 1997. (3-598-40380-1)

The **book** hunger. Ed. by Ronald Barker and Robert Escarpit / 1973. 155 p. / Soft (3-7940-5128-9) [Koprod. m. d. UNESCO, Paris]

The **Book** Trade of the World. Ed. by Sigfred Taubert (Vol. I - IV) and Peter Weidhaas (Vol. III & IV) / 4 vols. [1979 v. Vlg. für Buchmarkt - Forschung, Wiesbaden übernommen]
• Vol. I. Europe and International Section / 1972. 543 p. / Hard (3-598-10183-X)
• Vol. II. The Americas, Australia, New Zealand / 1976. 377 p. / Hard (3-598-10184-8)
• Vol. III. Asia. Introduction by Ingo-Eric M. Schmidt-Braul / 1981. 284 p. / Hard (3-598-10185-6)
• Vol. IV. Africa. Introduction by Hans M. Zell. With an index to volumes I-IV comp. by Carol Bundy / 1984. 391 p. / Hard (3-598-10354-9)

Books in other Languages. A Guide to Selection Aids and Suppliers. Comp. by Leonard Wertheimer / 4th ed. 1979. 140 p. (3-598-40010-1)

Boßmeyer, Christine / Ilse Jöstlein; Rainer Klar; Gottfried Mälzer; Klaus Sailer: Fortschritte des EDV-Verbundes in Bibliotheken der USA (OCLC und BALLOTS). Bericht über eine Studienreise. Hrsg.: Arbeitsstelle für Bibliothekstechnik bei der Staatsbibliothek Preußischer Kulturbesitz / 1976. 204 S. / Br (3-7940-3217-9)

Bouché, Reinhard: Datenerhebungskatalog für die Dokumentation von Literatur und Projekten. Herbert H. Engel und Hans-Jochen Schneider / 1975. 206 S. (3-7940-3275-6)

Bramley, Gerald: Outreach. Library Services for the Institutionalised, the Elderly and the Physicaly

Handicapped. / 1978. 232 S. (0-85157-254-5) [Clive Bingley]

Brauer, Werner: Graphik + (und) Design. Grundlagen, Werbung, Information, Gestaltung, Typographie, Druck, Photographie / 1976. 234 S. (3-7940-3310-8)

Breitkopf, Johann G.: Versuch den Ursprung der Spielkarten und die Einführung des Leinenpapiers zu erforschen. - Nachdr. d. Ausg. Leipzig 1784 / 1985. 144 S., 14 Taf. / Gb (3-598-07216-3) [Lizenzausg. m. Genehm. v. Zentralantiquariat d. DDR, Leipzig]

Brendel, Detlev / Bernd Grobe: Journalistisches Grundwissen. Darstellung der Formen und Mittel journalistischer Arbeit und Einführung in die Anwendung empirischer Daten in den Massenmedien. Mit einem Vorw. v. Kurt Koszyk / 1976. 264 S., 30 Abb. u. Tab. / Kst. (UTB 565) (3-7940-2654-3)

Brenneke, Adolf: Archivkunde. - Unveränderter Nachdr. d. Ausg. Leipzig: Koehler & Amelang 1953 / 1970. XIX, 542 S. / Ln (Vlg. – Nr. 04280) [Lizenzausg. m. Genehm. v. Zentralantiquariat d. DDR, Leipzig]

Brenneke, Adolf: Archivkunde. Ein Beitr. zur Theorie u. Geschichte d. europ. Archivwesens. - Nachdr. d. Orig.- Ausg. 1953. - Bearb. u. erg. von Wolfgang Leesch / 1988. XIX, 469 S. / Ln (3-598-10785-4)
- 2. völlig neubearb. u. erw. Aufl. Brenneke, Adolf / Wolfgang Leesch: Archivkunde / 2 Bde. 1993ff. (3-598-11074-X)
- Bd. 2: Internationale Archivbibliographie. Mit besonderer Berücksichtigung des deutschen und österreichischen Archivwesens / 1993. XXIV, 438 S. / Gb (3-598-11076-6)

Brewer, Gordon J.: The Literature of Geography / 2. überarb. Aufl. 1978. 264 S. (0-85157-280-4) [Clive Bingley]

Briefe Deutscher Philosophen (1750-1850). Mikrofiche-Edition. ca. 800 Quellenwerke. Vorw. u. Hrsg. v. Norbert Henrichs; Horst Weeland. Bearb. v. Ingo Rill; Martin Roether / 1990. 3141 Fiches. Lesefaktor 24x, Aufnahmefaktor 1: 20 / Silberfiche (3-598-33020-0) / Diazofiche (3-598-33010-3)

Briefwechsel deutschsprachiger Philosophen 1750-1850. Band 1: Register, Verfasser / Adressaten, Adressaten / Verfasser. Band 2: Nachweise, Briefe, Briefsammlungen. Hrsg. v. Norbert Henrichs; Horst Weeland / 2 Bde. 1987. Zus. 1120 S. / Ln zu (3-598-21910-5)

British Biographical Archive / Britisches Biographisches Archiv (BBA I). A one-alphabet cumulation of 324 of the most important English-language biographical reference works originally published between 1601 and 1929. Approximately 330. 000 entries covering 170. 000 individuals. Microfiche edition. Ed. by Paul Sieveking. Managing ed.: Laureen Baillie / 1984-1989. 1,236 fiches. Reader factor 24x / Silverfiche (3-598-30479-X) / Diazofiche (3-598-30467-6)
- British Biographical Index / Britischer Biographischer Index (BBI I). Ed.: Humanities Reference Unit University of Glasgow / 4 Vols. 1991. Cplt. 2,045 p. / Hard (0-86291-390-X)

British Biographical Archive. Series II / Britisches Biographisches Archiv. Neue Folge (BBA II). Microfiche edition Ed.: Humanities Reference Unit University of Glasgow 1991-1994. 632 fiches. Reader factor 24x / Silverfiche (3-598-33629-2) / Diazofiche (3-598-33628-4)
- British Biographical Index / Britischer Biographischer Index (BBI II). Comp. by David Bank; Theresa McDonald; University of Glasgow / 7 vols. 2nd cumulated and enl. ed. 1998. Cplt. CXCVI, 3179 p. / Hard (3-598-33630-6)

British Catalogue of Music 1957-1985. Ed. by Michael Chapman; Elizabeth Robinson / 10 vols. 1988 / 1989. Per Vol. 420 p. / Hard (3-598-31590-2)

The **British** Library General Catalogue of Printed Books to 1975 (BLC). Ed.: The British Library / 360 vols. 1979-1987. Cplt. 18,0360 p. / Hard (3-598-31000-5)
- Supplement. Ed. by Paul Guthrie / 6 vols. 1987 / 1988. Per vol 560 p. / Hard (3-598-31390-X)
- Bible Sequence / 4 vols. 1981. Per vol 520 p. (BLC vols. 28-31) / Hard (3-598-31028-5)
- England Sequence / 8 vols. 1982. (6 vols. BLC and 2 index vols.) / Hard (3-598-31360-8)
- England. Vol. 1: Titles Index. Vol. 2: Subheading Index. Ed. by Judi Vernau / 2 vols 1984. Cplt. 856 p. / Hard (3-598-31362-4)
- Liturgies Sequence. An international Bibliographical Compilation of Primary and Secondary Liturgical Publications in the British Library. Ed. by Judi Vernau / 2 vols. 1983. Per vol 460 p. (BLC vols. 194, 195) / Hard (3-598-31194-X)
- Russia / 1984. 447 p. / Hard (3-598-31285-7) [BLC Vol. 285, includes Index of Subheadings]

The **British** Library General Catalogue of Printed Books 1976-1982. Ed.: The British Library / 50 vols. 1983. Cplt. 27,850 p. / Hard (3-598-30500-1)

The **British** Library General Catalogue of Printed Books 1982-1985. Ed.: The British Library / 26 vols. 1986. Cplt. 14,610 p. / Hard (3-598-31363-2)

The **British** Library General Catalogue of Printed Books 1986-1987. Ed.: The British Library / 22 vols. 1988. Cplt. 8,500 p. / Hard (3-598-32740-4)

The **British** Library General Catalogue of Printed Books 1988-1989. Ed.: British Library / 28 vols. 1990 / 1991. Cplt. 1,3440 p. / Hard (3-598-33050-2)

The **British** Library General Catalogue of Printed Books 1990-1992. Ed.: The British Library / 27 vols. 1993. Cplt. 11,340 p. / Hard (3-598-33640-3)

The **British** Library General Catalogue of Printed Books 1993-1994. Ed.: The British Library / 27 vols. 1995. Cplt. 12,700 p. / Hard (3-598-33940-2)

The **British** Library General Catalogue of Printed Books 1995-1996. Ed.: The British Library / 27 vols. 1997. Cplt. CLXII, 12,714 p. / Hard (3-598-34370-1)

The **British** Library General Catalogue of Printed Books 1997-1998. Ed.: The British Library / 27 vols. 1999. Cplt. 11,232 p. / Hard (3-598-34660-3)

The **British** Library General Subject Catalogue 1975-1985. Including the systematic Ed.: The British Library / 75 vols. 1986. Cplt. 35,850 p. / Hard (3-598-32191-0)

The **British** Library General Subject Catalogue 1986-1990. Ed.: British Library. Project Consultant: Vernau, Judi / 42 vols. with the systematic 1991 / 1992. Cplt. 20,100 p. / Hard (3-598-31810-3)

British Museum: Catalogue of Western Manuscripts in the Old Royal and King's Collections Ed. by George F. Warner; Julius P. Gilson / 3 vols. and 3 microfiches 1997. Cplt. LXXIV, 1,146 p., 125 illus. / Hard (3-598-11314-5)

Broxis, Peter F.: Organising the Arts / 1978. 132 S. (0-85157-051-8) [Clive Bingley]

Bruce, Anthony: A Bibliography of the British Army, 1660-1914 / 1985. XII, 422 p. / Hard (3-598-10574-6)
- A Bibliography of the British Military History From the Roman Invasion to the Restoration 1660 / 1981. X, 350 p. / Hard (3-598-10359-X)

Bruderer, Herbert E.: Handbuch der maschinellen und maschinenunterstützten Sprachübersetzung. Automatische Übersetzung natürlicher Sprachen und mehrsprachige Terminologiedatenbanken. Hrsg.: Institut f. linguistische Datenverarbeitung / 1978. XVI, 864 S. / Lin (3-7940-7005-4)

Bruderer, Herbert E.: Sprache -Technik - Kybernetik. Aufsätze zur Sprachwissenschaft, maschinellen Sprachverarbeitung, künstlichen Intelligenz und Computerkunst / 1978. 187 S. / Br (3-597-09535-3) [Minerva Publikation]

Brunner, Klaus: Katalog der Ritter-Waldauf-Bibliothek. Eine ehemalige Predigerbibliothek in Hall / Tirol / 1983. XXXIV, 427 S., 9 Taf / Br (3-598-10489-8)

Das **Buch** als Quelle historischer Forschung. Dr. Fritz Juntke anlässlich seines 90. Geburtstages gewidmet. Hrsg. v. Joachim Dietze / 1977. 189 S. / Br (3-7940-7003-8)

Buch und Bibliothek (BuB). Fachzeitschr. d. Vereins d. Bibliothekare an Öffentlichen Bibliotheken. Hrsg. von Klaus Hohlfeld, Ilse Michaelis und Werner Mevissen / Jg. 27/1975 - Jg. 31/1979 (monatlich) (ISSN 0340-0301)

Buch und Bibliothekswissenschaft im Informationszeitalter. Internationale Festschrift für Paul Kaegbein zum 65. Geburtstag. Hrsg. v. Engelbert Plassmann, Wolfgang Schmitz und Peter Vodosek / 1990. 485 S. / Ln (3-598-10915-6)

Buch und Zeitschrift in Geistesgeschichte und Wissenschaft. Hrsg. von Paul Kaegbein
- 1: Krawehl, Otto-Ernst: Die „Jahrbücher für Nationalökonomie und Statistik" unter den Herausgebern Bruno Hildebrand und Johannes Conrad (1863 - 1915) / 1977. 127 S. (3-7940-6450-X)
- 2: Mettler, Dieter: Stefan Georges Publikationspolitik Buchkonzeption und verlegerisches Engagement / 1979. 131 S. (3-598-20352-7)

Buchanan, Brian: Bibliothekarische Klassifikationstheorie. Übers. v. Ute Reimer-Böhner / 1989. 151 S. / Br (3-598-10788-9)

Buchhandel. Internationale Bibliographie der Monographien. Bearb. v. Klaus Gerhard Saur / 1965. 148 S.

Buchwissenschaftliche Beiträge. Hrsg. v. Deutschen Bucharchiv München
- 1: Delp, Ludwig: Die Kulturabgabe / 1950. 203 S. (3-7940-3389-2)
- 2: Delp, Ludwig: Das Deutsche Bucharchiv 1947-67. Geleitw. von Hans Striedl / 1967. 48 S. / Br (Vlg.-Nr. 03147)
- 3: Werner, Andreas: Der Börsenverein des deutschen Buchhandels nach 1945 / 1971. 182 S. (3-7940-3138-5)

Bücher für die Wissenschaft. Bibliotheken zwischen Tradition und Fortschritt. Festschrift für Günter Gattermann zum 65. Geburtstag. Hrsg. v. Gert Kaiser in Verb. mit Heinz Finger und Elisabeth Niggemann / 1994. X, 562 S. / Ln (3-598-11205-X)

Bücher, Menschen und Kulturen. Festschrift für Hans-Peter Geh zum 65. Geburtstag. Hrsg. v. Birgit Schneider, Felix Heinzer und Vera Trost / 1999. XXXI, 432 S. / Gb (3-598-11399-4)

Bulletin of Reprints (ISSN 0303-4550) [Erscheinen eingestellt. Früher im Verlag Gröner, Amsterdam]
- Vol. XI (1974) / 1976. 264 p. / Hard (3-7940-3477-5)
- Vol. XII (1975) / 1977. 217 p. / Hard (3-7940-3478-3)
- Vol. XIII (1976) / 1977. 210 p. / Hard (3-7940-3479-1)
- Vol. XIV (1977) / 1978. 259 p. / Hard (3-7940-3464-3)
- Vol. XV (1978) / 1979. 240 p. / Hard (3-7940-0130-3)
- Vol. XVI (1979) / 1981. 250 p. / Hard (3-598-00095-2)
- Vol. XVII (1980) / 1981. 210 p. / Hard (3-598-00101-0)

Die **Bundesrepublik** Deutschland und Frankreich: Dokumente 1949-1963. Hrsg.: Historische Kommission bei d. Bayerischen Akademie d. Wissenschaften u. d. Institut f. Zeitgeschichte; Horst Möller und Klaus Hildebrand / 3 Bde. u. Registerbd. 1996-99 / Gb (3-598-23680-8)
- Bd. 1. Aussenpolitik und Diplomatie. Bearb. v. Ulrich Lappenküper / 1996. 21, 1002 S. / Gb (3-598-23681-6)
- Bd. 2. Wirtschaft. Bearb. v. Andreas Wilkens / 1997. 6, 1095 S. / Gb (3-598-23682-4)
- Bd. 3. Parteien, Öffentlichkeit, Kultur. Bearb. v. Herbert Elzer / 1997. 6, 1012 S. / Gb (3-598-23683-2)
- *Bd. 4. Register [i.Vb.]*

Burghardt, Anton: Allgemeine Wirtschaftssoziologie. Eine Einführung / 1974. 231 S. / Kst. (UTB 349) (3-7940-2624-1)

Business Archives. Studies on International Practices. Ed.: Committee on Business Archives of the International Council Archives / 1983. 167 p. / Soft (3-598-10516-9)

Busshoff, Heinrich: Kritische Rationalität und Politik. Eine Einführung in die Philosophie des Politischen und die Wissenschaftslehre der Politischen Wissenschaft / 1976. 284 S. / Kst. (UTB 623) (3-7940-2656-X)

Busshoff, Heinrich: Systemtheorie als Theorie der Politik. Eine Studie über politische Theorie als Grundlagendisziplin der politischen Wissenschaft 1975. 284 S. / Kst. (UTB 467) (3-7940-2641-1)

Calvo, Gabriel / Eberhard Sauppe: Diccionario de Biblioteconomía / Wörterbuch des Bibliothekswesens / 1997. XV, 410 S. / Gb (3-598-11269-6)

The **Capitals** of Europe / Les Capitales de l'Europe. A Guide to the Sources for the History of their Architecture and Construction / Guide des sources de l'architecture et de l'urbanisme. Ed.: International Council on Archives. Ed. in Chief: Agnes Ságvári. Ass.: Erzsébet C. Harrach / 1980. 359 p., numerous ills. / Hard (3-598-10093-0)

Capurro, Rafael: Information. Ein Beitrag zur etymologischen und ideengeschichtlichen Begründung des Informationsbegriffs / 1979. 320 S. (3-598-07089-6)

Card Catalog of the Harvard University Music Library to 1985. Microfiche edition with a Guide (printed volume) / 1986. 224 fiches. Reader factor 42x / Silverfiche (3-598-40150-7) / Diazofiche (3-598-40151-5)

Card Catalogs of the Harvard Law School Library 1817-1981. Microfiche edition / 6 sections 1984 / 1985. ca. 4,015,000 cards on 2422 fiches. Reader factor 42x (International Law and Relations 24x) / Diazofiche (3-598-40100-0)
- 1: Author-Title Catalog / ca. 1750,960 cards on 1,085 fiches (3-598-40101-9)
- 2: Anglo-American Subject Catalog / ca. 470,958 cards on 288 fiches (3-598-40105-1)
- 3: Foreign and Comparative Law Subject Catalog / ca. 520,966 cards on 330 fiches (3-598-40106-X)
- 4: Catalog of International Law and Relations / 2 Parts ca. 525,000 cards on 342 fiches (3-598-40107-8)
- Part I to 1963 / ca. 360,963 cards on 231 fiches (3-598-30750-0)
- Part II 1964-1981 / ca. 16,5000 cards on 111 fiches (3-598-40108-6)
- 5: Jurisdictional Shelflist / ca. 550,970 cards on 288 fiches (3-598-40109-4)
- 6: International Shelflist / ca. 209,680 cards on 89 fiches (3-598-40110-8)
- Guide to the card catalogs of the Harvard Law School Library 1817-1981. Prepared by the Harvard Law School Library / 1986. 190 S. (3-598-40111-6)

Card Catalogs of the Harvard University Fine Arts Library 1895-1981. Microfiche edition / 4 sections 1984. 516 fiches. Reader factor 42x / Diazofiche (3-598-40130-2)
- 1: Dictionary Catalog. Titles Catalogued Through February 1981 / 1984. 355 fiches (3-598-40131-0)
- 2: Catalog of Auction Sales Catalogs. Titles Catalogued Through December 1983 / 21 fiches (3-598-40132-9)
- 3: The Catalog of the Rübel Asiatic Research Collection. Titles Catalogued Through December 1983 / 33 fiches (3-598-40133-7)

- 4: Shelflist Catalog. Titles Catalogued Through December 1983 / 107 fiches (3-598-40134-5)
- Guide to the Card Catalogs of the Harvard University. Fine Arts Library 1859-1981. Prepared by the Harvard University Fine Arts Library / 1984. III, 96 p. / Soft (3-598-40140-X)

La **carte** manuscrite et imprimée du XVI au XIX siécle. Journée d'Etude sur 'Histoire du Livre et des Documents Graph., Vallenciennes, 17 novembre 1981. Sous la dir. de Frédéric Barbier / 1983. 132 S. (3-598-10478-2)

Casada, James A.: An Annoted Bibliography of Exploration in Afrika / 1988. 400 p. (3-598-10640-8)

Catalog of the Hebrew Collection of the Harvard College Library. Ed. by Charles Berlin / 11 vols. in 3 parts 1995. Cplt. LXV, 6,369 p. / Hard (3-598-22620-9)

The **Catalog** of the Rübel Asiatic Research Collection. Titles Catalogued Through December 1983 / 7 vols. 1989. Cplt. 3,680 p. / Hard (3-598-40141-8)

Catalogue collectif des ouvrages en langue arabe acquis par les Bibliothèques Françaises 1952-1983 / Union Catalogue of Arab Books in French Libraries 1952-1983. Ed.: Bibliothèque Nationale, Catalogue Collectif des Ouvrages Etrangers (C.C.O.E.). Préparé par Georges Haddad; Mohamed Said avec la participation de l'Institut du monde arabe / 4 vols. 1984. 2.667 S. / Gb (3-598-10510-X)

Catalogue des manuscrits et xylographes orientaux de la Bibliothèque Impériale Publique de St. Petersbourg. - Nachdr. d. Ausg. 1852. - Hrsg. v. Bernhard Dorn u. Reinhold Rost / 1978. XLIV, 719 S., 2 Taf. / Ln (3-598-05015-1) [Lizenzausg. m. Genehm. v. Zentralantiquariat d. DDR, Leipzig]

Catalogue des ouvrages imprimés au XVIe Siècle. Bibliothèque Sainte Geneviève, Paris: Science, Techniques, Médecine. Redigé par Jacqueline Linet et Denise Hillard avec la collaboration de Xavier Lavagne / 1980. XVI, 493 S. / Gb (3-598-10119-8)

Catalogue Général Auteurs, des Livres Imprimés jusqu' à 1900 / General Catalogue of Printed Books up to 1900. Bibliothèque Municipale de Grenoble / Grenoble Public Reference Library / 12 vols. 1980/ 1981. Cplt. 11.600 S. / Gb (3-598-10160-0)

Catalogue général des ouvrages en langue française 1926-1929. Publié sous la direction de Bernard Dermineur / 9 vols. 1987-1989. Cplt. 4.712 S. / Gb (3-598-30990-2)
- Auteurs / 3 vols. 1987. Cplt. 2385 S. (3-598-30991-0)
- Matières / 4 vols. 1989. Cplt. 1495 S. (3-598-30995-3)
- Titres / 2 vols. 1988. Cplt. 832 S. (3-598-30992-9)

Catalogue général des ouvrages en langue française 1930-1933. Ed. par Bernard Dermineur / 15 vols. 1993 / 1995. Cplt. XXXIII, 6080 S. / Gb (3-598-32860-5)
- Auteurs / 7 vols. 1993. Cplt. XXXVIII, 3.022 S. (3-598-32861-3)
- Matières / 5 vols. 1995. Cplt. XXXVIII, 1.872 S. (3-598-32872-9)
- Titres / 3 vols. 1994. Cplt. VII, 1.186 S. (3-598-32868-0)

Catalogue of Cartographic Materials in the British Library 1975-1988. Ed.: The British Library / 3 vols. 1989. Cplt. 1,500 p. / Hard (3-598-32774-9)

Catalogue of Dated and Datable Manuscripts. c. 700-1600 in the Department of Manuscripts. The British Library Comp. by Andrew G. Watson / 2 vols. 1996. Cplt. XII, 627 p. / Hard (3-598-11313-7)

The **Catalogue** of Printed Music in the British Library to 1980 (CPM). Preface by Tim Neighbor. Ed.: The British Library, London; Laureen Baillie / 62 vols. 1980-1987. Cplt. 2,5170 p. / Hard (3-598-31400-0)

Catalogue of Reproductions of Paintings / Catalogue de reproductions de peintures / Catálogo de reproducciones de pinturas / Katalog der Reproduktionen von Gemälden [Koprod. m. d. UNESCO, Paris]
- Catalogue of Reproductions of Paintings Prior to 1860 / Catalogue de reproductions de peintures antérieures à 1860 / Catálogo de reproducciones de pinturas anteeriiores a 1860 / Katalog der Reproduktionen von Gemälden vor 1860 / 9., erw. Aufl. 1972. 501 S., 1391 einfarb. Reprod. / Br (3-7940-5122-X)
- Catalogue of Reproductions of Paintings 1860-1973 / Katalog der Reproduktionen von Gemälden 1860-1973 / 1974. 343 S. / Br (3-7940-5125-4)

Catalogue thématique des sources du grand motet français (1663-1792). Rédigé sous la direction de Jean R. Mongrédien / 1984. 234 S. / Br (3-598-10561-4)

Český biografický archiv a Slovenský biografický archiv / Tschechisches und Slowakisches Biographisches Archiv / Czech and Slovakian Biographical Archive (CSBA). Eine Kumulation von Einträgen aus 214 biographischen Nachschlagewerken vom 9. Jahrhundert bis 1992. Mikrofiche-Edition Bearb. v. Ulrike Kramme und Zelmíra Urra Muena / 1993-1998. 687 Fiches, Lesefaktor 24x / Silberfiche (3-598-33430-3) / Diazofiche (3-598-33431-1)

Chancen und Grenzen moderner Technologie im Museum. Bericht über ein internationales Symposium veranstaltet von den ICOM-Nationalkomitees der Bundesrepublik Deutschland, Österreich und der Schweiz vom 16.-18. Mai 1985 am Bodensee. Hrsg. v. Hermann Auer / 1986. 241 S. / Br (3-598-10631-9)

Charlatanarie der Buchhandlung, welche den Verfall derselben durch Pfuschereyen, Praenumerationes, Auctiones, Nachdrucken, Trödeleyen u.a. m. befördert / 1988. 96 S. / Soft (3-598-07237-6)

Children's Authors and Illustrators. A Guide to Manuscript Collections in United States Research Libraries. Compiled by James H. Fraser / 1980. XI, 119 p. (3-598-40504-9)

Chinese Biographical Archive / Chinesisches Biographisches Archiv (CBA). Microfiche edition. Comp. by Stephan von Minden / 1996-1999. 12 installments on 453 fiches. Reader factor 24x / Silverfiche (3-598-33911-9) / Diazofiche (3-598-33910-0)

Chirgwin, F. John / Phyllis Oldfield: The Library Assistant's Manual / 1978. 118 S. (0-85157-263-4) [Clive Bingley]

Chitnis, A. C.: Scotland's Age of Equilibrium. Society, Politics, Economics, and Ideas 1600 - 1800 / 1990. 250 p. (3-598-10792-7)

Cigánik, Marek: Informationsfonds in Wissenschaft, Technik und Wirtschaft. Übers. aus d. Slowak. / 2., neuverf. Aufl. 1973. 624 S. (3-7940-4185-2)

Clain-Stefanelli, Elvira E.: Numismatic Bibliography / 1985. XXII, 1,848 p. / Hard (3-598-07507-3) [Koprod. mit Battenberg]

Clodius, Heinrich J.: Primae liniae Bibliothecae Lusoriae sive notitia scriptorum de ludis praecipue domesticis ac privatis ordine alphabetico digesta. Bibliographie alter Drucke über Spiele und Sport aller Art. Nachdr. der Ausgabe Leipzig 1761. / 1971. 166 S. (3-7940-5016-9)

The **College** Library. A Collection of Essays. Hrsg. von George Jefferson; G. C. Smith-Burnett / 1978. 208 S. (0-85157-252-9) [Clive Bingley]

Comenius, Johann A.: Informatorium Maternum, Mutter Schul. - Nachdruck d. Orig.-Ausg. 1636 / 1988. XI, 167 S. / Gb (3-598-07252-X) [Lizenzausg. m. Genehm. v. Zentralantiquariat der DDR, Leipzig]

Comics and Visual Culture / La bande dessinée et la culture visuelle / Comics und visuelle Kultur International research studies from 10 different countries / Contributions internationales de recherche scientifique des 10 pays differentes / Internationale Forschungsbeiträge aus 10 verschiedenen Ländern Ed. by Alphons Silbermann; H. D. Dyroff / Internationales Institut für Zeichentrickfilm und Comicsforschung (IZC) / 1986. 264 p. / Hard (3-598-10604-1)

Commercial Radio in Africa. Hrsg. Deutsche Afrika-Gesellschaft / 1970. IX, 307 S. (3-7940-5120-3)

Communication of Scientific Information. Hrsg. von Stacey B. Day / 1975. 240 S., 8 Abb., 10 Tab. / Br (3-7940-5153-X)

Communication Research and Broadcasting. Ed.: Internationales Zentralinstitut für das Jugend- u. Bildungsfernsehen
- 1: School radio in Europe / Schulfunk und Europa. A documentation with contributions given at the European School Radio Conference Munich 1977 / 1979. 198 p. / Soft (3-598-20200-8)
- 2: Effects and Functions of Television: Children and Adolescents. A bibliography of selected research literature 1970-1978. Comp. by Manfred Meyer; Ursula Nissen / 1979. 172 p. / Soft (3-598-20201-6)
- 3: Women, Communication, and Careers. Ed. by Marianne Grewe-Partsch; Gertrud J. Robinson / 1980. 138 p. / Soft (3-598-20202-4)
- 4: Sturm, Hertha / Sabine Jörg: Information Processing by Young Children Piaget's Theory of Intellectual Development Applied to Radio and Television German ed.: Informationsverarbeitung durch Kinder / 1981. 91 p. / Soft (3-598-20204-0)
- 5: Health Education by Television and Radio. Contributions to an International Conference with a Selected Bibliography. Ed. by Manfred Meyer / 1981. 476 p. / Soft (3-598-20203-2)
- 6: Children and Formal Features of Television. Approaches and Findings of Experimental and Formative Research. Ed. by Manfred Meyer / 1983. 333 p. / Soft (3-598-20205-9)
- 7: Children and Families watching Television. A Bibliography of Research on Reception Processes. Rev. by Werner Müller; Manfred Meyer / 1985. 159 p. / Soft (3-598-20206-7)
- 8: Television and Young People. A Bibliography of International Literature 1969-1989. Comp. by Kurt Aimiller; Paul Löhr; Manfred Meyer / 1989. 225 p. / Soft (3-598-20207-5)
- 9: Media Communication in Everyday Life. Interpretative Studies on Children's and Young People's Media Actions. Ed. by Michael Charlton; Ben Bachmair / 1990. 224 p. / Soft (3-598-20208-3)
- 10: Aspects of School Television in Europe. A Documentation. Ed. by Manfred Meyer / 1992. 595 p. / Soft (3-598-20209-1)
- 11: Educational Programmes on Television - Dificiencies, Support, Chances. Contributions to an International Symposium. Ed. by Manfred Meyer. Organized by Int. Zentralinstitut für das Jugend- und Bildungsfernsehen (IZI), München; Bayerische Landeszentrale für neue Medien (BLM) / 1993. 283 p. / Soft (3-598-20210-5)

Communications. The European Journal on Communication / Le Journal Européen de la Communication. Die Deutsche Zeitschrift für Kommunikation / Ed. on behalf of the German Association of Communication Research by Alfons Silbermann; Heinz Neubert; Walter Nutz / Vol. 17-19. 1992-94 (3 issues annually). (ISSN 0341-2059)

Le **Communisme**. Catalogue des Livres et Brochures des XIXe et XXe Siècles Ed.: Institut Français d'Histoire Sociale, Paris. Elaboré, sous la direction de Denise Fauvel-Rouif, par de nombreux rédacteurs / 1989. VIII, 378 p. / Gb (3-598-10800-1)

A **Comprehensive** Handbook of the United Nations. Compiled by Min-Chuan-Ku / 2 vols.
- Vol. 1. 1978. 700 p. / Hard (3-7940-7026-7)
- Vol. 2. 1979. 700 p. / Hard (3-7940-7027-5)

Computergestützte Bibliotheksarbeit. Das Verfahren BASIS im Einsatz bei der Stadtbücherei Bochum. Hrsg. v. Gertrud König, unter Mitarb. v. Günther Nowak; Susanne Rauhut; Jutta Seibert / 1987. 176 S., 74 Abb. / Br (3-598-10696-3)

Computer-based aids to parliamentary work. Datenverarbeitung für die Parlamentsarbeit. Report of the Committee on Science and Technology of the Council of Europe. Ed. by Leitstelle Politische Dokumentation, Berlin / 1976. 85 S. (3-7940-5154-8)

Computers in the Humanities and the Social Sciences Achievements of the 1980s - Prospects for the 1990s. Proceedings of the Cologne Computer Conference 1988: Uses of the Computer in the Humanities and Social Sciences held at the University of Cologne, September 1988. Ed. by Heinrich Best; Ekkehard Mochmann; Manfred Thaller / 1991. XVII, 520 p. / Hard (3-598-11041-3)

Computersimulation in der regionalen und in der Stadtentwicklungsplanung. Anwendung und neuere Ansätze. Bd. 1: Kolloquium. Bd. 2: Anlagen / 2 Bde. 1978. 296 , 108 S. (3-7940-5423-7)

Conceptual Basis of the Classification of Knowledge. Proceedings of the Ottawa Conference on the Conceptual Basis of the Classification of Knowledge. Actes du Colloque d'Ottawa sur les fondements de la Classification des savois. In englisch, französisch und russisch. Ed. by Jercy A. Wojciechowski / 2nd ed. 1978. 503 p. / Hard (3-7940-3649-2)

Conservation in Archives: Proceedings of the International Symposium Ottawa, Canada May 10-12, 1988. Ed.: National Archives of Canada / 1990. IV, 310 p. / Hard (3-598-07564-2)

Continuing Education: Issues and Challenges. Papers from th Conference held at Moraine Valley Community College, Palos Hills, Illinois, U.S.A. August 13-16, 1985. Ed. by Esther E. Horne, School of Library and Information Science, The Catholic University of America, Washington D.C. / 1985. 434 p. / Hard (3-598-10600-9)

Cordell, Helene: Directory of South-East Asian Library Collections in the United Kingdom and Western Europe / 3rd rev. ed. 1988. 240 p. (3-598-10796-X)

Un **Corpus** des Liturgies Chrétiennes sur Micro-fiches / A Corpus of Christian Liturgy on Microfiche. Liturgies Latines, Liturgies Issues de la Reforme, Liturgies Orientales, Subsidia. Publié par le CIPOL-centre international des publications oecuméniques des liturgies / Microfiche edition / 1973-1983. 523 Diazofiches (3-598-30255-X) [Lizenzveröffentlichung seit 1984]

Corpus Librorum Emblematum
• Series B. Secondary Literature on the Emblem (3-598-22280-7)
• No. 1. Daly, Peter M. / Mary V. Silcox: The Modern Critical Reception of the English Emblem / 1991. XVI, 337 p. / Hard (3-598-22281-5)
• No. 2. Daly, Peter M. / Mary V. Silcox: The English Emblem Bibliography of Secondary Literature / 1990. XIX, 179 p. / Hard (3-598-22282-3)

Countryside Planning in Practice: the Scottish Experience Ed. by Paul H. Selman / 1988. XXI, 279 p. / Hard (3-598-10691-2)

Crass, Hanns M.: Bibliotheksbauten des 19. Jahrhunderts in Deutschland Kunsthistorische und architektonische Gesichtspunkte und Materialien. Withan English Summary: Library Buildings in Germany during the 19th Century / 1976. 179 S., 224 Abb. auf Taf / Ln (3-7940-3177-6)

Crime & Justice in American History. Historical Articles on the Origins and Evolution of American Criminal Justice. Ed. and Intro. by Eric H. Monkkonen , University of California, Los Angeles / 11 vols. in 20 parts 1991 / 1992 Cplt. 7745 p. / Hard (3-598-41407-2) [1991 von Meckler Publishing, Westport, CT übernommen]
• Vol. 1. The Colonies and Early Republic / 2 vols. 1991. Cplt. XXIV, 911 p. / Hard (3-598-41408-0)

• Vol. 2. Courts and Criminal Procedure / 1991. X, 558 p. / Hard (3-598-41410-2)
• Vol. 3. Delinquency and Disorderly Behavior / 1991. X, 398 p. / Hard (3-598-41411-0)
• Vol. 4. The Frontier / 1991. X, 574 p. / Hard (3-598-41412-9)
• Vol. 5. Policing and Crime Control / 3 vols. 1992. Cplt. XXXIV, 1113 p. / Hard (3-598-41413-7)
• Vol. 6. Prisons and Jails / 2 vols. 1992. Cplt. XXIV, 774 p. / Hard (3-598-41415-3)
• Vol. 7. The South / 2 vols. 1992. Cplt. XXIII, 505 p. / Hard (3-598-41417-X)
• Vol. 8. Prostitution, Drugs, Gambling and Organized / 2 vols. 1992. Cplt. XXV, 794 p. / Hard (3-598-41418-8)
• Vol. 9. Violence and Theft / 3 vols. 1992. Cplt. XXXVIII, 1,019 p. / Hard (3-598-41420-X)
• Vol. 10. Reform / 1992. X, 445 p. / Hard (3-598-41422-6)
• Vol. 11. Theory and Methods in Criminal Justice History / 2 vols. 1992. Cplt. XXV, 657 p. / Hard (3-598-41423-4)

Cumulative Microform Reviews, 1972-76 / 1977. 619 p. / Hard (3-598-10937-7) [1990 von Meckler Publishing, Westport, CT übernommen]

Current contents Africa. Ed. by Irmtraud D. Wolcke-Renk / Vol. 6-18. / 1981-93 (4 issues annually) (ISSN 0340-7632)

Czayka, Lothar: Grundzüge der Aussagenlogik. Mit Anwendungsbeispielen aus den Wirtschaftswissen-schaften / 2., wesentl erw. Aufl. 1972. 110 S., zahlr. Abb. / Kst. (UTB 124) (3-7940-2608-X)
1. Aufl. → *Studiengruppe für Systemforschung Nr. 98*

Czayka, Lothar: Systemwissenschaft. Eine kritische Darstellung mit Illustrationsbeispielen aus den Wirtschaftswissenschaften / 1974. 110 S. / Kst. (UTB 185) (3-7940-2610-1)

Data documentation. Some Principles and applica-tions in science and industry. Proceedings of the workshop on Data Documentation organized by the School for Medical Documentation of the Univ. of Ulm , Reisensburg Castle, July 7-11, 1975. Commission of the European Communities. Ed. by Wilhelm Gaus and Rolf Henzler / 1977. 189 S. (3-7940-3022-2)

Datenbasen - Datenbanken - Netzwerke. Praxis des Information Retrieval. Hrsg. v. Rainer Kuhlen / 3 Bde. 1979 / Lin (3-598-10032-9)
• Bd. 1. Datenbasen - Datenbanken - Netzwerke Aufbau von Datenbasen / 1979. 376 S. / Lin (3-598-10033-7)
• Bd. 2. Konzepte von Datenbanken / 1979. 353 S. / Lin (3-598-10034-5)
• Bd. 3. Erfahrungen mit Retrievalsystemen / 1980. 380 S. / Lin (3-598-10035-3)

Dauenhauer, Erich: Curriculumforschung. Eine Einführung mit Praxisbeispielen aus der Berufs-pädagogik / 1976. 168 S., 56 Tab. / Kst. (UTB 566) (3-7940-2647-0)

Dauenhauer, Erich: Einführung in die Arbeitslehre / 1975. 168 S. / Kst. (UTB 471) (3-7940-2639-X)

Davies, Helen: Libraries in West Africa. A Bibliog-raphy / 1982. XIX, 170 p. / Soft (3-598-10440-5) [Hans Zell Publishers]

Day, Alan Edwin: Archaeologie. A Reference Handbook. / 1978. 319 S. (0-85157-242-1) [Clive Bingley]

Day, Alan Edwin: Discovery and Exploration. A Reference Handbook. The Old World. / 1980. 295 S.

Debes, Dietmar: Das Figurenalphabet / 1968. 176 S., 32. S., Abb. (Vlg. -Nr. 04158)

DeCoster, Jean de: Dictionary for automotive en-gineering / Dictionaire du génie automobile / Wör-terbuch für Kraftfahrzeugtechnik English-French-German / anglais-français-allemand avec definitions des termes français / Englisch-Französisch-Deutsch. Mit Definitionen der deutschen Begriffe / 1982. 290 p. (3-598-10430-8)
• 2nd enl. ed. 1986. 620 p. / Hard (3-598-10591-6)
• 3rd rev. and enl. ed. 1990. 620 p. / Hard (3-598-10881-8)
• 4th rev. and enl. ed. DeCoster, Jean de / Otto Vollnhals / 1998. 691 p. / Hard (3-598-11370-6)

Della Santa, Leopoldo: Della Costruzione e del Regolamento di una Pubblica Universale Biblioteca: con la Pianta dimonstrativa / Über den Bau und die Verwaltung einer öffentlichen Universalbibliothek: mit einem veranschaulichenden Plan. Traktat / Trattato. Band 1: Italienisches Originalwerk, Band 2: deutsche Übersetzung / 2 Bde. 1984. Je 76 S. / Br i. Sch. (3-598-07225-2)

Demoskopie und Aufklärung. Ein Symposium. Norman Bradburn - Renate Köcher - Helmut Kohl - Hermann Lübbe - Heinz Maier-Leibnitz - Hubert Markl - Elisabeth Noelle-Neumann - Helmut Sihler. Vorw. v. Elisabeth Noelle-Neumann. Hrsg.: Institut f. Demoskopie, Allensbach / 1987. 76 S. / Br (3-598-10752-8)

Der **Deutsch**-israelische Dialog. Dokumentation eines erregenden Kapitels deutscher Aussenpolitik. Teil I: Politik (3 Bände), Teil II: Wirtschaft / Land-wirtschaft (2 Bände), Teil III: Kultur (3 Bände). Vorw. v. Jitzhak Ben-Ari. Geleitw. v. Helmut Kohl. Eingel. u. Hrsg. v. Rolf Vogel / 1987/1990. Zus. CLXXXVIII, 3966 S. / Ln (3-598-21940-7)
• Teil I: Politik / 3 Bde. 1987-1988
  - Bd. 1. 1987. XL, 523 S., 8 Abb.-S. (3-598-21941-5)
  - Bd. 2. 1988. S. XIII, 525-1058 (3-598-21942-3)
  - Bd. 3. 1988. S. XIII, 1059-1699 (3-598-21943-1)
• Teil II: Wirtschaft / Landwirtschaft / 2 Bde. 1989
  - Bd. 4. 1989. XXXVII, 451 S. (3-598-21944-X)
  - Bd. 5. 1989. S. XIX, 453-1010 (3-598-21945-8)
• Teil III: Kultur / 3 Bde. 1989-1990
  - Bd. 6. 1989. XIII, 433 S. (3-598-21946-6)
  - Bd. 7. 1990. IX, 432 S. (3-598-21947-4)
  - Bd. 8. 1990. X, 392 S. (3-598-21948-2)

Deutsche Biographische Enzyklopädie (DBE). Hrsg. v. Walther Killy und Rudolf Vierhaus unter Mitarbeit von Dietrich von Engelhardt, Wolfram Fischer, Franz Georg Kaltwasser u. Bernd Moeller / 1995 ff. / Hld (3-598-23160-1)
• Bd. 1 / 1995. XXV, 642 S. / Hld (3-598-23161-X)
• Bd. 2 / 1995. XXI, 680 S. / Hld (3-598-23162-8)
• Bd. 3 / 1996. XXI, 680 S. / Hld (3-598-23163-6)
• Bd. 4 / 1996. XXIII, 679 S. / Hld (3-598-23164-4)
• Bd. 5 / 1996. XXIV, 680 S. / Hld (3-598-23165-2)
• Bd. 6 / 1997. XXIII, 679 S. / Hld (3-598-23166-0)
• Bd. 7 / 1997. XXIII, 695 S. / Hld (3-598-23167-9)
• Bd. 8 / 1998. XXIV, 712 S. / Hld (3-598-23168-7)
• Bd. 9 / 1998. XXIII, 695 S. / Hld (3-598-23169-5)
• Bd. 10 / 1999. XXIV, 680 S. / Hld (3-598-23170-9)

Deutsche Drucke des Barock 1600-1720. Katalog der Herzog August Bibliothek Wolfenbüttel. Begr. v. Martin Bircher,. Bearb. v. Thomas Bürger / 4 Abt. 46 Bde. Haupttl u. Reg. 1977-1996 / Ln (3-598-32099-X) [1985 übernommen. Vorher im Vlg. Kraus International Publications, München]
- Abt. A: Bibliotheca Augusta 1977-1996 (3-598-32100-7)
  - Bd. 1. Ethica, Grammatica, Poetica, Rhetorica / 1977. XV, 346 S. (3-598-32101-5)
  - Bd. 2. Arithmetica, Astronomica, Bellica, Geographica, Geometrica, Medica, Oeconomica, Physica / 1979. VIII, 408 S. (3-598-32102-3)
  - Bd. 3. Politica 1: 1600-1628 / 1980. VIII, 311 S. (3-598-32103-1)
  - Bd. 4. Politica 2: 1629-1697 / 1980. V, 348 S. (3-598-32104-X)
  - Bd. 5. Quodlibetica 1 / 1987. VIII, 271 S. (3-598-32105-8)
  - Bd. 6. Quodlibetica 2 / 1987. V, 290 S. (3-598-32106-6)
  - Bd. 7. Quodlibetica 3 / 1987. V, 326 S. (3-598-32107-4)
  - Bd. 8. Historica 1: A-S / 1989. VIII, 293 S. (3-598-32108-2)
  - Bd. 9. Historica 2: T-Z. Juridica / 1990. V, 295 S. (3-598-32109-0)
  - Bd. 10. Theologica A-C / 1994. VIII, 261 S. (3-598-32110-4)
  - Bd. 11. Theologica D-G / 1995. V, 244 S. (3-598-32111-2)
  - Bd. 12. Theologica H-L / 1995. VIII, 283 S. (3-598-32112-0)
  - Bd. 13. Theologica / 1995. VIII, 291 S. (3-598-32113-9)
  - Bd. 14. Theologica / 1995. VIII, 259 S. (3-598-32114-7)
  - Bd. 15. Theologica / 1996. VIII, 341 S. (3-598-32115-5)
- Abt. B: Mittlere Aufstellung 1982-1992 (3-598-32130-9)
  - Bd. 1. Literatur 1: A-N / 1982. X, 322 S. (3-598-32131-7)
  - Bd. 2. Literatur 2: O-Z / 1982. V, 339 S. (3-598-32132-5)
  - Bd. 3. Theologie: 1 A-E / 1986. IX, 445 S. (3-598-32133-3)
  - Bd. 4. Theologie: 2 F-L / 1986. IX, 572 S. (3-598-32134-1)
  - Bd. 5. Theologie: 3 M-S / 1986. IX, 491 S. (3-598-32135-X)
  - Bd. 6. Theologie: 4 T-Z, Register / 1986. IX, 303 S. (3-598-32136-8)
  - Bd. 7. Allgemeines, Buchwesen, Geographie, Wissenschaftskunde, Historische Hilfswissenschaften / 1989. IX, 272 S. (3-598-32137-6)
  - Bd. 8. Biographische Schriften / 1989. VIII, 321 S. (3-598-32138-4)
  - Bd. 9. Historische Hilfswissenschaften. Neuere Geschichte (A-N) / 1990. VIII, 265 S. (3-598-32139-2)
  - Bd. 10. Neuere Geschichte O-Z. Geschichte einzelner Länder / 1990. V, 279 S. (3-598-32140-6)
  - Bd. 11. Deutsche Geschichte / Allgemeines / 1990. VIII, 272 S. (3-598-32141-4)
  - Bd. 12. Deutsche Geschichte / Territorien und Städte A-M / 1990. VII, 271 S. (3-598-32142-2)
  - Bd. 13. Deutsche Geschichte / Territorien und Städte N-Z / 1990. VIII, 284 S. (3-598-32143-0)
  - Bd. 14. Deutsche Geschichte (Braunschweig-Lüneburg) / 1990. VIII, 297 S. (3-598-32144-9)
  - Bd. 15. Geschichte einzelner Länder / 1991. VIII, 319 S. (3-598-32145-7)
  - Bd. 16. Kulturgeschichte, Militaria, Medizin, Naturwissenschaften / 1991. VIII, 294 S. (3-598-32146-5)
  - Bd. 17. Ökonomie, Naturwissenschaften / 1991. VIII, 379 S. (3-598-32147-3)
  - Bd. 18. Recht / 1992. VIII, 334 S. (3-598-32148-1)
  - Bd. 19. Quodlibetica Nova 1: A-L / 1992. IX, 285 S. (3-598-32149-X)
  - Bd. 20. Quodlibetica Nova 2: M-Z / 1992. V, 299 S. (3-598-32150-3)
- Abt. C: Helmstedter Bestände 1983-1989 (3-598-32160-0)
  - Bd. 1. M (Medica), N (Mathematica), O (Philosophica), P (Philologica), Q (Historia litteraria), S (Historia ecclesiastica) / 1983. IX, 440 S. (3-598-32161-9)
  - Bd. 2. C 1183-C 2242 A-K Helmstedter (Theologica 1: A-Q) / 1988. VIII, 369 S. (3-598-32162-7)
  - Bd. 3. C 2243-C 3116 A-K Helmstedter (Theologica 2: R-Z) L Helmstedter (Juridica) / 1988. VI, 365 S. (3-598-32163-5)
  - Bd. 4. T Helmstedter (Geschichte) / 1988. IX, 320 S. (3-598-32164-3)
  - Bd. 5. Miscellanea 1: A-L / 1989. IX, 262 S. (3-598-32165-1)
  - Bd. 6. Miscellanea 2: M-Z Einblattdrucke / 1989. VI, 266 S. (3-598-32166-X)
- Abt. D: Sonderbestände 1993/94 (3-598-32180-5)
  - Bd. 1. Bibelsammlung. Bibliothek Alvensleben / 1993. VII, 351 S. (3-598-32181-3)
  - Bd. 2. Musiksammlung / 1994. IX, 310 S. (3-598-32182-1)
- Gesamtregister
  - Namenregister / 1996. XI, 425 S. (3-598-32184-8)
  - Titelregister / 1996. XI, 360 S. (3-598-32185-6)
  - Register der Drucker, Verleger und Orte / 1996. XI, 333 S. (3-598-32186-4)
- Register zu den Bänden A1-A7, B1-B6, C1-C3 [Interimsregister] / 1988. IX, 306 S. (3-598-32800-1)

Deutsche Drucke des Barock 1600-1720. Mit den Ergänzungen Polnische Drucke und Polonica 1501-1700 sowie Ungarische Drucke und Hungarica 1480-1720. Kataloge der Herzog August Bibliothek Wolfenbüttel. Mikrofiche-Edition. Begr. v. Martin Bircher. Bearb. v. Thomas Bürger, Malgorzata Goluszka, Marian Malicki, S. Katalin Németh / 1996. 157 Fiches, Lesefaktor 24x u. 4 Indexbde. / Silberfiche (3-598-32187-2)
- Index: Polonica und Hungarica / 1997. IX, 262 S. / Ln (3-598-32188-0)

Deutsche Kunstbibliotheken / German Art Libraries. Berlin, Florenz, Köln, München, Nürnberg, Rom. Hrsg.: Arbeitsgemeinschaft der Kunstbibliotheken / 1975. 101 S. / Lin (3-7940-3424-4)

Deutsche Literatur in Titelblättern. Reproduktionen aus Büchern d. Stadt- und Universitätsbibliothek Frankfurt a. M. Hrsg. v. Alfred Estermann u. Hans-Albrecht Koch / 1978. XVI, 212 S. / Ln (3-7940-7058-5)

Deutsche literarische Zeitschriften 1880-1945. Ein Repertorium. Hrsg.: Deutsches Literaturarchiv, Marbach. Bearb. v. Thomas Dietzel; Hans O. Hügel / 4 Bde. u. 1 Reg.-Bd. 1988. Zus. 1.623 S. / Ln (3-598-10645-9)

Deutsche literarische Zeitschriften 1945-1970. Ein Repertorium. Hrsg.: Deutsches Literaturarchiv Marbach. Bearb. v. Fischer, Bernhard; Dietzel, Thomas / 3 Bde. u. 1 Reg.-Bd. 1992. Zus. 1304 S. / Ln (3-598-22000-6)

Deutsche Nationalbibliographie
- Ergänzung 1. Verzeichnis der Schriften, die 1933-1945 nicht angezeigt werden durften. Bearb. u. hrsg. v. d. Deutsche Bücherei, Leipzig. - Nachdr. d. Ausg. Leipzig 1949 / 1974. 433 S. / Ln (3-7940-5043-6) [Lizenzausg. m. Genehm. v. Zentralantiquariat der DDR, Leipzig]
- Ergänzung 2. Verzeichnis der Schriften, die infolge von Kriegseinwirkungen vor dem 8. Mai 1945 nicht angezeigt werden konnten. Bearb. u. hrsg. v. d. Deutsche Bücherei, Leipzig. - Nachdr. d. Ausg. Leipzig 1949 / 1974. 664 S. / Ln (3-7940-5044-4) [Lizenzausg. m. Genehm. v. Zentralantiquariat der DDR, Leipzig]

Deutsche Presseforschung. Hrsg. v. Gebhardt Hartwig [Vorher im Schünemann Universitätsvlg., Bremen]
- 25: Henkel, Martin / Rolf Taubert: Die deutsche Presse 1848-1850. Eine Bibliographie / 1986. 634 S., 37 Abb. / Lin (3-598-21626-2)
- 26: Presse und Geschichte II. Neue Beiträge zur historischen Kommunikationsforschung. Von Hartwig Gebhardt; Hans J. Köhler; Helmut W. Lang. Hrsg.: Deutsche Presseforschung / 1987. 426 S. / Br (3-598-21627-0)
- 27: Petrat, Gerhardt: Einem besseren Dasein zu Diensten. Die Spur der Aufklärung im Medium Kalender zwischen 1700 und 1919 / 1991. XIII, 242 S., 24 Abb. / Br (3-598-21628-9)
- 28: Französische Revolution und deutsche Öffentlichkeit. Wandlungen in Presse und Alltagskultur am Ende des achtzehnten Jahrhunderts. Hrsg. v. Holger Böning / 1992. XII, 549 S., 117 Abb. / Br (3-598-21629-7)

Deutsche und Polen zwischen den Kriegen. Minderheitenstatus und „Volkstumskampf" im Grenzgebiet 1920-1939 / Polacy i Niewcy między wojnami. Status mniejszos'ci i walka graniczna. Amtliche Berichterstattung aus beiden Ländern / Reporty wadz polskich i niemieckich z lat 1920-1939. Hrsg. im Auftr. d. Instituts f. Zeitgeschichte, München u. d. Generaldirektion d. polnischen Staatsarchive, Warschau v. Rudolf Jaworski; Marian Wojciechowski. Bearb. v. Przemyslaw Hauser; Mathias Niendorf. (Texte und Materialien zur Zeitgeschichte Bd. 9) / 2 Bde. 1996. Zus. XXXI, 1.156 S. / Ln (3-598-22810-4)

Deutscher Dokumentartag. Hrsg.: Deutsche Gesellschaft für Dokumentation e.V. (DGD)
- 1973. Frankfurt am Main vom 22. bis 26. 10. 1973. Informationswissenschaften und Informatik. Der Mikrofilm in der Dokumentation / Datenverarbeitung für die Informationspraxis. / Bearb. von Peter Port; Mathilde Schelzinger / 1974. 256 S. (3-7940-3641-7)
- 1974. Bonn-Bad Godesberg vom 7.-11. 10. 1974. Bd. 1: Staatliche Informationssysteme, Informationspolitik und Informationsrecht, öffentliche Komiteesitzungen. (3-7940-3642-5). Bd. 2: Probleme der Terminologiearbeit in Information und Dokumentation. (3-7940-3643-3) / 2 Bde. / 1975. VIII, 472 S.; VI, 242 S.
- 1975. Bad Kreuznach vom 29.9. bis 02. 10. 1975. Information und Dokumentation in der Wirtschaft. Bearb. von Mathilde von der Laake; Peter Port / 1976. 375 S. (3-7940-3647-6)
- 1976. Münster vom 4. 10. bis 7. 10- 1976. Information und Dokumentation zum Umweltschutz / Das IuD-Programm der Bundesregierung. Bearb. von Mathilde von der Laake; Peter Port / 1977. VIII, 416 S. (3-7940-3655-7)

- 1977. Datendokumentation, Benutzerforschung, Mehrspachigkeit, Tariffragen, Indexierung. Bearb. von Mathilde von der Laake; Peter Port / 1978. IV, 486 S. (3-7940-3656-5)
- 1978. Frankfurt am Main vom 02.10. bis 07. 10. 1978.; 30 Jahre DGD 1948 - 1978.; IuD und Verlagswesen; Finanzierung von IuD-Dienstleistungen über Preise?; Ausbildung und Tarifprobleme;. Forschungsprogramm für die Informationswissenschaft / 1979. 294 S. (3-598-03657-4)
- 1979. Willingen, Hoch-Sauerland, vom 01.10. bis 07.10.1979. Das IuD-Programm heute; Online-Benutzergruppe; Bibliometrie; Scientometrie; Terminologiearbeit; Datenschutz; Tariffragen, Berufsbilder; Informationsmarkt; Gesprächskreise. Bearb. von Mathilde von der Laake; Hilde Strohl-Goebel / 1980. 594 S. (3-598-20259-8)
- 1980. Berlin, den 29-9. bis 3. 10. 1980. IuD und Normung. Neue Kommunikationstechnologie / Berufspolitik / Datenbank / Betriebsstatistik / Infometrie / Parlamentsdokumentation / Arbeitskreis Senioren Gesprächskreise / 1981. 438 S. (3-598-20260-1)
- 1981. Kleincomputer in Information und Dokumentation. Bearb. von Hilde Strohl-Goebel / 1982. 735 S. (3-598-20261-X)
- 1982. Fachinformation im Zeitalter der Informationsindustrie Lübeck - Travemünde vom 27. 9. bis 30. 9. 1982 Bearb. v. Hilde Strohl-Goebel / 1982/1983. 477 S. / Lin (3-598-20262-8)
- 1983. Fachinformation und Bildschirmtext Göttingen, vom 3. bis 7. 10. 1983 Bearb. v. Hilde Strohl-Goebel / 1983/1984. 511 S. / Lin (3-598-20263-6)
- 1984. Fachinformation: Methodik, Management, Markt Neue Entwicklungen, Berufe, Produkte Bearb. v. Hilde Strohl-Goebel / 1985/1986. 479 S. / Ln (3-598-20265-2)
- 1985. Perspektiven der Fachinformation, Programme - Praxis - Prognosen Darmstadt, vom 09. bis 12. 10. 1984. Hrsg. von d. Dt. Gesell. für Dokumentation e.V. (DGD), Bearb. von Hilde Strohl-Goebel 1985. 559 S. (3-598-20264-4)

Deutsches Bibliotheksadressbuch. Verz. von Bibliotheken i. d. Bundesrepublik Deutschland einschl. Berlin (West). Hrsg. von d. Dt. Bibliothekskonferenz
- Deutsches Bibliotheksadressbuch 1. Ausgabe / 1974. XIX, 603 S. (3-7940-2239-4)
- 2. Ausgabe / 1976. XI, 498 S. (3-7940-224084)

Deutsches Biographisches Archiv / German Biographical Archive (DBA I). Eine Kumulation aus 265 der wichtigsten biographischen Nachschlagewerke für den deutschen Bereich bis zum Ausgang des 19. Jahrhunderts. Mikrofiche-Edition. Hrsg. v. Bernhard Fabian. Bearb. unter Leitung v. Willi Gorzny / 1982-1985. 623.000 S. auf 1.447 Fiches. Lesefaktor 24x / Silberfiche (3-598-30421-8) / Diazofiche (3-598-30410-2)
- Deutscher Biographischer Index / German Biographical Index (DBI I). Ein Register zum Deutschen Biographischen Archiv. Hrsg. v. Willi Gorzny. Bearb. v. Uta Koch, Hans A. Koch und Angelika Koller / 4 Bde. 1986. Zus. XXVI, 2.298 S. / Ln (3-598-30432-3)

Deutsches Biographisches Archiv, Neue Folge bis zur Mitte des 20. Jahrhunderts / German Biographical Archive, A Sequel to the midtwentieth Century (DBA II). Die Neue Folge des Deutschen Biographischen Archivs kumuliert 301 biographische Lexika erschienen bis zur Mitte des 20. Jahrhunderts mit insgesamt ca. 493.000 Personen.

Mikrofiche-Edition. Bearb. v. Willi Gorzny / 1989-1993. 1.457 Fiches, Lesefaktor 24x / Silberfiche (3-598-32820-6) / Diazofiche (3-598-32834-6)
- Deutscher Biographischer Index / German Biographical Index (DBI II). Der Index kumuliert in einem Alphabet die biographischen Kurzinformationen zu den ca. 4801249 Artikeln im DBA I und II Bearb. v. Willi Gorzny / 8 Bde. 2. kumulierte u. erw. Ausg. 1997. Zus. CXCII, 4.017 S. / Ln (3-598-32848-6)
- Deutscher Biographischer Index. CD-ROM Ausgabe / German Biographical Index. CD-ROM-Edition / 1998. (3-598-40284-8)

Deutsches Biographisches Jahrbuch. Register zu Band 1 bis 5, 10 und 11 Bearb. v. Heinrich Ihme / 1986. 99 S. / Ln (3-598-10656-4)

Deutsches Literatur-Lexikon. Biographisch-bibliographisches Handbuch. Begr. v. Wilhelm Kosch / 3. völlig neubearb. Aufl. (3-907820-00-2) [1991 übernommen. Bis Bd. 12 im Francke Vlg., Bern]
- Bd. 1. Aal-Bremeneck. Hrsg. v. Bruno Berger u. Heinz Rupp / 1968. XII S., 1.024 Spalten / Ln (3-907820-01-0)
- Bd. 2. Bremer-Davidis. Hrsg. v. Bruno Berger u. Heinz Rupp / 1969. VIII S., 1.024 Spalten / Ln (3-907820-02-9)
- Bd. 3. Davidis-Eichendorff. Fortgeführt v. Bruno Berger. Hrsg.: Heinz Rupp (Ältere Abt.) / 1971. VIII S., 1.048 Spalten / Ln (3-907820-03-7)
- Bd. 4. Eichenhorst-Filchner. Hrsg. v. Heinz Rupp (Ältere Abt.) u. Hildegard Emmel (Neuere Abt.) / 1972. X S., 1.024 Spalten / Ln (3-907820-04-5)
- Bd. 5. Filek-Fux. Hrsg. v. Heinz Rupp (Ältere Abt.) u. Hildegard Emmel (Neuere Abt.) / 1978. X S., 926 Spalten (3-907820-05-3)
- Bd. 6. Gaa-Gysin. Hrsg. v. Heinz Rupp (Ältere Abt.) u. Carl Ludwig Lang (Neuere Abt.). Red.: Carl Ludwig Lang / 1978. XIII S., 1.092 Spalten / Ln (3-907820-06-1)
- Bd. 7. Haab-Hogrebe. Hrsg. v. Heinz Rupp (Ältere Abt.) u. Carl Ludwig Lang (Neuere Abt.). Red.: Carl Ludwig Lang / 1979. XII S., 1.440 Spalten / Ln (3-907820-07-X)
- Bd. 8. Hohberg-Kober. Hrsg. v. Heinz Rupp (Mittelalter) u. Carl Ludwig Lang (Neuzeit). Red.: Carl Ludwig Lang / 1981. XII S., 1.443 Spalten / Ln (3-907820-08-8)
- Bd. 9. Kober-Lucidarius. Hrsg. v. Heinz Rupp (Mittelalter) u. Carl Ludwig Lang (Neuzeit). Red.: Carl Ludwig Lang / 1984. XIV S., 1.716 Spalten / Ln (3-907820-09-6)
- Bd. 10. Lucius-Myss. Hrsg. v. Heinz Rupp (Mittelalter) u. Carl Ludwig Lang (Neuzeit). Red.: Carl Ludwig Lang / 1986. XIV S., 1.714 Spalten / Ln (3-907820-10-X)
- Bd. 11. Naaff-Pixner. Hrsg. v. Heinz Rupp (Mittelalter) u. Carl Ludwig Lang (Neuzeit). Red.: Carl Ludwig Lang / 1988. XIV S., 1348 Spalten / Ln (3-907820-11-8)
- Bd. 12. Plachetka-Rilke. Hrsg. v. Heinz Rupp (Mittelalter) u. Carl Ludwig Lang (Neuzeit). Red.: Carl Ludwig Lang / 1990. XIV S., 1.290 Spalten / Ln (3-907820-12-6)
- Bd. 13. Rill-Salzmann. Hrsg. v. Heinz Rupp (Mittelalter) u. Carl Ludwig Lang (Neuzeit). Red.: Carl Ludwig Lang / 1991. XIV S., 768 Spalten / Ln (3-907820-13-4)
- Bd. 14. Salzmesser-Schilling. Hrsg. v. Heinz Rupp (Mittelalter) u. Carl Ludwig Lang (Neuzeit). Red.: Carl Ludwig Lang / 1992. XIV S., 663 Spalten / Ln (3-907820-14-2)
- Bd. 15. Schilling-Schnydrig. Hrsg. v. Heinz Rupp

(Mittelalter) u. Carl Ludwig Lang (Neuzeit). Red.: Carl Ludwig Lang / 1993. XIV S., 708 Spalten / Ln (3-907820-15-0)
- Bd. 16. Schobel-Schwaiger. Hrsg. v. Hubert Herkommer (Mittelalter) u. Carl Ludwig Lang (Neuzeit). Red.: Carl Ludwig Lang / 1995. XIV S., 744 Spalten / Ln (3-907820-18-5)
- Bd. 17. Schwalb-Siewert. Hrsg. v. Hubert Herkommer (Mittelalter) u. Carl Ludwig Lang (Neuzeit). Red.: Carl Ludwig Lang / 1997. XIV S., 338 S. / Ln (3-907820-20-7)
- Bd. 18. Siff-Speorri. Hrsg. v. Hubert Herkommer (Mittelalter) u. Carl Ludwig Lang (Neuzeit). Red.: Carl Ludwig Lang / 1998. XIV S., 616 Spalten / Ln (3-907820-23-1)
- Ergänzungsband I: A-Bernfeld. Hrsg. v. Heinz Rupp (Mittelalter) u. Carl Ludwig Lang (Neuzeit). Red.: Carl Ludwig Lang / 1994. XIV S., 768 Spalten / Ln (3-907820-16-9)
- Ergänzungsband II: Bernfeld-Christen. Hrsg. v. Heinz Rupp (Mittelalter) u. Carl Ludwig Lang (Neuzeit). Red.: Carl Ludwig Lang / 1994. XIV S., 606 Spalten / Ln (3-907820-17-7)
- Ergänzungsband III: Christener-Fowelin. Hrsg. v. Hubert Herkommer (Mittelalter) u. Carl Ludwig Lang (Neuzeit). Red.: Carl Ludwig Lang / 1997. XVI S., 756 Spalten / Ln (3-907820-19-3)
- Ergänzungsband IV: Fraenkel-Hermann. Hrsg. v. Hubert Herkommer (Mittelalter) u. Carl Ludwig Lang (Neuzeit). Red.: Carl Ludwig Lang / 1997. XVI S., 644 Spalten / Ln (3-907820-21-5)
- Ergänzungsband V: Hermann-Lyser. Hrsg. v. Hubert Herkommer (Mittelalter) u. Carl Ludwig Lang (Neuzeit). Red.: Carl Ludwig Lang / 1998. XIV S., 770 Spalten / Ln (3-907820-22-3)
- Ergänzungsband VI: Maag-Ryslavy. Hrsg. v. Hubert Herkommer (Mittelalter) u. Carl Ludwig Lang (Neuzeit). Red.: Carl Ludwig Lang / 1998. XIV S., 636 Spalten / Ln (3-907820-76-2)

Deutsches Museum - Bildarchiv. Hrsg.: Deutsches Museum, München / Mikrofiche-Edition 1987-1996. Bildtl: 370 Photo-Ordner u. 110 Ordner mit Rückvergrösserungen v. techn. Zeichnungen auf 480 Silberfiches. Lesefaktor 24x u. Register (3-598-30403-X)
- Register zur Mikrofiche-Edition. Bearb. v. Wilhelm Füßl u. Christiane Hennet / 1997. 248 S. / Br (3-598-30405-6)

Deutsches Theater. Hrsg. v. Ludwig Tieck - Nachdr. d. Ausg. Berlin, 1817 / 2 Bde. 1980 Bd. 1: XXXII, 407 S., Bd. 2: XXII, 344 S. / Ln (3-598-07234-1) [Lizenzausg. m. Genehm. v. Zentralantiquariat d. DDR, Leipzig]

Deutsches Theater-Lexikon. Biographisch-bibliographisches Handbuch. Begr. v. Wilhelm Kosch. Fortgef. v. Hanspeter Bennwitz, ab Lfg. 22 v. Ingrid Bigler-Marschall (3-907820-26-6) [1991 übernommen. Bis Bd. II im Francke Vlg., Bern]
- Bd. I: A-Hurka / 1953. VIII, 864 S. / Ln (3-907820-27-4)
- Bd. II: Hurka-Pallenberg / 1960. VI, 864 S. / Ln (3-907820-28-2)
- Bd. III: Pallenberg-Singer / 1992. 479 S. / Ln (3-907820-29-0)
- Bd. IV: Singer-Tzschoppe / 1998. VIII, 460 S. / Ln (3-907820-30-4)

Deutsches Universitäts-Handbuch. Bearb. von Karl-Otto Saur / 2 Bde. 1967. 1360 S.
- 2. Ausg. Deutsches Universitäts-Handbuch (BRD und DDR). Bd. 1: Aachen - Hannover; Bd. 2:

Heidelberg - Würzburg; Bd. 3: Personen- und Fachregister / 3 Bde. 1969/70. LVI, 2331 S. (3-7940-2309-9)
weitere Aufl → **Handbuch** *der Universitäten*

Deutschland im ersten Nachkriegsjahr. Berichte von Mitgliedern des Internationalen Sozialistischen Kampfbundes (ISK) aus dem besetzten Deutschland 1945/46. Hrsg. u. bearb. v. Martin Rüther; Uwe Schütz; Otto Dann. Hrsg. v. Institut f. Zeitgeschichte (Texte und Materialien zur Zeitgeschichte Bd. 10. Red.: Werner Röder u. Christoph Weisz) / 1998. IX, 648 S. / Gb (3-598-11349-8)

Deutschland, Armenien und die Türkei 1895-1925. Dokumente und Zeitschriften aus dem Dr. Johannes-Lepsius-Archiv an der Martin-Luther-Universität Halle-Wittenberg. Mikrofiche-Edition. Hrsg. v. Hermann Goltz / 3 Tle. (3-598-34406-6)
- Tl. 1. Katalog. Dokumente und Zeitschriften aus dem Dr. Johannes-Lepsius-Archiv. Zsgst. u. bearb. v. Hermann Goltz; Axel Meissner / 1998. XXVIII, 622 S. / Gb (3-598-34407-4)
- Tl. 2. Mikrofiche-Edition. Bearb. V. Hermann Goltz und Axel Meissner unter Mitarb. v. Ute Blaar, Jana Büttner und Leonore Kratzsch / 1999. Ca. 38.000 S. auf ca. 325 Fiches, Lesefaktor 24x / Silberfiche (3-598-34408-2)
Tl. 3 [i.Vb.]

Deutschsprachige Bilderbücher: ein Verzeichnis 1945-1975 erschienener Titel / Children's Picture Books in German: a Checklist of Titles 1945-1975. Bearb. v. Willi Weismann, unter Mitarb. v. Ruth Jansen und Sylvia Stegner / 1980. XVI, 488 S. / Lin (3-598-10078-7)

Deutschsprachige Exilliteratur seit 1933 (3-907820-43-6) [1991 vom Francke Vlg., Bern übernommen]
- Bd. 1. Kalifornien. Hrsg. v. John M. Spalek, Joseph Strelka / 2 Tle. 1976. Zus. 1096 S. / Br (3-907820-44-4)
- Bd. 2. New York. Hrsg. v. John M. Spalek, Joseph Strelka / 2 Tle. 1989. Zus. XXX, 1817 S. / Ln (3-907820-45-2)
Bd. 3. USA [i.Vb.]
- Bd. 4. Bibliographien. Schriftsteller, Publizisten und Literaturwissenschaftler in den USA. Hrsg. v. John M. Spalek, Konrad Feilchenfeldt, Sandra H. Hawrylchak / 3 Tlbde. 1994. Zus. LIX, 2.110 S. / Ln (3-907820-47-9)

Deutschsprachige Lehrprogramme. Eine Bibliographie. Stand: Mai 1973. Hrsg.: Pädagogisches Zentrum Berlin, Referat Dokumentation. Bearb. von Heidemarie Bassen. / 6., neubearb. u. erw. Aufl. 1973. 142 S. / Br [Vorher im Eigenverlag erschienen]
- 7., erw. Aufl. Stand: April 1975 / 1975. 142 S. / Br (3-7940-3212-8)

Deutschsprachige Literatur Prags und der böhmischen Länder 1900-1925. Chronologische Übersicht und Bibliographie. Hrsg. u. eingel. v. Jürgen Born. Bearb. unter Mitw. v. Waltraud John und Jon Shepherd / 1991. VII, 227 S. / Lin (3-598-10855-9)
- 2. überarb. u. erw. Aufl. Unter Mitw. v. Martina Dickert und Klaus P. Wahner / 1993. X, 326 S. / Lin (3-598-11091-X)

Die **deutschsprachige** UNESCO-Literatur. Bibliographie 1946-72. Hrsg.: Deutsche UNESCO-Kommis-

sion. Zsgst. von Willi Gorzny u. Christiane Koschwitz, unter Mitarb. von Werner Kaupert / 2. Aufl. 1973. 222 S. / Lin (3-7940-3056-7)

Developments in collection building in university libraries in Western Europe. Papers presented at a symposium of Belgian, British, Dutch and German University Librarians; Amsterdam, 31. March - 2. April, 1976. Ed. by Willem R. H. Koops and Johannes Stellingwerff / 1977. 109 S. (3-7940-7020-8)

DGD-Schriftenreihe. Hrsg.: Deutsche Gesellschaft f. Dokumentation e.V., Frankfurt (ISSN 0344-5372)
- 1: Laisiepen, Klaus / Ernst Lutterbeck; Karl-Heinrich Meyer-Uhlenried: Grundlagen der praktischen Information und Dokumentation. Eine Einführung. Mit Geleitw. d. Bundesministers d. Innern u.e. Vorw. d. Präsidenten d. DGD / 1972. XIII, 652 S. (3-7940-3600-X)
  - 2., völlig neubearb. Aufl. 1980. XX, 826 S. / Lin (3-598-21251-8)
  gleichzeitig als **Handbuch** *der Internationalen Dokumentation und Information* **Bd. 1** *(3-598-20518-X) erschienen*
- 2: Meyer-Uhlenried, Karl-Heinrich: Der Entwurf von Informationssystemen Grundlagen und Hilfsmittel / 1973. XVI, 308 S., Tab. / Lin (3-7940-3622-0)
- 3: Dahlberg, Ingetraut: Grundlagen universaler Wissensordnung Probleme eines universalen Klassifikationssystems des Wissens 1974 XVIII, 366 S., 44 Taf / Lin (3-7940-3623-9)
- 4: Terminologie der Information und Dokumentation. Redaktion: Ulrich Neveling; Gernot Wersig. Hrsg.: Komitee Terminologie und Sprachfragen der Deutschen Gesellschaft für Dokumentation / 1975. X, 307 S. / Lin (3-7940-3625-5)
- 5: Kuhlen, Rainer: Experimentelle Morphologie in der Informationswissenschaft / 1977. 237 S. (3-7940-3624-7)
- 6: Kaminsky, Reiner: Information und Informationsverarbeitung als ökonomisches Problem / 1977. 207 S. (3-7940-3626-3)
- 7: Meyer-Uhlenried, Karl-Heinrich: Methodische Grundlagen für die Planung von Informationssystemen / 1977. XVI, 520 S. (3-7940-1)
- 8: Wersig, Gernot: Thesaurus - Leitfaden. Eine Einführung in das Thesaurus-Prinzip in Theorie und Praxis Unter Mitarb. v. Petra Schuck-Wersig / 2. erg. Aufl. 1985. 394 S. / Lin (3-598-21252-6)
- 9: Grundlagen der praktischen Information und Dokumentation. Ein Handbuch zur Einführung in die fachliche Informationsarbeit. Begr. v. Klaus Laisiepen, Ernst Lutterbeck und Karl Heinrich Meyer-Uhlenried. Hrsg. v. Marianne Buder, Werner Rehfeld und Thomas Seeger / 2 Bde. 3., völlig neu gefasste Ausg. (*1. u. 2. Aufl.:* → *Bd. 1 d.DGD-Schriftenreihe*) 1990. Zus. XLI, 1.230 S. / Gb (3-598-21253-4) / 1991. Zus. XLI, 1.230 S. / Gb (3-598-11048-0)
  4. Aufl.: → **Grundlagen** *d. praktischen Information*

Diamantidis, Demetrius / Willi Hahn; Karl Franz Stock: Bestellprogramm für eine Lehrbuchsammlung / 1971. 79 S. (3-7940-3199-7)

Dictionarium Bibliothecarii Practicum. Wörterbuch des Bibliothekars in 20 Sprachen. Hrsg. von Zoltán Pipics / 2. Aufl. 1968. 420 S. / Ln
- 3. Aufl. 1969. 375 S. / Ln (Vlg.-Nr. 04108)
- 4. Aufl., correcta et aucta 1970. 375 S. / Ln (Vlg.-Nr. 04108)

- 5. Aufl. 1971. 375 S. (3-7940-4108-9)
- 6., verb. und erw. Auflage. Dictionarium Bibliothecarii Practicum ad usum internationalem in XXII linguis. The Librarian's Practical Dictionary in 22 Languages / Wörterbuch des Bibliothekars in 22 Sprachen (Ausgangssprache Englisch). Hrsg. von Zoltán Pipics. Mitarb.: Jósef Bödey / 1974. 385 S. / Gemeinschaftsausg. mit Akadémiai Kiadó, Budapest (3-7940-4109-7)
- 7. Aufl. 1977. 385 S. / Gb (3-7940-4110-0)

Dictionarium Museologicum. by ICOM - National Centre of Museums, Budapest (CIDOC). Ed. by I. Eri and B. Végh / 1986. 774 p. / Ln (3-598-07530-8)

Diels, Herrmann A.: Colloquium über antikes Schriftwesen (1908) / Vorlesungen über Herodot (1907-08). Einl. v. Jürgen Dummer und Wolfgang Rösler. Nachw. v. Hildebrecht Hommel: Berliner Erinnerungen 1920-21, Herrmann Diels zum Gedächtnis. - Reprint / 1984. 56 S. / Br (3-598-10319-0)

Dietze, Joachim: B. Travens Wortschatz. Ein Frequenzwörterbuch zu seinen drei Schaffensperioden / 1998. 705 S. (3-598-11383-8)
- Einführung in die Informationslinguistik. Die Linguistische Datenverarbeitung in der Informationswissenschaft (Linguistische Studien) / 1989. 195 S. / Br (3-598-07044-6)
- Texterschliessung: Lexikalische Semantik und Wissensrepräsentation / 1993. 102 S. / Br (3-598-11179-7)

Dietzel, Armin: Geistige Freiheit und das Buch. Aufsätze zum Buch in seinem Beziehungsgeflecht / 1989. 160 S. (3-598-10743-9)

Digitales Informationssystem für Kunst und Sozialgeschichte (DISKUS) CD-ROM-Edition. Hrsg.: Department Computer & Letteren d. Rijksuniversiteit in Utrecht, in Zus.-Arb. mit d. Bildarchiv Foto Marburg - Deutsches Dokumentationszentrum f. Kunstgeschichte Philipps-Universität Marburg (3-598-40301-1)
- Tl. 001: Russische Avantgarde. Sammlung Ludwig. Hrsg.: Museum Ludwig, Köln / CD-ROM-Edition 1995. ca. 600 Abb. (3-598-40300-3)
- Tl. 002: Gedruckte Porträts 1500-1618 aus der Graphischen Sammlung des Germanischen Nationalmuseums. Hrsg.: Germanisches Nationalmuseum, Nürnberg / CD-ROM-Edition 1995. 2900 Abb. (3-598-40305-4)
- Tl. 003: Das politische Plakat der DDR (1945-1970) Aus dem Bestand des Deutschen Historischen Museums in Berlin. Hrsg.: Deutsches Historisches Museum, Berlin / CD-ROM-Edition 1995. 2800 Abb. (3-598-40303-8)
- Tl. 004: Photographische Perspektiven aus den 20er Jahren Aus dem Bestand des Museums für Kunst und Gewerbe in Hamburg. Hrsg.: Museum f. Kunst u. Gewerbe, Hamburg / CD-ROM-Edition 1995. 2300 Abb. (3-598-40302-X)
- Tl. 005: Italienische Zeichnungen vom 14. bis 18. Jahrhundert im Berliner Kupferstichkabinett. Hrsg.: Kupferstichkabinett, Berlin / CD-ROM-Edition 1995. 2800 Abb. (3-598-40304-6)
- Tl. 006: Die Gemälde der Nationalgalerie / The Paintings of the Nationalgalerie. Hrsg.: Nationalgalerie d. Staatlichen Museen zu Berlin / CD-ROM-Edition 1996. 3100 Abb. (3-598-40308-9)
- Tl. 007: Wallraf-Richartz-Museum Köln. Gemälde- und Skulpturenbestand / Wallraf-Richartz-Mu-

seum Cologne. Collection of Paintings and Sculptures. Hrsg.: Wallraf-Richartz-Museum, Köln / CD-ROM-Edition 1996. 2500 Abb. (3-598-40309-7)
- Tl. 008: Plakate des ersten Weltkrieges 1914-1918 / Posters from World War One 1914-1918. Hrsg.: Deutsches Historisches Museum, Berlin / CD-ROM-Edition 1996. 800 Abb. (3-598-40306-2)
- Tl. 009: Politische Allegorien und Satiren aus der Graphischen Sammlung des Germanischen Nationalmuseums / Political Allegories and Satires from the Prints Collection of the Germanisches Nationalmuseum. Hrsg.: Germanisches Nationalmuseum, Nürnberg / CD-ROM-Edition 1996. 2500 Abb. (3-598-40307-0)
- Tl. 010: Politische Abzeichen der Kaiserzeit und der Weimarer Republik / Political Badges from the Imperial Age and the Weimarer Republic. Hrsg.: Deutsches Historisches Museum, Berlin / CD-ROM-Edition 1996. 800 Abb. (3-598-40310-0)
- Tl. 011: Filmplakate der Österreichischen Nationalbibliothek (1910-1955) / Film Posters of the Austrian National Library (1910-1955). Hrsg.: Österreichische Nationalbibliothek, Wien / CD-ROM-Edition 1998. (3-598-40311-9)
- Tl. 012: 1848 - Politik, Propaganda, Information und Unterhaltung aus der Druckerpresse / 1848 - Politics, Propaganda, Information and Entertainment from the Printing Press. Hrsg.: Deutsches Historisches Museum, Berlin / CD-ROM-Edition 1998. (3-598-40314-3)
- Tl. 013: Gemäldegalerie Berlin / Picture Gallery Berlin. Hrsg.: Gemäldegalerie, Staatliche Museen zu Berlin - Preußischer Kulturbesitz / CD-ROM-Edition 1998. (3-598-40316-X)
- Tl. 014: Plakate der SBZ / DDR - Politik, Wirtschaft, Kultur. Sammlung des Deutschen Historischen Museums. Hrsg.: Deutsches Historisches Museum, Berlin / CD-ROM-Edition 1999. (3-598-40315-1)

Directories and associations of the book trade and librarianship Adreßbücher und Verbände des Buch- und Bibliothekswesens An internationale Guide / 2. ed. 1971. 96 S. (3-7940-2017-0)

Directory of Administrative ans Judical Authorities in EC Countries. Rev. by Walter Duic / 2nd ed. 1990. 1.000 p. (3-598-10882-6)

Directory of continuing education opportunities for library, information, media, personnel 1979. Comp. by CLENE, inc., the Continuing Library Education Network and Exchange / 1979. XIX, 292 S. (3-598-40009-8)

Directory of European Political Scientists. Ed. by Jean Blondel and Carol Walker / 3. fully rev. ed. 1979. XXIV, 461 p. / Hard (3-598-10055-8) [Hans Zell Publishers]
- 4th fully rev. ed. Ed.: European Consortium for Political Research, University of Essex / 1985. XXVI, 627 p. / Hard (3-598-10417-0)

Directory of International Cooperation in Science and Technology Index of Institutions in the Federal Republic of Germany Cooperating with Developing Countries. Second enlarged and revised Edition Ed. by Klaus Gottstein; Research Unit Gottstein in the Max Planck Society. Comp. by Jörn Behrmann; Werner Bock-Werthmann, in Zus.-Arb. mit Georg Menache. 2nd ed / 1989. 256 p. / Hard (3-598-10626-2)

Dissertationen Katalog der Universität Tübingen. Mikrofiche-Edition / 1983. 7.013.600 Ktn. auf 370 Fiches. Lesefaktor 42x / Silberfiche (3-598-30449-8)

Dittmar, Peter: Die Darstellung der Juden in der populären Kunst zur Zeit der Emanzipation. Hrsg.: Zentrum f. Antisemitismusforschung d. Technischen Universität Berlin / 1992. 482 S. / Gb (3-598-10971-7)

Documentation et Recherches. Publiées par l'Institut Historique Allemand
- Hartmann, Peter C.: Archives, bibliothèques et centres de documentation à Paris pour l'histoire des XIXe et XXe siècles. Guide pratique pour historiens, politologues et journalistes / 1978. 136 S. / Br (3-7940-3320-5)
- Hildebrand, Klaus: Le Troisième Reich. Traduit de l'Allemand par François Manfrass-Sirjacques / 1985. 244 S. / Br (3-598-20313-6)
- L'histoire medievale et les ordinateurs / Medieval History and Computers. Rapports d'une Table ronde internationale, Paris 1978. Publiés par Karl F. Werner / 1981. 126 S. / Br (3-598-10204-6)
- Kaiser, R. / M. T. Kaiser-Guyot: Documentation numismatique de la France médiévale collections de monnaies et sources de l'histoire monetaire / 1982. 113 S. / Br (3-598-10330-1)

Documents for the History of Collecting. Ed. by Carol T. Dowd; Anna C. Sones. The Provenance Index of the Getty Art History Information Program (3-598-21690-4)
- Italian Inventories / 2 vols. 1992-96 (3-598-21691-2)
- Vol. 1: Labrot, Gérard: Italian Inventories Collections of Paintings in Naples 1600-1780. Ed. by Carol Dowd; Anna C. Sones, with the assistance of Antonio Delfino / 1992. XIII, 806 p. / Hard (3-598-21692-0)
- Vol. 2: Safarik, Eduard: Italian Inventories, Collezione dei dipinti Colonna Inventari 1611-1795 The Colonna Collection of Paintings. Inventories 1611-1795 Ed. by Anna C. Sones, with the assistance of Cinzia Pujia / 1996. XXI, 1111 p. / Hard (3-598-21693-9)

Documents that move and speak: Audiovisual Archives in the New Information Age Proceedings of a symposium organized for the International Council of Archives held by the National Archives of Canada, Ottawa, April 30, 1990 - May 3, 1990. Ed.: National Archives of Canada / Archives Nationales du Canada / 1991. 318 p. / Hard (3-598-11043-X)

Döring, Detlef: Die Bestandsentwicklung der Bibliothek der Philosophischen Fakultät der Universität zu Leipzig von ihren Anfängen bis zur Mitte des 16. Jahrhunderts. Ein Beitrag zur Wissenschaftsgeschichte der Leipziger Universität in ihrer vorreformatorischen Zeit. (Beiheft 99 zum Zentralblatt für Bibliothekswesen) / 1990. 177 S. / Br (3-598-07033-0)

Dokumentation der Wissenschaft: Medizin. Fakultäten, Institutionen, Personen / 1970. XVI, 437 S. (3-7940-2419-2)
- Naturwissenschaften. Fakultäten, Institutionen, Personen / 1970. XVI, 507 S. (3-7940-2460-5)
- Technik. Fakultäten, Institutionen, Personen / 1970. (XIII,) 441 S. (3-7940-2139-8)
- Wirtschafts-, Rechts- und Sozialwissenschaften. Fakultäten, Institutionen, Personen / 1970. (XI,) 345 S. (3-7940-2129-0)

Dokumentation - ein Rationalisierungsfaktor? Bericht über 5jährige Erfahrungen. Ergänzung zum Taschenbuch der Dokumentation und Technik / 1955. 17 S. [→ *Taschenbuch 1954*]

Die **Dokumentation** und ihre Probleme. Vorträge, gehalten auf d. 1. Tagung d. Deutschen. Gesellschaft für Dokumentation vom 21.-24. Sept 1942 in Salzburg. Dt. Ges. für Dokumentation. Hrsg. u. mit e. Nachw. u. Anm. vers. von Erich Pietsch. - Unveränd. Nachdr. d. Ausg. Leipzig 1943 / 1975. VII, 205, 17 Abb., 38 S. Anhang / Ln (3-7940-3640-9)

Dokumentation Westeuropa. Hrsg.: Deutsches Historisches Institut, Paris
- 1: Hartmann, Peter Klaus: Pariser Archive, Bibliotheken und Dokumentationszentren zur Geschichte des 19. und 20. Jahrhunderts. Eine Einführung in Benützungspraxis u. Bestände für Historiker, Politologen u. Journalisten / 1976. 131 S., 1 Kte / Lin (3-7940-3330-2)
- 2: Menges, Franz: Quellen zur westeuropäischen Geschichte und Kultur in bayerischen Bibliotheken / 1977. 171 S. / Lin (3-7940-3331-0)
- 3: Manfrass, Klaus: Politik und politische Wissenschaft in Frankreich Politische Organisationen, Publikationen / Presseorgane, Dokumentationsstätten, Forschungseinrichtungen / 1979. IV, 234 S. / Lin (3-598-20238-5)
- 4: Paravicini, Werner: Das Nationalarchiv in Paris. Ein Führer zu den Beständen aus dem Mittelalter und der Frühen Neuzeit Vorw. v. J. Favier. Hrsg. v. Karl F. Werner / 1980. 199 S. / Lin (3-598-20239-3)
- 5: Paravicini, Werner: Die Nationalbibliothek in Paris Ein Führer zu den Beständen aus dem Mittelalter und der Frühen Neuzeit Vorw. v. Georges LeRider / 1981. 133 S. / Lin (3-598-20240-7)

Dokumentation zur jüdischen Kultur in Deutschland 1840-1940. Die Zeitungsausschnittsammlung Steininger. Mikrofiche-Edition. Hrsg.: Archiv Bibliographia Judaica e.V. / 3 Tle. 1995-1999 (3-598-33316-1)
- Abt. I. Teil 1: Bildende Künstler. Teil 2: Darstellende Künstler / 1995/1996. 101 Fiches. Lesefaktor 24x / Silberfiche (3-598-33317-X)
- Abt. II. Musik / 1996/1997. 102 Fiches. Lesefaktor 24x / Silberfiche (3-598-33322-6)
- Abt. III. Schriftsteller / 1999. Ca. 120 Fiches, Lesefaktor 24x / Silberfiche (3-598-33326-9)

Dokumente zur Geschichte der kommunistischen Bewegung in Deutschland. Reihe 1945-46. Hrsg. v. Günter Benser; Hans J. Krusch / 6 Bde. 1993-1997. Zus. LVI, 3.464 S. (3-598-11114-2)
- Bd. 1. Protokolle des Sekretariats des Zentralkomitees der KPD Juli 1945 bis April 1946 / 1993. 577 S. / Gb (3-598-11115-0)
- Bd. 2. Protokolle der erweiterten Sitzungen des Sekretariats des Zentralkomitees der KPD Juli 1945 bis Februar 1946 / 1994. XII, 708 S. / Gb (3-598-11116-9)
- Bd. 3. Protokoll der Reichsberatung der KPD 8 / 9. Januar 1946 / 1994. XIII, 553 S. / Gb (3-598-11117-7)
- Bd. 4. Stenographisches Protokoll der Reichskonferenz der KPD 2 / 3. März 1946 / 1995. XIV, 769 S. / Gb (3-598-11118-5)
- Bd. 5. Stenographisches Protokoll des 15. Parteitages der KPD 19 / 20. April 1946 / 1996. XVIII, 694 S. / Gb (3-598-11119-3)
- Bd. 6. Register / 1997. 163 S. / Gb (3-598-11123-1)

Domanovszky, Ákos: Functions and Objects of Author and Title Cataloguing. A contribution to cataloguing theory. English text ed. by Anthony Thompson / 1975. 173 p. / Hard (3-7940-4113-5)

Dortmunder Beiträge zur Zeitungsforschung. Hrsg.: Institut für Zeitungsforschung der Stadt Dortmund / Kurt Koszyk. Ab Bd. 26: Hans Bohrmann.
- 1: Behrbalk, Erhard: Die „Westfälische Zeitung". Ein Beitrag zur Geschichte der westfälischen Tagespresse im 19. Jahrhundert (1848-1883). - Koszyk, Kurt: Aktenstücke zum. „Westfälischen Anzeiger". Dokumentation / 1958. 179 S., 4 Abb.-S. / Br (3-7940-2501-6)
- 2: Koszyk, Kurt: Das „Dampfboot" und der Rhedaer Kreis.- Wand, Albert: Die „Westfälische Provinzialzeitung" (1873-1882).- Lindemann, Margot: Gerhard Wilhelm von Eicken - ein Arzt als Journalist / 1958. 90 S. / Br (3-7940-2502-4)
- 3: Lindemann, Margot: Pressefrühdrucke als Spiegel französischer Geschichte. Neue Zeitungen und Pamphlete aus zwei Jahrhunderten / 1959. 108 S., 14 Abb. / Br (3-7940-2503-2)
- 4: D'Ester, Karl / Heinz O. Sieburg; Heinrich Hofmeier: Die Zeitungswissenschaft im Dienste der deutsch-französischen Beziehungen. Die Politik Richelieus und Renaudots „Gazette". Der Zahnarzt Georges Fattet in der Karikatur von 1848 / 1959. 75 S. / Br (3-7940-2504-0)
- 5: Sandgathe, Günter: Der „Westfälische Anzeiger" und die politischen Strömungen seiner Zeit 1798-1809 / 1960. 214 S. / Br (3-7940-2505-9)
- 6: Hess, Ulrich: Louis Viereck und seine Münchner Blätter für Arbeiter 1882-1889. - Lindemann, Margot: Der Münchner Pressebandit / 1961. VI, 101 S. u. 6 Taf / Br (3-7940-2506-7)
- 7: Grosskopff, Rudolf: Die Zeitungsverlagsgesellschaft Nordwestdeutschland GmbH 1922-1940. Beispiel einer Konzentration in der deutschen Provinzpresse / 1963. X, 231 S. / Br (3-7940-2507-5)
- 8: Märthesheimer, Peter: Publizistik und gewerkschaftliche Aktion. Das Bild der IG Metall in westdeutschen Zeitungen dargestellt an der Tarifauseinandersetzung 1961/62 in der Metallindustrie / 1964. 88 S. / Br (3-7940-2508-3)
- 9: Stein, Elisabeth: Der Freimüthige an der Haar als Organ des politischen Katholizismus 1849-1850 / 1965. 105 S. / Br (3-7940-2509-1)
- 10: Klutentreter, Wilhelm: Die „Rheinische Zeitung" von 1842 / 43 in der politischen und geistigen Bewegung des Vormärz / 2 Bde. Bd. 1: 1966 Bd. 2: Dokumente / 1967. 266 S. / Br (3-7940-2510-5)
- 11: Grote, Bernd: Der deutsche Michel. Ein Beitrag zur Publizistischen Bedeutung der Nationalfiguren / 1967. 89 S. / Br (3-7940-2511-3)
- 12: Scholand, Hildegard: Presse und Beamtenpolitik. Eine Inhaltsanalyse / 1968. 165 S. / Br (3-7940-2512-1)
- 13: Kothenschulte, Uwe: Hermann Löns als Journalist / 1968. 184 S. / Br (3-7940-2513-X)
- 14: Fischer, Heinz D.: Publizistik in Suburbia. Strukturen und Funktionen amerikanischer Vorortzeitungen. Hrsg. v. Kurt Koszyk / 1971. 270 S., 16 Abb. / Br (3-7940-2514-8)
- 15: Schütze, Peter: Die Entwicklungsgeschichte lokaler Wechselseiten im deutschen Pressewesen bis 1945 / 1971. 178 S. / Br (3-7940-2515-6)
- 16: Bieber, Horst: Paul Rohrbach. Ein konservativer Publizist und Kritiker der Weimarer Republik / 1972. 270 S. / Br (3-7940-2516-4)
- 17: Richter, Rolf: Kommunikationsfreiheit - Verlegerfreiheit? Zur Kommunikationspolitik der Zeitungsverleger in der Bundesrepublik Deutschland 1945-1969 / 1973. 429 S. / Br (3-7940-2517-2)
- 18: Petzke, Ingo: Journalistische Mitbestimmung in schwedischen Zeitungsverlagen / 1974. 101 S. / Br (3-7940-2518-0)
- 19: Overesch, Manfred: Presse zwischen Lenkung und Freiheit. Preussen und seine offiziöse Zeitung von der Revolution bis zur Reichsgründung (1848 - 1871 / 72) / 1974. 210 S. / Br (3-7940-2519-9)
- 20: Haase, Amine: Katholische Presse und die Judenfrage. Inhaltsanalyse katholischer Periodika am Ende des 19. Jahrhunderts / 1975. 262 S. / Br (3-7940-2520-2)
- 21: Verzeichnis und Bestände westfälischer Zeitungen. Hrsg. v. Kurt Koszyk im Auftr. d. Histor. Komm. f. Westfalen / 1975. 112 S. / Br (3-7940-2521-0)
- 22: Tölcke, Carl Wilhelm: Carl Wilhelm Tölckes Presseberichte zur Entwicklung der deutschen Sozialdemokratie 1848 - 1893. Quellen zur Geschichte der deutschen Arbeiterbewegung Bearb. von Arno Herzig / 1976. 278 S. (3-7940-2522-9)
- 23: Schaber, Will: B. F. Dolbin. Der Zeichner als Reporter / 1976. 177 S., zahlr. Abb. / Ln (3-7940-2523-7)
- 24: Taubert, Rolf: Autonomie und Integration. Das Arbeiter-Blatt Lennep. Eine Fallstudie zur Theorie und Geschichte von Arbeiterpresse und Arbeiterbewegung 1848-1850 / 1977. 215 S. / Br (3-7940-2524-5)
- 25: Goldstein, Moritz: Berliner Jahre. Erinnerungen 1880-1933 / 1978. 269 S. / Br (3-7940-2525-3)
- 26: Lindemann, Margot: Nachrichtenübermittlung durch Kaufmannsbriefe Brief"Zeitungen" in d. Korrespondenz Hildebrand Veckinchussens (1398 - 1428) / 1978. 116 S. (3-7940-2526-1)
- 27: Kunze, Christine: Journalismus in der UdSSR. Eine Untersuchung über Aufgaben und Funktionen sowjetischer Journalisten unter besonderer Berücksichtigung der Struktur der Massenmedien in der UdSSR und der Diskussion des Berufsbildes in der Zeitung „Zurnalist" / 1978. 342 S. / Br (3-7940-2527-X)
- 28: Weischenberg, Siegfried: Die Elektronische Redaktion. Publizistische Folgen der neuen Technik. Unter Mitarb. von Kurt P. Christophersen / 1987. 144 S. (3-7940-3055-9)
- 29: Kisker, Klaus P. / Manfred Knoche; Axel Zerdick: Wirtschaftskonjunktur und Pressekonzentration in der Bundesrepublik Deutschland / 1979. 230 S. / Br (3-598-02529-7)
- 30: Presse im Exil Beiträge zur Kommunikationsgeschichte des deutschen Exils 1933-1945. Hrsg. v. Hanno Hardt; Elke Hilscher; Winfried B. Lerg / 1979. 516 S. / Br (3-598-02530-0)
- 31: Heinrich-Jost, Ingrid: Literarische Publizistik Adolf Glaßbrenners (1810-1876). Die List beim Schreiben der Wahrheit / 1980. 399 S. / Br (3-598-21281-X)
- 32: Jarren, Otfried: Stadtteilzeitung und Kommunikation / 1980. 162 S. / Br (3-598-21282-8)
  - 2. durch ein Nachw. erg. Aufl. 1983. 165 S. / Br (3-598-21294-1)
- 33: Pressefrühdrucke aus der Zeit der Glaubenskämpfe (1514-1648). Bestandsverzeichnis des Instituts für Zeitungsforschung / 1980. 247 S. / Br (3-598-21283-6)
- 34: Schaber, Will: Der Gratgänger Welt und Werk Erich Schairers (1887-1956) / 1981. 204 S., 26 Abb. / Ebr (3-598-21286-0)
- 35: Hagelweide, Gert: Literatur zur deutschsprachigen Presse. Eine Bibliographie (Von den Anfängen bis 1970) / 14 Bde, 3 Reg.-Bde. 1985-2002 / Lin (3-598-21284-4)
  - Bd. 1. Handbücher, Lexika, Bibliographien, Pressesammlung und -dokumentation, Organisation der Presse (Verbände), Zeitungs-, Publizistik- und Kommunikationswissenschaft 1-13132. Presse im Wechselspiel der Medien und der Öffentlichkeit / 1985. XXXVI, 464 S. / Lin (3-598-21288-7)
  - Bd. 2. Presseverlag - Träger der Aussage - Presseinhalt 13133-23742. Formgebung und Gestaltung - Inhaltsbeschaffung und -vermittlung - Nachrichtenwesen / 1989. XXII, 372 S. / Lin (3-598-21289-5)
  - Bd. 3. Technische Herstellung und Vertrieb - Der Rezipient 23743-33164 / 1989. XX, 311 S. / Lin (3-598-21290-9)
  - Bd. 4. Wesen und Funktion periodischer Druckpublizistik 33165-47705. Tageszeitung (Presse), Die Zeitschrift, Almanache und Kalender, Die Presse (Tageszeitung) in Geschichte und Gegenwart / 1993. XXIV, 546 S. / Lin (3-598-21291-7)
  - Bd. 5. Deutschsprachige Länder. Literatur zur deutschsprachigen Presse 44706-58007. Teil I: Deutschland. Lokale Pressegeschichte - Druck-, Verlags- und Vertriebsorte. Aachen-Eutin / 1995. XIX, 386 S. / Lin (3-598-21292-5)
  - Bd. 6. Deutschsprachige Länder. Literatur zur deutschsprachigen Presse 58008-69708. Teil I. Deutschland. Lokale Pressegeschichte, Druck-, Verlags- und Vertriebsorte. Fellbach-Lyck / 1995. XXI, 410 S. / Lin (3-598-21304-2)
  - Bd. 7. Deutschsprachige Länder. Literatur zur deutschsprachigen Presse 69709-80619. Teil I: Deutschland. Lokale Pressegeschichte, Druck-, Verlags- und Vertriebsorte. Magdeburg-Zwönitz / 1996. XXII, 390 S. / Lin (3-598-21305-0)
  - Bd. 8. Deutschsprachige Länder. Literatur zur deutschsprachigen Presse 80620-89198. Teil II: Liechtenstein-Österreich-Schweiz / 1997 XIX, 321 S. / Lin (3-598-21306-9)
  - Bd. 9. Länder ausserhalb des deutschen Sprachraums 89199-98384. Afrika - Amerika - Asien - Australien - Europa. Deutschsprachige Literatur zur Presse des Auslands. Literatur zur Presse der deutschen Minderheiten (deutsch und fremdsprachig) / 1998. XVIII, 348 S. / Lin (3-598-21307-7)
- 36: Studnitz, Cecilia von: Kritik des Journalisten. Ein Berufsbild in Fiktion und Realität / 1983. 238 S. / Ebr (3-598-21287-9)
- 37: Hoefer, Frank T.: Pressepolitik und Polizeistaat Metternichs. Die Überwindung von Presse und politischer Öffentlichkeit in Deutschland und den Nachbarstaaten durch das Mainzer Informationsbüro (1833-1848) / 1983. 240 S. / Br (3-598-21293-3)
- 38: Uphaus-Wehmeier, Annette: Zum Nutzen und Vergnügen - Jugendzeitschriften des 18. Jahrhunderts. Ein Beitrag zur Kommunikationsgeschichte / 1984. 294 S., 4 Ill., 10 Taf / Br (3-598-21295-X)
- 39: Marktzutritt bei Tageszeitungen zur Sicherung der Meinungsvielfalt durch Wettbewerb. Hrsg. v. Gerd Kopper / 1984. 185 S. / Ebr (3-598-21296-8)
- 40/41: Struktur und Organisation des Pressevertriebs / 2 Tle.
  - Tl. 1 (40). Brummund, Peter: Der deutsche Zeitungs- und Zeitschriftengrosshandel / 1985. 502 S. / Ebr (3-598-21297-6)
  - Tl. 2 (41). Schwindt, Peter: Zeitungen und Zeitschriften im Einzelhandel / 1985. 133 S. / Ebr (3-598-21298-4)
- 42: Stein, Peter: Die NS-Gaupresse 1925-1933. Forschungsbericht - Quellenkritik, neue Bestandsaufnahme / 1987. 275 S. / Ebr (3-598-21299-2)

• 43: Maoro, Bettina: Die Zeitungswissenschaft in Westfalen 1914-45. Das Institut für Zeitungswissenschaften in Münster und die Zeitungswissenschaft in Dortmund. Vorw. v. Hans Bohrmann / 1987. 491 S. / Ebr (3-598-21300-X)
• 44: Hagelweide, Gert: Quellenkunde zur Pressegeschichte Dortmunds und der Grafschaft Mark. Bibliographie, Standortnachweis, Archivalien und Literatur / 1990. XVIII, 242 S., 20 Abb. / Ebr (3-598-21301-8)
• 45: Klose, Hans G.: Die Zeitungswissenschaften in Köln Ein Beitrag zur Professionalisierung der deutschen Zeitungswissenschaft inder ersten Hälfte des 20. Jahrhunderts / 1989. XX, 239 S. / Ebr (3-598-21302-6)
• 46: Ludwig Wronkow Berlin - New York, Journalist und Karikaturist bei Mosse und beim „Aufbau". Eine illustrierte Lebensgeschichte. Von Michael Groth; Barbara Posthoff / 1989. 231 S. / Ln (3-598-21303-4)
• 47: Wilking, Thomas: Strukturen lokaler Nachrichten. Eine empirische Untersuchung von Text- und Bildberichterstattung Vorw. v. Otfried Jarren / 1990. X, 263 S. / Br (3-598-21308-5)
• 48: Obermeier, Karl M.: Medien im Revier. Die Entwicklung der Zeitungslandschaft des Ruhrgebiets seit dem Zweiten Weltkrieg. Vorw. v. Hans Bohrmann / 1991. 497 S. / Ebr (3-598-21309-3)
• 49: Macat, Andreas: Die Bergische Presse. Bibliographie und Standortnachweis der Zeitungen und zeitungsähnlichen Periodika seit 1769. Vorw. v. Hans Bohrmann / 1991. 249 S. / Gb (3-598-21310-7)
• 50: Plakatsammlung des Instituts für Zeitungsforschung der Stadt Dortmund Kommentar- und Registerband zur Mikrofiche-Ausgabe. Bearb. v. Barbara Posthoff (3-598-21313-1)
  - Bd. 1. Teil A: Deutsche Plakate / 1992. 174 S. / Gb (3-598-21311-5)
  - Bd. 2. Teil B und C: Ausländische und deutschsprachige Plakate / 1992. 139 S. / Gb (3-598-21312-3)
• 51: Ros, Guido: Adalbert von Bornstedt und seine Deutsche - Brüsseler Zeitung. Ein Beitrag zur Geschichte der deutschen Emigrantenpublizistik im Vormärz. Vorw. v. Kurt Koszyk / 1993. 301 S. / Ebr (3-598-21314-X)
• 52: Gerasch, Sabine: Prozesswirklichkeit und Gerichtsberichterstattung. Eine Untersuchung der Lokalberichterstattung zu Strafprozessen bei Gewalt- und Sexualdelikten vor dem Dortmunder Land- und Amtsgericht / 1995. 196 S. / Ebr (3-598-21315-8)
• 53: Müsse, Wolfgang: Die Reichspresseschule - Journalisten für die Diktatur? Ein Beitrag zur Geschichte des Journalismus im Dritten Reich / 1995. 299 S. / Ebr (3-598-21316-6)
• 54: Hüffer, Jürgen B.: Vom Lizenzpressesystem zur Wettbewerbspresse Lizenzverleger und Altverleger im Rheinland und in Westfalen 1945-1953/54 / 1995. 376 S. / Ebr (3-598-21317-4)
• 55: Neumann, Sieglinde: Redaktionsmanagement in den USA. Fallbeispiel „Seattle Times" / 1997. XVI, 273 S. / Ebr (3-598-21318-2)
• 56: Klammer, Bernd: Pressevertrieb in Ostdeutschland. Die wirtschaftlichen und politischen Interessen beim Aufbau eines Pressegrosshandelssystems nach der Oktoberwende 1989 / 1997. XI, 255 S. / Ebr (3-598-21319-0)

Drube, Herbert: Pullach im Isartal. Unsere Heimat in Vergangenheit und Gegenwart. Mit Beitr. von Franz Schauer u. Walter Fuchs / 1982. 192 S.

Duerden, Denis: A Concise Handbook of African Art. Ancient, Modern, and Contemporary / 1986. 360 p. (3-598-10544-4)

Duic, Walter Z.: Africa-Administration. Handbuch des öffentlichen Lebens, der Verwaltung und Justiz der afrikanischen Staaten / Directory of Public Life, Administration and Justice for the African States / Repertoire des Organismes officiels et professionels de l'Administration et de la Jus / 1978. XXXII, 1285 S. / Lin (3-7940-7010-0)
• Europa-Administration. Handbuch d. Verwaltung u. Justiz für d. Europ. Gemeinschaften. Als Ms. gedr. / 1976. 1161 S. (3-7940-3017-6)

Dußler, Sepp / Fritz Kolling: Moderne Setzerei. Unter Mitwirk. namhafter Fachleute / 1967. 144 S. / Lin (Vlg.-Nr. 08703) [1968 v. Ullstein Vlg, Berlin übernommen]
• 2., erw. u. überarb. Aufl. 1971. 191 S. / Lin (3-7940-8703-8)
• 3. überarb. Aufl. 1973. 188 S. / Lin (3-7940-8703-8)
• 4. Aufl. 1974. 188 S. / Lin (3-7940-8708-8) [Anschließend im Vlg. Beruf + Schule, Itzehoe]

Edhofer, Ingrid: Die Biogenese Chl-bindender Proteine. Regulation der Apoprotein-Akkumulation durch Chlorophyll und Licht / 1998. XV, 179 S. / Br (3-598-11405-2)

Egenolff, Christian: Sprichwörterlexikon. - Nachdruck d. Orig.Ausg. Frankfurt a. M. 1552. - Mit e. Nachw. von Hans Henning / 1968. 864 S. / Ln (Vlg.-Nr. 04158)

Eichhoff, Jürgen: Wortatlas der deutschen Umgangssprachen (3-907820-48-7) [1991 vom Francke Vlg., Bern übernommen]
• Bd. 1. 1977. 50 S., 54 Ktn. / Gb (3-907820-49-5). Studienausgabe / Kt (3-907820-50-9)
• Bd. 2. 1978. 50 S., 71 Ktn, Fragebogen, 1 Beil / Gb (3-907820-51-7). Studienausgabe / Kt (3-907820-52-5)
• Bd. 3. 1993. 58 S., 62 Ktn. / Gb (3-907820-53-3). Studienausgabe / Kt (3-907820-54-1)
  Bd. 4 [i.Vb.]

Eichstätter Hochschulreden. Hrsg.: Katholische Universität, Eichstätt [Minerva Publikation]
• 1: Fina, Kurt: Der polare Aufbau des geschichtsdidaktischen Feldes / 1978. 34 S. / Gh (3-597-30001-4)
• 2: Wehner, Ernst G.: Entwicklungstendenzen moderner Psychologie / 1978. 17 S. / Gh (3-597-30002-2)
• 3: Wellnhofer, Peter: Archaeopteryx und Probleme der Evolutionstheorie / 1978. 18 S. / Gh (3-597-30003-0)
• 4: Kaiser, Philipp: Ist der Marxismus notwendig atheistisch? / 1978. 19 S. (3-597-30004-9)
• 5: Renk, Herta E.: Literatur und Schule - Zur Situation der Literaturdidaktik / 1978. 25 S. / Gh (3-597-30005-7)
• 6: Baumgartner, Konrad: Der Wandel des Priesterbildes zwischen dem Konzil von Trient und dem II. Vatikanischen Konzil / 1978. 21 S. / Gh (3-597-30006-5)
• 7: Hunfeld, Hans: Didaktik als Wissenschaft? / 1978. 23 S. / Gh (3-597-30007-3)
• 8: Hischer, Erhard: Das Kind im Krankenhaus. Eine Herausforderung der Sozialpädagogik / 1978. 36 S. (3-597-30008-1)
• 9: Knopp, Werner: Professoren und Studenten 1977. Was denken sie? Was wünschen sie? Wie handeln sie? / 1978. 26 S. / Gh (3-597-30009-X)

• 10: Pannenberg, Wolfhart: Die Auferstehung Jesu und die Zukunft des Menschen / 1978. 18 S. (3-597-30010-3)
• 11: Stone, Margaret: Das Deutschstudium an britischen Schulen und Universitäten / 1978. 18 S. / Gh (3-597-30011-1)
• 12: Schleissheimer, Bernhard: Sein und Sollen. Zur Frage der theoretischen Begründbarkeit praktischer Normen / 1978. 18 S. / Gh (3-597-30012-X)
• 13: Müller, Hubert: Das Gesetz in der Kirche zwischen amtlichem Anspruch und konkretem Vollzug. Annahme und Ablehnung universalkirchlicher Gesetze als Anfrage an die Kirchenrechtswissenschaft / 1978. 26 S. / Gh (3-597-30013-8)
• 14: Peters, Dieter S.: Biologische Einsicht und ethische Entscheidung. Ill. v. Renate Klein-Rödder / 1978. 28 S. / Gh (3-597-30014-6)
• 15: Ratzinger, Joseph: Zum Begriff des Sakraments / 1979. 20 S. (3-597-30015-4)
• 16: Hübner, Reinhard M.: Der Gott der Kirchenväter und der Gott der Bibel. Zur Frage der Hellenisierung des Christentums / 1979. 32 S. (3-597-30016-2)
• 17: Hürten, Heinz: Vorurteil und Vernunft. Leitbegriffe der politischen Diskussion in der Epoche der demokratischen Revolution / 1979. 23 S. / Gh (3-597-30017-0)
• 18: Blaicher, Günther: Vorurteil und literarischer Stil Zur Interaktion von Autor und zeitgenössischem Leserpublikum in Byrons Don Juan / 1979. 26 S. / Gh (3-597-30018-9)
• 19: Gsell, Otto: Strukturelle Semantik und Wortschatzunterricht / 1980. 27 S. / Gh (3-597-30019-7)
• 20: Gutmann, Wolfgang F. / Klaus Bonik: Die Dynamik von Selbstorganisation und Destruktion im heutigen Evolutionsverständnis / 1980. 39 S. / Gh (3-597-30020-0)
• 21: Sutor, Bernhard: Die Kardinaltugenden - Erziehungsziele Politischer Bildung? / 1980. 18 S. / Gh (3-597-30021-9)
• 22: Wehrle, Paul: Die Bedeutung des Symbols für die religiöse Erziehung / 1980. 40 S. (3-597-30022-7)
• 23: Huber, Wolfgang: Simuliertes Verstehen, Künstliche Intelligenz und natürlicher Schwachsinn / 1980. 22 S. / Gh (3-597-30023-5)
• 24: Luthe, Heinz O.: Soziale Welt - Soziologische Welt / 1981. 19 S. / Gh (3-597-30024-3)
• 25: Thomaneck, Jürgen K.: Fremdsprachenunterricht und Soziolinguistik / 1981. 22 S. / Gh (3-597-30025-1)
• 26: Paulig, Peter: Die Schulverwaltung: Funktionsanalyse, Reformnotwendigkeit und Kriterien eines Alternativmodells / 1981. 31 S. / Gh (3-597-30026-X)
• 27: Prokop, Ernst: Universität und Erwachsenenbildung / 1981. 29 S. / Gh (3-597-30027-8)
• 28: Bammesberger, Alfred: Englische Sprachwissenschaft. Ein Neuansatz in der Textkritik der altenglischen Dichtung / 1981. 17 S. / Gh (3-597-30028-6)
• 29: Krämer, Peter: Kirche der freien Gefolgschaft. Kirchenrechtliche Überlegungen zu einem umstrittenen Kirchenmodell / 1981. 18 S. / Gh (3-597-30029-4)
• 30: Maas-Ewerd, Theodor: Liturgische Einheit in Vielfalt. Die Einheit des Gottesdienstes der Kirche und der ausgeträumte Traum von einer „Welt-Einheitsliturgie" / 1981. 31 S. / Gh (3-597-30030-8)
• 31: Nossol, Alfons: Das Phänomen Kirche in Polen / 1982. 12 S. / Gh (3-597-30031-6)

- 32: Edmaier, Alois: Dient einander selbstlos in Wahrheit und Liebe / 1982. 24 S. / Gh (3-597-30032-4)
- 33: Grötzbach, Erwin: Das Hochgebirge als menschlicher Lebensraum / 1982. 26 S. / Gh (3-597-30033-2)
- 34: Rosen, Klaus: Über heidnisches und christliches Geschichtsdenken in der Spätantike / 1982. 29 S. / Gh (3-597-30034-0)
- 35: Sawicki, Stefan: Das Ethos der polnischen Literatur / 1982. 14 S. / Gh (3-597-30035-9)
- 36: Krings, Hermann: Die Wissenschaft und ihre Kritiker / 1983. 17 S. / Gh (3-597-30036-7)
- 37: Gärtner, Hans: Lesenlernen durch Faszination? Auf dem Wege zu einem psychoanalytisch und psycholinguistisch begründeten Leselernkonzept / 1983. 31 S. / Gh (3-597-30037-5)
- 38: Gnilka, Joachim: Jesu ipsissima mors. Der Tod Jesu im Lichte seiner Martyriumsparänese / 1983. 16 S. / Gh (3-597-30038-3)
- 39: Unverricht, Hubert: Kammermusik im 20. Jahrhundert. Zum Bedeutungswandel des Begriffs / 1983. 28 S. / Br (3-597-30039-1)
- 40: Zottl, Anton: Existenz in der Alternative. Ein pastoraltheologisches Gespräch mit Jean Jacques Rousseau / 1983. 35 S. / Gh (3-597-30040-5)
- 41: Kohut, Karl: Die spanische und lateinamerikanische Literatur im französischen Exil / 1984. 36 S. / Gh (3-597-30041-3)
- 42: Geiger, Willi: Wissenschaftsfreiheit als Problem der politischen Ordnung / 1984. 23 S. / Gh (3-597-30042-1)
- 43: Bucher, Alexius J.: Warum sollen wir gut sein? Zur Möglichkeit einer vernünftigen Letztbegründung sittlicher Normen / 1984. 24 S. / Gh (3-597-30043-X)
- 44: Wimmer, Ruprecht: Ferdinand Raimunds Zauberspiele / 1984. 29 S. / Gh (3-597-30044-8)
- 45: Jendrowiak, Hans W.: Benötigt die Pädagogik ein Weltbild? / 1985. 18 S. / Gh (3-597-30045-6)
- 46: Tschiedel, Hans J.: Lucan und die Tränen Caesars / 1985. 26 S. / Gh (3-597-30046-4)
- 47: Steindl, Michael: Wenn die Worte fehlen. In unterdeterminierten Texten von Ausländern und Deutschen / 1985. 28 S. / Gh (3-597-30047-2)
- 48: Clauss, Manfred: Gesellschaft und Staat in Juda und Israel / 1985. 32 S. / Gh (3-597-30048-0)
- 49: Steinbach, Josef: Zur Bewertung von Lebenschancen / 1986. 27 S. / Gh (3-597-30049-9)
- 50: Giessner, Klaus: Sahara „Die grosse Wüste" als Forschungsprojekt der Physiogeographie / 1988. 70 S., 19 Abb. / Gh (3-597-30050-2)
- 51: Kimura, Naoji: Das Christentum als sprachliches Problem in Japan / 1986. 21 S. / Gh (3-597-30051-0)
- 52: Plaum, Ernst: Psychologie der Werte - antiquiert oder aktuell? Beiträge zur Entwicklung einer objektiven Hierarchie der Zielvorstellungen und Sinnbezüge / 1986. 29 S. / Gh (3-597-30052-9)
- 53: Ballestrem, Karl: Macht und Moral. Ein Grundproblem der politischen Ethik / 1986. 18 S. / Gh (3-597-30053-7)
- 54: Kreuzer, Karl J.: Die Pädagogik im Spannungsfeld der Kultur / 1988. 23 S. / Gh (3-597-30054-5)
- 55: Wilke, Jürgen: Massenmedien und sozialer Wandel / 1986. 23 S. / Gb (3-597-30055-3)
- 56: Spölgen, Johannes: Jesus Christus glauben lernen. Zur Diskussion um eine Didaktik des Glaubens / 1987. 21 S. / Gh (3-597-30056-1)
- 57: Gössmann, Elisabeth: Wie könnte Frauenforschung im Rahmen der Katholischen Kirche aussehen? / 1987. 25 S. / Gh (3-597-30057-X)
- 58: Röller, Wolfgang: Die Auslandsverschuldung Latein-Amerikas. Ihre Ursachen, Probleme und

mögliche Lösungen / 1987. 19 S. / Gh (3-597-30058-8)
- 59: Deininger, Jürgen: Die antike Welt in der Sicht Max Webers / 1987. 26 S. / Gh (3-597-30059-6)
- 60: Bubb, Heiner: Energie, Entropie, Ergonomie. Eine arbeitswissenschaftliche Betrachtung / 1987. 35 S. / Gh (3-597-30060-X)
- 61: Ratzinger, Joseph: Abbruch und Aufbruch. Die Antwort des Glaubens auf die Krise der Werte / 1988. 19 S. / Gh (3-597-30061-8)
- 62: Felix, Rainer: Geräusch, Klang, Musik - Ein spektraltheoretischer Zugang / 1988. 18 S., 13 Abb. / (3-597-30062-6)
- 63: Manzanera, Miguel: Lateinamerikanische kirchliche Basisgemeinschaften. Eine Herausforderung für die Kirche / 1988. 32 S. / Gh (3-597-30063-4)
- 64: Betten, Anne: Lancelot-Roman, Luther-Bibel, Lessing-Dramen. Beispiele neuer sprachhistorischer Arbeitsweisen / 1988. 26 S. / Gh (3-597-30064-2)
- 65: Stolte, Dieter: Die Rolle der Medien in einer freiheitlichen Gesellschaft / 1988. 19 S. / Gh (3-597-30065-0)
- 66: Gross, Engelbert: Der Anspruch der Öffentlichkeit und das „Geheimnis des Glaubens". Erörterungen zu einem Dilemma des Religionsunterrichts in der öffentlichen Schule / 1989. 32 S. / Gh (3-597-30066-9)
- 67: Wickert, Ulrich: Einheit in der Vielfalt. Die Rekapitulation der Kirchengeschichte im Mysterium Marianum / 1989. 30 S. / Gh (3-597-30067-7)
- 68: Pylak, Boleslaw: Maria und das Geheimnis der Eucharistie / 1989. 19 S. / Gh (3-597-30068-5)
- 69: Greca, Rainer: Können soziale Organisationen sinnvoll handeln? / 1989. XXXX, 34 S. / Gh (3-597-30069-3)
- 70: Stöckl, Günther: 1000 Jahre christliches Russland und wir / 1989. 24 S. / Gh (3-597-30070-7)
- 71: Fell, Margret: Katholische Erwachsenenbildung zwischen gesellschaftspolitischer Verantwortung und kirchlichem Interesse / 1989. 24 S. / Gh (3-597-30071-5)
- 72: Becher, Ursula A.: Der implizierte Leser der Historiographie. Zur didaktischen Dimension der Geschichtswissenschaft / 1989. 23 S. / Gh (3-597-30072-3)

Eid, Mohammad Salahuddin: Die blockfreien Staaten in den Vereinten Nationen / 1970. 267 S. (3-7940-4015-5)

Einführung in die Kommunikationswissenschaft. Der Prozess der politischen Meinungs- und Willensbildung. Ein Kurs im Medienverbund Loseblattausg. Erarb. v. einer Projektgruppe am Institut f. Politikwissenschaft d. Universität München. Redakt.: Walter Hömberg / 2 Tle. Teil 1: (3-7940-3231-4). Teil 2: (3-7940-3234-9) / 2 Bde. 1976. 472 S. (3-7940-3231-4 / 3-7940-3234-9)
- 2., aktualisierte und erweiterte Aufl. 1981. 510 S. (3-598-10410-3)
- 3. verb. Aufl. 1983. Zus. 510 S. (3-598-10519-3)

Der **Einsatz** von Kleincomputern in Bibliotheken unter Berücksichtigung von Verbundsystemen. Bericht e. Syposiums, veranst. vom Deutschen Bibliotheksinstitut Berlin u. d. Gesamthochschule Essen am 9. / 10. Okt. 1978. Dt. Bibliotheksinst. Berlin; Gesamthochschulbibliothek Essen. Red. Detlef Schwarz / 1979. 161 S. (3-598-10031-0)

Einzelveröffentlichungen der Historischen Kommission zu Berlin. Hrsg.: Historische Kommission zu

Berlin [Vorher im Colloquium Vlg., Berlin]
- 76: Wirtschaft im geteilten Berlin 1945-90. Forschungsansätze und Zeitzeugen. Hrsg. v. Wolfram Fischer; Johannes Bähr / 1993. XII, 369 S. / Gb (3-598-23220-9)
- 77: Scarpa, Ludovica: Gemeinwohl und lokale Macht. Honoratioren und Armenwesen in der Berliner Luisenstadt im 19. Jahrhundert / 1995. XII, 389 S. / Gb (3-598-23221-7)
- 78: Sadmon, Zeev W.: Die Gründung des Technions in Haifa im Lichte deutscher Politik 1907-1920 / 1994. XVI, 312 S. / Gb (3-598-23222-5)
- 79: Ratz, Ursula: Zwischen Arbeitsgemeinschaft und Koalition. Bürgerliche Sozialreformer und Gewerkschaften im Ersten Weltkrieg / 1994. XII, 575 S. / Gb (3-598-23223-3)
- 80: Vom Generalplan Ost zum Generalsiedlungsplan. Dokumente. Hrsg. v. Czeslaw Madajczyk, unter Mitarb. v. Stanislaw Biernacki; Karin Borck; Hans H. Hahn; Eligiusz Janus; Blanka Meissner; Michael G. Müller / 1994. XXXVI, 576 S. / Gb (3-598-23224-1)
- 81: Strenge, Barbara: Juden im preußischen Justizdienst 1812-1918. Der Zugang zu den juristischen Berufen als Indikator der gesellschaftlichen Emanzipation / 1995. XIV, 393 S. / Gb (3-598-23225-X)
- 82: Die Juden und die jüdischen Gemeinden Preußens in amtlichen Enquêten des Vormärz. Enquête des Ministeriums des Inneren und der Polizei über die Rechtsverhältnisse der Juden in den preußischen Provinzen 1842 - 1843. Enquête des Ministeriums der geistlichen, Unterrichts- und Medizinal-Angelegenheiten überdie Kultus-, Schul- und Rechtsverhältnisse der jüdischen Gemeinden in den preußischen Provinzen 1843 - 45. Bearb. u. Hrsg. v. Manfred Jehle. Mit e. Beitr. v. Herbert A. Strauss / 4 Bde. 1998. Zus. XCIII, 1.672 S. / Gb (3-598-23226-8)

Die **Emigration** der Wissenschaften nach 1933. Disziplingeschichtliche Studien. Hrsg. v. Herbert A. Strauss, Klaus Fischer, Christhard Hoffmann und Alfons Söllner / 1991. 282 S. / Gb (3-598-11044-8)

Empirische Sozialforschung / Empirical Social Research. Eine Dokumentation / An Inventory. Hrsg.: Zentralarchiv für Empirische Sozialforschung der Universität Köln, in Zus.-Arbeit mit d. Informationszentrum Sozialwissenschaften (IZ)
- Bd. 1968: Von Thomas Aage Herz; Hagen Stegemann. Unter Mitarb. von Michael Bauer u.a / 1969. XXVIII, 441 S. / Lin (3-7940-3390-6)
- Bd. 1969: Von Thomas Aage Herz; Hagen Stegemann. Unter Mitarb. von Sylvia Witte / 1970. XXIII, 442 S. / Lin (3-7940-3392-2)
- Bd. 1970: Von Thomas Aage Herz; Hagen Stegemann. Unter Mitarb. von Sylvia Witte / 1971. XIX, 538 S. / Lin (3-7940-3400-7)
- Bd. 1971: Von Thomas Aage Herz; Hagen Stegemann; Sylvia Witte / 1972. XXIII, 486 S., 5 S., 4 Tab. / Lin (3-7940-3437-6)
- Bd. 1972: Von Thomas Aage Herz; Hagen Stegemann; Sylvia Witte / 1973. XIX, 562 S. / Lin (3-7940-3438-4)
- Bd. 1973: Von Thomas Aage Herz; Hagen Stegemann; Sylvia Witte / 1974. XXIII, 578 S. / Lin (3-7940-3449-2)
- Bd. 1974: Von Thomas Aage Herz; Hagen Stegemann; Sylvia Witte / 1975. XXII, 482 S. / Lin (3-7940-3443-0)
- Bd. 1975: Von Thomas A. Herz; Hagen Stegemann / 1976. XXI, 644 S. / Lin (3-7940-3444-9)

- Bd. 1976: Von Thomas A. Herz; Hagen Stegemann / 1977. XXI, 656 S. / Lin (3-7940-3445-7)
- Bd. 1977: Von Karl Heinz Reuband; Hagen Stegemann / 1978. XVII, 922 S. / Lin (3-7940-3446-5)
- Bd. 1978: Von Karl Heinz Reuband; Hagen Stegemann / 1979. XXIX, 728 S. / Lin (3-598-20978-9)
- Bd. 1979: Von Gerhard Held; Karl Heinz Reuband / 1980. XIX, 702 S. / Lin (3-598-20979-7)
- Bd. 1980: Von Gerhard Held; Karl Heinz Reuband / 1981. XXXI, 668 S. / Lin (3-598-20980-0)
- Bd. 1981:Von Gerhard Held; Karl Heinz Reuband / 1982 XXXI, 610 S. / Lin (3-598-20981-9)

Encyclopedia of Film Directors in the United States of America and Europe.
- Vol. 1. Comedy Films to 1991. Ed. by Alfred Krautz, in cooperation with Hille Krautz; Joris Krautz. Comp. with the assistance of Eberhard Spiess. Preface by Siodmak, Curt / 1993. 408 p. / Hard (3-598-21501-0)
- Vol. 2. Crime Films to 1995. Ed.by Alfred Krautz, in cooperation with Hille Krautz; Joris Krautz. Comp. with the assistance of Eberhard Spiess / 1997. 314 p. / Hard (3-598-21502-9)

Encyclopedia of World Problems and Human Potential. Ed.: Union of International Associations / 2nd ed. 1986. 1,400 p. / Hard (3-598-21864-8)
- 3rd ed. 2 vols. 1991. Cplt. 2136 p. / Hard (3-598-10842-7)
- 4th ed. 3 vols. 1994 / 1995 / Hard (3-598-11165-7)
  - Vol. 1. World Problems / 1994. 1264 p. / Hard (3-598-11225-4)
  - Vol. 2. Human Potential-Transformation and Values / 1994. 931 p. / Hard (3-598-11226-2)
  - Vol. 3 Actions-Strategies-Solutions / 1995. 976 p. / Hard (3-598-11227-0)

Encyclopedia of World Problems and Human Potential - Encyclopedia PLUS. CD-ROM-Edition. Ed.: Union of International Associations / 1995 (3-598-40294-5)

Encyclopedic Dictionary of Electronic, Electronical Engineering and Information Processing / Enzyklopädisches Wörterbuch der Elektronik, Elektrotechnik und Informationsverarbeitung. English-German / German-English / Englisch-Deutsch / Deutsch-Englisch. Comp. by Michael Peschke / 8 vols. 1990-1997. Cplt. CXXI, 3,148 p. / Hard (3-598-10680-7)
- Vol. 01. 1990. XV, 431 p. / Gb (3-598-10681-5)
- Vol. 02. 1993. XVI, 402 p. / Gb (3-598-10843-5)
- Vol. 03. 1996. XV, 396 p. / Gb (3-598-11122-3)
- Vol. 04. 1996. XV, 381 p. / Gb (3-598-10845-1)
- Vol. 05. 1997. XV, 373 p. / Gb (3-598-10846-X)
- Vol. 06. 1997. XV, 407 p. / Gb (3-598-10847-8)
- Vol. 07. 1997. XV, 387 p. / Gb (3-598-10848-6)
- Vol. 08. 1997. XV, 371 p. / Gb (3-598-10849-4)

Energie und Gerechtigkeit. Hrsg. v. Reiner Kümmel und Monika Suhrcke / 1984. XIV, 255 S. / Br (3-597-10259-X)

Englert, Ludwig / Otto Mair; Siegfried Mursch: Georg Kerschensteiner. Bibliographie / Bibliography / 1976. 164 S. / Br (3-7940-3276-4)

English Bibliography / Englische Bibliographie / Bibliographie anglaise / Bibliografia Generale Inglese / Bibliografi General Inglesa . CD-ROM-Edition / 1998
- 15th Century to 1901 / 15. Jahrhundert - 1901 / du XVe siècle à 1901 / dal XV secolo al 1901 / Siglo XV - 1901. CD-ROM-Edition / 4 discs (3-598-40374-7)
- 1901 to 1945. CD-ROM-Edition / 2 discs (3-598-40375-5)

Enzyklopädische Information im 19. Jahrhundert. Die Ergänzungswerke zum Brockhaus Konversationslexikon: „Zeitgenossen“ - „Die Gegenwart“ - „Unsere Zeit“ mit einem Gesamtregister, herausgegeben von Otmar Seemann. Mikrofiche-Edition / 1985-1995 / Silberfiche (3-598-32295-X) / Diazofiche (3-598-32294-1)
- Die Gegenwart, Leipzig 1848-1856 Eine enzyklopädische Darstellung der neuesten Zeitgeschichte für alle Stände. Verlag: F. A. Brockhaus. Mikrofiche-Edition / 1985. 12 Bde. ca. 9.737 S. auf 26 Fiches. Lesefaktor 42x / Diazofiche (3-598-30671-7)
- Gesamtindex „Zeitgenossen“ - „Die Gegenwart“ - „Unsere Zeit“ Die Ergänzungswerke zum Brockhaus Konversationslexikon: „Zeitgenossen“ - „Unsere Zeit“ - „Gegenwart“ mit einem Gesamtindex. Hrsg. u. Einl. v. Otmar Seemann / 1995. 344 S. / Br (3-598-23610-7)
- Unsere Zeit, Leipzig 1857-1891 Jahrbuch zum Conversations-Lexikon 1857-64, Deutsche Revue der Gegenwart. Monatsschrift zum Conversations-Lexikon. Neue Folge 1865-74, Deutsche Revueder Gegenwart. Neue Folge 1875-79, Deutsche Revue der Gegenwart 1880-91. Verlag: F. A. Brockhaus. Mikrofiche-Edition / 1994. Ca. 52.000 S. auf 440 Fiches. Lesefaktor 24x / Silberfiche (3-598-32293-3) / Diazofiche (3-598-32292-5)
- Zeitgenossen, Leipzig und Altenburg 1816-1841 Biographien und Charakteristiken 1816-1828. Ein biographisches Magazin fürdie Geschichte unserer Zeit 1829-1841. Verlag: F. A. Brockhaus. Mikrofiche-Edition / 1994. 14.308 S. auf 157 Fiches. Lesefaktor 24x / Silberfiche (3-598-32291-7) / Diazofiche (3-598-32290-9)

Enzyklopädisches Wörterbuch Kartographie in 25 Sprachen / Encyclopedic Dictionary of Cartography in 25 Languages. Definitionen in Deutsch, Englisch, Französisch, Spanisch, Russisch - mit Äquivalenten in algerischem und marokkanischem Arabisch, Bulgarisch, Dänisch, Chinesisch, Finnisch, Hindi, Italienisch, traditionellem und transliteriertem Japanisch, Niederländisch, Norwegisch, Polnisch, Portugiesisch, Schwedisch, Slowakisch, Thai, Tschechisch und Ungarisch / Definitions in German, English, French, Spanish and Russian with Equivalent Terms in Algerian and Moroccan Arabic, Bulgarian, Danish, Chinese, Finnish, Hindi, Italian, Classical and transliterated Japanese, Dutch, Norwegian, Polish, Portuguese, Swedish, Slovakian, Thai, Czech and Hungarian. Hrsg. v. J. Neumann / 2. Aufl. 1997. 586 S. / Gb (3-598-10764-1)

Ersch, Johann S. / Johann G. Gruber: Allgemeine Encyclopädie der Wissenschaften und Künste Erste Section, Band 1-99, Leipzig 1818-1882. Zweite Section, Band 1-43, Leipzig 1827-1889. Dritte Section, Band 1-25, Leipzig 1830-1850. Mikrofiche-Edition / 1996. 81.613 S., 300 Ktn. u. Abb. auf 310 Fiches, Lesefaktor 24x / Silberfiche (3-598-33511-3)

Esbeck, Friedrich J. von: Reden. Ein Beitrag zur deutschen Kultur- und Theatergeschichte. - Nachdr. d. Ausg. Leipzig 1881. - Nachw. u. Erg.-Bibliogr. v. Wolfram Günter / 1985. 358, 34 S., 2 Illus. / Lin

(3-598-07232-5) [Lizenzausg. m. Genehm. v. Zentralantiquariat der DDR, Leipzig]

Estermann, Alfred: Die deutschen Literatur-Zeitschriften 1815-1850. Bibliographien - Programme - Autoren / 11 Bde. 2. überarb. u. verb. Aufl. 1991. Zus. 5.180 S. / Ln (3-598-10723-4)

Estermann, Alfred: Die deutschen Literatur-Zeitschriften 1850-1880. Bibliographien - Programme / 5 Bde. mit Reg. 1987 / 1989. Zus. 3.013 S. / Ln (3-598-10708-0)

Estermann, Alfred: Inhaltsanalytische Bibliographien deutscher Kulturzeitschriften des 19. Jahrhunderts - IBDK / 10 Bde. in 16 Tlbdn 1995/96 / Ln (3-598-23310-8)
- Bd. 1. Deutsches Museum (1851-1867) A-Z / 2 Bde. 1995. Zus. X, 606 S. / Ln (3-598-23311-6)
- Bd. 2. Telegraph für Deutschland (1837-1848) / 1995. VI, 378 S. / Ln (3-598-23312-4)
- Bd. 3. Die Gartenlaube (1853-1880 (-1944)) A-Z / 2 Bde. 1995. Zus. X, 660 S. / Ln (3-598-23313-2)
- Bd. 4. Berliner Conversationsblatt (1827-1829) / Der Freihafen (1838-1944) / Hallische Jahrbücher (1838-1844) / Königsberger Literatur-Blatt (1841-1845) / 1995. VI, 408 S. / Ln (3-598-23314-0)
- Bd. 5. Phoenix (1835-1838) / Frankfurter Museum (1855-1859) / Neues Frankfurter Museum (1861) / Weimarer Sonntags-Blatt (1855-1857) / 1995. VI, 557 S. / Ln (3-598-23315-9)
- Bd. 6. Deutsche Roman-Zeitung (1864-1880 (-1925)) / 1995. VI, 592 S. / Ln (3-598-23316-7)
- Bd. 7. Das Jahrhundert (1856-1859) / Deutsches Magazin (1861-1863) / Freya (1861-1867) / Orion (1863-1864) / Deutsche Warte (1871-1875) / Der Salon (1868-1890) / 1995. VI, 566 S. / Ln (3-598-23317-5)
- Bd. 8. Westermanns Monatshefte (1856-1880 (-1986)) / 1995. VI, 446 S. / Ln (3-598-23318-3)
- Bd. 9. Blätter für literarische Unterhaltung (1826-1850 (-1898)) A-Z / 5 Bde. 1996. Zus. XXII, 1.868 S. / Ln (3-598-23319-1)
- Bd. 10. Gesamtregister / 1996. VII, 216 S. / Ln (3-598-23320-5)

Estermann, Alfred: Kontextverarbeitung. Buchwissenschaftliche Studien. Hrsg. v. Klaus D. Lehmann und Klaus Gerhard Saur, in Verb. mit d. Stadt- u. Universitätsbibliothek Frankfurt am Main / 1998. 486 S. / Ln (3-598-11371-4)

EUDISED. European Educational Research Yearbook. Project Reports - People - Contacts. Ed. by Council of Europe, Directorate of Education, Culture and Sports (ISSN 0947-5826)
- 1993 / 1995. XVIII, 701 p. / Hard (3-598-23460-0)
- 1994/95 / 1996. XVII, 516 p. / Hard (3-598-23461-9)
- 1995/96 / 1996. XIX, 494 p. / Hard (3-598-23462-7)
- 1996/97 / 1997. XIX, 480 p. / Hard (3-598-23463-5)
- 1997/98 / 1998. XIX, 442 p. / Hard (3-598-23464-3)

EUDISED R & D Bulletin. Ed.: Council of Europe. Directorate of Education, Culture and Sport / Vol. 25-28(1985)/1986 - Vol. 45-48(1992)/1993-1994 (4 issues and 1 Cumulative Index-Vol. annually) (ISSN 0378-7192)

Europäischer Kongreß über Dokumentationssysteme und –netze. Hrsg.: Kommission der Europäischen Gemeinschaften, Generaldirektion „Wissenschaftliche und Technische Information und Informationsmanagement“

- (1.) Erste Europäische Tagung über Dokumentationssysteme und -netze. Luxemburg, 16., 17. und 18. Mai 1973 / 1974. 403 S. / Br (3-7940-5146-7)
- First European Congress on Documentation Systems and Networks / 1974. 397 S. / Br (3-7940-5147-5)
- Premier Congres Europeen sur les Systemes et Eseaux Documentaires / 1974. 406 S. / Br (3-7940-5148-3)
- (2.) Zweiter Europäischer Kongreß über Dokumentationssysteme und -netze, Luxemburg, 27.-30. Mai 1975 / 1976. 272 S. / Br (3-7940-5164-5)
- Second European Congress on Information Systems and Networks / 1976. 232 S. / Br (3-7940-5164-5)
- Deuxième congrès européen sur les systèmes et réseaux documentaires / 1976. 255 S. / Br (3-7940-5165-3)
- (3.) Dritter Europäischer Kongreß über Dokumentationssysteme und -netze. Die Überwindung der Spachbarrieren. Luxemburg, 3. - 6. Mai 1977 / 2 Bde. 1977. 610 S; 291 S. (3-7940-5183-1)
- Third European Congress on Information Systems and Network. Overcoming the language barrier / 1977. 675 S., 213 S. / Br (3-7940-5184-X)
- Troisième congrès européen sur les systèmes et réseaux documentaires. Franchir la barrière linguistique / 1977. 730 S., 180 S. / Br (3-7940-5185-8)

Europäisches Bibliotheks- Adressbuch. Bearb. v. Klaus Gerhard Saur. Sonderdr. d. 2. Ausgabe 1968 des Internationalen Bibliotheks-Handbuches (Handbuch d. technischen Dokumentation u. Bibliographie, Bd. 8, 2. Ausg., Tl. 1) / 1968. XXX, 861 S. / Lin (Vlg.-Nr. 02238)

European Music Directory. Music Industry, Music Publishers, Agencies, Associations and Foundations, Radio and Television, Orchestras, Teaching and Instruction, Competitions and Prizes, Festivals / 2 vols. 3rd. new rev. ed. of Stadler's Musikhandbuch / 1998. XII, 858 p. / Hard (3-598-11347-1)

Fabian, Ruth / Corinna Coulmas: Die deutsche Emigration in Frankreich nach 1933. / 1979. 136 S. / Br (3-598-07076-4)

Fachbegriffe und Sinnbilder der Datenverarbeitung. Hrsg.: Institut für Datenverarbeitung, Dresden / 1968. 196 S. / Br (Vlg.-Nr. 04178)
- 2. Aufl. 1976. 166 S. (3-7940-4179-8)

Fachkatalog Afrika / Subject Catalog Africa / Catalogue matières Afrique. Hrsg.: Stadt- und Universitätsbibliothek Frankfurt a. M. Bearb. v. Irmtraud D. Wolcke-Renk.
- Bd. 1/1. Geschichte I / History I / Histoire I. Stand Juni 1986 / 2. überarb. u. erw. Aufl. 1987. IX, 349 S., 7 S. Reg / Lin (3-598-20921-5)
- Bd. 1/2. Geschichte II / History II / Histoire II. Stand Juni 1986 / 2. überarb. u. erw. Aufl. 1987. V, 416 S., 9 S. Reg / Lin (3-598-20922-3)
- Bd. 2. Politik / Politics / Politique. Stand: September 1977 / 1979. X, 269 S. / Lin (3-598-02852-0)
- Bd. 3. Literatur, Literaturwissenschaft / Literature / Litterature Lettres. Stand: Dezember 1977 / 1979. X, 358 S. / Lin (3-598-02853-9)
- Bd. 4. Kulturanthropologie (Völkerkunde, Kulturgeschichte, Religion, Kunst) / Social and cultural anthropology / Anthropologie social et culturelle. Stand: Dezember 1978 / 1980. VIII, 589 S. / Lin (3-598-20924-X)

- Bd. 5. Geowissenschaften, Sozialwissenschaften / Geography, Social Sciences / Géographie, Sciences Sociales. Stand: Dezember 1980 / 1981. X, 579 S. / Lin (3-598-20925-8)
- Bd. 6. Sprachen, Linguistik / Languages, Linguistics / Langues, Linguistique. Stand Dezember 1981 / 1982. X, 460 S. / Lin (3-598-20926-6)
- Bd. 7. Buch- und Bibliothekswesen, Erziehung, Wissenschaft / Library Science, Education, Science / Bibliotheconomie, Education, Sciences. Stand Dezember 1982 / 1983. X, 414 S. / Lin (3-598-20927-4)
- Bd. 10: Ostafrika / East Africa / Afrique de l'Est. Stand Dezember 1983 / 2 Bde. 1984. Zus. XVI, 1.343 S., 25 S. / Lin (3-598-20929-0)
  - Bd. 10/1: Ostafrika I (Kenya) / East Africa I (Kenya) / Afrique de l'Est I (Kenya) / X, 712, 4 S.
  - Bd. 10/2 Ostafrika II (Tanzania Uganda) / East Africa II (Tanzania Uganda) / Afrique de l'Est II (Tanzanie, Ouganda) / VI, 631, 21 S.
- Bd. 11. Indischer Ozean / 1988. 300 S. / Lin (3-598-20931-2)
- Bd. 12/2. Westafrika II (Binnenstaaten) / West Africa II (Land Locked Countries) / Afrique de l'Ouest II (Pays Intérieurs). Stand Dezember 1990 / 1992. IX, 259 S. / Lin (3-598-20933-9)
- Bd. 12/3. Westafrika III (Nigeria) / West Africa III (Nigeria) / Afrique de l'Ouest III (Nigeria). Stand Oktober 1994 / 1995. IX, 414 S. / Lin (3-598-20934-7)

Fachkatalog Film / Subject Catalog Film. Hrsg.: Stadt- und Universitätsbibliothek Frankfurt am Main.
- Bd. 1. Literatur zu Personen. Bestandsverzeichnis bis Juli 1981 / Literature on Persons. Holdings as of Juli 1981. Bearb. v. Norbert Ruecker, Norbert und Thomas Siedhoff / 1982. XII, 385 S. / Lin (3-598-10414-6)

Faust-Bibliographie. Bearb. v. Hans Henning. Hrsg.: Nationale Forschungs- und Gedenkstätten der klassischen deutschen Literatur in Weimar [Lizenzausg. m. Genehm. v. Aufbau Vlg., Berlin]
- Tl. I: Allgemeines, Grundlagen, Gesamtdarstellungen. Das Faust-Thema vom 16. Jahrhundert bis 1790 / 1966. XVIII, 512 S. / Ln (3-7940-5036-3)
- Tl. II: Goethes Faust
- Bd. 1: Ausgaben und Übersetzungen / 1968. IX, 233 S. / Ln (3-7940-5037-1)
- Bd. 2 / Tlbd. 1: Sekundärliteratur zu Goethes Faust / 1968. VII, 320 S. / Ln (3-7940-5038-X)
- Bd. 2 / Tlbd. 2: Sekundärliteratur zu Goethes Faust / 1968. 320 S. / Ln (3-7940-5039-8)
- Tl. III: Faust neben und nach Goethe / 1976. IX, 472 S. / Ln (3-7940-5040-1)

Fenlon, Iain: Catalogue of the printed music and music manuscripts before 1801 in the Music Library of the University of Birmingham, Barber Institute of Fine Arts / 1976. XIII, 140 S. (3-7940-7001-1)

Fernsehen und Bildung. Internationale Zeitschrift für Medienpsychologie und Medienpraxis. Hrsg. f. d. Internationale Zentralinstitut für das Jugend und Bildungsfernsehen (IZI) in München v. Helmut Oeller; Hertha Sturm; Redakt.: Marianne Grewe-Partsch / Jg. 9/1975 bis Jg. 16/1982 (2 Hefte jährlich) (ISSN 0015-0150)

Fertig, Eymar: Forschungsobjekt Buch. Internat. Bibliographie zur Soziologie u. Psychologie des Lesens. Hrsg. v. Heinz Steinberg / 3. Aufl. 1971. 230 S. (3-7940-3192-X)

Feuereisen, Fritz / Ernst Schmacke: Die Presse in Afrika. Ein Handbuch für Wirtschaft und Werbung / 1968. 254 S. (Vlg.-Nr. 03058)
- 2., überarb. Ausg. Die Presse in Afrika / The Press in Africa. Ein Handbuch für Wirtschaft und Werbung / A Handbook for Economics and Advertising / 1973. 279 S. / Lin (3-7940-3058-3)

Feuereisen, Fritz / Ernst Schmacke: Die Presse in Asien und Ozeanien. Ein Handbuch für Wirtschaft und Werbung / 1968. 303 S.
- 2. Aufl. 1973. 376 S. (3-7940-3078-8)

Feuereisen, Fritz / Ernst Schmacke: Die Presse in Lateinamerika. Ein Handbuch für Wirtschaft u. Werbung / 1968. 272 S.
- 2. Aufl. 1973. 268 S. (3-7940-3068-0)

Feuereisen, Fritz: Die Presse in Europa / The Press in Europe. Ein Handbuch für Wirtschaft u. Werbung / 1971. 328 S. (3-7940-3299-3)

Film, Television, Sound Archives Series. Film - Television - Sound Archives Series. Papers and Reference Tools for Film Archivists. Dealing with Audiovisual Material. Ed. by Wolfgang Klaue; Eva Orbanz. Ed. Board: Michelle Aubert; Harriet W. Harrisson; Henning Schou.
- 1: The FIAF Cataloguing Rules for Film Archives Comp. and ed. for the FIAF Cataloguing. Commission by Harriet W. Harrisson / 1991. XI, 239 p. / Hard (3-598-22590-3)
- 2: Posters of GDR-Films 1945-1990. Ed. by Babett Stach; Helmut Morsbach / 1991. 149 p. / Hard (3-598-22591-1)
- 3: German Film Posters 1895-1945. Comp. and ed. by Babett Stach; Helmut Morsbach / 1992. 152 p. / Hard (3-598-22593-8)
- 4: Terms and Methods for Technical Archiving of Audiovisual Materials / Termes et méthodes d'archivage technique de documents audiovisuels / Sachwörter und Methoden für die technische Archivierung audiovisueller Materialien / Terminologia y méthodos para a English - French - German - Spanish - Russian. Comp. and ed. by Günter Schulz for the FIAF Cataloguing Commission / 1992. 88 p. / Hard (3-598-22592-X)
- 5: World Directory of Moving Image and Sound Archives. Comp. and ed. by Wolfgang Klaue / 1993. 192 p. / Hard (3-598-22594-6)

Film Semiotik / Semiotics of Films / Sémiologie du Cinéma Eine Bibliographie. A Bibliography / Une Bibliographie. Hrsg. v. Achim Eschbach und Wendelin Rader / 1978. XXVIII, 203 S. / Lin (3-7940-7059-3)

Filmkultur zur Zeit der Weimarer Republik. Beiträge zu einer internationalen Konferenz vom 15. bis 18. Juni 1989 in Luxemburg. Hrsg. v. Uli Jung; Walter Schatzberg, im Auftr. v. Cinémathèque Municipale de Luxembourg; Clark University, Worcester, MA, USA; Goethe Institut München; Thomas-Mann-Bibliothek, Luxemburg / 1991. 322 S. / Gb (3-598-11042-1)

Fink, Gottfried W.: Wesen und Geschichte der Oper. Ein Handbuch für alle Freunde der Tonkunst. - Nachdr. d. Ausg. Leipzig, 1838 / 1982. 335 S. / Ln (3-598-07233-3)

The **First** Century of German Language Printing in the United States of America 1728-1830. A Bibliography Based on the Studies of Oswald Seidensticker and Wilbur H. Oda. Ed.: Niedersächsische Staats- und Universitätsbibliothek Göttingen. Karl J. Arndt, Reimer C. Eck. Rev. by Gerd J. Bötte and Werner Tannhof using a preliminary compilation by Annelies Müller / 2 vols. 1989. Cplt. XXIV, 1245 p. / Hard (3-598-07578-2)

Fischer, Engelbert: Die Grossmacht der Jugend- und Volksliteratur. Vom patriotischen, religiösen u. pädagogisch-didaktischen Standpunkte kritisch beleuchtet. - Nachdr. d. Ausg. Neustift am Walde bei Wien u. Stoizendorf bei Eggenburg 1877-1886 / 12 Bde. in 6 Bdn 1979. 4529 S. / Ln (3-598-07056-X)

Fischer, Erika J.: The Inauguration of „Oscar". Sketches and Documents from the Early Years of the Hollywood Academy of Motion Picture Arts and Sciences and the Academy Awards, 1927-1930 / 1988. 333 p., 12 photos and 33 drawings / Soft (3-598-10753-6)

Fischer, Konrad: Geschichte des Deutschen Volks- schullehrerstandes. - Nachdr. der Ausg. Hannover 1892 / 1972. 818 S. / Ln (3-7940-5027-4) [Lizenz- ausg. m. Genehm. v. Zentralantiquariat d. DDR. Leipzig]

Fischer, Paul: Goethe-Wortschatz. Ein sprachge- schichtliches Wörterbuch zu Goethes sämtlichen Werken. - Nachdr. d. Ausg. Leipzig 1929] / 1983. XI, 905 S. / Ln (3-598-07209-0)

Fjällbrant, Nancy / Malcolm Stevenson: User Education in Libraries. / 1978. 173 S. (0-85157-251-0) [Clive Bingley]

Fladt, Philipp Wilhelm Ludwig: Anleitung zur Registratur-Wissenschaft und von Registratoribus deren Amt und Pflichten: wobey zugl. d. dahin gehörige Nachricht von Canzley-Wesen u. sonstigen nöthigen Erfordernüßertheil wird. Nebst e. Erl. einiger hierin befindlichen Stellen. - Nachdr. d. Ausg. Franckfurt und Leipzig 1765 / 1975. 18, 254, 16 S. / Ln (3-7940-3015-X)

Flemig, Kurt: Karikaturisten-Lexikon / 1993. XIV, 325 S., 14 Abb. / Ln (3-598-10932-6)

Flemmer, Walter: Verlage in Bayern. Geschichte und Geschichten. Mit einem einführenden Kapitel über die Frühgeschichte des bayerischen Verlags- wesens v. Fritz Schmitt-Carl / 1974. 493 S. / Ln (3-7940-3418-X)

Flögel, Ute: Pressekonzentration im Stuttgarter Raum / 1971. 232 S., 5 Abb., 29 Tab., 2 Übers., 9 Ktn. / Br (3-7940-3399-X)

Flugschriftensammlung Gustav Freytag. Vollstän- dige Wiedergabe der 6265 Flugschriften aus dem 15.-17. Jahrhundert sowie des Katalogs von Paul Hohenemser auf Mikrofiche. Mikrofiche-Edition. Hrsg.: Stadt- und Universitätsbibliothek Frankfurt, Gustav-Freytag-Sammlung / 1980/1981. 6.265 Flugschriften auf 746 Fiches. Lesefaktor 42x / Silberfiche (3-598-21189-9) / Diazofiche (3-598-21190-2)

Föderalismus. Bauprinzip einer freiheitlichen Grundordnung in Europa. Hrsg. v. Karl Assmann u.

Thomas Goppel. Mit e. Beitr. von Alfons Goppel / 1978. 190 S. / Br (3-598-07088-4)

Fohrbeck, Karla / Andreas Johannes Wiesand: Bi- bliotheken und Bibliothekstantieme. Materialbericht und Erhebungen zu Bestand, Ausleihe und Entwick- lungstendenzen in den Bibliothekssystemen der BRD. E. Untersuchung d. Instituts für Projektstudien, Hamburg / 1974. 152 S., zahlr. Tab. / Br (3-7940-3043-5)

Fokus. Praxis Information und Kommunikation. Hrsg. v. Hans Meuer; Hans G. Kruse.
- 1: Rechtsstandpunkte in der Datenverarbeitung. Hrsg. v. Michael Bartsch / 1991. VIII, 116 S. / Gb (3-598-22400-1)
- 2: Schnell, Christoph: Grenzen des Computers / 1991 81 S., 3 Abb. / Gb (3-598-22404-4)
- 3: Supercomputer 1986-90. Anwendungen, Architekturen, Trends. Hrsg. v. Hans Meuer / 1992. X, 395 S., 194 Abb. u. 44 Tab. / Gb (3-598-22401-X)
- 4: Datenkommunikation 1986-91. Aspekte und Entwicklungen. Hrsg. v. Wolfgang Effelsberg / 1993. X, 449 S. / Gb (3-598-22402-8)
- 5: Verteilte Multimedia-Systeme. Hrsg. v. Wolfgang Effelsberg; Kurt Rothermel / 1993. VIII, 227 S. / Gb (3-598-22407-9)
- 6: Das wissenschaftliche Rechenzentrum der Zukunft. Hrsg. v. Helmut Weberpals / 1993. 84 S. / Gb (3-598-22408-7)
- 7: Basissoftware für Vektor- und Parallelrechner. Hrsg. v. Hans Meuer / 1993. 171 S. / Gb (3-598-22406-0)
- 8: Formale Methoden für verteilte Systeme. Hrsg. v. Hartmut König / 1993. 178 S. / Gb (3-598-22409-5)
- 9: Haan, Oswald: Vektorrechnung Architektur - Programmierung - Anwendung / 1993. VIII, 168 S. / Gb (3-598-22405-2)
- 10: Facing the New World of Information Technology Proceedings of the 6th International Siemens Nixdorf IT User Conference June 8-10, 1994 in Copenhagen. Ed. by Hans Meuer / 1994. XIV, 691 p. / Hard (3-598-22410-9)
- 11: Supercomputer 1994. Anwendungen, Architekturen, Trends. Hrsg. v. Hans Meuer / 1994. 240 S. / Gb (3-598-22411-7)
- 12: DV-Organisation. Hrsg. v. Werner Dirlewanger / 1995. 229 S. / Gb (3-598-22403-6)
- 13: Supercomputer 1995. Anwendungen, Architekturen, Trends. Hrsg. v. Hans Meuer / 1995. VIII, 271 S. / Gb (3-598-22412-5)
- 14: Supercomputer 1996. Anwendungen, Architekturen, Trends. Hrsg. v. Hans Meuer / 1996. VIII, 246 S. / Gb (3-598-22413-3)
- 15: Supercomputer 1997. Anwendungen, Architekturen, Trends. Hrsg. v. Hans Meuer / 1997. IV, 191 S. / Gb (3-598-22414-1)
- 16: Supercomputer 1998. Anwendungen, Architekturen, Trends. Hrsg. v. Hans Meuer / 1998. 217 S. / Gb (3-598-22415-X)

Forschungsarbeiten in den Sozialwissenschaften. Dokumentation. Hrsg.: Informationszentrum für Sozialwissenschaftliche Forschung.
- Forschungsarbeiten 1969. Red.: Ernst August Jüres / 1971. 384 S. / Br (3-7940-3387-6)
- Forschungsarbeiten 1970. Red.: Matthias Herfurth [u.a.] Arbeitskreis für die Dokumentation Sozialwissenschaftlicher Forschung / 1971. 493 S. / Br (3-7940-3440-6)
- Forschungsarbeiten 1971. Red. u. Vorw. Matthias Herfurth; Ernst August Jüres; Peter Tewes / 1972. XV, 364n 4 S. / Br (3-7940-3441-4)

- Forschungsarbeiten 1972. Bearb. von Matthias Herfurth; Klaus Moske; Rolf Schmauch / 1973. XIV, 576 S. / Br (3-7940-3442-2)

Forsyth, Michael: Bauwerke für Musik. Konzertsäle und Opernhäuser, Musik und Zuhörer vom 17. Jahrhundert bis zurGegenwart. Übers. v. Michael Dickreiter; Regine Dickreiter / 1992. 374 S., 262 Abb. / Gb (3-598-11029-4)

Fortegnelse over ikke-statslige landbrugsorganisa- tioner inden for de Europaeiske Faellesskaber. Ver- zeichnis der im Rahmen der Europäischen Gemein- schaften zusammengeschlossenen land- und ernäh- rungswirtschaftlichen Verbände. Hrsg.: Kommission der Europäischen Gemeinschaften / 5. Ausg. 1977. 464 S. / Br (3-7940-3021-4)

Fothergill, Richard / Ian Butchard: Non-Book Materials in Libraries. A Practical Guide. / 1978. 256 S. (0-85157-253-7) [Clive Bingley]

France – Allemagne / Deutschland – Frankreich. Relations internationales et interdépendances bilatérales / Internationale Beziehungen und gegen- seitige Verflechtung. Ed. par Deutsch-Französisches Institut Ludwigsburg. Publié avec le concours de Deutsch-Französisches Jugendwerk.
- I: Menyesch, Dieter / Bérénice Manac'h: Une Bi- bliographie 1963-1982 / Eine Bibliographie 1963- 1982. / 1984. XXX, 793 S. / Br (3-598-10490-1)
- II: Manac'h, Bérénice / Dieter Menyesch; Joachim Schild: Une Bibliographie 1983-1990 / Eine Bibliographie 1983-1990 / 1993. XXX, 422 S. / Br (3-598-11049-9)

François Villon - Bibliographie und Materialien 1489-1988 / François Villon- Bibliographie et matériaux littéraires 1489-1988. Hrsg. v. Rudolf Sturm / 2 Bde. 1990. Zus. 691 S. / Gb (3-598-10892-3)
- Bibliographie / Bibliographie 1990 344 S. / Gb (3-598-10893-1)
- Materialien zu Werk und Wirkung / Matériaux Littéraires pour les Etudes sur l'Œuvre / 1990. 347 S. / Gb (3-598-10894-X)

Frank Wedekind. A Bibliographic Handbook. Comp. and annotated by Robert A. Jones; Leroy R. Shaw / 2 vols. 1996. Cplt. XLIV, 759 p. / Hard (3-598-11306-4)

Franz Kafka. Eine kommentierte Bibliographie der Sekundärliteratur. (1955-1980, mit einem Nachtrag 1980-1985). Hrsg. v. Marie L. Caputo-Mayr; Julius M. Herz / 1987. 692 S. (3-317-01569-1) [1991 vom Francke Vlg, Bern übernommen]

Franz Kafkas Werke. Eine Bibliographie der Pri- märliteratur (1908-1980). Hrsg. v. Marie L. Caputo- Mayr; Julius M. Herz / 1982. 94 S. / Gb (3-907820-66-5) [1991 vom Francke Vlg, Bern übernommen]

Frasch, Gisela: Kommunale Politik und Öffentliche Bibliothek. Ein Planspiel / 1984. 233 S. / Br (3-598-10428-6)

Die **Frauenfrage** in Deutschland. Bibliographie. Hrsg.: Deutscher Akademikerinnenbund e.V. Bearb. v. Ilse Delvendahl; Doris Marek (ISSN 0344-1415) [Bis Bd. 7. Bibliographie 1976 (1977) im Selbstverlag erschienen]

- Bd. 1: Strömungen und Gegenströmungen 1790-1930. Sachlich geordnete und erläuterte Quellenkunde. Hrsg. v. Hans Sveistrup und Agnes Zahn-Harnack. - Nachdr. d. unveränd. 2. Aufl. Tübingen 1961 / 3. Aufl. 1984. XVI, 800 S. / Lin (3-598-20189-3)
- Bd. 8. 1977. Deutscher Akademikerinnenbund / 1979. 111 S. / Br (3-598-20185-0)
- Bd. 9. 1978. Bearb. von Ilse Delvendahl u. Doris Marek. Dt. Akademikerinnenbund / 1980. VIII, 134 S. / Br (3-598-20186-9)
- Bd. 10. 1931-1980 / 1981. XV, 957 S. / Lin (3-598-20187-7)

Die **Frauenfrage** in Deutschland. Neue Folge. Bibliographie. Hrsg.: Deutscher Akademikerinnenbund (Bd. 1-3); Institut „Frau und Gesellschaft" (IFG) (Bd. 4 ff.)
- Bd. 1. 1981. Bearb. v. Ilse Delvendahl / 1983. XIII, 229 S. / Lin (3-598-20188-5)
- Bd. 2. 1982-1983. Bearb. v. Ilse Delvendahl / 1985. X, 458 S. / Lin (3-598-20190-7)
- Bd. 3. 1984. Bearb. v. Ilse Delvendahl / 1987. 343 S. / Lin (3-598-20191-5)
- Bd. 4. 1985. Bearb. v. Gisela Ticheloven, unter Mitarb. v. Susanne Rudolph; Susanne Urban / 1990. VI, 413 S. / Lin (3-598-20192-3)
- Bd. 5. 1986. Bearb. v. Marion Göhler / 1991. IV, 371 S. / Lin (3-598-20193-1)

Freimann, Aron / Moses Marx: Thesaurus Typographiae Hebraicae Saeculi XV. The Hebrew Printing during the fifteenth century. - Nachdruck d. Ausg. Berlin, 1924-31/ 1970. 334 Bl. / Ln (Vlg.-Nr. 06502)

Fritzsche, Gottfried: Theoretische Grundlagen der Nachrichtentechnik / Bd. 1 u. 2 / 1974 / Kst. (UTB 235 u. 236) (3-7940-2617-9 u. 3-7940-2618-7)

From Haven to Conquest. Readings in Zionism and the Paletine Problem until 1948. Ed. by Walid Khalidi / 1971. 914 S. (3-7940-5155-6)

25 Jahre Amerika-Gedenkbibliothek. Berliner Zentralbibliothek. Hrsg. von Peter K. Liebenow / 1979. 286 S., 16 S. Abb. / Gb (3-598-10914-5)

Fürsten-Postkarten: Sammlung Frühsorge. Mikrofiche-Edition Bearb. v. Christoph Behrens; Gotthardt Frühsorge / 1988. 256 S. Textbd, 5288 Postktn. auf 86 Fiches, davon 5 farb. u. 81 schw.-w. Silberfiches, Lesefaktor 24x (3-598-32532-0)

Fugger, Wolfgang: Ein nutzlich und wolgegrundt Formular manncherley schöner schriefften. Mit e. Nachw. von Friedrich Pfäfflin. - Vollst. Faksimile-Ausg. d. Schreibmeisterbuchs von 1553 / 1967. 224 S. (3-7940-3047-8)

Fuhlrott, Rolf: Deutschsprachige Architektur-Zeitschriften. Enstehung u. Entwicklung der Fachzeitschriften für Architektur in der Zeit von 1789-1918; mit Titelverz. u. Bestandsnachweisen / 1975. VIII, 373 S. / Lin (3-7940-3653-0)

Function and Organization of a National Documentation Centre in a Developing Country. Von einer Arbeitsgruppe der Fédération International du documentation (FID) u. d. Ltg. von Harald Schütz / 1975. 218 S. / Br (3-7940-5151-3) [Koprod. m. UNESCO, Paris]

Funke, Fritz: Buchkunde. Ein Überblick über die Geschichte des Buch- und Schriftwesens / 3. Aufl., unveränd. Nachdr. d. 2., verb. u. erw. Aufl. 1963. 1969. 324 S. / Gb (3-7940-4219-0) [Lizenzausg. m. Genehm. v. VEB Vlg. f. Buch- u. Bibliothekswesen, Leipzig]
- 4. Aufl., unveränd. Nachdruck der 2. verb. und erw. Aufl. 1963 / 1978. 324 S. / Gb (3-7940-4219-0)
- 5., neubearb. Aufl. Buchkunde. Ein Überblick über die Geschichte des Buches / 1992. 393 S., 170 Abb. / Gb (3-598-11051-0) [Vorher im VEB Bibliographisches Institut, Leipzig]
- 6., überarb. u. erg. Aufl. 1998. 396 S. / Gb (3-598-11390-0)

Funke, Klaus-Detlef: Innere Pressefreiheit. Zu Problemen d. Organisation v. Journalisten. Mit einem Beitrag v. Werner Brede: Von d. IG Druck u. Papier zur Mediengewerkschaft / 1972. 259 S. / Br (3-7940-3290-X)

Gaspary, Udo: Tabellen und Formeln für Investitionsrechnungen. Mit Erläuterungen und Beispielen / 1979. 80 S., 52 Tab. / Kst. (UTB 891) (3-598-02664-1)

Geimer, Hildegard / Reinhold Geimer: Research Organisation and Science Promotion in the Federal Republic of Germany / 1981. XX, 197 p. / Hard (3-598-10357-3)

Gekeler, Otto / Klaus-Dieter Herdt; Walter Oberender: Warenkatalogisierung und Kommunikation über die Ware. Hrsg. vom AWV - Ausschuß für wirtschaftliche Verwaltung in Wirtschaft und Öffentlicher Hand e.V. Mit e. Geleitw. von Joachim Böttger (AWV-Schriftenreihe 125) / 1974. XV, 232 S., 63 Abb. u. Tafeln / Lin (3-7940-444198-4)

Genealogische Bestände der Universitätsbibliothek Düsseldorf. Hrsg. v. Günter Gattermann / 1991. XVI, 541 S. / Ln (3-598-10925-3)

Eine **Generation** später. Bundesrepublik Deutschland 1953-1979. Hrsg.: Institut f. Demoskopie Allenbach / Elisabeth Noelle-Neumann; Edgar Piel / 1982. 272 S., 120 Abb. / Lin (3-598-10475-8)

Genzmer, Fritz: Das Buch des Setzers. Kurzgefaßtes Lehr- und Handbuch für den Schriftsetzer / 9., völlig neu bearbeitete Aufl. 1967. 266 S. / Ln (Vlg.-Nr. 08701) [1968 v. Ullstein Fachvlg., Frankfurt a. M. übernommen]

German-Americans in the World Wars. Ed. by Don H. Tolzmann / 5 vols. in 9 parts Cplt. LXXX, 3,255 p., 12 photos black/white / Hard (3-598-21530-4)
- Vol. I: The Anti-German Hysteria of World War One. German-Americans and the World War, with Special Emphasis on Ohio's German-Language Press, by Carl Wittke. (Reprint) The Anti-German Hysteria: The Case of Robert Paul Prager. Selected Documents. Ed. by Franziska Ott / 1995. X, 365 p. (3-598-21531-2)
- Vol. II: The World War One Experience. The U.S. Senate Hearings Against the National German-American Alliance (Reprint) / 2 parts 1995 Cplt. X, 708 p. (3-598-21532-0)
- Vol. III: Research on the German-American Experience of World Ware One: A Case Study. The Survival of an Ethnic Community: The

Cincinnati Germans, 1918 through 1932, by Don H. Tolzmann. - Reprint / 1995 X, 422 p. (3-598-21533-9)
- Vol. IV: The World War Two Experience. The Internment of German-Americans Documents. Ed. by Arthur D. Jacobs; Joseph E. Fallon / 3 vols. in 4 parts 1995 (3-598-21534-7)
  - Section 1: From Suspicion to Internment. U.S. Government Policy Toward German-Americans. 1939-48 / 2 parts 1995. Cplt. X, 751 p. (3-598-21535-5)
  - Section 2: Government Preparation for and Implementation of the Repatriation of German-Americans 1943-48 / 1995. X, 391 p. (3-598-21536-3)
  - Section 3: German-American Camp Newspapers Internees' View of Life in Internment / 1995. XII, 255 p. (3-598-21537-1)
- Vol. V: Germanophobia in the U.S. The Anti-German Hysteria and Sentiment of the World Wars. Supplement and General Index / 1998. XVIII, 354 p. (3-598-21538-X)

German Yearbook on Business History. Ed. by The German Society for Business History, Cologne. In cooperation with the Institute for Bank-Historical Research, Frankfurt a. M. / Ab Jg. 1981 (ISSN 0722-2416) [Vorher im Vlg. Springer, Berlin]
- 1981. Ed. by Wolfram Engels; Hans Pohl / 1981. 127 p. / Hard (3-598-22962-3)
- 1982. Ed. by Wolfram Engels; Hans Pohl / 1982. 186 p. / Hard (3-598-22963-1)
- 1984. Ed. by Wolfram Engels; Hans Pohl / 1985. 158 p. / Hard (3-598-22964-X)
- 1985. Ed. by Hans Pohl; Bernd Rudolph / 1986. VIII, 162 p. / Hard (3-598-22965-8)
- 1986. Ed. by Hans Pohl; Bernd Rudolph / 1987. VIII, 168 p. / Hard (3-598-22966-6)
- 1987. Ed. by Hans Pohl; Bernd Rudolph / 1990. VIII, 152 p. / Hard (3-598-22967-4)
- 1988. Ed. by Hans Pohl; Bernd Rudolph / 1990. VIII, 186 p. / Hard (3-598-22968-2)
- 1989-92. Ed. by Hans Pohl; Bernd Rudolph. Redaktion: Gabriele Jachmich; Manfred Pohl / 1992. 243 p. / Hard (3-598-22980-1)
- 1993. Ed. by Hans Pohl; Bernd Rudolph / 1993. 207 p. / Hard (3-598-22981-X)
- 1994. Ed. by Hans Pohl; Bernd Rudolph / 1994. 186 p. / Hard (3-598-22982-8)
- 1995. Ed. by Hans Pohl; Bernd Rudolph / 1995. 135 p. / Hard (3-598-22983-6)

Gesammt-Verlags-Katalog des Deutschen Buchhandels und des mit ihm im direkten Verkehr stehenden Auslandes Mikrofiche-Edition der Ausgabe 1881-1893. Münster 1894. Hrsg. v. Adolph Russell; Johannes Basch / 1986. 28 Bde, 38541, CCLXX Spalten auf 217 Fiches. Lesefaktor 24x / Silberfiche (3-598-10625-4) / Diazofiche (3-598-10624-6)

Gesamtkatalog der Malik-Buchhandlung A.G. Mit einem Nachw. v. Wieland Herzfelde. -Nachdr. d. Ausg. Berlin 1925 / 1976. 304 S., 8 Taf., mehrere Abb. / Br (3-7940-4107-0)

Gesamtplan für das wissenschaftliche Bibliothekswesen. Von der Arbeitsgruppe Bibliotheksplan Baden-Württemberg. Red.: Elmar Mittler. Hrsg. i. Auftr. d. Kultusministeriums Baden-Württemberg
- Bd. 1: Universitäten. Vorwort v. Wolfgang Kehr / 2. Aufl. 1973. 680 S, 8 Abb., 87 Tab. u. Übersichten / Lin (3-7940-3041-9) / Br (3-7940-3042-7)
- Bd. 2: Hochschulen, ADV-Einsatz, Kooperation. Vorwort v. Wolfgang Kehr / 1975. XII, 468 S. / Br (3-7940-3045-1)

Gesamttabelle für die Mehrwertsteuer. Zum Ablesen aller Steuerbeträge, Brutto- u. Nettowerte bis DM 100.000
- 5,5% / 1968. 80 S. / Kt
- 11% / 1968. 80 S. / Kt

Gesamtverzeichnis der Kongress-Schriften (GKS) in Bibliotheken der Bundesrepublik Deutschland einschließlich Berlin (West) / Union List of Conference Proceedings in Libraries of the Federal Republic of Germany including Berlin (West). Hrsg. u. bearb. v. d. Staatsbilbiothek Preussischer Kulturbesitz, Abteilung Gesamtkatalog und Dokumentation
- Schriften von und zu Kongressen, Konferenzen, Kolloquien, Symposien, Tagungen, Versammlungen und dergleichen vor 1971
- Hauptband mit Besitznachweisen. Stand: Januar 1976. Publications from and about congresses, conferences, colloquia, symposia, meetings / 1976. XIV, 563 S. / Lin (3-7940-3001-X)
- Registerband / 1976. VI, 344 S. / Lin (3-7940-3002-8)
- Supplement für Kongresse bis 1976 mit Register / Supplement for conferences through 1976 with index / 1977. VIII, 297 S. / Lin (3-7940-3003-6)
- Supplement für Kongresse bis 1978 mit Register. Supplement for conferences through 1977 with index / 1980. VIII, 834 S. / Lin (3-7940-3006-0)
- Anschluss-Supplement für Kongresse bis 1981 mit Register. Stand März 1981 / Additional supplement for Conferences through 1981 with index. Holdings as of March 1981 / 1982. XII, 608 S. / Lin (3-598-10434-0)

Gesamtverzeichnis der Übersetzungen deutschsprachiger Werke / Bibliography of Translations of German-Language Publications (GvÜ) Berichtszeitraum 1954-1990. Bearb. v. Willi Gorzny / 12 Bde. 1991-1994. Zus. 5.170 S. / Gb (3-598-22536-9)
- Bd. 1 / 1991. VIII, 369 S. / Gb (3-598-22537-7)
- Bd. 2 / 1992. IV, 312 S. / Gb (3-598-22538-5)
- Bd. 3 / 1992. IV, 311 S. / Gb (3-598-22538-3)
- Bd. 4 / 1992. IV, 323 S. / Gb (3-598-22540-7)
- Bd. 5 / 1992. IV, 317 S. / Gb (3-598-22541-5)
- Bd. 6 / 1992. IV, 372 S. / Gb (3-598-22542-3)
- Bd. 7 / 1993. IV, 393 S. / Gb (3-598-22543-1)
- Bd. 8 / 1993. IV, 662 S. / Gb (3-598-22544-X)
- Bd. 9 / 1994. IV, 506 S. / Gb (3-598-22545-8)
- Bd. 10. Registerbd. 1 / 1994. V, 426 S. / Gb (3-598-22546-6)
- Bd. 11. Registerbd. 2 / 1994. V, 463 S. / Gb (3-598-22547-4)
- Bd. 12. Registerbd. 3 / 1994. V, 479 S. / Gb (3-598-22548-2)

Gesamtverzeichnis der Zeitschriften und Serien in Bibliotheken der Bundesrepublik Deutschland einschließlich Berlin (West). Stand: April 1974 Union list of serials in libraries of the Federal Republic of Germany including Berlin (West) Neue u. geänd. Titel seit 1971 mit Besitznachweisen. Bearb. u. Hrsg. von d. Staatsbibliothek Preuß. Kulturbesitz, Abt. Gesamtkataloge u. Dokumentation in Zusammenarbeit mit d. Arbeitsstelle für Bibliothekstechnik / 1974. XII, 227 S. (37940-2990-9)
- Stand: Nov. 1974 / 1975. X, 436 S. (3-7940-2991-7)
- Stand: Nov. 1975. Ausg. 5. / 1976. X, 608 (3-7940-2992-5-)
- Stand: Juni 1976 / 1976. VIII, 606 S. (3-7940-2993-3)
- Stand: April 1977 / 1977. VIII, 917 S. (3-7940-2994-1)
- Stand: April 1978 / 1978. 1351 S. (3-7940-2995-X)

Gesamtverzeichnis des deutschsprachigen Schrifttums (GV)

- 1700-1910. Vorw. v. Hans Popst und Rainer Schöller. Bearb. unter d. Leit. v. Hilmar Schmuck, Willi Gorzny und Peter Geils / 160 Bde. 1979-1987 / Ln (3-598-301804-X)
  - Nachträge. Bearb. v. Hilmar Schmuck / 1987. 294 S. / Ln (3-598-30158-8)
  - Mikrofiche-Edition. Bearb. unter Leitung v. Hilmar Schmuck und Willi Gorzny / 1986. 160 Bde. u. Nachtragbd. auf 795 Fiches. Lesefaktor 24x / Diazofiche (3-598-30590-7)
- 1911-1965. Hrsg. v. Reinhard Oberschelp mit einem Geleitw. v. Wilhelm Totok. Bearb. v. Willi Gorzny / 150 Bde. 1976-1981. Zus. 79.560 S. / Ln (3-7940-5600-0)
  - Mikrofiche-Edition. Hrsg. v. Reinhard Oberschelp. Bearb. v. Willi Gorzny / 1984. 150 Bde. auf 400 Fiches. Lesefaktor 42x / Silberfiche (3-598-30456-0) / Diazofiche (3-598-30455-2)

Gesamtverzeichnis des deutschsprachigen Schrifttums außerhalb des Buchhandels 1966-1980 / 45 Bde. 1988 / 1990. Zus. 20.160 S. / Ln (3-598-31630-5)

Gesamtverzeichnis Deutschsprachiger Hochschulschriften 1966-1980. Hrsg. v. Willi Gorzny / 40 Bde. (24 Bde. Haupttl., 16 Bd. Reg.) 1984-1990. Zus. 205.505 S. / Lin (3-598-30600-8)

Gesamtverzeichnis deutschsprachiger Zeitschriften und Serien in Bibliotheken der Bundesrepublik Deutschland einschließlich Berlin (West). GDZS. Titel vor 1971 mit Besitznachweisen. Bearb. u. Hrsg.: Staatsbibliothek Preußischer Kulturbesitz, Abt. Gesamtkataloge u. Dokumentation. Stand: Dez 1976 / 1. A-H. 1977. X, 757 S. [Nebent.:] Union list of German language serials in libraries of the Federal Republic of Germany including Berlin (West)
- 2. I-Z. 1977. S. 758-1617.
- 2. Ausgabe. 1978. XIV, 612 S. / (Titel vor 1971 mit Besitznachweisen) (3-598-02801-6)

Geschichte und Kultur der Juden in Bayern. Aufsätze. Hrsg. v. Manfred Treml; Josef Kirmeier, unter Mitarb. v. Evamaria Brockhoff (Veröffentlichung zur Bayerischen Geschichte und Kultur, Bd. 17) / 1988. 614 S., 24 Abb. / Gb (3-598-07543-X)
- Lebensläufe. Vorw. v. Manfred Treml; Bernward Deneke. Hrsg. v. Manfred Treml; Wolf Weigand, unter Mitarb. v. Evamaria Brockhoff (Veröffentlichung zur Bayerischen Geschichte und Kultur, Bd. 18) / 1988. 328 S., 40 Abb. / Gb (3-598-07544-8)

Geschichts- und Romanen-Litteratur der Deutschen. - Nachdr. d. Ausg. Breslau 1798. - Mit e. Nachwort von Hans-Joachim Koppitz / 1973. VIII, 220S. , 1 Bl., 20 S. Nachwort / Ln (3-7940-5031-2)

Gesellschaft für Exilforschung: Nachrichtenbrief 1984 bis 1994 mit Gesamtregister / Society for Exil Studies: Newsletter 1984 to 1994 with General Index. Hrsg.: Society for Exile Studies, Inc / Gesellschaft f. Exilforschung e.V. Redakt.: Ernst Loewy, unter Mitarb. v. Patrik von ZurMühlen; Barbara Seib; Eva Tiedemann; Elsbeth Wolffheim u.a / 3 Bde. 1995. Zus. X, 1.794 S. / Ln (3-598-11243-2)

Gillespie, Paul D. / Paul Katzenberger; John Page: Problems of document delivery for the EURONET user. Technical report. Prepared for the Commission of the European Communities, Directorate General

for Scientific and Technical Information Management. Franklin Institut, Munich / 1979. XIV, 228 S. (3-598-10096-8)

Gittig, Heinz: Bibliographie der Tarnschriften 1933 bis 1945 / 2. wesentl erw. Aufl. 1996. XXIV, 260 S. / Ln (3-598-11224-6)

Global Books in Print PLUS. Complete English-language bibliographic information from the United States, United Kingdom, Continental Europe, Australia, New Zealand, Africa, Asia, Latin America, Canada and the Oceanic States. CD-ROM-Edition / Bowker Electronic Publishing since 1993 (12 updates annually) (3-598-40273-2) [incl. the database of *International* Books in Print]

Global Guide to Media and Communications. Ed. by John Lent / 1987. XII, 145 p. / Hard (3-598-10746-3)

Glossarium Artis. Dreisprachiges Wörterbuch der Kunst / Dictionnaire des Termes d'Art / Dictionary of Art Terms. Sous le patronage du Comité International d'Histoire de l'Art. Redakt.: Renate Rieth und Rudolf E. Huber / 10 Bde. [1982 vom Max Niemeyer Vlg., Tübingen üvbernommen]
- Bd. 1. Burgen und Feste Plätze / Chateauxforts et places fortes. Der Wehrbau vor Einführung der Feuerwaffen. Mit Anhang: Kriegsgeräte und schwere Waffen / L'architecture militaire avant l'introduction des armes à feu. Avec supplément: attirail de guerre et armes lourdes / 2. Aufl. 1977. 280 S., 173 Abb. / Br (3-598-10452-9)
  - 3. Aufl. 1995. 315 S. / Gb (3-598-11183-5)
- Bd. 2. Liturgische Geräte, Kreuze und Reliquiare der christlichen Kirchen / Objets liturgiques, croix et reliquaires des églises chrétiennes Systematisches Fachwörterbuch / Dictionnaire spécialisé et systématique / 2. unveränd. Aufl. 1982. 159 S., 108 Abb. / Sn (3-598-10453-7)
  - 3. vollst. neubearb. u. erw. Aufl. Kirchengeräte, Kreuze und Reliquiare der christlichen Kirchen / Objets liturgiques, croix et reliquaires des églises chrétiennes / Ecclesiastical utensils, crosses and reliquaries of the christian churches / 1992. 365 S., 267 Abb. / Gb (3-598-11079-0)
- Bd. 3. Bogen und Arkaden / Arcs et Arcades. Systematisches Fachwörterbuch / Dictionnaire spécialisé et systématique. M. Beitr. v. J. Courvoisier, R. Gemtos, L. Grodecki, H. R. Hahnloser, G. Janneau, R. Lehni, M. Masciotta, J. S. Meseguer, W. Meyer, P. Moisy, R. Recht u. d. Mitgl. d. Wissenschaftl. Kommission / 2. unveränd. Aufl. 1982. 167 S., 164 Abb. / Sn (3-598-10454-5)
  - 3., neu bearb. u. erw. Aufl. Bogen und Mauerwerk / Arcs et maconnerie / Arches and Masonry. Systematisches Fachwörterbuch / Dictionnaire spécialisé et systématique / Specialized and Systematic Dictionary. Unter Mitarb. u. Beratung v. Günther Binding, S. Muthesius, H. Pflüger, Michel Procès una A. Vivis / 1998. 253 S. / Ln (3-598-11252-1)
- Bd. 4. Paramente und Bücher der christlichen Kirchen / Parements liturgiques et livres des églises chrétiennes. Systematisches Fachwörterbuch / Dictionnaire spécialisé et systématique. Anhang: Kirchenfeste, katholische und griechisch-orthodoxe Geistlichkeit / Supplement: Fêtes de l'église, clergé catholique et grecorthodoxe / 2. unveränd. Aufl. 1982. 203 S., 73 Abb. / Sn (3-598-10455-3)

- Bd. 5. Treppen / Escaliers / Staircases. Systematisches Fachwörterbuch / Dictionnaire specialisé et systématique / Specialized and Systematic Dictionary / 2. vollst. neubearb. u. erw. Aufl. 1985. 278 S., 300 Ill / Gb (3-598-10456-1)
- Bd. 6. Gewölbe und Kuppeln / Voutes et coupoles. Systematisches Fachwörterbuch / Dictionnaire spécialisé et systématique / 2. unveränd. Aufl. 1982. 249 S., 139 Abb. / Sn (3-598-10457-X)
  - 3. vollst. neubearb. u. erw. Aufl. Gewölbe / Voutes / Vaults. Systematisches Fachwörterbuch / Dictionnaire spécialisé et systématique / Specialized and Systematic Dictionary. Unter Mitarb. v. R. Lecourt, K. Lodge, E. Anderson, E. Troll-Kraft, Torsten Gebhard, F. Hart, W. Meyer, S. Muthesius und H. Reuther / 1988. 312 S., 338 Abb. / Gb (3-598-10758-7)
- Bd. 7. Festungen / Forteresses / Fortresses. Der Wehrbau nach Einführung der Feuerwaffen / L'architecture après l'introduction des armes à feu / Military architecture after the introduction of firearms / 1979. 298 S., 190 Abb. / Br (3-598-10458-8)
  - 2., vollst. neu bearb. u. erw. Aufl. Festungen / Fortifications / Fortifications. Der Wehrbau nach Einführung der Feuerwaffen. Systematisches Fachwörterbuch / L'Architecture militaire après l'introduction des Armes à Feu. Dictionnaire spécialisé et systématique / Military Architecture after the Introduction of Firearms. Spezialized and Systematic Dictionary. Unter Mitarb. v. Helmut Pflüger, Philippe Truttmann und Quentin Huges / 1990. 407 S., 265 Abb. / Gb (3-598-10806-0)
- Bd. 8. Das Baudenkmal / Le monument historique / The historic monument Zu Denkmalschutz und Denkmalpflege / Termes concernant la protection et la conservation des monuments historiques / Terminology concerned with the protection and preservation of historic monuments / 1981. 326 S., 135 Abb. / Br (3-598-10459-6)
  - 2. völlig neubearb. u. erw. Aufl. 1994. 331 S., 220 Abb. / Ln (3-598-11113-4)
- Bd. 9. Städte / Villes / Towns. Stadtpläne, Plätze, Strassen, Brücken. Systematisches Fachwörterbuch / Plans, Places, Voies, Ponts. Dictionnaire spécialisé et systématique / Plans, Squares, Roads, Bridges. Specialized and Systematic Dictionary / 1987. 408 S., 248 Abb. / Gb (3-598-10460-X)
- Bd. 10. Holzbaukunst / Architecture en Bois / Architecture of Wood. Fachwerk - Dachgerüst - Zimmermannswerkzeug. Systematisches Fachwörterbuch / Construction en Pan de Bois - Charpente de Toit - Outils du Charpentier. Dictionnaire spécialisé et systématique / Timber-frame Construction - Roof Frame - Carpenter's Tools. Spezialized and Systematic Dictionary. Unter Mitarb. u. Beratung v. Jean-Pierre Anderegg, Maurice Berthoud, Günther Binding, Torsten Gebhard, Richard Harris, Günther Heine, Michel Procès, Peter Smith und Philip Walker / 1997. 331 S., 266 Abb. / Gb (3-598-10461-8)

Glossary on Educational Technologie. Hrsg. Internationales Zentralinstitut für das Jugend- und Bildungsfernsehen / 1973. 140 S. (3-7940-5134-3)

Gluschkow, Viktor Michaijlovič: Einführung in die technische Kybernetik. Übers. von Manfred Peschel. Überarb. von G. Wunsch / 2 Bde. 1970. 126 u. 173 S. / Br (3-7940-4275-1 u. 3-7940-4276-X)

unveränd. Ausg. 2 Bde. 1971. 126 u. 173 S. / Kst. (UTB 45 u. 46) (3-7940-2603-9 u. 3-7940-2604-7)

Görner, Franz: Osteuropäische bibliographische Abkürzungen. East European bibliographical abbreviations

- Bd. 1: Ost- und südosteuropäische Sprachen einschließlich der Sprachen im Westen der Sowjetunion / East and Southeast European Languages including the languages in the Western Soviet Union / 3., neu bearb. u. erg. Aufl. 1975. XVII, 301 S. / Lin (3-7940-3196-2)

Goethes Bibliothek. Katalog. Hrsg.: Nationale Forschungs- und Gedenkstätten der klassischen deutschen Literatur in Weimar. Bearb. v. Hans Ruppert. (Nachdr. d. Ausg. v. 1958) / 1978. 825 S. / Gb (3-7940-5034-7)

Göttinger Atomrechtskatalog / Atomic Energy Law Catalogue. Hrsg.: Institut für Völkerrecht der Universität Göttingen. Redakt.: Gottfried Zieger / 29 Bde. 1960-1976. Zus. 11.850 S. / Lin (3-598-30360-2) [1982 v. d. Zentralstelle für Atomernergie Dokumentation, Karlsruhe übernommen]

- Internationale Bibliographie des Atomenergierechts 1976-1987 / International Atomic Energy Law Bibliography 1976-1987 Internationales Schrifttum und Quellen 1976-1987 / International Bibliography and Sources 1976-1987 Bearb. v. Gertie Bauer; Werner Bischof; Peter Morgenstern; Norbert Pelzer; Dietrich Rauschning / 4 Bde. 1988. Zus. 2.193 S. / Gb (3-598-10797-8)

Götz von Olenhusen, Albrecht / Christa Gnirss:: Handbuch der Raubdrucke Bd. 2: Theorie und Klassenkampf. Sozialisierte Drucke u. proletarische Reprints. Eine Bibliographie / 1973. 509 S., 50 Abb. / Lin (3-7940-3419-8) [Bd. 1 ist nicht erschienen]

Gorman, Gary / Maureen Mahoney: Guide to Current National Bibliographies in the Third World / 1983. XVIII, 328 p. / Hard (3-598-10446-4) [Hans Zell Publishers]

Goslich, Lorenz: Zeitungs-Innovationen / 1987. X, 198 S. / Br (3-598-10695-5)

Der **Gotha**

- Mikrofiche-Edition / 1982-1984. 475 Bde, ca. 415.000 S. auf 496 Fiches. Lesefaktor 24x / Silberfiche (3-598-30330-0)
- Supplement: Der „Österreich-Gotha". Mit Ergänzungswerken zum deutschen Adel / 1997. ca. 22.000 S. auf 140 Fiches / Silberfiche (3-598-30359-9)

Gottschalk, Günther: Hesse Lyrik Konkordanz. Mit Wortindex und Wortfrequenzlisten / 1987. VI, 837 S. / Ln (3-598-10686-6)

Griechisches Biographisches Archiv / Greek Biographical Archive (GBA). Mikrofiche-Edition. Bearb. v. Hilmar Schmuck / 1998 ff. ca. 300 Fiches in 12 Lfgn, Lesefaktor 24x / Diazofiche (3-598-34180-6) / Silberfiche (3-598-34181-4)

Gronemeyer, Reimer: Zigeuner in Osteuropa. Eine Bibliographie zu den Ländern Polen, Tschechoslowakei und Ungarn. Mit einem Anhang über ältere sowjetische Literatur / 1983. 280 S. / Lin (3-598-10506-1)

Grosse deutsche Lexika. Aufklärung und frühes neunzehntes Jahrhundert. Mikrofiche-Edition. Hrsg. v. Walther Killy / 1992. 422 Fiches, Lesefaktor 24x (3-598-40620-7)

- Allgemeine deutsche Real-Encyklopädie für die gebildeten Stände. (Conversations-Lexikon). In 12 Bänden. 7. Originalaufl. Leipzig: Brockhaus 1827. Mikrofiche-Edition / 1992. 113 Fiches, Lesefaktor 24x / Silberfiche (3-598-40622-3)
- Hederich, Benjamin: Gründliches Antiquitäten-Lexicon. Worinne die merckwürdigsten Alterthümer der Jüden, Griechen. zulänglich beschrieben. Leipzig: Gleditsch 1743. Mikrofiche-Edition / 1992. 17 Fiches, Lesefaktor 24x / Silberfiche (3-598-40627-4)
- Hederich, Benjamin: Gründliches Lexicon Mythologicum. Worinne sowohl die fabelhafte, als wahrscheinliche und eigentliche Historie der alten und bekannten Römischen, Griechischen und Egyptischen Götter. 2. und verb. Aufl. Leipzig: Gleditsch 1741. Mikrofiche-Edition / 1992. 11 Fiches, Lesefaktor 24x / Silberfiche (3-598-40626-6)
- Hederich, Benjamin: Reales Schul-Lexicon. Worinne nicht allein von den Ländern, Städten, Schlössern. wie auch von den Zeiten, Völckern, Geschlechten, Personen. zur Geographie, Chronologie. eine nöthige Nachricht gegeben. Leipzig: Gleditsch 1717. Mikrofiche-Edition / 1992. 16 Fiches, Lesefaktor 24x / Silberfiche (3-598-40625-8)
- Hübner, Johann: Reales Staats- Zeitungs- und Conversations-Lexicon. Darin so wohl die Religionen, die Reiche und Staaten, Meere, Seen. als auch andere in Zeitungen und täglichen Umgang vorkommende, ingleichen juristische und Kunstwörter beschrieben werden. Nebst acht Kupfertafeln. Neue verb. Ausg. Leipzig: Gleditsch 1795. Mikrofiche-Edition / 1992. 15 Fiches, Lesefaktor 24x / Silberfiche (3-598-40624-X)
- Hübner, Johann: Reales Staats-, Zeitungs- und Conversations-Lexicon. Darinnen so wohl die Religionen und geistlichen Orden, die Reiche und Staaten, Meere, Seen. als auch andere in Zeitungen und täglicher Conversationvorkomme aus fremden Sprachen entlehnte Wörter. beschrieben werden. Neue mit Kupfern versehene Aufl. Nebst vollständigem Register und einer ausführlichehn Vorrede. Leipzig: Gleditsch 1737. Mikrofiche-Edition / 1992. 14 Fiches, Lesefaktor 24x / Silberfiche (3-598-40623-1)
- Universal-Lexikon oder vollständiges encyclopädisches Wörterbuch. Herausgegeben von H. A. Pierer. Bd. 1-26. Altenburg: Literatur-Comptoir 1835-1836. Mikrofiche-Edition / 1992. 211 Fiches, Lesefaktor 24x / Silberfiche (3-598-40621-5)
- Walch, Johann Georg: Philosophisches Lexicon. Worinnen die in allen Theilen der Philosophie, vorkommende Materien und Kunstwörter erkläret. wie auch mit einer kurzen kritischen Geschichte der Philosophie aus dem Bruckerischen grossen Werke versehen von Justus Christian Hennings. 4. Aufl. in 2 Theilen. Leipzig: Gleditsch 1775. Mikrofiche-Edition / 1992. 25 Fiches, Lesefaktor 24x / Silberfiche (3-598-40628-2)

Groth, Michael: The Road to New York: The Emigration of Berlin Journalists 1933-1945 / 2nd ed. 1988. X, 384 p. / Hard (3-598-10782-X)

Grund, Uwe: Indices zur sprachlichen und literarischen Bildung in Deutschland / 5 Bde. 1991-1995 / Gb (3-598-22569-5)

- Bd. 1. Zeitschrift für Deutschkunde 1920-1943. Beiträger - Themen - Textprofile / 1991. 285 S. / Gb (3-598-22570-9)
- Bd. 2. Zeitschrift für Deutsche Bildung 1925-1944. Beiträger - Themen - Textprofile / 1995. 222 S. / Gb (3-598-22571-7)
- Bd. 3. Amtsblätter Preussen, Bayern, Deutsches Reich 1911-1945. Erlasse - Bekanntmachungen - Erhebungen / 1995. 242 S. / Gb (3-598-22572-5)
- Bd. 4. Deutschunterricht 1948-1970. Beiträger - Themen - Textprofile / 1995. 320 S. / Gb (3-598-22573-3)
- Bd. 5. Der Deutschunterricht 1947-1970 / Wirkendes Wort 1950-1970. Beiträger - Themen - Textprofile / 1995. 296 S. / Gb (3-598-22574-1)

Grundlagen der praktischen Information und Dokumentation. Ein Handbuch zur Einführung in die fachliche Informationsarbeit. Begr. v. Klaus Laisiepen; Ernst Lutterbeck; Karl H. Meyer-Uhlenried. Hrsg. v. Marianne Buder; Werner Rehfeld; Thomas Seeger; Dietmar Strauch / 4. vollst. neugefasste Ausg. in 2 Bdn 1996. Zus. XLVII, 1069 S. / Gb (3-598-11309-9)
- Unveränd. Ausg. in 1 Broschurband / 1997. XLVII, 1069 S. / Br (3-598-11310-2)
1.u. 2. Aufl.: → **DGD**-*Schriftenreihe Bd. 1*
3: Aufl.: → **DGD**-*Schriftenreihe Bd. 9*

Grundlagen des Bibliotheksbaus. Bibliotheksgebäude / 1986. 384 S., 96 Abb. / Ln (3-598-07226-0)

Grundwissen Buchhandel - Verlage. Hrsg. v. Klaus W. Bramann; Joachim Merzbach
- Bd. 1. Göhler, Wolfgang: Kaufmännisches Rechnen. Statistik / 2. verb. Aufl. 1983. X, 322 S. / Br (3-598-20055-2)
  - 3. überarb. u. erw. Ausg. Schmelzle, Wolfgang / Wolfgang Göhler: Kaufmännisches Rechnen - Buchführung / 1997. IX, 368 S. / Ln (3-598-20063-3)
- Bd. 2. Bramann, Klaus W. / Joachim Merzbach; Roger Münch: Sortiments- und Verlagskunde / 1993. 382 S. / Ln (3-598-20052-8)
  - 2. überarb. u. erw. Aufl. 1995. 413 S. / Ln (3-598-20065-X)
- Bd. 3. Göhler, Wolfgang: Buchführung. Unter Mitarb. v. Joachim Merzbach / 1984. XII, 322 S. / Br (3-598-20053-6)
- Bd. 4. Paulerberg, Herbert: Marketing und Werbung der Sortimentsbuchhandlung. Lehr- und Praxisbuch / 1986. VIII, 139 S. / Br (3-598-20054-4)
- Bd. 5. Blana, Hubert: Die Herstellung. Ein Handbuch für die Gestaltung, Technik und Kalkulation von Buch, Zeitschrift und Zeitung / 1986. XIX, 354 S., 250 Abb. / Br (3-598-20056-0)
  - 2. verb. Ausg. 1991. 454 S., 310 Abb. / Gb (3-598-20058-7)
  - 3. überarb. Aufl. 1993. 472 S., 310 Abb. / Gb (3-598-20062-5)
  - 4. überarb. Aufl. 1998. 447 S., 340 Abb. / Gb (3-598-20067-6)
- Bd. 6. Göhler, Wolfgang: Kostenrechnung im Verlag / 1989. 300 S. (3-598-20057-9)
- Bd. 7. Schiller, Manfred: EDV für Sortimentsbuchhändler / 1992. XII, 288 S. / Br (3-598-20059-5)
- Bd. 8. Heimler, Gerhard: Kartographie - Touristik - Bildbände / 1990. 200 S. (3-598-20061-7)

Grundzüge des Marketing. Von Franz Böcker; Dudo von Eckardstein; Rainer Hauzeneder u.a. / 3. überarb. Aufl. 1976. 182 S., 56 Tab. / Kst. (UTB 564) (3-7940-2648-9)

Grune, Siegfried: Bildschirmarbeitsplätze. Eine Bibliographie / 1985. XII, 456 S. / Lin (3-598-10601-7)

Güdter, Bernd: Verstehen üben. Bilden in Interaktion und Kommunikation / 1976. 179 S. / Kst. (UTB 626) (3-7940-2657-8)

Guggenheimer, Eva / Heinrich Guggenheimer: Etymologisches Wörterbuch der jüdischen Familiennamen / 1996. XLI, 522 S. / Gb (3-598-11260-2)

Guide to Microforms in Print. Incorporating International Microforms in Print [Seit 1990. Vorher bei Meckler Publishing, Westport, CT]
- 1990
  - Author-Title / 1990. 695 p. / Hard (3-598-10928-8)
  - Subject Guide / 1990. 1,440 p. / Hard (3-598-10929-6)
  - Supplement / 1991. XV, 106 p. / Hard (3-598-10930-X)
- 1991. Ed. by Barbara Hopkinson
  - Author-Title / 1991. XIII, 1,651 p. / Hard (3-598-11047-2)
  - Subject Guide / 1991. XXI, 1,408 p. / Hard (3-598-11005-7)
  - Supplement / 1991. XXI, 150 p. / Hard (3-598-11006-5)
- 1992. Ed. by Barbara Hopkinson. Assistent ed.: Darin Laracuente
  - Author-Title / 1992. XIII, 1,704 p. / Hard (3-598-11082-0)
  - Subject Guide / 1992. XXI, 1,434 p. / Hard (3-598-11083-9)
  - Supplement / 1992. XXI, 127 p. / Soft (3-598-11095-2)
- 1993. Ed. by Barbara Hopkinson. Assistant ed.: Darin Laracuente
  - Author-Title / 1993. XIII, 1,706 p. / Hard (3-598-11136-3)
  - Subject Guide / 1993. XXI, 1,444 p. / Hard (3-598-11137-1)
  - Supplement / 1993. XXI, 107 p. / Soft (3-598-11138-X)
- 1994. Ed. by Barbara Hopkinson
  - Author-Title / 1994. XIV, 1,877 p. / Hard (3-598-11172-X)
  - Subject Guide / 1994. XXI, 1,632 p. / Hard (3-598-11173-8)
  - Supplement / 1994. XXI, 139 p. / Soft (3-598-11174-6)
- 1995. Ed. by Barbara Hopkinson
  - Autor-Title / 2 vols. 1995. Cplt. XIX, 1,869 p. / Hard (3-598-11231-9)
  - Subject Guide / 2 vols. 1995. Cplt. XXIV, 1,652 p. / Hard (3-598-11232-7)
  - Supplement / 1995. XX, 130 p. / Soft (3-598-11233-5)
- 1996. Ed. by Barbara Hopkinson
  - Author-Title / 2 vols. 1996. Cplt. XIV, 2,015 p. / Hard (3-598-11291-2)
  - Subject Guide / 2 vols. 1996. Cplt. XX, 1,767 p. / Hard (3-598-11290-4)
  - Supplement / 1996. XX, 130 p. / Soft (3-598-11312-9)
- 1997 Ed. by Barbara Hopkinson and Irene Izod
  - Author-Title / 2 vols. 1997. Cplt. XVII, 2,143 p. / Hard (3-598-11325-0)
  - Subject Guide / 2 vols. 1997. Cplt. XXIV, 1,922 p. / Hard (3-598-11326-9)
  - Supplement / 1997. XX, 121 p. / Soft (3-598-11338-2)
- 1998. Ed. by Irene Izod
  - Author-Title / 2 vols. 1998. Cplt. XVII, 2,036 p. / Hard (3-598-11363-3)
  - Subject Guide / 2 vols. 1998. Cplt. XXIV, 1,593 p. / Hard (3-598-11365-X)

- Supplement / 1998. XX, 123 p. / Soft (3-598-11366-8)

Guide to the Archival Materials of the German-speaking Emigration to the United States after 1933 / Verzeichnis der Quellen und Materialien der deutschsprachigen Emigration in den USA seit 1933 / 3 vols. 1991-1996 (3-907820-92-4) [1991 vom Francke Vlg, Bern übernommen]
- Vol. 1. Ed. by John M. Spalek, in collab. with Adrienne Ash; Sandra H. Hawrylchak / 2 vols. 1978. XXV, 1113 p. / Hard (3-907820-93-2)
  - Nachdr. Orig.-Ausg. / 2 vols 1997. XXV, 1113 p. / Hard (3-907820-93-2)
- Vol. 2. Ed. by John M. Spalek; Sandra H. Hawrylchak / 1992. XVII, 847 p. / Hard (3-907820-94-0)
- Vol. 3. Ed. by John M. Spalek; Sandra H. Hawrylchak / 2 vols. 1996. Cplt. XVI, 970 p. (3-907820-95-9)

Guide to the Sources of United States Military History. Ed. by Robin Higham / 1975. XIII, 559 p., numerous ills. / Hard (3-598-10432-4)

Guides to International Organization. Ed.: Union of International Associations
- 1: African International Organization. Directory and African Participation in other International Organizations / 1984. 598 p. / Hard (3-598-21650-5)
- 2: Arab and Islamic International Organization. Directory and Arab / Islamic Participation in other International Organizations / 1984. 477 p. / Hard (3-598-21651-3)
- 3: International Organization. Abbreviations and Adresses / 1985. 529 p. / Hard (3-598-21652-1)
- 4: Intergovernmental Organization. Directory / 1985. IV, 754 p. / Hard (3-598-21653-X)

Guides to the Sources for the History of the Nations / Guides des Sources de l'Histoire des Nations / Quellenführer zur Geschichte der Nationen. Ed.: International Council on Archives / Conseil International des Archives [1978 übernommen]
2nd Series: Africa South of the Sahara
- Vol. 9: Africa South of the Sahara in the Netherlands. Ed.: Netherland State Archive Service. Compiled by M. P. Roessingh; W. Visser / 1978. 241 p. / Hard (3-7940-3819-3)
3rd Series: North Africa, Asia and Oceania / 3ème Série: Afrique du Nord, Asie et Océanie (3-598-21460-X)
- Vol. 1: Guide des Sources de l'Histoire d'Afrique du Nord, d'Asie et d'Oceanie conservees en Belgique. Ed.: Archives générales du Royaume. Par Emile Vandewoude; André Vanrie / 1972. 622 p. (3-7940-3801-0)
- Vol. 2: Sources de l'Histoire de l'Asie et de l'Océanie dans les Archives et Bibliothèques françaises. Ed.: Commission Française du Guide des Sources de l'Histoire des Nations
  - Part 1. Archives / 1981. XXIII, 593 p. / Hard (3-598-21472-3)
  - Part 2. Bibliothèque Nationale / 1981. XI, 316 p. / Hard (3-598-21473-1)
  - Part 3. Autres bibliothèques / 1992. 330 p. / Hard (3-598-21488-X)
- Vol. 3: Sources of the History of North Africa, Asia and Oceania in Scandinavia
  - Part 1. Sources of the History of North Africa, Asia and Oceania in Denmark. Ed.: The Danish National Archives. Compiled by C. Rise Hansen / 1980. 842 p. / Hard (3-598-21474-X)

- Part 2. Sources of the History of North Africa, Asia and Oceania in Finland, Norway, Sweden. Ed.: The National Archives of Finland, The National Archives of Norway, The National Archives of Sweden. Compiled by B. Federley; Y. Kihlberg; D. Jörgensen; S. Söderlind; L. Näslund; F. Ludwigs / 1981. 233 p. / Hard (3-598-21475-8)
• Vol. 4: Sources of the History of Asia and Oceania in the Netherlands. Ed.: Netherlands State Archives Service Royal Institute of Linguistics and Anthropology
  - Part 1. Sources up to 1796 Comp. by Marius P. Roessingh / 1982. 337 p. / Hard (3-598-21476-6)
  - Part 2. Sources 1796-1949 Comp. by Frits G. Jaquet / 1983. 547 p. / Hard (3-598-21477-4)
• Vol. 5: Sources de l'Histoire du ProcheOrient et de l'Afrique du Nord dans les Archives et Bibliothèques françaises. Ed.: Commission française du Guide des Sources de l'Histoire des Nations
  - Part 1. Archives / 3 vols. 1996. Cplt. 1,365 p. / Hard (3-598-21478-2)
  - Part 2. Bibliothèque Nationale / 1984. X, 480 p. / Hard (3-598-21479-0)
• Vol. 6: Quellen zur Geschichte Nordafrikas, Asiens und Ozeaniens in der Bundesrepublik Deutschland bis 1945. Hrsg. v. Ernst Ritter / 1984. XLVI, 386 p. / Hard (3-598-21480-4)
• Vol. 7: Guía de Fuentes para la Historia de Asia en España. Ed. by the Spanish State Archives. Comp. by Luis S. Belda / 1987. 242 p. / Hard (3-598-21482-0)
• Vol. 8: Quellen zur Geschichte Afrikas, Asiens und Ozeaniens im Österreichischen Staatsarchiv bis 1918 / 1986. XII, 273 p. / Hard (3-598-21484-7)
• Vol. 9: Sources of the History of Africa, Asia, Australia and Oceania in Hungary. With a Supplement: Latin America. Ed.: National Archives of Hungary / 1991. XVII, 451 p. / Hard (3-598-21485-5)
• Vol. 10: Sources of the History of Africa, Asia and Oceania in Yugoslavia. Ed.: Union of Societies of Archivists of Yugoslavia / 1991. VII, 164 p. / Hard (3-598-21486-3)

Gurnsey, John: The Information Professions in the Electronic Age. / 1985. 256 p. (3-85157-380-0)

Hacker, Rupert: Bibliothekarisches Grundwissen
• Unter Mitarbeit v. Hans Popst; Rainer Schöller / 1971. 368 S., 19 Abb. U. Tab. / Lin (3-7940-3191-1)
• 2., unveränderte Aufl. Unter Mitarbeit v. Hans Popst; Rainer Schöller / 1973. 368 S., 19 Abb. U. Tab. / Lin (3-7940-3207-1) / Kst. (UTB 148) (3-7940-2607-1)
• 3., neubearb. Aufl. Unter Mitarbeit v. Hans Popst; Rainer Schöller / 1976. 432 S. / Kst. (UTB 148) (3-7940-2658-6)
• 4. völlig überarb.Aufl. 1982. 340 S., mehrere Tab., Reg / Kst. (UTB 148) (3-598-02671-4)
• 5. durchges. Aufl. 1989. 340 S. mehrere Tab., Reg / Br (3-598-10867-2)
• 6., völlig neubearb. Aufl. 1992. 406 S. / Br (3-598-11078-2)

Hackforth, Josef: Fernsehen, Programm, Programmanalyse. Bibliographie 1960-1969 / 1981. XXV, 542 S. / Lin (3-598-10329-8)

Hackforth, Josef / Ulrich Steden; Ute Alte-Teigeler: Fernsehen Programm Programmanalyse. Auswahlbibliographie 1970-1977 / 1978. XXIV, 621 S. / Lin (3-7940-7084-4)

Hadamitzky, Wolfgang: Japanese, Chinese, and Korean Surnames and How to Read Them. 125. 947 Japanese, 594 Chinese, and 259 Korean Surnames written with Kanji as they appear in Japanese texts. Vol. 1: From Characters to Readings. Vol. 2: From Readings to Characters / 2 vols. (Vol. 1 in 2 parts) 1998. Cplt. XIX, 1550 p. / Hard (3-598-11334-X)

Hadamitzky, Wolfgang / Marianne Rudat-Kocks: Japan-Bibliografie / Bibliography of Japan. Verzeichnis deutschsprachiger japanbezogener Veröffentlichungen / German-Language Publications on Japan / 6 Bde. in 13 Teilbänden 1990 ff. (3-598-22145-2)
• Reihe A: Monographien, Zeitschriften, Karten
  - Bd. 1: 1477-1920 / 1990. XXVI, 419 S. / Gb (3-598-22146-0)
  - Bd. 2: 1921-1950 / 1993. XVII, 258 S. / Gb (3-598-22149-5)
  - Bd. 3: 1951-1985 / 2 Tlbde. 1995 / 1996 Teil 1 / 1995. XVIII, 282 S. / Gb (3-598-22152-5) Teil 2 / 1996. XV, 434 S. / Gb (3-598-22156-8)
• Reihe B: Aufsätze
  - Bd. 1: Bis 1920 / 3 Tlbde. 1998 / 1999 Teil 1 / 1998. XXI, 409 S. / Gb (3-598-22147-9) Teil 2 / 1998. XIX, 384 S. / Gb (3-598-22148-7) *Teil 3 sowie Bde. 2 u. 3 [i.Vb.]*

Hänsel, Markus: Die anonym erschienenen autobiographischen Schriften des neunzehnten Jahrhunderts. Bibliographie. Mit einem Nachweis für die Bibliotheken Deutschlands / 1986. VI, 292 S. / Lin (3-598-10643-2)

Haller, Klaus: Katalogkunde. Formalkataloge und formale Ordnungsmethoden / 1980. 314 S. (3-598-10126-0)
• 2. Aufl. 1983. 315 S. / Lin (3-598-10431-6)
• 3. erw. Aufl. Katalogkunde. Eine Einführung in die Formal- und Sacherschliessung / 1998. 269 S. / Br (3-598-11364-1)

Haller, Klaus: Titelaufnahme nach RAK. Eine Einführung in die „Regeln für die Alphabetische Katalogisierung". Hrsg. Generaldirektion der Bayerischen Staatlichen Bibliotheken / 1976. X, 229 S. (3-7940-3036-2)
• 2., überarb. und erw. Aufl. In Zus.-Arbeit mit Mechthilde Bonse; Ursula Gaiser; Rupert Hacker; Hans Popst; Rainer Schöller / 1978. (3-7940-3036-2)
• 3., überarb. Aufl. 1979. 251 S., zahlr. Abb. / Lin (3-598-03046-0)

Haller, Klaus / Hans Popst: Katalogisierung nach den RAK-WB. Eine Einführung in die „Regeln für die alphabetische Katalogisierung in wissenschaftlichen Bibliotheken (RAK-WB)" / 1981. 278 S. / Lin (3-598-10363-8)
• 2. überarb. Aufl. 1982. 278 S. / Lin (3-598-10436-7)
• 3. überarb. Aufl. 1984. 293 S. / Lin (3-598-10543-6)
• 4. geänd. u. aktualis. Aufl. 1991. 303 S., mehrere Abb. / Br (3-598-10926-1)
• 5. überarb. Aufl. 1996. 327 S. / Br (3-598-11305-6)

Hamann, Richard: Theorie der bildenden Künste / 1983. 276 S., 75 Fotos, davon 2 farb., Reg / Kst. (UTB 1213) (3-598-02669-2)

Hamberger, G. C. / J. G. Meusel: Das gelehrte Teutschland oder Lexikon der jetzt lebenden teutschen Schriftsteller. Registerband zur 5. Ausgabe von 1796 bis 1834. Geleitw. v. Reinhard Oberschelp. Bearb. v. Maria Th. Kirchberg und Rainer Pörzgen / 1979. 448 S. / Lin (3-598-10059-0)

Hamburg-Bibliographie. Hrsg.: Staats- u. Universitätsbibliothek Hamburg - Carl von Ossietzky
• 1992 / 1995. XXI, 746 S. / Ln (3-598-23620-4)
• 1993 / 1997. XXI, 798 S. / Ln (3-598-23621-2)

A **Handbook** for African Writers. Comp. a. ed. by James Gibbs / 1986. VIII, 218 p. / Hard (3-598-10639-4) [Hans Zell Publishers]

Handbuch der baubezogenen Bedarfsplanung. Bemessung des Flächenbedarfs im Hochschulbereich. Mit einem Beiheft Tabellensätze / Hrsg.: Zentralarchiv für Hochschulbau, Stuttgart / 1974. 372 S. (3-598-10016-7)
• 1. Ergänzung zur Ausgabe 1974 / 1976. VI, 57 S. (3-598-10017-5)

Handbuch der Bayerischen Bibliotheken. Hrsg.: Generaldirektion der Bayerischen Staatlichen Bibliotheken / 1983. 371 S. / Br (3-598-10500-2)

Handbuch der Bibliotheken Bundesrepublik Deutschland, Österreich, Schweiz. Hrsg.: Helga Lengenfelder / 1984. XIV, 329 S. / Lin (3-598-10522-3)
• 2. Aufl. Hrsg. v. Helga Lengenfelder. Bearb. v. Bettina Bartz; Ruth Lochar; Helmut Opitz / 1989. 540 S. / Lin (3-598-10827-3)
• 3. Aufl. Red.: Bettina Bartz; Helmut Opitz; Elisabeth Richter / 1993. X, 650 S. / Lin (3-598-11156-8)
• 4. Aufl. Red.: Bettina Bartz; Helmut Opitz / 1996. X, 670 S. / Lin (3-598-11288-2)
• 5. Ausg. Red.: Willemina von der Meer; Helmut Opitz / 1998. XII, 633 S. / Lin (3-598-11351-X)

Handbuch der Internationalen Dokumentation und Information / Handbook of International Documentation and Information (ISSN 0939-1959) bis1972 u.d.T.: Handbuch der technischen Dokumentation und Bibliographie. Hrsg. von Karl-Otto Saur / 10 Bde. in 22 Teilen / 1957 - 1971 (3-7940-1100-7)

~ **Bd. 1**: Die technische Dokumentation, ihre Träger, Verfahren und Mittel. Bearb. von Karl Otto Saur und Grete Gringmuth / 1957. 231 S.
• 2. Aufl. 1961. IV, 354 S.
• 3. Aufl. 1964. VI, 540 S.
• 4. erw. u. neubearb. Ausg. Die Internationale wissenschaftliche Dokumentation und Information / 1969. XVI, 663 S., 4 Übersichten / Lin (3-7940-1019-1)
• Laisiepen, Klaus / Ernst Lutterbeck; Karl H. Meyer Uhlenried: Grundlagen der praktischen Information und Dokumentation. Eine Einführung / 2. völlig neubearb. Aufl. 1980. XX, 826 S. / Lin (3-598-20518-X)
gleichzeitig als Bd. 1 der DGD-Schriftenreihe (3-598-21251-8) erschienen
1. u. 3. Aufl.: → DGD-Schriftenreihe
4. Aufl.: → Grundlagen d. praktischen Information

~ **Bd. 2** / Vol. 2: Die Fachliteratur für den Dokumentar. Bibliographien für Technik und Wirtschaft / 1957 / Lbl.
• 2. erw. Aufl. Fachliteratur für den Dokumentar und Bibliothekar in 1300 Literaturnachweisen / 1958 / Lbl.
• 3. erw. Aufl. Fachliteratur für den Dokumentar und Bibliothekar in 1600 Literaturnachweisen / 1959 / Lbl.

- 4. Aufl. Die Fachliteratur für den Bibliothekar, Buchhändler, Dokumentar, Literaturingenieur und Verleger / 1961. LV, 322 S.
- 5. Aufl. Die Fachliteratur für den Autor, Bibliothekar, Buchhändler, Dokumentar, Literaturingenieur und Verleger / 1963. XLII, 394 S.
- 6. Ausg. Die Fachliteratur zum Buch- und Bibliothekswesen / 1965. CXII, 621 S.
- 7. Ausg. 1966.
- 8. Ausg. 1967. VIII, 636 S.
- 9. Ausg. 1970. 650 S. (3-7940-1027-2)
- 10. Ausg. International Bibliography of the Book Trade and Librarianship / Fachliteratur zum Buch- und Bibliothekswesen. Ed. by Helga Lengenfelder; Gitta Hausen. T.1. Europa / 1973. XIX, 406 S. (3-7940-1127-9)
- 11. Ausg. 1973-75 [nebst Nachtr.] / 1976. XXIV 704 S. [Mehr als 9000 Titel.]
- 12th ed. (1976-1979) / 1981. XXIX, 692 p. / Hard (3-598-20516-3)

~ **Bd. 3** 2700 technische Bibliographien des In- und Auslandes, nach 56 Fachgebieten geordnet / 1959. 653 S. / Lbl.
- 4. Ausg. Fach- und Spezialbibliographie sowie Dokumentationsdienste für Technik, Wissenschaft, Wirtschaft / 1962. LIII, 392 S.
  - Ergänzungsbd.: Internationale Bibliographie der Fachbibliographien für Technik, Wissenschaft, Wirtschaft / 1963. XXXVI, 283 S.
- 5. Ausg. Bd. 3 A: Internationale Bibliographie der Fachbibliographien für Technik, Wissenschaft u. Wirtschaft. Nachweise der einmalig erscheinenden Bibliographien, Dokumentationen, Kataloge, Verzeichnisse für Technik, Wissenschaft, Wirtschaft und ihre Einzelgebiete. 1964. XLVIII, 470 S.
- 7. Ausg. Stand: Nov 1965 / 1965. L, 329 S.
- 8. Ausg. Stand: Nov 1966 / 1966. XXXVIIII, 305 S.
- 9. Ausg. Einmalig erscheinende Bibliographien / 1969. VIII, 887 S. / Lin (3-7940-1039-6)

~ **Bd. 3 B**: Internationale Bibliographie der Fachbibliographien für Technik, Wissenschaft u. Wirtschaft. Ausgabe B. Nachweise der periodisch erscheinenden Bibliographien, Dokumentationen, Kataloge, Verzeichnisse für Technik, Wissenschaft, Wirtschaft und ihre Einzelgebiete. Stand: Juni 1965 / 1965. IL, 513 S.
- 2. Ausg. 1969. VIII, 825 S. / Lin (3-7940-1139-2)

~ **Bd. 4**: Technik und Wirtschaft in fremden Sprachen. Internationale Bibliographie der Fachwörterbücher, Fachencyklopädien, Fachsprachlehrmittel, der Abhandlungen zur Übersetzungstechnik, zur Wörterbuchherstellung, der Bibliographien von Wörterbüchern und Übersetzungsdiensten / 1960. XII, 449 S.
  - 1. Ergänzungsbd. / 1961. VIII, 181 S.
  - 2. Ergänzungsbd. / 1962. CIX, 224 S.
- 2., erw. Aufl. 1963. LIV, 338 S.
  - Ergänzungsbd. 1964. XLVI, 244 S.
- 3. Ausg. 1966. CXLVI, 304 S.
  - Ergänzungsbd. 1967. XCII, 229 S.
- 4. Ausg. Technik, Wissenschaft und Wirtschaft in fremden Sprachen. Internationale Bibliographie der Fachwörterbücher / Techniques, science and economics in foreign languages. International bibliography of dictionaries / 1969. VIII, 633 S.
- 5., neu bearb. Ausg. Fachwörterbücher und Lexika. Ein internationales Verzeichnis / International bibliography of dictionaries. Red. Leitung: Helga Lengenfelder / 1972. XXVI, 511 S.
- 6. Ausg. International Bibliography of Specialized Dictionaries / Fachwörterbücher und Lexika.

Hrsg. von Helga Lengenfelder unter Mitarb. von Henri van Hoof / 1979. XXI, 470 S.

~ **Bd. 5** / Vol. 5: Internationale Bibliographie der Fachadressbücher Wirtschaft, Wissenschaft, Technik / International bibliography of economics-, science- and technique-directories / 1962. LXVII, 309 S.
- Ergänzungsbd. 1963. XLVI, 264 S.
- 2. Ausg. 1964. LXXII, 318 S.
- 3. Ausg. 1966. 422 S.
- 4. Ausg. 1970. X, 652 S. (3-7940-1056-6)
- 5. Ausg. International Bibliography of Special Directories / Internationale Bibliographie der Fachadressbücher. Ed. by Helga Lengenfelder / 1973. XV, 535 S.
- 6. Ausg. 1978. XVI, 472 S.
- 7th ed. 1982. XX, 474 p. / Hard (3-598-20520-1)

~ **Bd. 6** / Vol. 6: Internationaler Generalkatalog der Zeitschriften für Technik und Wirtschaft. Vollst. bibliograph. Nachweise d. Periodika, getr. nach 50 Fachgebieten u. d. 78 Ländern, wo sie erschienen sind. Stand September 1962. Bearb. v. Karl-Otto Saur; Grete Gringmuth; Bärbel Saur / 1. Ausg. 1962. CXX, 482 S.
  - Zeitschriftenverzeichnis zum 6. Band 1962 / 1962. S. CXXI-CCLIII
- 2. Ausg.: Ergänzung 1963 zur 1. Ausgabe 1962 mit Stand Oktober 1963. Bearb. v. Karl-Otto Saur; Bärbel Saur / 1963. CXXVII, 324 S.
- 3. Ausg. Internationale Bibliographie der Fachzeitschriften für Technik und Wirtschaft / 1964
- 4. Ausg. 1965. V, 495 S.
- 5. Ausg. Internationale Bibliographie der Fachzeitschriften / World Guide. to Periodicals. Bearb. von Karl-Otto Saur / 3 Teile 1967. LXII, 2272 S. (3-7940-1367-0)

~ **Bd. 7** / Vol. 7: Internationales Verlagsadressbuch Verz. v. 14100 Buch- u. Zeitschriftenverlagen aus 100 Ländern sowie d. Anschriften v. 208 Verleger- u. Sortimenterverbänden / 1964. XIV S. 338 Bl.
- 2. Ausg. aus 140 Ländern mit Sachreg. zu d. Verlagsgebieten sowie d. Anschriften buchhändlerischer Organisationen, Verleger- u. Sortimenter-Verbände / 1965. XXVI, 668 S.
- 3. Ausg. 1967. XXXII, 764 S.
- 4. Ausg. 1969. 3 Tle. 1.337 S. (3-7940-1080-9)
- 5., überarb. Ausg. Internationales Verlagsadreßbuch mit ISBN-Register / Publisher's International Directory with ISBN Index / 2 Tle. 1972. XXIII, 1168 S. / Lin (3-7940-1170-8)
  - Teil 1: Europa / Europe / 1972. XVI, 543 S. / Lin (3-7940-1171-6)
  - Teil 2: Afrika, Amerika, Asien, Ozeanien / Africa, America, Asia, Oceania. Register, Index / 1972. XVII-XXIII, S. 545-1168 / Lin (3-7940-1172-4)
- 6. Ausg. 1974. XV, 1182 S., 213 S. Anhang / Lin (3-7940-1270-4) [Mit einem Anhang: Internationales ISBN-Verlagsregister. Bearb. u. Hrsg.: Internationale ISBN-Agentur Staatsbibliothek Preußischer Kulturbesitz, Berlin / 1975. → *Internationales ISBN-Verlagsregister*]
- 7. Ausg. 1977. XI, 620 S., 270 S. Anhang / Lin (3-7940-1371-9) [Mit einem Anhang: International ISBN Publishers' Index / Internationales ISBN-Verlagsregister. Bearb u. Hrsg.: Internationale ISBN-Agentur Staatsbibliothek Preußischer Kulturbesitz / 2nd. Ed. 1977. → *Internationales ISBN-Verlagsregister*]
- 8th ed. Publishers' International Directory with ISBN Index / Internationales Verlagsadressbuch mit ISBN-Register. Bearb u. Hrsg.: Internationale

ISBN-Agentur Staatsbibliothek Preußischer Kulturbesitz / 2 vol. 1979. 890 p. / Hard (3-7940-1372-7)
- 9th ed. 2 parts 1982. 1190 p. / Hard (3-598-20519-8)
- 10th ed. 1983. 2 vols. 1357 p. / Hard (3-598-20521-X)
- 11th ed. / 1984. 2 vols. XIX, 1603 p. / Hard (3-598-20525-2)
- 12th ed. / 1985. 2 vols.: Vol. I: XIX, 1452 p.; Vol. II: V, 406 p. / Hard (3-598-20529-5)
- 13th ed. / 1986. 2 vols.: Vol. I: XIX, 1158 p.; Vol. II: V, 1068 p. / Hard (3-598-20532-5)
- 14th ed. / 2 vols. 1987. Vol. I: XIX, 1221 p., Vol. II: VI, p. 1223-2192 / Hard (3-598-20535-X)
- 15th ed. / 2 vols. 1988. Vol. I: XIX, 1770 p., Vol. II: V, 1771-2250 / Hard (3-598-20537-6)
- 16th ed. Publishers' International ISBN-Directory (PIID). Ed.: International ISBN Agency, Berlin / 2 vols. 1989. Cplt. XXXXIII, 1925 p. / Hard (3-598-20538-4)
- 17th ed. Ed. in collab. with International ISBN Agency, Berlin / 3 vols. 1990. Cplt. 3800 p. / Lin (3-598-20540-6)
- 18th ed. 3 vols. 1991. Cplt. XXXVII, 4624 p. / Hard (3-598-21600-9)
- 19th ed. 3 vols. 1992. Cplt. XXXXV, 5073 p. / Hard (3-598-21601-7)
- 20th ed. 3 vols. 1993. Cplt. L, 5662 p. / Hard (3-598-21602-5)
- 21st ed. 3 Vols. 1994. Cplt. L, 3847 p. / Hard (3-598-21603-3)
- 22nd ed. 3 vols. 1995. Cplt. IL, 4138 p. / Hard (3-598-21604-1)
- 23rd ed. 3 vols. 1996. Cplt. 4481 p. / Hard (3-598-21605-X)
- 24th ed. 3 vols. 1997. Cplt. III, 5046 p. / Hard (3-598-21606-8)
- 25th ed. 3 vols. 1998. Cplt. LII, 4436 p. / Hard (3-598-21607-6)

CD-ROM-Ausgabe: → *Publishers'* International ISBN-Directory PLUS (PIID PLUS)

~ **Bd. 8** / Vol. 8: Internationales Bibliotheksadreß-buch. Bearb. von Klaus Gerhard Saur / 2 Bde. 1966. 1524 S.
- 2. Ausg. 1968. 1952 S.
- 3. Ausg. Internationales Bibliotheks-Handbuch / 1970. 4 Tle. 2281 S. (3-7940-1488-X)
- 4., erw. Aufl. Internationales Bibliotheks-Handbuch / World Guide to Libraries. Hrsg. v. Helga Lengenfelder / 2 Teile 1974. XXIX, 1603 S. (3-7940-1588-6)
  - Tl. 1: Europa, Amerika (3-7940-1688-2)
  - Tl. 2: Afrika, Asien, Ozeanien, Register (3-7940-1788-9)
- 5th ed. World Guide to Libraries / Internationales Bibliotheks-Handbuch. Hrsg. v. Helga Lengenfelder / 1980. XXV, 1030 p. / Hard (3-598-205033-1)
- 6th ed. Hrsg. v. Helga Lengenfelder / 1983. XXXII, 1186 p. / Hard (3-598-20523-6)
- 7th ed. Hrsg. v. Helga Lengenfelder / 1985. XXXIII, 1203 p. / Hard (3-598-20531-7)
- 8th ed. Editorial Director: Helga Lengenfelder. Ed by Barbara Fischer; Ruth Lochar; Bettina Pautler / 1987. XXXIV, 1279 p. / Hard (3-598-20536-8)
- 9th ed. Ed.: Helga Lengenfelder. Ed. by Bettina Bartz; Ruth Lochar; Helmut Opitz / 1989. XXXVIII, 1001 p. / Hard (3-598-20539-2)
- 10th ed. Ed. by Bettina Bartz; Helmut Opitz; Elisabeth Richter / 1991. XXVIII, 1039 p. / Hard (3-598-20541-4)
- 11th ed. Ed. by Bettina Bartz; Helmut Opitz; Elisabeth Richter / 1993. XXVIII, 1178 p. / Hard (3-598-20720-4)

- 12th ed. Ed. by Bettina Bartz; Helmut Opitz / 2 vols. 1995. Cplt. XXXII, 1082 p. / Hard (3-598-20721-2)
- 13th ed. Ed. by Willemina van der Meer; Helmut Opitz / 2 vols. 1998. Cplt. XXXVI, 1187 p. / Hard (3-598-20723-9)
- 14th ed. Ed. by Willemina van der Meer / 2 vols. 1999. Cplt. XXXIV, 1210 p. / Hard (3-598-20725-5)

~ **Bd. 9** / Vol. 9: Internationales Verzeichnis von Abkürzungen. Bearb. von Klaus Gerhard Saur / 2 Teile 1968. 1.148 S.
- Spillner, Paul: Internationales Wörterbuch der Abkürzungen von Organisationen / International Dictionary of Abbreviations of Organizations
  - Tl. 1: A-H / 2 Bde. 1970. 527 S. (3-7940-1098-1)
  - Tl. 2: I-R / 3 Bde. 1971. XXXII, 430 S. (3-7940-1198-8)
  - Tl. 3: S-Z / 3 Bde. 2. Ausg. 1972. XXXIX, 334 S. (3-7940-1296-4)
- 2. Aufl. Internationales Wörterbuch der Abkürzungen von Organisationen / International Dictionary of Abbreviations of Organizations. Bearb. von Paul Spillner / 3 Teile 1970 / 1972. Zus. LV, 1.295 S. (3- 7940-1398-0)

~ **Bd. 10** / Vol. 10: Internationales Universitäts-Handbuch / World Guide to Universities / 1971/72. 4 Tle. Zus. CXXXIII, 3609 S. (3-7940-1060-4)
  - Tl. 1: Europa. Albanien-Großbritannien / 1971. XXII, 1.040 S. / Lin (3-7940-1010-8)
  - Tl. 2: Europa. Irland-Vatikan / 1971. S. IL-LXVII, 1041-1928 / Lin (3-7940-1020-5)
  - Tl. 3: Europa. Amerika: Kanada, Vereinigte Staaten, Lateinamerika / 1971. S. LXXXIX-CXXVI, 1929-2705 / Lin (3-7940-1040-X)
  - Tl. 4: Afrika, Asien, Ozeanien, Gesamtregister / 1972. S. CXXVII-CXLIII, 2707-3609 / Lin (3-7940-1050-7)
- 2., erw. Ausg. Bearb. v. Michael Zils.
  - Tl. I / 1-2: Europa / 1976. 2 Bde. LXXXIX, 1253 S. / Lin (3-7940-1130-9)
  - Tl. II / 1-2: Afrika, Amerika, Asien, Ozeanien Afrika, Amerika, Asien, Ozeanien / 1977. 2 Bde. LVI, 1994 S. / Lin (3-7940-1145-7)

~ **Bd. 11** / Vol. 11: Hoof, Henry van: Internationale Bibliographie der Übersetzung / International Bibliography of Translation / 1973. XVI, 591 S. / Lin (3-7940-1011-6)

~ **Bd. 12** / Vol. 12: Internationales Verzeichnis der Wirtschaftsverbände / World Guide to Trade Associations. Bearb. von Michael Zils / 2 Teile 1973. XXII, 1.429 S. / Lin (3-7940-1032-9)
  - Tl. 1: Europa / 1973. IX, 659 S. / Lin (3-7940-1012-4)
  - Tl. 2: Afrika, Amerika, Asien, Ozeanien, Register / 1973. S. XI-XX, 661-1429 / Lin (3-7940-1022-1)
- 2nd ed. World Guide to Trade Associations / Internationales Verzeichnis der Wirtschaftsverbände. Ed. by Michael Zils / 1980. XVI, 845 p. / Hard (3-598-20513-9)
- 3rd ed. Ed. by Michael Zils / 1985. XXXIII, 1,259 p. / Hard (3-598-20527-9)
- 4th ed. World Guide to Trade Associations. Compiled by Michael Zils / 1995. XX, 536 p. / Hard (3-598-20722-0)
- 5th ed. Ed. by Michael Zils / 1998/99. 2 Vol. / Hard (3-598-20724-7)
  - Vol. I: Trade Associations / 1998. XVIII, 606 p. / Hard (3-598-20726-3)
  - *Vol. II: Chambers of Industry and Commerce [i.Vb.]*

~ **Bd. 13** / Vol. 13: Verbände und Gesellschaften der Wissenschaft. Ein internationales Verzeichnis / World Guide to Scientific Associations. Bearb. v. Michael Zils / 1974. XIV, 481 S. / Lin (3-7940-1013-2)
- 2. Ausg. World Guide to Scientific Associations and Learned Societies / Internationales Verzeichnis wissenschaftlicher Verbände und Gesellschaften. Bearb. v. Michael Zils / 1978. XII, 510 S. (3-7940-1213-5)
- 3rd ed. Ed. by Michael Zils / 1982. 619 S. (3-598-20517-1)
- 4th ed. Ed. by Barbara Verrel. Comp. by Helmut Opitz / 1984. XIX, 947 p. / Hard (3-598-20522-8)
- 5th ed. Ed. by Michael Sachs / 1990. XVI, 672 p. / Hard (3-598-20530-9)
- 6th ed. Ed. by Michael Zils / 1994. XIV, 542 p. / Hard (3-598-20580-5)
- 7th ed. Ed. by Michael Zils / 1997. XIV, 529 p. / Hard (3-598-20581-3)

~ **Bd. 14** / Vol. 14: Wennrich, Peter: Anglo-amerikanische und deutsche Abkürzungen in Wissenschaft und Technik / Anglo-American and German Abbreviations in Science and Technology / 4 Bde.: Tl. 1-3 u. Nachtr. / Lin (3-7940-1014-0)
- Tl. 1: A - E / 1976. VII, 607 S. / Lin (3-7940-1024-8)
- Tl. 2: F - O / 1977. V, S. 609 - 1448 / Lin (3-7940-1024-8)
- Tl. 3: P - Z / 1978. V, S. 1449 - 2276 / Lin (3-7940-1044-2)
- Tl. 4: Nachträge / 1980. VI, 618 S. / Lin (3-598-20512-0)

~ **Bd. 15** / Vol. 15: Internationales Buchhandels-Adressbuch. International Directory of Booksellers. Bearb. v. Michael Zils / 1978. XII, 948 S. / Lin (3-7940-1115-5)
- 2nd edition / 2. Ausgabe. International Book Trade Directory: Europe, Australia, Oceania, Latin America, Africa and Asia / Internationales Buchhandelsadressbuch: Europa, Australien, Ozeanien, Lateinamerika, Afrika und Asien 1989 Comp. by Michael Sachs / 1989. XII, 739 p. / Hard (3-598-10755-2)
- 3rd ed. 1997. XII, 495 p. / Hard (3-598-22236-X)

~ **Bd. 16** / Vol. 16: Museums of the World / 3rd rev. ed. / 1981. VIII, 623 p. / Hard (3-598-10118-X)
- 4th rev. and enl. Ed. by Bettina Bartz; Helmut Opitz; Elisabeth Richter / 1992. IV, 642 p. / Hard (3-598-20533-3)
- 5th rev. and enl. ed. Ed. by Elisabeth Richter / 1995. IX, 672 p. / Hard (3-598-20604-6)
- 6th ed. Ed. by Bettina Bartz / 1997. IX, 673 p. / Hard (3-598-20605-4)
1. u. 2. Aufl. → Museums of the World

~ **Bd. 17** / Vol. 17: World Guide to Special Libraries / Internationales Handbuch der Spezialbibliotheken. Ed. by Helga Lengenfelder / 1983. XXX, 990 p. / Hard (3-598-20528-7)
- 2nd ed. Ed. by Bettina Bartz; Ruth Lochar; Helmut Opitz / 2 vols. 1990. Cplt. XXXVI, 1196 p. / Hard (3-598-22230-0)
- 3rd ed. Ed. by Helmut Opitz; Elisabeth Richter / 2 vols. 1995. Cplt. XXXV, 1260 p. / Hard (3-598-22234-3)
- 4th ed. Ed. by Willemina van der Meer; Helmut Opitz / 2 vols. 1998. Cplt. XLII, 1275 p. / Hard (3-598-22249-1)

~ **Bd. 18** / Vol. 18: Herbote, Burkhard: Handbuch für deutsch-internationale Beziehungen / Handbook of German-International Relations. Verzeichnis deutscher und ausländischer Vertretungen, Verbindungsbüros und Informationsstellen in Politik, Wirtschaft, Kultur, Medienwirtschaft und Tourismus / A Directory of German and Foreign Representation Abroad, Liason Offices and Information Center / 1994 XVI, 660 S. / Gb (3-598-22231-9)
- 2. überarb. Aufl. Herbote, Burkhard: Handbuch für deutsch-internationale Beziehungen / Handbook of German-International Relations. Verzeichnis deutscher und ausländischer Vertretungen, Verbindungsbüros und Informationsstellen aus den Bereichen Politik, Aussenwirtschaft und Banken, Kulturaustausch, Medien, Entwicklungszusammenarbeit, humanitärer Hilfe, Wissenschaft und Forschung. Hrsg. in Zus.-Arb. mit d. Institut f. Auslandsbeziehungen, Stuttgart / 3 Bde. 1997. Zus. LIV, 1203 S. / Gb (3-598-22235-1)
  - Bd. 1. Europa / Europe / 1997 XVIII, 415 S. / Gb (3-598-22237-8)
  - Bd. 2. Afrika und America / Africa and America / 1997 XVIII, 401 S. / Gb (3-598-22238-6)
- Bd. 3. Asien und Ozeanien / Asia and Oceania / 1997 XVIII, 387 S. / Gb (3-598-22239-4)

~ **Bd. 19** / Vol. 19: World Guide to Foundations. Ed. by Michael Zils / 1997. XIV, 559 p. / Hard (3-598-11315-3)

Handbuch der katholisch-theologischen Bibliotheken. Hrsg.: Arbeitsgemeinschaft Katholisch-Theologischer Bibliotheken.
- 2., neu bearb. u. wesentl. erw. Aufl. u.d.T.: Handbuch der kirchlichen katholisch-theologischen Bibliotheken in der Bundesrepublik Deutschland und in West-Berlin. Bearb. v. Franz Rudolf Reichert. Mit e. Beitr. über kathol.-öffentl. Büchereien von Erich Hodik / 1979. 175 S., 1 Kte / Gb (3-598-07072-1)
- 3. völlig neubearb. Ausg. Handbuch der katholisch-theologischen Bibliotheken. Bearb. v. Franz Wenhardt. Mit einem Beitrag über katholische öffentliche Büchereien v. Erich Hodick / 1991. 175 S., 1 Kte / Gb (3-598-10919-9)

Handbuch der Massenkommunikation. Hrsg. v. Kurt Koszyk und Karl H. Pruys / 1981. 368 S. / Lin (3-598-10406-5) [Lizenzausg. m. Genehm. v. Deutschen Taschenbuch Vlg. (dtv), München]

Handbuch der Museen / Handbook of Museums Bundesrepublik Deutschland, Deutsche Demokratische Republik, Österreich, Schweiz, Liechtenstein
- Bd. 1: Bundesrepublik Deutschland (3-7940-3411-2); Bd. 2: DDR, Österreich, Schweiz (3-7940-3412-0) / Bearb. von Gudrun B. Kloster / 2 Bde. / 1971. XXII, 1300 S. (3-7940-3413-9)
- 2. verb. Aufl. 1981. VII, 750 S. / Lin (3-598-10345-X)

Handbuch der Pressearchive Mit einem Anhang Presse-, Rundfunk-, Fernseh-, Filmarchive. InternationaleAuswahlbibliographie 1971-1982. Hrsg. v. Hans Bohrmann; Marianne Englert. Anh. Zsgst. v. Wilbert Ubbens / 1984. 265 S. / Lin (3-598-10361-1)

Handbuch der technischen Dokumentation und Bibliographie: → Handbuch der Internationalen Dokumentation und Information / Handbook of International Documentation and Information

Handbuch der Tonstudiotechnik. Hrsg.: Schule für Rundfunktechnik. Bearb. v. Michael Dickreiter / 2., verb. Aufl. 1978. XXXII, 861 S. (3-7940-7065-8)

- 3., durchges. Aufl. 1979. XXXI, 862 S.
  (3-598-07065-9)
- 4., durchgesehene Aufl. 1980. 862 S.
  (3-598-10199-6)
- 5. völlig neubearb. Aufl. 2 Bde. / Lin
  (3-598-10588-6)
  - Bd. 1. Raumakustik, Schallquellen, Schallwahr-
    nehmung, Schallwandler, Beschallungstechnik,
    Aufnahmetechnik, Klanggestaltung / 1987. XIV,
    429 S. / Lin (3-598-10589-4)
  - Bd. 2. Tonregieanlagen, Hörfunk-Betriebstech-
    nik, Schallspeicherung, Digitaltechnik / 1990.
    XV, 453 S., 400 Abb. / Lin (3-598-10590-8)
- 6. vollst. neubearb. Aufl. 2 Bde. 1997 / Lin
  (3-598-11320-X)
  - Bd. 1. Raumakustik, Schallquellen, Schallwahr-
    nehmungen, Schallwandler, Beschallungstech-
    nik, Aufnahmetechnik, Klanggestaltung / 1997.
    XIV, 430 S. / Lin (3-598-11321-8)
  - Bd. 2. Tonregieanlagen, Hörfunk-Betriebstech-
    nik, Schallspeicherung, Digitaltechnik / 1997.
    XVI, 520 S. / Lin (3-598-11322-6)

Handbuch der Universitäten und Fachhochschulen.
Deutschland, Österreich, Schweiz (ISSN 1430-4635)
Bis zur 6. Ausg. 1993 u. d. T.: Handbuch der Universitäten und Fachhochschulen. Bundesrepublik
Deutschland, Österreich, Schweiz (ISSN 0933-4831)
- 3. Aufl. Handbuch der Universitäten und Fachhochschulen. Bundesrepublik Deutschland,
  Österreich, Schweiz. Hrsg. v. Helga Lengenfelder.
  Redakt.: Gertrud Conrad-Bergweiler, Barbara
  Fischer und Angelika Riedel / 1985. XXII, 329 S.
  (3-598-10534-7)
- 4. Aufl. Red.: Barbara Fischer, Ruth Lochar und
  Bettina Pautler / 1988. XX, 513 S. / Lin
  (3-598-10718-8)
- 5. Ausg. Red.: Bettina Bartz, Helmut Opitz und
  Elisabeth Richter / 1990. XX, 487 S. / Lin
  (3-598-10895-8)
- 6. Aufl. Red.: Bettina Bartz, Helmut Opitz und
  Elisabeth Richter / 1993. XXII, 546 S. / Lin
  (3-598-11155-X)
- 7. Aufl. Handbuch der Universitäten und Fachhochschulen. Deutschland, Österreich, Schweiz.
  Red.: Bettina Bartz, Helmut Opitz / 1996. XXIII,
  578 S. / Lin (3-598-11287-4)
- 8. Ausg. Mit Registern zu den Hochschulleitern,
  Professoren und Sachgebieten. Red.: Helmut
  Opitz, Axel Schniederjürgen / 1998. XXIV, 718 S. /
  Ln (3-598-11352-8)
- 9. Ausg. Red.: Peter Nouwens; Helmut Opitz /
  1999. XXVI, 767 S. / Ln (3-598-11388-9)
1. u. 2. Aufl → **Deutsches** *Universitäts-Handbuch*

Handbuch der Wasserzeichenkunde. Hrsg. u. bearb.
v. K. T. Weiss. - Unveränd. Nachdr. / 1983. X, 327 S. /
Ln (3-598-07208-2)

Handbuch des deutschsprachigen Exiltheaters
1933-1945. Hrsg. v. Frithjof Trapp; Werner
Mittenzwei; Henning Rischbieter; Hansjörg
Schneider / 2 Bde. in 3 Tlbdn 1998. XLVI, 1617 S. /
Gb (3-598-11373-0)
- Bd. I. Verfolgung und Exil deutschsprachiger
  Theaterkünstler. Bearb. v. Ingrid Maaß; Michael
  Philipp / 1998 X, 525 S. / Gb (3-598-11374-9)
- Bd. II. Biographisches Lexikon der
  Theaterkünstler. Bearb. v. Frithjof Trapp; Bärbel
  Schrader; Dieter Wenk; Ingrid Maaß / 2 Tlbde.
  1998. Zus. XXXVI, 1092 S. / Gb (3-598-11375-7)

Handbuch Lesen. Im Auftr d. Stiftung Lesen u. der
Deutschen Literaturkonferenz hrsg. v. Bodo
Franzmann, Klaus Hasemann, Dietrich Löffler und
Erich Schön, u. Mitarb. v. Georg Jäger, Wolfgang R.
Langenbucher und Ferdinand Melichar / 1999. XII,
690 S. u. 16 Abb.-S. / Ln (3-598-11327-7)

Handbuch Vereinte Nationen. Eine Veröffentlichung der Forschungsstelle der Deutschen Gesellschaft für die Vereinten Nationen, Bonn / Berlin.
Hrsg. v. Rüdiger Wolfrum; Norbert J. Prill; Jens A.
Bückner / 1977. XXV, 577 S. / Lin (3-7940-2248-3)

Handbuch zur „Völkischen Bewegung" 1871-1918.
Hrsg. v. Uwe Puschner, Walter Schmitz und Justus
H. Ulbricht / 1996. XXVII, 978 S. / Gb
(3-598-11241-6)
- Unveränd. Broschurausg. 1999. XXVIII, 978 S. / Br
  (3-598-11421-4)

Handbücher und Lexika zur Militärgeschichte.
Militärhistorische Nachschlagewerke aus dem 19.
und beginnenden 20. Jahrhundert. Mikrofiche-
Edition / 1997. ca. 23.300 S. auf 198 Fiches.
Lesefaktor 24x / Silberfiche (3-598-33871-6) /
Diazofiche (3-598-33870-8)

Hauke, Hermann / Hermann Köstler: Das Eichstätter Stundenbuch. Die lateinische Handschrift 428
der Bibliothek des Bischöflichen Seminars Eichstätt
/ 1977. 133 S., Teilfaks. u. Kommentar mit Bibliogr /
Ln (3-7940-7044-5)

Health Information for All: A Common Goal / In-
formation Biomédicale pour tous: Un Objectif
Commun / L'Informazione Biomedica per tutti: un
obiettivo comune Second European Conference of
Medical Libraries. Bologna, Italy, Nov. 2-6, 1988 /
Deuxième Conférence Européenne des Biblio-
thèques Médicales. Bologne, Italie, 2-6 Nov. 1988 /
Seconda Conferenza Europea delle Biblioteche
Biomediche. Bologna, Italia 2-6 N Ed. by David W.
Stewart; Derek J. Wright on behalf of the European
Association for Health Information and Libraries /
pour L'Association Européen pour l'Information et
les Bibliothèques de Santé / per L'Associazione
Europea per l'Informazione e le Biblioteca / 1990.
584 p. / Hard (3-598-10860-5)

Hebrew Books from the Harvard College Library.
Microfiche edition. Ed. by Charles Berlin / 1990-
1992. 4934 vols. on 11,453 fiches. Reader factor 24X
/ Silberfiche (3-598-41200-2) / Diazofiche
(3-598-41160-X)
- Part I. Rabbinical Works. Responsa, Homiletics,
  Religious Ethics, Codes of Jewish Law, Rabbinic
  Texts and Commentaries, Biblical Commentaries,
  Kabbalah and Hasidism, Haggadoth and Liturgy /
  3,198 vols. on 8,647 fiches. Reader factor 24X /
  Silverfiche (3-598-41201-0) Diazofiche
  (3-598-41161-8)
- Part II. Library Secular Works. Philosophy and
  Theology, Belles Lettres, Biography, History,
  Ladino, Judeo-Persian, Hebrew Language and
  Grammar, Judeo-Arabic / 1,736 vols. on 2,806
  fiches. Reader factor 24X / Silverfiche
  (3-598-41202-9) / Diazofiche (3-598-41162-6)
- Index to the Microfiche edition Ed. by Charles
  Berlin / 1996. X, 219 p. / Hard (3-598-41220-7)

Hefele, Bernhard: Drogenbibliographie. Verzeichnis
der deutschsprachigen Literatur über Rauschmittel
und Drogen von 1800 bis 1984 mit einer Übersicht
über internationale Bibliographien / 2 Bde. 1988.
924 S. / Lin (3-598-10671-8)

Hefele, Bernhard: Jazz Bibliography / Jazz Biblio-
graphie. International Literature on Jazz, Blues,
Spirituals, Gospel and Ragtime Music with a
selected List of Works on the Social and Cultural
Background from the Beginning to the Present /
Verzeichnis des internationalen Schrifttums über
Jazz, Blues, Spirituals, Gospel, ragtime mit einer
Auswahlbibliographie über den sozialen und
kulturellen Hinetergrund / 1981. VIII, 368 p. / Hard
(3-598-10205-4)

Hefele, Bernhard: Jazz, Rock, Pop / Jazz, Rock, Pop.
Eine Bibliographie deutschsprachiger Literatur 1988
und 1989 / A Bibliography of German-Language
Literature from 1988 to 1989 / 1991. XII, 457 S. / Gb
(3-598-10914-8)

Hegel Bibliography. Background Material on the
International Reception of Hegel within the Context
of the History of Philosophy / Hegel Bibliographie.
Materialien zur Geschichte der internationalen
Hegel-Rezeption und zur Philosophie-Geschichte.
- Part I / Teil I: Ed. by Kurt Steinhauer. Stichw.-Reg.
  von Gitta Hausen / 1980. XVI, 894 p. / Hard
  (3-598-03184-X)
- Part II / Teil II: Comp. by Kurt Steinhauer. Under
  the assist. of Hans-Dieter Schlüter. Keywood index
  by Anton Sergl / 2 parts 1998. Cplt. XVIII, 1,128 p.
  / Hard (3-598-10787-0)

Hehl, Hans: Die elektronische Bibliothek / 1999.
185 S. / Br (3-598-11416-8)

Heiber, Helmut: Universität unterm Hakenkreuz / 2
Tle. in 3 Bdn (3-598-22628-4)
- Tl. I: Der Professor im Dritten Reich. Bilder aus
  der akademischen Provinz / 1991. 652 S. / Ln
  (3-598-22629-2)
- Tl. II: Die Kapitulation der Hohen Schulen. Das
  Jahr 1933 und seine Themen. Bd. 1 / 1992.. 668 S.
  / Gb (3-598-22630-6). Bd. 2 1994. 858 S. / Gb
  (3-598-22631-4)

Heidtmann, Frank: Die bibliothekarische
Berufswahl. Eine empirische Untersuchung der
Berufswahl d. Bibliothekars d. gehobenen Dienstes
an öffentlichen u. wissenschaftl. Bibliotheken (Veröffentlichungen des Instituts für Bibliothekarausbil-
dung der Freien Universität Berlin Bd. 3) / 1974. 496
S. / Br (3-7940-3197-0)

Heidtmann, Frank: Zur Theorie und Praxis der
Benutzerforschung Unter besonderer Berücksichtigung der Informationsbenutzer von Universitätsbibliotheken / 1971. 283 S. (3-7940-3200-4)

Heimeran, Ernst: Von Büchern und von Bücher-
machern. Mit e. Nachw. von Herbert G. Göpfert /
1969. 196 S. / Lin (3-7940-3158-X)

Heinsius, Wilhelm: Alphabetische Verzeichnis der
von 1700 bis zu Ende 1810 erschienenen Romane
und Schauspiele, welche in Deutschland und in den
durch Sprache und Literatur damit verwandten
Ländern gedruckt worden sind. - Nachdruck d. Orig.
Ausg. 1813 / 1972. 376 Spalten

Helmrich, Herbert / Bernd Ahrndt: Schüler in der
DDR. Zwei Erlebnisberichte (1945-1952 / 1964-1977)
/ 1980. 157 S. / Br (3-598-10121-X)

Henker, Michael / Karlheinz Scherr; Elmar Stolpe: Von Senefelder zu Daumier. Die Anfänge der lithographischen Kunst. (Veröffentlichung zur Bayerischen Geschichte und Kultur, Nr. 16 / 88). Hrsg. v. Claus Grimm / 1988. 260 S., 217 Abb. / Ln (3-598-10803-6) / Br (3-598-10804-4)
- De Senefelder á Daumier. Les débuts de l'art lithographique. (Veröffentlichung zur Bayerischen Geschichte und Kultur Nr. 16 / 88) Ed. by Claus Grimm / 1988. 260 S., 217 illus. / Br (3-598-07540-5)

Henning, Hans: Faust-Variationen. Beiträge zur Editionsgeschichte vom 16. bis zum 20. Jahrhundert / 1993. 420 S. / Ln (3-598-11108-8)

Hentschke, Th.: Allgemeine Tanzkunst Theorie und Geschichte, antike und moderne (gesellschaftliche und theatralische) Tanzkunst und Schilderung der meisten National- und Charaktertänze. - Reprint d. Orig.-Ausg. 1836 / 1986. XVI, 278, XLV S. / Ln (3-598-07242-2)

Herbote, Burkhard: World Tourism Directory. 3rd ed. '95 / 96 / 3 Vols. 1995. Cplt. LX, 1945 p. / Hard (3-598-11181-9)
- Part 1: Europe / 1995. XX, 579 p. / Hard (3-598-11202-5)
- Part 2: Herbote, Burkhard / Marian Goldberg: The Americas / 1995. XX, 660 p. / Hard (3-598-11203-3)
- Part 3: Africa, Middle East, Asia and Oceania / 1995. XX, 706 p. / Hard (3-598-11204-1)

Hermann Hesse. Personen und Schlüsselfiguren in seinem Leben. Ein alphabetisches annotiertes Namensverzeichnis mit sämtlichen Fundstellen in seinen Werken und Briefen. Band I: A-I. Band II: J-Z. Hrsg. v. Ursula Apel, mit Unterstützung d. Komitees d. Internationalen Hermann-Hesse-Kolloquien in Calw / 2 Bde. 1989. Zus. XXII, 1.057 S. / Ln (3-598-10841-9)
- Supplement A - Z / 1993. XXVII, 512 S. / Ln (3-598-11158-4)

Hermann, Peter: Informationsrecherchesysteme / 1973. 280 S. / Ln (3-7940-4286-7)

Hermann, Peter: Praktische Anwendung der Dezimalklassifikation. Klassifizierungstechnik / 6., erw. u. verb. Aufl. 1970. 115 S. / Br (Vlg.-Nr. 04283)

Hermanns, Hartmut / Cay Lienau; Peter Weber: Arbeiterwanderungen zwischen Mittelmeerländern und den mittel- und westeuropäischen Industrieländern. Eine annotierte Auswahlbibliographie unter geographischen Aspekten / 1979. XV, 244 S. / Lin (3-598-10057-4)

Herzog, Hanna: Contest of Symbols. The Sociology of Election Campaigns through Israeli Ephemera. Frwd. by Sidney Verba. Ed.: Harvard University Library / 1988. 191 p. / Hard (3-598-10749-8)

Hessische Bibliographie. Hrsg.: Stadt- und Universitätsbibliothek Frankfurt am Main, in Zus.-Arb. mit den wissenschaftlichen Bibliotheken des Landes Hessen
- Bd. I. Berichtsjahr 1977 / 1979. XVI, 463 S. / Ln (3-598-20170-2)
- Bd. II. Berichtsjahr 1978. Mit Nachträgen aus 1977 / 1980. XVI, 528 S. / Ln (3-598-20171-0)
- Bd. III. Berichtsjahr 1979 / 1981. XV, 668 S. / Ln

(3-598-20172-9)
- Bd. IV. Berichtsjahr 1980 mit Nachträgen aus 1977-1979 / 1982. XV, 726 S. / Ln (3-598-20173-7)
- Bd. V. Berichtsjahr 1981 / 1983. XIV, 710 S. / Lin (3-598-20174-5)
- Bd. VI. Berichtsjahr 1982 mit Nachträgen aus 1977-1981 / 1984. XIV, 856 S. / Lin (3-598-20175-3)
- Bd. VII. Berichtsjahr 1983 mit Nachträgen ab 1965 / 1985. XVII, 749 S. / Ln (3-598-20176-1)
- Bd. VIII. Berichtsjahr 1984 / 1986. 759 S. / Ln (3-598-20177-X)
- Bd. IX. Berichtsjahr 1985 mit Nachträgen ab 1965 / 1987. XVII, 839 S. / Ln (3-598-20178-8)
- Bd. X. Berichtsjahr 1986 mit Nachträgen ab 1965 / 1988. XVII, 799 S. / Ln (3-598-20179-6)
- Bd. XI. Berichtsjahr 1987 mit Nachträgen ab 1965 / 1989. XVII, 844 S. / Ln (3-598-20180-X)
- Bd. XII. Berichtsjahr 1988 mit Nachträgen ab 1965 / 1990. XVII, 853 S. / Ln (3-598-20161-3)
- Bd. XIII. Berichtsjahr 1989 mit Nachträgen ab 1965 / 1992. XVII, 831 S. / Ln (3-598-20162-1)
- Bd. XIV. 1990 mit Nachträgen ab 1965 / 1993. XVII, 858 S. / Ln (3-598-20163-X)
- Bd. XV. Berichtsjahr 1991 mit Nachträgen ab 1965 / 1994. XVIII, 757 S. / Ln (3-598-20164-8)
- Bd. XVI. Berichtsjahr 1992 / 1995. XVIII, 845 S. / Ln (3-598-20165-6)
- Bd. XVII. Berichtsjahr 1993 / 1996. XVIII, 912 S. / Ln (3-598-20166-4)
- Bd. XVIII. Berichtsjahr 1994 / 1996. LVIII, 895 S. / Ln (3-598-20167-2)
- Bd. XIX. Berichtsjahr 1995 / 1997. XVIII, 862 S. / Ln (3-598-20168-0)
- Bd. XX. Berichtsjahr 1996 / 1998. XVIII, 956 S. / Ln (3-598-20169-9)

Hessische Bibliographie. Berichtszeitraum 1 / 1997-19 / 1995. CD-ROM-Edition. Hrsg.: Stadt- u. Universitätsbibliothek Frankfurt am Main, in Zus.-Arb. mit d. wissenschaftlichen Bibliotheken d. Landes Hessen Kumulierte Ausg. 1998 (3-598-40410-7)

Heuschele, Otto: Umgang mit dem Genius. Essays und Reden / 1974. 244 S. (3-7940-3031-1)

Hildreth, Charles R.: Library Automation in North America. A Reassessment of the Impact of New Technologies on Networking. Ed.: Commission of the European Communities / 1987. IX, 196 p. / Soft (3-598-10735-8)

Hink, Wolfgang: Die Fackel. Herausgeber Karl Kraus. Bibliographie und Register 1899 bis 1936 / 2 Bde. 1994. Zus. XXII, 678 S. / Gb (3-598-11175-4)

HIS-Brief. Hrsg.: Hochschul-Informations-System-GmbH, Hannover [Bis Nr. 33 im Eigenverlag]
- 34: Neuwirth, Wolfgang: ERZUL- Organisations-Handbuch. Ein Programmsystem zur Automatisierung d. Zulassg in Numerus-Clausus-Fächern an d. Univ. Erlangen-Nürnberg / 1973. 73 S. (3-7940-4334-0)
- 35: Heindlmeyer, Peter: Eine quantitative Analyse des Bedarfs an Hochschulabsolventen bis 1991 / 1973. 71 S. (3-7940-4335-9)
- 36: Organisation der Hochschule und des Studiums. Eine sozialwissenschaftliche Effizienzuntersuchung zur Planung v. Kapazitätserweiterungen. Zsfassg d. Projektberichts anhand ausgew. Ergebnisse. Hrsg.v. d. Forschungsgruppe Hochschulkapazität, Univ. Mannheim / 1973. 129 S. (3-7940-4336-7)

- 37: Ausbildungskapazität und Approbationsordnung Bericht über ein gemeinsames Seminar der Hochschul-Informations-System GmbH u. d. Univ. d. Saarlandes 1972 in Saarbrücken / 1973. 130 S. (3-7940-4337-5)
- 38: Flächenbedarf für das Aufbau- und Verfügungszentrum der Gesamthochschule Kassel. Raumprogramm als Ergebnis e. Flächenbilanzierg f. d. beiden ersten Baustufen. Horst Gerken u.a / 1973. 165 S. (3-7940-4338-3)
- 39: Personal-Verwaltungs-System (PVS). Ein EDV-gestütztes System zur Verwaltung d. Personals u. d. Stellen d. Hochschulen. Von Friedrich Blahusch u.a / 1973. XII, 202 S. (3-7940-4339-1)
- 40: Hochschulforschung Erfahrungen zur Datenerfassung im Forschungsbetrieb. Pilotstudien an der Univ. Hamburg. Von Ingo Hartmann u.a. Arbeitsgruppe FOPRODA / 1973. VIII, 133, 52 S. (3-7940-4340-5)
- 41: Frey, Hans Werner: VERSTAT - Veranstaltungsstatistik auf EDV-Basis. Unter Mitarb. von Gunther Riderer; Rainer Deisenroth / 1974. 132 S. (3-7940-4341-3)
- 42: Hempel, Joachim / Arndt Kehler: Probleme der Kosten-Nutzen-Analyse für Informationssysteme in öffentlichen Verwaltungen / 1974. VI, 78 S. (3-7940-4342-1)
- 43: Cloes, Henning: Einsatzmöglichkeiten des Informations- und Verwaltungssystems HIS-LVS-I im Lehrbetrieb von Hochschulen / 1974. IX, 130 S. (3-7940-4342-1)
- 44: Berg, Gunnar: Modell eines Dokumentationssystems Hochschulplanung. Vorschlag zur Errichtung e.kooperativen Informationsverbundes Meindl, Ulrich / 1974. XI, 125 S. (3-7940-4344-8)
- 45: Siefert, Joachim: Das Studium der Stadt-, Regional- und Landesplanung in der Bundesrepublik Deutschland WS 1972/73 - SS 1973/1974. IV, 368 S. (3-7940-4345-6)
- 46: Boberg, Gisela / Kirsten Ebeling: Organisationsreform der Hochschulbibliothek. Eine Bibliographie / 1974. VII, 58 S. (3-7940-4346-4)
- 47: Raum-, Gebäude- und Grundstücksdateien im Hochschulbereich. Bericht d. Aussschuß für Baubestandserfassung. Redaktion: Horst Gerken / 1974. XII, 171 S. (3-7940-4347-2)
- 48: Keller, Arno / Manfred Weichhold: Beiträge zur Hochschulentwicklungsplanung. Unter Mitw. von Helmut Berndt / 1974. X, 121 S. (3-7940-4348-0)
- 49: Hansen, Hans-Peter / Karl Heinrich Mylius: Die soziale Situation der Studenten. Ergebnisse einer Befragung d. Studentenschaft an d. Univ. Münster / 1974. XIX, 217 S. (3-7940-4349-9)
- 50: Prüfungs-Operations-System (POS). Bericht über e. Befragung bei Prüfungsämtern u. Studienberatungsstellen Von Alma Bischof-Peters u.a / 1974. X, 88 S. (3-7940-4350-2)
- 51: Brockstedt, Jürgen / Christa Deneke: Planung und Organisation in den Hochschulen. Organe und Institutionen / 1975. XVI, 305 S. (3-7940-4351-0)
- 52: Groh, Helmut / Ulrich Lange; Roland Schröder: EDV-gestützte Raumbelegung durch das Programmsystem ZUOP (Zuordnungsoptimierung). Bericht über die Anwendung an d. Gesamthochschule Kassel / 1975. IX, 80 S. (3-7940-4352-9)
- 53: Vorwerk, Claus: Stand der Datenverarbeitung in Verwaltung und Planung von Universitäten in den USA. Bericht über e. Studienreise durch die USA im Sept. 1974 / 1975. XIII, 125 S. (37940-4353-7)

- 54: Bonin, Hinrich / Rosemarie Wenzel: Datenschutz und Datensicherung im Rahmen d. Haushalts-, Kassen- und Rechnungswesens e. Hochschule bei Verwendung e. ADV-Systems. Dargest. am Beispiel d. Systems HISKAM / 1975. XII, 105 S. (3-7940-4354-5)
- 55: Brockstedt, Jürgen: EDV-Einsatz in Hochschulverwaltungen. Ausstattung u. Anwendungsbereiche Reiner Reissert. Unter Mitarb. von Christa Deneke / 1975. VIII, 347 S. (3-7940-4355-3)
- 56: Hochschulberichtssystem (HBS). Entwurf e. period., standarisierten Berichtssystems für wiss. Hochschulen Projektgruppe Erlangen / 1975. VII, 94 S. / Br (3-7940-4356-1)
- 57: Fischer, Jürgen / Christoph Oehler; Jochen Pohle: Hochschulentwicklungsplanung Konzepte, Verfahren, Arbeitshilfen / 1975. XIV, 309 S. (3-7940-4357)
- 58: Kautzmann, Alfred: Aufbau und Verwendung eines Gebäudeatlasses zur Ergänzung einer Hochschulraumdatei Manuelle u. EDV-mäßige Erstellung von schematischen Grundrissen / 1976. XI, 81 S. (3-7940-4358-8)
- 59: Giseler, Hinrich / Joachim Hempel; Jörg Winterhoff: Personalverwaltungs-System (PVS II). Bericht über die Weiterentwicklung eines EDV-gestützten Systems zur Verwaltung des Personals und der Stellen der Hochschulen / 1976. 102 S. (3-7940-4359-6)
- 60: Verwaltung, Steuerung und Prognose des Studentendurchlaufs in der Medizin Studenten-Informationssystem Medizin. Unter Mitarb. von Renate Foerst / 1975. IX, 103 S. (3-7940-4360-X)
- 61: Unterstützung der Lehrorganisation durch Datenverarbeitung. Hrsg.: Projektgruppe Lehrverwaltung u. -organisation / 1977. V, 337 S. (3-7940-4361-8)
- 62: Meindl, Ulrich: Zur Situation der Studienreform Eine kommentierte Dokumentation / 1977. VIII, 289 S. (3-7940-4362-6)
- 63: Kleinefenn, Andreas: Betriebskosten an Hochschulen / 1977. XI, 157 S. (3-7940-4363-4)
- 64: Schnitzer, Klaus. Raumbezogene Nutzungsuntersuchungen Handbuch zur Raumnutzungsuntersuchung. Dokumentation u. Entscheidungshilfe für d. Anwendung von Verfahren zur Unters. d. räuml. Nutzung in Hochschulen. Unter Mitarb. von H. Gerken u. G. Schneider / 1978. VII, 199 S. (3-7940-4364-2)

History of Women in the United States. Ed. by Nancy F. Cott, Yale University / 20 vols. in 28 parts 1992 / 1994. Cplt. 11,564 p. / Hard (3-598-41454-4) [1991 von Meckler Publishing, Westport, CT übernommen]
- Vol. 1. Theory and Method in Women's / 2 vols. 1992. Cplt. XXIV, 526 p. / Hard (3-598-41455-2)
- Vol. 2. Household Constitution and Family Relationships / 1992. XI, 430 p. / Hard (3-598-41456-0)
- Vol. 3. Domestic Relations and Law / 1992. XI, 449 p. / Hard (3-598-41457-9)
- Vol. 4. Domestic Ideology and Domestic Work / 2 vols. 1992. Cplt. XXIV, 591 p. / Hard (3-598-41458-7)
- Vol. 5. The Intersection of Work and Family Life / 2 vols. 1992. Cplt. XXVI, 704 p. / Hard (3-598-41459-5)
- Vol. 6. Working on the Land / 1993. XIV, 484 p. / Hard (3-598-41460-9)
- Vol. 7. Industrial Wage Work / 2 vols. 1993. Cplt. XXVIII, 656 p. / Hard (3-598-41461-7)
- Vol. 8. Professional and White-Collar Employments / 2 vols. 1993. Cplt. XXVIII, 668 p. / Hard (3-598-41462-5)
- Vol. 9. Prostitution / 1993. X, 379 p. / Hard (3-598-41463-3)
- Vol. 10. Sexuality and Sexual Behavior / 1993. XIV, 577 p. / Hard (3-598-41464-1)
- Vol. 11. Women's Bodies: Health and Childbirth / 1993. XIV, 392 p. / Hard (3-598-41465-X)
- Vol. 12. Education / 1994. XIV, 467 p. / Hard (3-598-41466-8)
- Vol. 13. Religion / 1994. XIV, 485 p. / Hard (3-598-41467-6)
- Vol. 14. Intercultural and Interacial Relations / 1994. XIV, 442 p. / Hard (3-598-41468-4)
- Vol. 15. Women and War / 1994. XIV, 504 p. / Hard (3-598-41469-2)
- Vol. 16. Women Together: Organizational Life / 1994. XIV, 515 p. / Hard (3-598-41470-6)
- Vol. 17. Social and Moral Reform / 2 vols. 1994. Cplt. XXXII, 795 p. / Hard (3-598-41471-4)
- Vol. 18. Woman and Politics / 2 vols. 1994. Cplt. XXVIII, 844 p. / Hard (3-598-41472-2)
- Vol. 19. Woman Suffrage / 2 vols. 1994. Cplt. XXVIII, 743 p. / Hard (3-598-41473-0)
- Vol. 20. Feminist Struggles for Sex Equality / 1994. XIV, 537 p. / Hard (3-598-41474-9)

Hitler. Reden, Schriften, Anordnungen Februar 1925 - Januar 1933. Hrsg.: Institut für Zeitgeschichte / 6 Bde. in 12 Tlbdn sowie Register u. 1 Ergänzungsbd. in 4 Tlbdn / 1991 ff. (3-598-21930-X)
- Bd. I: Die Wiedergründung der NSDAP. Februar 1925-Juni 1926. Hrsg. u. komment. v. Clemens Vollnhals / 1991. XXIX, 496 S. / Ln (3-598-21931-8)
- Bd. II: Vom Weimarer Parteitag bis zur Reichstagswahl Juli 1926 - Mai 1928
  - Teil 1: Juli 1926 - Juli 1927. Hrsg.: Bärbel Dusik / 1991. XVI, 437 S. / Ln (3-598-21932-6)
  - Teil 2: August 1927 - Mai 1928. Hrsg.: Bärbel Dusik / 1991. XVI, 445 S. / Ln (3-598-21937-7)
- Bd. II/A: Aussenpolitische Standortbestimmung nach der Reichstagswahl Juni-Juli 1928. Eingel. v. Gerhard L. Weinberg; Christian Hartmann; Klaus A. Lankheit. Hrsg. v. Weinberg, Gerhard L. / 1995. XXIV, 212 S. / Ln (3-598-22004-9)
- Bd. III: Zwischen den Reichstagswahlen. Juli 1928-September 1930
  - Teil 1: Juli 1928 - Februar 1929 / Hrsg. Bärbel Dusik u. Klaus A. Lankheit, unter Mitw. v. Christian Hartmann / 1993. XV, 464 S. / Ln (3-598-21934-2)
  - Teil 2: März 1929 - Dezember 1929 / Hrsg.: Klaus A. Lankheit / 1994. XV, 574 S. / Ln (3-598-21938-5)
  - Teil 3: Januar 1930 - September 1930 / Hrsg.: Christian Hartmann / 1994. XV, 513 S. / Ln (3-598-21939-3)
- Bd. IV: Von der Reichstagswahl bis zur Reichspräsidentenwahl. Oktober 1930-März 1932
  - Teil 1: Oktober 1930 - Juni 1931. Hrsg.: Constantin Goschler / 1993. XV, 445 / Ln (3-598-21935-0)
  - Teil 2: Juli 1931 - Dezember 1931. Hrsg.: Christian Hartmann / 1996. XIII, 341 S. / Ln (3-598-22001-4)
  - Teil 3: Januar bis März 1932. Hrsg.: Christian Hartmann / 1997. XIII, 306 S. / Ln (3-598-22005-7)
- Bd. V: Von der Reichspräsidentenwahl bis zur Machtergreifung. April 1932-Januar 1933
  - Teil 1: April 1932 - September 1932. Hrsg. v. Klaus A. Lankheit / 1996. XV, 392 S. / Ln (3-598-21936-9)
  - Teil 2: Oktober 1932 - Januar 1933. Hrsg. v. Klaus A. Lankheit; Christian Hartmann / 1998. XVII, 436 S. / Ln (3-598-22002-2)
- *Bd. VI: Register [i.Vb.]*

- Der Hitler-Prozess 1924. Wortlaut der Hauptverhandlung vor dem Volksgericht München I. Hrsg. v. Lothar Gruchmann; Reinhard Weber, unter Mitarb. v. Otto Gritschneder / 4 Tlbde. 1997-1999 / Ln (3-598-11238-6)
  - Teil 1 / 1997. LXV, 366 S. / Ln (3-598-11317-X)
  - Teil 2 / 1997. XI, 413 S. / Ln (3-598-11318-8)
  - Teil 3 / 1998. XI, 439 S. / Ln (3-598-11319-6)
  - Teil 4 / 1999. XI, 441 S. / Ln (3-598-11355-2)

Hitzeroth, Christiane / Dagmar Marek; Isa Müller: Leitfaden für die formale Erfassung von Dokumenten in der Literaturdokumentation / 1976. IX, 493 S. (3-7940-7001-X)

Hochschulbibliotheken. Alternative Konzepte und ihre Kosten. Bericht zu einem Forschungsprojekt. Durchgef. v. Horst Höfler; Lutz Kandel; Achim Linhardt / 1984. 252 S. / Br (3-598-10550-9)

Hochschulplanung. Hrsg.: Hochschul-Informations-System GmbH, Hannover. [Bis Bd. 12 im Beltz Verlag, Weinheim]
- 13: Heindlmeyer, Peter: Berufsausbildung und Hochschulbereich. Eine quantitative Analyse f. d. Bundesrepublik Deutschland / 1973. 182 S. (3-7940-3113-X)
- 14: Bessai, Burghardt: Der Aufbau einer Informationsbank, insbesondere einer Datenbank als Voraussetzung für die Lösung von Managementproblemen im Hochschulbereich / 1973. VI, 333 S. (3-7940-3114-8)
- 15: Beckmann, Jochen: Gravitationstheoretischer Ansatz zur Ermittlung des regionalen Studentenaufkommens in NRW. Eine Analyse der Studienortwahl / 1973. VIII, 134 S. (3-7940-3115-6)
- 16: Rischkowsky, Franziska: Thesaurus Hochschulplanung. Ein aufgabenbezogener Thesaurus f. d. Literaturdokumentation d. Hochschul-Informations-System GmbH (HIS), Hannover. Unter Mitarb. von Gisela Boberg / 1973. IX, 205 S. (3-7940-3116-4)
- 17: Hussain, Khateeb M. / Hans Ludwig Freytag: Resource, costing and planning models in higher education / 1973. XVIII, 144 S. (3-7940-3117-2)
- 18: Das Verfahren der Flächenbedarfsplanung für die Universität Bielefeld. Beispiele zur Planung universitärer Forschung mit Anwendungsorientierung. Von Einhard Schrader u.a / 1974. XVI, 316 S. (3-7940-3118-0)
- 19: Anwendung des HIS-Simulationsmodells B an der Universität Karlsruhe. Von Hans Frey u.a / 1975. 119 S. (3-7940-3119-9)
- 20: Bonin, Hinrich: Hiskam: ein computergestütztes Informationssystem zur Abwicklung des Haushalts-, Kassen- und Rechnungswesens an Hochschulen. Ein Sachstandsbericht zur Erstimplementation / 1975. XVI, 371 S. (3-7940-3120-2)
- 21: Foerst, Renate / Hans W. Frey: Organisation der Lehre und Ausbildungskapazität in der klinischen Medizin Unters. in Freiburg u. überregional. Auslastungsvergleich Unter Mitarb. von H.-D. Bilsky / 1975. XI, 221 S. (3-7940-3121-0)
- 22: Ipsen, Detlev / Gerhard Portele: Organisation von Forschung und Lehre an westdeutschen Hochschulen / 1976. XIV, 285 S. (3-7940-3122-9)
- 23: Kort, Ute: Akademische Bürokratie. E. empir. Unters. über d. Einfluß von Organisationsstrukturen auf Konflikte an westdt. Hochschulen / 1976. VII, 172 S. (3-7940-3123-7)
- 24: Albert, Willi / Christoph Oehler: Die Kulturausgaben der Länder, des Bundes und der Gemeinden einschließlich Strukturausgaben zum Bildungswesen / 1976. VII, 505 S. (3-7940-3124-5)

- 25: Studienplanung und Organisation der Lehre. Ergebnisse einer empir. Untersuchungin d. Hochschulregionen Frankfurt und Darmstadt. Unter Mitarb. von Christoph Oehler / 1976. XVII, 538 S.
- 26: Foerst, Renate / Elke Korte: Organisation der Lehre und Ausbildungskapazität in der Zahnmedizin / 1976. X, 159 S. (3-7940-3126-1)
- 27: Griebach, Heinz / Karl Lewin; Martin Schacher: Studienverlauf und Beschäftigungssituation von Hochschulabsolventen. Bd. 1. Textteil; Bd. 2. Tabellenteil / 1977. XV, 257 S. u. XIX, 125 S. (3-7940-3127-X)
- 28: Birk, Lothar: Abiturienten zwischen Schule, Studium und Beruf Wirklichkeit und Wünsche / 1978. XV, 115 S. (3-7940-3128-8)
- 29: Organisation und Reform des Studiums Eine Hochschullehrerbefragung. Unter Mitarb. von Christoph Oehler / 1978. XI, 102 S. (3-7940-3129-6)
- 30: Rau, Einhard: Hochschulreform in Schweden. Ein Überblick / 1978. X, 95 S. (3-7940-3130-X)
- 31: Foerst, Renate / Elke Korte: Pharmazie in Freiburg. Studiengang u. Curricularrichtwert / 1978. IX, 120 S. (3-7940-3131-8)
- 32: Studenten zwischen Hochschule und Arbeitsmarkt. E. Unters. über Einstellungen u. Meinungen von Studenten an wissenschaftlichen Hochschulen / 1980. 187 S. in getr. Zählung (3-598-20442-6)
- 33: Lewin, Karl / Martin Schacher: Studium oder Beruf? Studienberechtigte 1976, 2 Jahre nach Erwerb d. Hochschulreife / 1979. XIX, 209 S. (3-598-20443-4)

Das **Hochschulwesen** in der UdSSR 1917-1967. Hrsg. von V. E. Eljutin. Übersetzung von Olga Meier-Kraut / 1969. 103 S. (3-7940-3238-1)

Hodnet, Grey: Leadership in the Soviet National Republics. A Quantitative Study of Recruitment Policy. / 1978. 410 p. (3-598-40023-3)

Höfig, Willi: Die Behandlung von Tageszeitungen an wissenschaftlichen Bibliotheken. Eine bibliothekarische Leitstudie / 1975. 163 S. (3- 7940-3274)

Höhne, Heinz: Die internationale Entwicklung auf dem Gebiete der alphabetischen Katalogisierung seit der Internationalen Katalogisierungskonferenz von Paris 1961 / 1979. II, 155, 57 S. (3-598-10151-1)

Hörning, Karl: Gesellschaftliche Entwicklung und soziale Schichtung. Vergleichende Analyse gesellschaftlichen Strukturwandels / 1976. 208 S. / Kst. (UTB 624) (3-7940-2655-1)

Holl, Oskar: Fremdsprache Deutsch. Deutschunterricht, Germanistik u. dt. Image in d. USA. E. Erfahrungsbericht / 1974. 160 S. / Lin (3-7940-3097-4)

Holl, Oskar: Wissenschaftskunde. Bd. 1-2 / 2 Bde. 1973. 363 S. / Kst. (UTB 286 u. 287) (3-7940-2620-9 u. 3-7940-2621-7)
- 2., unveränd. Aufl. 1976. 184 S.; VIII, 178 S. (3-7940-2620-9 / 3-7940-2621-7)

Holstein, Hermann: Schultheorie - Wozu? Überlegungen und Aspekte einer Theorie „der Schultheorie" / 1975. (3-597-24398-8) [Minerva Publikation]

Holtz, Christina: Comics, ihre Entwicklung und Bedeutung / 1980. 328 S. (3-598-10154-6)

Horn, Reinhard / Wolfram Neubauer: Fachinformation Politikwissenschaft. Literaturhinweise, Informationsbeschaffung und Informationsverarbeitung / 1987. 198 S., Reg / Kst. (UTB 1475) (3-598-02676-5)

Hüfner, Klaus / Jens Naumann: The United Nations System. An International Bibliography / Das System der Vereinten Nationen. Eine internationale Bibliographie. Eine Veröffentlichung der Forschungsstelle der Deutschen Gesellschaft für die Vereinten Nationen, Bonn / Berlin / 1976-79.
- Vol. 1: Learned Journals and Monographs 1945-1965 / Wissenschaftliche Zeitschriften und Monographien 1945-1965 / Unveränd. Sonderausg. 1976. LV, 519 S. / Hard (3-7940-2250-5) [Koprod. mit Walter de Gruyter, Berlin]
- Vol. 2 A: Learned Journals 1965-1970 / Wissenschaftliche Zeitschriften 1965-1970 / 1977. XCII, 286 p. / Hard (3-7940-2251-3)
- Vol. 2 B: Learned Journals 1971-1975 / Wissenschaftliche Zeitschriften 1971-1975 / 1977. LX, 436 p. / Hard (3-7940-2252-1)
- Vol. 3 A: Monographs and Articles in Collective Volumes 1965-1970 / Monographien und Artikel in Sammelbänden 1965-1970 / 1978. XCII, 492 p. / Hard (3-7940-2253-X)
- Vol. 3 B: Monographs and Articles in Collective Volumes 1971-1975 / Monographien und Artikel in Sammelbänden 1971-1975. Grusswort v. Kurt Waldheim / 1979. LVI, 692 S. / Hard (3-7940-2254-8)
- Vol. 4 A: Learned Journals 1976-1980 / Wissenschaftliche Zeitschriften 1976-1980. Address by Javier Pérez de Cuéllar. Frwd by Klaus. A. Hüfner. Publication of the Research Unit of the German United Nations Ass., Bonn / Berlin / Eine Veröffentlichung d. Forschungsstelle d. Deutschen Gesellschaft f. d. Vereinten Nationen, Bonn / Berlin / 1991. XCII, 566 p. / Hard (3-598-10993-8)
- Vol. 4 B: Learned Journals 1981-1985 / Wissenschaftliche Zeitschriften 1981-1985 Address by Javier Pérez de Cuéllar. Frwd by Klaus. A. Hüfner. Publication of the Research Unit of the German United Nations Ass., Bonn / Berlin / Eine Veröffentlichung d. Forschungsstelle d. Deutschen Gesellschaft für die Vereinten Nationen, Bonn / Berlin / 1991. LXI, 498 p. / Hard (3-598-10994-6)
- Vol. 5 A: Monographs and Articles in Collective Volumes 1976-1980 / Monographien und Artikel in Sammelbänden 1970-1980 Frwd by Klaus Hüfner. Address by Boutros Boutros Ghali. A Publication of the Research Unit of the German United Nations Ass., Bonn / Berlin / Eine Veröffentlichung d. Forschungsstelle d. Deutschen Gesellschaft f. d. Vereinten Nationen, Bonn / Berlin / 1994. LXXXIII, 301 p. / Hard (3-598-11111-8)

Hunter, Eric: AACR 1978. An Introduction / 1979. 128 S. (0-85157-282-0) [Clive Bingley]

Hury, Carlo: Luxemburgensia. Eine Bibliographie der Bibliographien / 2., erw. völlig neu bearb. Aufl. 1978. XX, 352 S. / Lin (3-598-07090-X)

ICA Handbooks Series. Published by the ICA publications Committee
- 1: Duchein, Michel: Archive Buildings and Equipment / 1977. 201 S., 51 ills. / Soft (3-7940-3780-4)
- 2: Taylor, Hugh A.: The Arrangement and Description of Archival Materials. With a Contribution: Les Instruments de Recherches dans les Archives by Etienne Taillemite / 1980. 181 p. / Soft (3-598-20272-5)

- 3: Dictionary of Archival Terminology / Dictionnaire de terminologie archivistique English-French. With Equivalents in Dutch, German, Italian. Ed. by Peter Walne. Comp. by Frank B. Evans, François S. Himly, Peter Walne / 1984. 226 p. / Hard (3-598-20275-X)
- 4: Glossary of Basic Archival and Library Conservation Terms English with Equivalents in Spanish, German, Italian, French and Russian Frwd by Peter Walne. Ed. by Carmen C. Nogueira. Comp. by the Committee on Conservation, International Council on Archives 1988 151 p. / Hard (3-598-20276-8)
- 5: Manual of Archival Reprography. Ed. by Lajos Körmendy. Comp. by James A. Keene, Lajos Körmendy, Ted F. Powell, Georges Weill / 1989. 223 p., 77 illus. / Hard (3-598-20277-6)
- 6: Duchein, Michel: Archive Buildings and Equipment. Ed. by Peter Walne,. Trans. by David Thomas / 2nd enl. rev. ed. 1988. 232 p. / Hard (3-598-20278-4)
- 7: Dictionary of Archival Terminology / Dictionnaire de Terminologie Archivistique English and French. With Equivalents in Dutch, German, Italian, Russian and Spanish. Ed. by Peter Walne / 2nd rev. ed. 1988. 212 p. / Hard (3-598-20279-2)
- 8: The Management of Business Records. Ed. by Anna Ch. Ulfsparre / 1988. 72 p. / Hard (3-598-20280-6)
- 9: Ermisse, Gérard: Les Services de Communication des Archives au Public / 1994. X, 306 p. / Hard (3-598-20281-4)

IFLA / Saur Professional Library (ISSN 0946-0624)
- 1: Cataloging of the Hand Press. A Comparative and Analytical Study of Cataloging Rules and Formats Employed in Europe. Prepared by Henry L. Snyder; Heidi L. Hutchinson / 1994. XII, 200 p. / Hard (3-598-23400-7)

IFLA Annual. Annual Reports / Rapport Annuels [Bis Bd. 1971: Scandinavian Library Center, Copenhagen]
- 1972. Proceedings of the 38th General Council Meeting, Budapest, 1972. Ed. by Peter Havard-Williams; Willem R. H. Koops; Wilhelmina E. S. (Milisa) Coops / 1973. 252 p. / Soft (3-7940-4298-0)
- 1973. Proceedings of the 39th General Council Meeting, Grenoble, 1973. Ed. by Peter Havard-Williams; Willem R. H. Koops; Wilhelmina E. S. (Milisa) Coops / 1974. 255 p. / Soft (3-7940-4299-9)
- 1974. Proceedings of the 40th General Council Meeting, Washington, 1974. Ed. by Peter Havard-Williams; Willem R. H. Koops; Wilhelmina E. S. (Milisa) Coops / 1975. 314 p. / Soft (3-7940-4300-6)
- 1975. Proceedings of the 41st General Council Meeting, Oslo, 1975. Ed. by Peter Havard-Williams; Willem R. H. Koops; Wilhelmina E. S. (Milisa) Coops / 1976. 232 p. / Soft (3-7940-4301-6)
- 1976. Proceedings of the 42nd General Council Meeting, Lausanne, 1976. Ed. by Peter Havard-Williams; Willem R. H. Koops; Wilhelmina E. S. (Milisa) Coops / 1977. 266 p. / Soft (3-7940-4302-2)
- 1977. Proceedings of the 43rd General Council Meeting, Brussels, 1977. Ed. by Peter Havard-Williams; Willem R. H. Koops; Wilhelmina E. S. (Milisa) Coops / 1978. 275 p. / Soft (3-7940-4303-0)

- 1978. Proceedings of the 44th General Council Meeting, Strebske Pleso, 1978. Ed. by Peter Havard-Williams; Willem R. H. Koops; Wilhelmina E. S. (Milisa) Coops / 1979. 157 p. / Soft (3-598-20659-3)
- 1979. Proceedings of the 45th General Council Meeting, Copenhagen, 1979. Ed. by Willem R. H. Koops; Wilhelmina E. S. (Milisa) Coops / 1980 232 p. / Soft (3-598-20661-5)
- 1980. Proceedings of the 46th General Council Meeting, Manila, 1980. Ed. by Willem R. H. Koops; Wilhelmina E. S. (Milisa) Coops / 1980. 219 p. / Soft (3-598-20660-7)
- 1981. Proceedings of the 47th General Council Meeting, Leipzig, 1981. Ed. by Willem R. H. Koops; Wilhelmina E. S. (Milisa) Coops / 1981. 230 p. / Soft (3-598-20662-3)
- 1982. Proceedings of the 48th General Council Meeting, Montreal, 1982. Ed. by Willem R. H. Koops; Wilhelmina E. S. (Milisa) Coops / 1983. 259 p. / Soft (3-598-20663-1)
- 1983. Proceedings of the 49th Council and General Conference. Munich, 1983. Ed. by Willem R. H. Koops; Wilhelmina E. S. (Milisa) Coops / 1984. 286 p. / Soft (3-598-20664-X)
- 1984. Proceedings of the 50th Council and General Conference. Nairobi, 1984. Ed. by Willem R. H. Koops; Wilhelmina E. S. (Milisa) Coops / 1985. 194 p. / Soft (3-598-20665-8)
- 1985. Proceedings of the 51st Council and General Conference, Chicago, 1985. Ed. by Willem R. H. Koops; Carol Henry / 1986. 260 p. / Soft (3-598-20666-6)
- 1986. Proceedings of the 52nd Council and General Conference, Tokyo 1986. Ed. by Willem R. H. Koops; Carol Henry / 1987. 234 p. / Soft (3-598-20667-4)
- 1987. Proceedings of the 53rd Council and General Conference, Brighton, 1987. Ed. by Willem R. H. Koops; Carol Henry / 1988. 256 p. / Soft (3-598-20668-2)
- 1988. Proceedings of the 54th Council and General Conference Sydney, 1988. Ed. by Willem R. H. Koops; Carol Henry / 1989. 213 p. / Soft (3-598-20669-0)
- 1989. Proceedings of the 55th Council and General Conference, Paris, 1989. Ed. by Carol Henry / 1990. 277 p. / Soft (3-598-20670-4)
- 1990. Proceedings of the 56th Council and General Conference, Stockholm, 1990. Ed. by Carol Henry / 1991. 249 p. / Soft (3-598-20671-2)
- 1991. Proceedings of the 57th Council and General Conference, Moscow, 1991. Ed. by Carol Henry / 1992. 262 p. / Soft (3-598-20672-0)
- 1992. Proceedings of the 58th General and General Conference, New Delhi, 1992. Ed. by Carol Henry / 1993. 262 p. / Soft (3-598-20673-9)
- 1993. Proceedings of the 59th Council and General Conference, Barcelona, 1993. Ed. by Carol Henry / 1994. 372 p. / Soft (3-598-20674-7)
- 1994. Proceedings of the 60th Council and General Conference, Havana, 1994. Ed. by Carol Henry / 1995. 214 p. / Soft (3-598-20675-5)
- 1995. Proceedings of the 60th Council and General Conference, Istanbul, 1995. Ed. by Carol Henry / 1996. 211 p. / Soft (3-598-20676-3)

IFLA Journal. Official quarterly Journal of the International Federation of Library Associations and Institutions. Ed.: International Federation of Library Associations and Institutions / since Vol. 1/1974 (ISSN 0340-0352)

IFLA Publications. Ed. by International Federation of Library Associations and Institutions (IFLA);

Willem R. H. Koops and Peter Havard-Williams (1 - 13); Willem R. H. Koops (14 - 51); Carol Henry (since 52/53)

- 1: Special Libraries - Worldwide. A collection of papers prepared for the Section of Special Libraries of the International Federation of Library Associations. Ed. by Günther Reichardt / 1974. 360 p. / Hard (3-7940-4421-5)
- 2: National Library Buildings. Proceedings of the Colloquium held in Rome, 3-6 September 1973. Ed. by Anthony Thompson / 1975. 144 p. / Hard (3-7940-4422-3)
- 3: Le contrôle bibliographique dans les pays en développement. Table Ronde sur le Contrôle Bibliographique Universel dans les Pays en Développement, Grenoble 22 25 ao-t 1973. Ed. by Marie L. Bossuat; Genevieve Feuillebois; Monique Pelletier / 1975. 165 S. / Gb (3-7940-4423-1)
- 4: National and International Library Planning. Key papers presented at the 40th Session of IFLA General Council, Washington, DC, 1974. Ed. by Vosper, Robert; Newkirk, Leone I. / 1976. 162 p. / Hard (3-7940-4424-X)
- 5: Reading in a Changing World. Papers presented at the 38th Session of the IFLA General Council, Budapest, 1972. Hrsg. v. Foster M. Mohrhardt / 1976. 134 p. / Hard (3-7940-4425-8)
- 6: The Organization of the Library Profession. A Symposium based on Contributions to the 34th Session of the IFLA GeneralCouncil, Liverpool, 1971. Ed. by A. H. Chaplin, / 2nd ed. 1976. 132 p. / Hard (3-7940-4309-X)
- 7: World Directory of Administrative Libraries. A guide of libraries serving national, state, provincial, and Länder-bodies, prepared for the Sub-Section of Administrative Libraries / IFLA. Ed. by Otto Simmler / 1976. 474 p. / Hard (3-7940-4427-4)
- 8: World directory of map collections. Comp. by the Geography and Map Libraries Sub-Sect. Ed. by Walter W. Ristow / 1976. 326 S. (3-7940-4428-2)
- 9: Standards for public libraries Prepared by the Sect. of Public Libraries / 2., corr. ed. 1977. 53 S. (3-7940-4429-0)
- 10: IFLA's First Fifty Years. Achievement and challenge in international librarianship. Ed. by Willem R. Koops; Joachim Wieder / 1977. 158 p. / Hard (3-7940-4430-4)
- 11: The International Federation of Library Associations and Institutions: Aselected list of references. Ed. by Edward P. Cambio / 2nd rev. and enl. ed. 1977 / Soft (3-7940-4431-2)
- 12: Library Service to Children. An International Survey. Ed. by Colin Ray / 1978. 158 S. (3-7940-4432-0)
- 2. ed. 1983. 168 p. / Hard (3-598-20392-6)
- 13: Allardyce, Alex: Letters for the International Exchange of Publications. A Guide to their Composition in English, French, German, Russian and Spanish. Ed. by Peter Genzel. Transl. by Jacques Lethére; Maria Razumovsky; Boris P. Kanyevsky; Adolfo Rodriguey / 1978. 148 p. / Hard (3-7940-4433-9)
- 14: Resource Sharing of Libraries in Developing Countries. Proceedings of the 1977 IFLA / UNESCO Pre-Session Seminar for Librarians from Developing Countries, Antwerp University, August 30 - September 4, 1977. Ed. by H. D. Vervliet / 1979. 286 p. / Hard (3-598-20375-6)
- 15: Libraries for All. A World of Books and their Readers / Bibliothéques pour tous. Le monde du livre et de ses lecteurs. Papers presented at the IFLA 50th Anniversary World Congress Brussels 1977. Ed. by W. R. Koops; R. Vosper / 1980. 163 p. / Hard (3-598-20376-4)

- 16: Library Service for the Blind and Physically Handicapped. An International Approach. Key Papers presented at the IFLA Conference 1978, Strbske Pleso. Ed. by Frank K. Cylke / 1979. 100 p. / Hard (3-598-20377-2)
- 17: Guide to the Availability of Theses. Compiled by Section of University Libraries and other General Research Libraries. Ed. by D. H. Borchardt; J. D. Thawley / 1981. 443 p. / Hard (3-598-20378-0)
- 18: Studies in the International Exchange of Publications. Ed. by Peter Genzel / 1981. 125 p. / Hard (3-598-20379-9)
- 19: Public Library Policy Proceedings of the IFLA / UNESCO Pre-Session Seminar, Lund, Sweden 1979. Ed. by K. C. Harrison / 1981. 152 p. / Hard (3-598-20380-2)
- 20: Library Education Programmes in Developing Countries with Special Reference to Asia UNESCO pre IFLA Conference Seminar, Manilla 1980. Ed. by Russell Bowden / 1982. X, 211 p. / Hard (3-598-20383-7)
- 21: Hébert, Françoise / Wanda Noel: Copyright and Library. Materials for the Handicapped A Study prepared for the International Federation of Library Associations and Institutions / 1982. 111 p. / Hard (3-598-20381-0)
- 22: Education of School Librarians. Some Alternatives Papers presented at the Seminar for the Education of School Librarians for Central America and Panama, at San José, Costa Rica 1978. Ed. and transl. by Sigrún K. Hannesdótir / 1982. 120 p. / Hard (3-598-20384-5)
- 23: Library Service for the Blind and Physically Handicapped. An international Approach. Volume 2. Ed. by Bruce E. Massis / 1982. 123 p. / Hard (3-598-20385-3)
- 24: Library Interior Layout and Design. Proceedings of the Seminar held in Frederiksdal, Denmark, 1980. Ed. by Rolf Fuhlrott; Michael Dewe / 1982. 145 p. / Hard (3-598-20386-1)
- 25: Line, Maurice / Steven Vickers: Universal Availability of Publications (UAP). A programme to improve the national and international provision and supply of information / 1983. 139 p. / Hard (3-598-20387-X)
- 26: Overton, David: Planning the Administrative Library / 1983. 264 p. / Hard (3-598-20388-8)
- 27: McDonald, Dennis D. / Eleanor J. Rodger; Jeffrey L. Squires: International Study of Copyright of Bibliographic Records in Machine-Readable Form. A Report prepared for the International Federation of Library Associationsand Institutions / 1983. 149 p. / Hard (3-598-20393-4)
- 28: Library Work for Children and Young Adults in Developing Countries / Les Enfants, les Jeunes et les Bibliothèques dans les Pays en voie de Développement Proceedings of the IFLA / UNESCO Pre-Session Seminar Leipzig, RDA, 10-15 ao-t, 1981 / Actes du Séminaire IFLA / UNESCO de Leipzig, GDR, 10-15 August, 1981. Ed. by Geneviève Patte; Sigrún K. Hannesdóttir / 1984. 283 p. / Hard (3-598-20389-6)
- 29: Guide to the Availability of Theses II: Non-University Institutions. Comp. by G. G. Allen; K. Deubert / 1985. VI, 124 p. / Hard (3-598-20394-2)
- 30: A Guide to Developing Braille and Talking Book Services. Ed. by Leslie Clark, in collab. with Dina N. Bedi; John M. Gill / 1984. 104 p. / Hard (3-598-20395-0)
- 31: World Directory of Map Collections. Compiled by the Section Geography and Map Libraries Sub-Section / IFLA. Ed. by John A. Wolter; Ronald E. Grim; David K. Carrington / 2nd ed. 1986. XLIII, 405 p. / Hard (3-598-20374-8)

- 32: International Guide to Library and Information Science Education. A Reference Source for Educational Programs in the Information Fields World-wide. Ed. by Josephine Riss-Fang; Paul Nauta, with the assist. of Anna J. Fang / 1985. 537 p. / Hard (3-598-20396-9)
- 33: University Libraries in Developing Countries. Structure and Function in Regard to Information Transfer for Science and Technology. Proceedings of the IFLA / UNESCO Pre-Session Seminar for Librarians from developing countries, München, August 16-19, 1983. Ed. by Anthony J. Loveday; Günter Gattermann / 1985. 183 p. / Hard (3-598-20397-7)
- 34: Pacey, Phillip: A Reader in Art Librarianship / 1985. XII, 199 p. / Hard (3-598-20398-5)
- 35: Inventaire Général des Bibliographies Nationales Rétrospectives / Retrospective National Bibliographies: an International Directory. Ed. by Marcelle Beaudiquez / 1986. 189 p. / Hard (3-598-20399-3)
- 36: Guidelines for Public Libraries. Prepared for the IFLA Section of Public Libraries / 3rd enl. and rev. ed. 1986. 90 p. / Hard (3-598-21766-8)
- 37: Baxter, Paula A.: International Bibliography of Art Librarianship. An Annotated Compilation / 1987. V, 94 p. / Hard (3-598-21767-6)
- 38: Automated Systems for Access to Multilingual and Multiscript Library Materials Problems and Solutions. Papers from the Pre-Conference held at Nihon Daigaku Kaikan Tokyo, Japan. August 21-22, 1986. Ed. for the Section on Library Services to Multicultural Populations and the Section on Information Technology by Christine Bossmeyer; Stephen W. Massil / 1987. 225 p. / Hard (3-598-21768-4)
- 39: Adaptation of Buildings for Library Use. Proceedings of the Seminar held in Budapest, 1985. Ed. by Michael Dewe / 1987. 254 p. / Hard (3-598-21769-2)
- 40/41: Preservation of Library Materials. Conference held at the National Library of Austria Vienna, Austria, April 7.-10. 1986. Sponsored by the Conference of Directors of National Libraries in Cooperation with IFLA and UNESCO. Ed. by Merrily A. Smith / 2 Vols.
  - Vol. 1 (40) / 1987. IX, 159 p. / Hard (3-598-21770-6)
  - Vol. 2 (41) / 1987. 155 p. / Hard (3-598-21771-4)
- 42: World Directory of Biological and Medical Sciences Libraries. Ed. by Ursula H. Poland / 1988. XII, 203 p. / Hard (3-598-21772-2)
- 43: Education and Research in Library and Information Science in the Information Age: Means of Modern Technology and Management. Ed. by Miriam H. Tees / 1988. 202 p. / Hard (3-598-21773-0)
- 44: Open Systems Interconnection. The Communications Technology of the 1990's. Papers from the Pre-Conference Seminar held at London, August 12-14, 1987. Ed. by Christine E. Smith / 1988. 254 p. / Hard (3-598-21774-9)
- 45/46: Newspaper Preservation and Access. Proceedings of the Symposium held in London, August 12.-15., 1987. Ed. for the Section on Serial Publications and the Working Group on Newspaper by Jan P. Gibb / 2 Vols.
  - Vol. 1 (45) / 1988. 230 p. / Hard (3-598-21775-7)
  - Vol. 2 (46) / 1988. p. VI, 231-449 / Hard (3-598-21776-5)
- 47: A l'Ecoute de l'Oeil. Les Collections Iconographiques et les Bibliothèques. Actes du Colloque organisé par la Section des Bibliothèques d'Art de l'IFLA Genève, 13-15 mars 1985 Rev. par Huguette Rouit; Jean P. Dubouioz / 1989. 348 S. / Gb (3-598-21777-3)
- 48: Library Buildings: Preparations for Planning. Proceedings of the Seminar held in Aberystwyth, August 10-14, 1978. Ed. by Michael Dewe / 1989. 278 p. / Hard (3-598-21778-1)
- 49/50: Harmonisation of Education and Training Programmes for Library, Information and Archival Personnel. Proceedings of an International Colloquium, London, August 9-15, 1987. Ed. by Ian M. Johnson; Fall A. Correa; Richard J. Neill; Martha B. Terry / 2 Vols.
  - Vol. 1 (49) / 1989. 164 p. / Hard (3-598-21779-X)
  - Vol. 2 (50) / 1989. p. 165-374 / Hard (3-598-21780-3)
- 51: International Directory of Libraries for the Blind. Ed. by Hiroshi Kawamura, under the auspices of the IFLA Section of Libraries for the Blind. Rev. by Carol Henry / 3rd ed. 1990. XXII, 257 p. / Hard (3-598-21781-1)
- 52/53: Riss-Fang, Josephine / Alice Songe: World Guide to Library, Archive and Information Science Associations. Ed.: International Federation of Library Associations and Institutions (IFLA) with the assistance of Anna J. Fang; Alexandra Herz / 1990. XXVII, 517 p. / Hard (3-598-10814-1)
- 54: Education and Training for Conservation and Preservation. Papers of an International Seminar on „The Teaching of Preservation Management for Librariens, Archivists and Information Scientists", sponsered by IFLA, FID and ICA, Vienna, April 11-13, 1986, with additional information Sources. Ed. by Josephine Riss-Fang; Ann Russel, with the assistance of Anna J. Fang / 1991. VII, 113 p. / Hard (3-598-21782-X)
- 55: Continuing Professional Education. An IFLA Guidebook. A Publication of the Continuing Professional Round Table (CPERT) of the International Federation of Library Associations and Institutions (IFLA). By Blanche Wooles. Reports from the Field ed. by Miriam H. Tess / 1991. X, 159 p. / Hard (3-598-21784-6)
- 56: Reference Service for Publications of Intergovernmental Organizations. Papers from an IFLA Workshop, Paris, August 24, 1989. Section on Government Information and Official Publications. Section on Bibliography. Section on Social Science Libraries. Ed. by Alfred Kagan; International Federation of Library Associations and Institutions / 1991. VI, 158 p. / Hard (3-598-21785-4)
- 57: Smith, Merrily A.: Managing the Preservation of Serial Literature / 1992 XI, 291 p. / Hard (3-598-21783-8)
- 58: La Presse de la Liberté Journée d'Etudes organisée par le Groupe de Travail IFLA sur les Journaux Paris le 24 ao-t 1989 / Seminar organised by the IFLA Working Group on Newspapers Paris 24 August 1989. Ed. par Eve Johansson; International Federation of Library Associations and Institutions / 1991. IV, 122 p., 10 illus. / Hard (3-598-21786-2)
- 59: Zielinska, Marie F. / Francis T. Kirkwood: Multicultural Librarianship. An International Handbook. Ed.: International Federation of Library Associations and Institutions / Section on Library Services to Multicultural Populations / 1992. XIV, 384 p. / Hard (3-598-21787-0)
- 60: Wellheiser, Johanna G.: Nonchemical Treatment Processes for Disinfectation of Insects and Fungi in Library Collections. Ed.: International Federation of Library Associations and Institutions / 1992. VIII, 118 p. / Hard (3-598-21788-9)
- 61: Will the Chain Break?: Differential Pricing as Part of a New Pricing Structure for Research Literature and its Consequences for the Future of Scholary. Communication Proceedings of the IFLA Workshop, Stockholm, 23 August 1990. Ed. by Ulrich Montag; Section on Acquisition and Exchange; Section on Serial Publications; UAP Office; LIBER / 1992. V, 90 p. / Hard (3-598-21789-7)
- 62: A la Recherche de la Mémoire. Le Patrimoine Culturel Actes du Colloque organisé par la Section des Bibliothèques d'Art de l'IFLA, Paris, 16-19 Ao-t 1989. Ed. by Huguette Rouit; Jean M. Humbert / 1993. 276 S. / Gb (3-598-21790-0)
- 63: World Directory of Map Collections. Ed. by Lorraine Dubreuil. Comp.: Section of geography and Map Libraries / 3rd ed. 1993. IX, 310 p. / Hard (3-598-21791-9)
- 64: Guidelines for Legislative Libraries. Ed. by Dermot Englefield, for the Parliamentary Libraries Section of the IFLA / 1993. XV, 123 p. / Hard (3-598-21792-7)
- 65: Documentation Nordic Art, Design, Bibliographies, Databases / Documentationl'Art des Pays Nordiques Design, Bibliographies, Bases de données Proceedings from the Art Libraries Satellite Meeting. Nationalmuseum, Stockholm, August 16-19, 1990. Ed. by Charlotte Hanner; Arlis Norden; IFLA Section of Art Libraries / 1993. 270 p., approx. 48 illus. / Hard (3-598-21793-5)
- 66/67: Continuing Professional Education and IFLA: Past, Present, and a Vision for the Future. Papers from the IFLA CPERT Second World Conference on Continuing Professional Education for Library and Information Science Professions. A Publication of the Continuing Professional Education Round Table (CPERT) of the International Federation of Libraries. Ed. by Blanche Woolls / 1993. XII, 365 p. / Hard (3-598-21794-3)
- 68: The Status, Reputation and Image of the Library and Information Profession. Proceedings of the IFLA Pre-Session Seminar Delhi, 24-28 August 1992. Under the Auspices of the IFLA Round Table for the Management of Library Associations. Ed. by Russell Bowden; Donald Wijasuriya / 1994. 228 p. / Hard (3-598-21795-1)
- 69: Global Perspectives on Preservation Education. Ed. by Michele V. Cloonan / 1994. X, 109 p. / Hard (3-598-21796-X)
- 70: Automated System for Access to Multilingual and Multiscript Library Materials. Proceedings of the Second IFLA Satellite Meeting, Madrid, August 18-19, 1993. Ed. by Sally M. McCallum; Monica Ertel / 1994. IV, 185 p. / Hard (3-598-21797-8)
- 71: The Image of the Library and Information Profession. How we see Ourselves - an Investigation. Ed. by Russell Bowden; Donald Wijasuriya / 1995. 86 p. / Hard (3-598-21798-6)
- 72/73: World Guide to Library Archive and Information Science Education. Ed. by Josephine Riss-Fang; Robert D. Stueart; Kulthida Tuamsuk / 2nd rev. and enl. ed. 1995. XIII, 585 p. / Hard (3-598-21799-4)
- 74: Bibliotecas de arte, arquitectura diseño / Art, Architecture and Design Libraries Perspectivas actuales / Current Trends / 1996. II, 441 p. / Hard (3-598-21801-X)
- 75: Multilingual Glossary for Art Librarians. English with Indexes in Dutch, French, German, Italien, Spanish and Swedish. Ed.: IFLA Section of Art Libraries / 2nd rev. and enl. ed. 1996. V, 183 p. / Hard (3-598-21802-8)

- 76: Measuring Quality International Guidelines for Performance Measurement in Academic Libraries / 1996. 171 p. / Hard (3-598-21800-1)
- 77: Szilvássy, Judith: Basic Serial Management Handbook. Ed.: IFLA Section on Serial Publications / 1996. VIII, 172 p. / Hard (3-598-21803-6)
- 78: Froehlich, Thomas J.: Survey and Analysis of the Major Ethical and Legal Issues Facing Library and Information Services. Ed.: IFLA Section on Serial Publications / 1997. VIII, 99 p. / Hard (3-598-21804-4)
- 79: Ressources pour les Bibliothèques et Centres Documentaires Scolaires / Resourcebook for School Libraries and Resource Centers. Ed. by Paulette Bernhard / 1997. 152 p. / Hard (3-598-21805-2)
- 80/81: Ward, Patricia L. / Darlene E. Weingand: Human Development: Competences for the Twenty-First Century Papers of the IFLA CPERT Third International Conference on Continuing Professional Education for the Library and Information Professions / 1997. XII, 400 p. / Hard (3-598-21806-0)
- 82: Hill, Thomas E.: International Directory of Art Libraries / Répertoire Internationale de Bibliotheques d'Art / Internationales Adressbuch der Kunstbibliotheken / Directorio Internationale de Bibliotecas de Arte / 1997. 2, XIII, 251 p. / Hard (3-598-21807-9)
- 83: Brian, Rob: Parliamentary Libraries and Informtion Services of Asia and the Pacific. Papers prepared for the 62nd Conference Beijing, China August 25-31, 1996 / 1997. IV, 106 p. / Hard (3-598-21808-7)
- 84: Library Preservation and Conservation in the '90s. Proceedings of the Satellite Meeting of the IFLA Section on Preservation and Conservation Budapest, Hungary 15-17 August, 1995. Ed. by Jean I. Whiffin; John Havermans / 1998. X, 181 p. / Hard (3-598-21809-5)
- 85: Multi-Script, Multilingual, Multi-character Issues for the Online Environment. Proceedings of a Workshop Sponsored by the IFLA Section on Cataloguing, Istanbul, Turkey, August 24, 1995. Ed. by John D. Byrum, Jr. and Olivia Madison / 1998. IV, 123 p. / Hard (3-598-21814-1)
- 86: Koenig, Michael E. D.: Information Driven management Concepts and Themes: A Toolkit for Librarians. With the assistance of Morgen MacIntosh / 1998. V, 73 p. / Hard (3-598-21815-X)
- 87: Parliamentary Libraries and Research Services in Central and Eastern Europe. Building More Effective Legislatures. Ed. by William H. Robinson and Raymond Gastelum / 1998. X, 238 p. / Hard (3-598-21813-3)
- 88: Intelligent Library Buildings. Proceedings of the 10th Seminar of the IFLA Section on Library Buildings and Equipment 1997. Ed. by Marie-Françoise Bisbrouck and Marc Chauveinc / 1999. VIII, 294 p. / Hard (3-598-21810-9)

An **Illustrated** Inventory of Famous Dismembered Works of Art - European Paintings / 1974 223 p., 363 ills. / Hard (3-7940-5126-2)

Imhof, Arthur E.: Historische Demographie I / CD-ROM-Edition 1995 (3-598-40330-5)

L'**impact** des techniques nouvelles sur l'industrie de l'edition. Compte rendu du symposium tenu à Luxembourg, 1979. Organisé par la Direction Générale „Information Scientifique et techniqueet gestion de l'information / 1981. 203 p. / Br (3-598-10129-5)

The **Impact** of New Technologies on Publishing. Proceedings of the Symposium organized by the Commission of the European Communities, Directorate General for Scientific and Technical Information Management, held in Luxembourg November 6-7, 1979. Ed.: Commission of the European Comunities / 1980. 195 p. / Hard (3-598-10128-7)

In der Gemeinschaft der Völker. Dokumente aus deutschen Archiven über die Beziehungen zwischen deutschen und anderen Nationen in elf Jahrhunderten. Zum X. Internationalen Archivkongress. Hrsg. v. Heinz Boberach; Eckhart G. Franz / 1984. 442 S., zahlr. Abb. / Lin (3-598-10575-4)

... **in** Szene gesetzt. Studien zur Inszenierenden Typografie. Komment. v. Christa Kochinke und Hans Peter Willberg. Hrsg.: Deutscher Werkbund Rheinland-Pfalz e.V. / 1990. 105 S. / Ebr (3-598-10968-7)

Index der Antiken Kunst und Architektur / Index of Ancient Art and Architecture. Denkmäler des griechisch-römischen Altertums in der Photosammlung des Deutschen Archäologischen Instituts Rom / Monuments of Greek and Roman cultural heritage in the photographic collection of the German Archaeological Institute in Rome. Mikrofiche-Edition. Hrsg.: Deutsches Archäologisches Institut, Abteilung Rom / German Archaeological Institute in Rome / 1988-1990. 2714 Photo-Fiches mit ca. 265.000 Photographien. Lesefaktor 24x / Silberfiche (3-598-32070-1)
- Teilausgaben / Special editions.
- Architekturteile / 1991. Fiche 1.005-1.031 (3-598-32089-2)
- Bronze / Unedle Metalle / 1991. Fiche 1.855-1.951 (3-598-32097-3)
- Figürliche Bauplastik / 1991. Fiche 890-1.004 (3-598-32088-4)
- Idealplastik / 1991. Fiche 1-279 (3-598-32084-1)
- Inschriften / 1991. Fiche 1.032-1.071 (3-598-32090-6)
- Kleinkunst / 1991. Fiche 1.952-1.994 (3-598-32098-1)
- Malerei / 1991. Fiche 1.081-1.167 (3-598-32092-2)
- Mosaik / 1991. Fiche 1.168-1.211 (3-598-32093-0)
- Portraitplastik / 1991. Fiche 280-514 (3-598-32085-X)
- Reliefs / 1991. Fiche 699-889 (3-598-32087-6)
- Sarkophage / 1991. Fiche 515-698 (3-598-32086-8)
- Stuck / 1991. Fiche 1.072-1.080 (3-598-32091-4)
- Terrakotta / 1991. Fiche 1.765-1.854 (3-598-32096-5)
- Topographie / 1991. Fiche 1.212-1.551 (3-598-32094-9)
- Vasen / 1991. Fiche 1.552-1.764 (3-598-32095-7)
- Begleitband - Register und Kommentar / 1991. 420 S. / Gb (3-598-32083-3)

Index of Conference Proceedings Received 1964-1988. Ed.: The Document Supply Centre of the British Library / 26 vols. 1989 / 1990. Cplt. 12,500 p. / Hard (3-598-31770-0)

Index of Jewish Art
- Vol. II, 3 parts: Iconographical Index of Hebrew Illuminated Manuscripts. Ed.: The Israel Academy of Sciences and Humanities. Institut de Recherche et d'Histoire des Textes. Compiled by Bezalel Narkiss; Gabrielle Sed-Rajna / 1981. 81 p., 566 index cards, 419 illus. / Soft (3-598-10207-0)
- Vol. IV, 3 parts: Illuminated Manuscripts of the Kaufmann Collection. Ed.: Library of the Hun-

garian Academy of Sciences, Budapest; The Israel Academy of Sciences and Humanities, Jerusalem; Institut de Recherche. Comp. by Bezalel Narkiss; Gabrielle Sed-Rajna / 1988. 119 p., 310 index cards, 150 illus. / Soft (3-598-10875-3)

The **Index** of Paintings Sold in the British Isles during the Nineteenth Century (3-598-22821-X)
- Vol. 1. 1801-1805 / 1989. XXVII, 1047 p. / Hard (3-598-22822-8)
- Vol. 2. 1806-1810 / 2 vols. 1989. Cplt. XXXI, 1409 p. / Hard
  - Part 1: A-N / 1989. XXXI, 689 p. (3-598-22823-6)
  - Part 2: O-Z and Anonymous / 1989. p. 693-1409 (3-598-22824-4)
- Vol. 3. 1811-1815 / 2 vols. 1989. 1993. Cplt. XXXI, 1455 p. / Hard
  - Part 1: A-N / 1989. XXXI, 704 p. (3-598-22825-2)
  - Part 2: O-Z / 1989. 751 p. (3-598-22826-0)

Index photographique de l'art en France. Microfiche edition. Ed.: Bildarchiv Foto Marburg im Forschungsinstitut f. Kunstgeschichte d. Philipps-Universität Marburg / 1979-1981 Approx. 93,696 reprod. on 976 fiches. Reader factor 24x / Silverfiche (3-598-30160-X)

Index to Festschriften in Librarianship. Ed. by Joseph Periam Danton and Jane F. Pulis / 1970. XII, 461 p. (3-7940-7033-X)

Index to Festschriften in librarianship 1967-1975. Ed. by Joseph Periam Danton and Jane F. Pulis. With assistance of Patiala Khoury Wallman / 1979. LXXXIV, 354 p. / Hard (3-598-07034-9)

Indexbücher der Technik / 1954-59
- Ultraschall / 1954
  - 1. Erg. - Bd. 1955
  - 2. Erg. - Bd. 1956
  - 3. Erg. - Bd. 1957
  - 4. Erg. - Bd. 1958
- Textilprüfung / 1954
  - 1. Erg. - Bd. 1955
  - 2. Erg. - Bd. 1956
  - 3. Erg. - Bd. 1957
  - 4. Erg. - Bd. 1958
- Schul- und Kirchenbau / 1954
  - 1. Erg. - Bd. 1955
  - 2. Erg. - Bd. 1956
  - 3. Erg. - Bd. 1957
  - 4. Erg. - Bd. 1958
- Human Relations / 1954
  - 1. Erg. - Bd. 1955
  - 2. Erg. - Bd. 1956
  - 3. Erg. - Bd. 1957
  - 4. Erg. - Bd. 1958
- Leistungslohn / 1955
  - 1. Erg. - Bd. 1956
  - 2. Erg. - Bd. 1957
  - 3. Erg. - Bd. 1958
  - 4. Erg. - Bd. 1959
- Innerbetrieblicher Transport / 2 Bde. 1955
  - 1. Erg. - Bd. / 2 Bde. 1956
  - 2. Erg. - Bd. / 2 Bde. 1957
  - 3. Erg. - Bd. / 2 Bde. 1958
  - 4. Erg. - Bd. / 2 Bde. 1959
- Regelungstechnik / 1954
  - 1. Erg. - Bd. 1955
  - 2. Erg. - Bd. 1956
  - 3. Erg. - Bd. 1957
  - 4. Erg. - Bd. 1957
  - 5. Erg. - Bd. 1958
  - 6. Erg. - Bd. 1958

Indian Biographical Archive / Indisches Biographisches Archiv (IBA). A Single-alphabet cumulation of approx. 180 sources in more than 260 volumes published from early 18th century to 1947 with a total of approx. 140,200 entries. Microfiche edition. Comp. by Laureen Baillie / 1997 ff. 12 instalments on approx. 500 fiches. Reader factor 24x / Silverfiche (3-598-34091-5) / Diazofiche (3-598-34090-7)

Information. Hrsg. v. Zentralarchiv für Hochschulbau, Stuttgart / Jg. 11/1978 bis Jg. 12/1979 (2 Hefte jährlich) (ISSN 0171-2314)

Information et documentation en matière de brevets en Europe occidentale. Inventaire des services offerts au public au sein de la Communaute europeenne. Ed. par la Commission des Communautes europeennes / 1976. 178 S. / Gb (3--7940-5168-8)
- 2nd éd. rév. et augmentée. Ed. par H. Bank, M. Fenat-Haessig et M. Roland, Commission des Communautes europeennes / 1980. 279 S. / Gb (3-598-10157-0)
- 3ième éd. rev. et augmentée. Ed. par Brenda Rimmer, Commission des Communautés européennes / 1989. 224 S. / Gb (3-598-10769-2)
Ed. allemand → **Patentinformation** *und Dokumentation in Westeuropa*
Ed. anglais → **Patentinformation** *and Documentation in Western Europe*

Information Handling in Offices and Archives. Ed. by Angelika Menne-Haritz / 1993. 197 p. / Gb (3-598-11146-0)

Information Retrieval und natürliche Sprache. Integrierte Verarbeitung von Daten und Texten im Modell Condor. Hrsg. v. H. G. Fischer / 1982. 507 S., zahlr. Schemata u. Taf / Br (3-598-10474-X)

Information Theory, Statistical Decision Functions, Random Prossesses. [Koproduktion mit Academia, Prag]
- Transactions of the Fourth Prague Conference 1965 / 1967. 726 S. / Ln (Vlg.-Nr. 06147)
- Transactions of the Sixth Prague Conference held at Prague from September. 19 to 25, 1971. Hrsg. von Jaroslav Konzesnik / 1973. 924 S. / Ln (3-7940-5046-0)

Information Trade Directory. An International Directory of Information Products and Services. Hrsg. : EUSIDIC (European Association of Information Services); James B. Sanders / 1979. 278 S. / Koporoduktion mit Learned Information Ltd. London (3-598-10072-8)

Information und Dokumentation im Aufbruch. Festschrift für Hans-Werner Schober. Hrsg. von Wilfried Kschenka / 1975. 256 S. (3-7940-3395-7)

Informationsdienste. Hrsg.: Gesellschaft für Information und Dokumentation mbH (GID), Frankfurt a. M. [Vorher im IDD Vlg. W. Flach, Frankfurt a. M.]
- 3: Forschungs- und Entwicklungsprojekte in Informationswissenschaft und -praxis. Stand Mai 1984 / 8. Ausg. 1985. 393 S. / Br (3-598-20828-6)
- 4: Wörterbuch der Reprografie. Deutsch mit Definitionen, Englisch, Französisch, Spanisch / 5., völlig neu bearb. Aufl. 1982. 353 S. / Br (3-598-10444-8)
  3. u. 6. Aufl. → **Wörterbuch** *der Reprographie*
- 5: Verzeichnis deutscher Informations- und Dokumentationsstellen. Bundesrepublik Deutschland und Berlin (West). Ausgabe 4 - 1982/83 / 1982 VII, 586 S. / Br (3-598-10437-5)

Ausg. 3 u. 5 → **Verzeichnis** *deutscher Informations- und Dokumentationsstellen*
- Internationale IuD-Gremien mit Beteiligung aus der Bundesrepublik Deutschland und Berlin (West). Ausgabe 1985. Stand: April 1985 / 1985. 324 S. / Br (3-598-20829-4)
- Verzeichnis deutscher Datenbanken, Datenbankbetreiber und Informationsvermittlungsstellen. Bundesrepublik Deutschland und Berlin (West). Ausgabe 1985. Stand: März 1985 / 1985. 255 S. / Br (3-598-10582-7)
Ausgabe 1988 → **Verzeichnis** *deutscher Datenbanken, Datenbankbetreiber und Informationsvermittlungsstellen*

Informationsmanagement. Hrsg.: Gesellschaft für Information und Dokumentation mbH (GID) Sektion für Systementwicklung der Fachinformation und -kommunikation / Robert Funk; Hans A. Koch; Werner Schwuchow; Hagen Stegemann
- 1: Augstin, Sonning: Wettbewerb versus Wettbewerbsbeschränkung. Der staatliche Eingriff in die marktwirtschaftl. Koordinierung des Informations- und Dokumentationsbereichs / 1978. 92 S. / Br (3-7940-3351-5)
- 2: Sauppe, Eberhard / Hartmut Müller; Rolf Westermann: Benutzerschulung in Hochschulbibliotheken. Ergebnisse e. von d. Dt. Forschungsgemeinschaft geförderten Grundlagenuntersuchung / 1980. 240 S. / Br (3-598-20942-8)
- 3: Schwuchow, Werner: Finanzierung und Preisgestaltung in Information und Dokumentation. Forschungsbericht Nr. 6 / 79 d. Projekts Wirtschaftlichkeit von Information und Dokumentation II. Hrsg. Ges. für Information u. Dokumentation mbH. (GID), Sekt. für Systementwicklung d. Fachinformation u. -Kommunikation (SfS) / 1979. VIII, 175 S. / Br (3-598-20944-4)
- 4: Ökonomische Aspekte der Fachinformation Internationale Fachtagung der Deutschen Gesellschaft für Dokumentation e.V. (DGD) Mai 1981, Garmisch-Partenkirchen. Tagungsbericht. Hrsg. v. Werner Schwuchow / 1981. 251 S. / Br (3-598-20945-2)
- 5: Wirtschaftlichkeit in Bibliotheken und IuD-Einrichtungen. Bibliographie 1970-1979. Hrsg. v. Hans J Bergmann; Robert Funk / 1981. 271 S. / Br (3-598-20946-0)
- 6: Nutzen der Fachinformation Internationale Fachkonferenz der Deutschen Gesellschaft für Dokumentation e.V. (DGD) Mai 1983 in Garmisch-Partenkirchen. Konferenzbericht. Hrsg. v. H. Stegmann / 1983. 237 S. / Br (3-598-20947-9)
- 7: Informationsverhalten und Informationsmarkt. Internationale Fachkonferenz der Deutschen Gesellschaft für Dokumentation e.V. (DGD). 1.-10. Mai 1985, Garmisch-Partenkirchen. Konferenzbericht. Hrsg. v. Werner Schwuchow; Hagen Stegemann / 1986. 311 S. / Br (3-598-20948-7)

Informationsströme in der Wirtschaft. Von W. Nemtschinow u.a. , Wissenschaftl. Bearb. d. Übers. a.d. Russ. von Wolfgang Schoppan / 1971. 330 S. (3-7940-4184-4)

Informationssysteme - Grundlagen und Praxis der Informationswissenschaften. Hrsg. v. Werner Kunz; Karl H. Meyer-Uhlenried; Horst W. Rittel
- 1: Gresser, Klaus / Herbert Paschen; Werner Schwuchow: Die Kosten der wissenschaftlichen und technischen Information. Ein Standartsystem für statistische Erhebungen / 1979. 136 S.

- 2: Wersig, Gernot: Das Krankenhaus-Informationssystem (KIS). Überlegungen zu Strukturen u. Realisierungsmöglichkeiten integrierter Krankenhaus-Informationssysteme. Mit e. Einl. von Günter Fuchs / 1971. 150 S. (3-7940-3452-X)
- 3: Berger, Albrecht: Die Erschließung von Verweisungen bei der Gesetzesdokumentation / 1971. 28, 204 S. (3-7940-3453-8)
- 4: Meyer-Uhlenried, Karl-Heinrich / Uta Krischker: Die Entwicklung eines Datenerfassungsschemas für komplexe Informationssysteme / 1971. XV, 194 S. (3-7940-3454-6)
- 5: Kunz, Werner / Horst W. J. Rittel: A systems analysis of the logic of research and information processes. Reasoning patters in organic chemistry / 1977. 74 S. (3-7940-3455-4)
- 6: Rolland, Maria Theresia: Thesaurusprobleme in Informationsverbundsystemes. Die Verwendung formalisierter Sprachen bei d. inhaltl. Erschließung von Dokumenten unter d. Gesichtspunkt d. Erfordernisse v. Informationsverbundstemen / 1973. 110 S. (3-7940-3456-2)
- 7: Gebhardt, Friedrich: Codierung der Texte und formalen Angaben für ein computergestütztes Dokumentationssystem / 1973. VII, 94 S. (3-7940-3457-0)

Informatisierung und Gesellschaft. Wie bewältigen wir die neuen Informations- und Kommunikationstechnologien. Hrsg. v. Gernot Wersig / 1983. 300 S. / Br (3-598-10503-7)

Infoterm Series. Ed.: International Information Centre for the Terminology (infoterm), Wien
- 1: Wüster, Eugen: The Road to Inform. Two reports prepared on behalf of UNESCO: Inventory of Sources of Scientific and Technical Terminology. A Plan for Estabishing an International Information Centre (Clearing House) for Terminology / 1974. IX, 141 S. / Br (3-7940-5501-2)
- 2: International bibliography of standardized vocabularies / Bibliographie internationale de vocabulaires normalisés / Internationale Bibliographie der Normwörterbücher. Hrsg. v. Eugen Wüster; Helmut Felber; Magdalena Krommer-Benz; Adrian Manu / 2nd enl. ed. 1979. XXIV, 540 p. / Soft (3-598-05502-1)
- 3: International co-operation in termninology Coopération internationale en terminologie 1. Infoterm-Symposium, Vienna, 9 to 11 April 1975 / 1976. 332 S. (3-7940-5503-9)
- 4: World Guide to Terminological Activities. Guide mondial des activites terminologiques. Organizations, Terminology, Banks, Committees. Organisations, Banques de terminologie, Comités / Hrsg. von Magdalena Krommer-Benz / 1977. 311 S. (3-7940-5504-7)
  - 2nd cpl. rev. and enl. ed. World Guide to Terminological Activities Organizations, Commissions, Terminology Banks. Prep. by Magdalena Krommer-Benz / 1985. X, 158 p. / Soft (3-598-21368-9)
- 5: Terminological Data-Banks. Proceedings of the First International Conference, Vienna 2nd and 3rd April 1979, convened by Infoterm. Introduction by Helmut Felber. Ed. by Christian Galinski / 1980. 207 p. / Soft (3-598-21365-4)
- 6: Theoretical and Methodological Problems of Terminology / Problémes théoriques et méthodologiques de la terminologie Proceeding of the International Symposium / Actes du colloque international, Moscow 1979-11-27 / 30. Ed. by

Magdalena Krommer-Benz / 1981. 608 p. / Soft (3-598-21366-2)
- 7: Terminologie for the eighties. With a special sect: 10 years of Infoterm. General ed.: Internat. Information Centre for Terminologie / 1982. 412 S.
- 8: Networking in Terminology - International Co-operation in Terminology Work / Travail dans le Cadre d'un Réseau de Terminologie - Coopéra-tion internationale dans le Travail terminologique Second Infoterm Symposium - Proceedings / Deuxième Symposium d'Infoterm - Actes Con-vened by Infoterm in co-operation with UNESCO. Issue ed. Magdalena Krommer-Benz / 1986. 642 p. / Soft (3-598-21369-7)

Innenpolitik in Theorie u. Praxis. Hrsg. v. Lutz R. Reuter; Rüdiger Voigt [Minerva Publikation]
- 1: Muszynski, Bernhard: Forschungspolitik und Humanisierung der Arbeit / 1982. 358 S. / Br (3-597-10374-X)
- 2: Seibel, Wolfgang: Die Nutzung verwaltungs-wissenschaftlicher Forschung für die Gesetzge-bung. Chancen und Risiken weniger komplexer Rechtssetzungen / 1984. 139 S. / Br (3-597-10392-8)
- 3: Reuter, Lutz R. / Thorsten Riedel; Rüdiger Schlott: Weiterbildung für Ausländer / 1982. 350 S. (3-597-10393-6)
- 3: Wohnen und kommunale Politik. Hrsg. v. Hiltrud Nassmacher / 1985. XII, 193 S. / Br (3-597-10393-6)
- 4: Billerbeck, Rudolf: Schutz für Kaliforniens Küste. Interessen und Instrumente in der ameri-kanischen Umweltpolitik / 1982. X, 276 S. / Br (3-597-10394-4)
- 5: Ebertzeder, Albrecht: Verrechtlichung des beruf-lichen Bildungswesens durch das Berufsbildungs-gesetz? / 1983. 122 S. / Br (3-597-10395-2)
- 6: Reeb, Hans J.: Erziehung in den Streitkräften. Eine interdisziplinäre Analyse geltender Bestim-mungen / 1983. II, 184 S. / Br (3-597-10396-0)
- 7: Soziale Selbsthilfegruppen in der Bundesrepu-blik Deutschland Aktuelle Forschungsergebnisse und Situationsdiagnosen. Hrsg. v. Walter H. Asam; Michael Heck / 2. unveränd. Aufl. 1987. II, 302 S. / Br (3-597-10397-9)
- 8: Wanders, Bernhard: Zwischen Dienstleistungs-unternehmen und politischer Basisbewegung. Mieterorganisation in der Bundesrepublik Deutschland. Empirische Untersuchung zur politi-schen Organisation wohnungsbezogener Ver-braucherinteressen in Mietervereinen und Deut-schem Mieterbund sowie in Mieterintitiativen / 1984. XIV, 285 S. / Br (3-597-10398-7)
- 9: Fälker, Margot: Schulpolitik als Resultat von Machtrelationen Computerunterstützte Datenana-lyse der Schulpolitik in Nordrhein-Westfalen von 1950-1966 / 1984. 395 S. / Br (3-597-10519-X)
- 10: Koch, Rainer: Berufliche Sozialisation öffent-licher Bediensteter. Zur Auswirkung eines inte-grierten Verwaltungsstudiums auf das berufliche Selbstverständnis / 1984. VI, 108 S. / Br (3-597-10520-3)
- 11: Zehn Jahre Hochschule der Bundeswehr Ham-burg Aufgaben, Entwicklungen, Perspektiven. Hrsg. v. Wolfgang Gessenharter; Harro Plander; Lutz R. Reuter / 1985. VIII, 370 S. / Br (3-597-10521-1)
- 12: Winkler-Haupt, Uwe: Gemeindeordnung und Politikfolgen. Eine vergleichende Untersuchung in vier Mittelstädten / 1988. XXIV, 247 S. / Br (3-597-10522-X)
- 13: Fuchs, Hans W. / Klaus P. Pöschl: Reform oder Restauration. Eine vergleichende Analyse der schulpolitischen Konzepte und Massnahmen der

Besatzungsmächte 1945-1949 / 1986. VI, 203 S. / Br (3-597-10523-8)
- 14: Damaschke, Kurt: Der Einfluss der Verbände auf die Gesetzgebung. Am Beispiel des Gesetzes zum Schutz vor gefährlichen Stoffen (Chemika-liengesetz) / 1986. XVI, 187 S. / Br (3-597-10524-6)
- 15: Schultz, Reinhard: Gewinn- und Kapitalbeteili-gung der Arbeitnehmer (Vermögensbildung) / 1987. 256 S. / Br (3-597-10525-4)
- 16: Posse, Achim U.: Föderative Politikverflech-tung in der Umweltpolitik / 1986. XIV, 181 S. / Br (3-597-10526-2)
- 17: Schlieckau, Jürgen: Politik und Offiziersbild. Die Offiziersausbildung in der Bundeswehr im Zeichen der Bonner Wende? / 1988. 355 S. / Br (3-597-10527-0)
- 18: Thomassen, Wolfgang: Politische Partizipation und Stadtentwicklungsplanung / 1988. II, 164 S. / Br (3-597-10528-9)
- 19: Reeb, Hans J.: Militär und Recht. Zur Rechts-ausbildung für Soldaten der Bundeswehr / 1988. 325 S. / Br (3-597-10631-5)
- 20: Hahn, Joachim: Stadtteilverfassung in Stutt-gart. Die Stuttgarter Bezirksbeiräte / 1989. II, 127 S. / Br (3-597-10632-3)
- 21: Frey, Klaus: Kommunale Umweltinformations-systeme. Strategien und Konflikte bei der Einfüh-rung kommunaler Umweltinformationssysteme in die kommunalpolitische Praxis am Beispiel der Städte Bielefeld, Düsseldorf, Wiesbaden und Würzburg / 1990. II, 162 S. / Br (3-597-10633-1)
- 22: Brettschneider, Frank: Wahlumfragen. Empi-rische Befunde zur Darstellung in den Medien und zum Einfluss auf das Wahlverhalten in der Bundesrepublik Deutschland und den USA / 1991. XII, 165 S. / Br (3-597-10634-X)

Integrale Anthropologie. Hrsg. v. Karel Mácha. [Minerva Publikation]
- 1: Die menschliche Individualität. Festschrift zum 85. Geburtstag von Prof. Dr. Herbert Cysarz. Hrsg. v. Karel Mácha / 1981. 239 S. / Br (3-597-10310-3)
- 2: Werz-Kovacs, Stephanie von: Heilige Mutter-schaft. Rekonstruktion matriarchalischer Elemen-te in der Religion und Mythologie der altindi-schen Bhil. Hrsg. v. Karel Mácha / 1982. II, 302 S. / Br (3-597-10311-1)
- 3: Der Nächste. Von Karel Mácha; Eugen Biser; Otakar Nahodil; W. L. Hohmann; Roland Pietsch; Theda Rehbock; Michael Rappenglück. Hrsg. v. Karel Mácha / 1983. 225 S. / Br (3-597-10312-X)
- 4: Kultur und Tradition Festschrift für Prof. Dr. Sc. Otakar Nahodil. Von K. Scalický; R. Preisner; E. Ducci; H. G. Kuttner; K. Vrána; M. Rappenglück; Karel Mácha. Hrsg. v. Karel Mácha / 1983. VIII, 361 S., 1 Abb. / Br (3-597-10313-8)
- 5: Geist und Erkenntnis. Zu spirituellen Grundla-gen Europas. Festschrift zum 65. Geburtstag von Prof. Dr. Tomas Spidlik SJ. Hrsg. v. Karel Mácha / 1985. II, 350 S. / Br (3-597-10314-6)
- 6: Das Wesen des Menschen. Festschrift zum 60. Geburtstag von Prof. Dr. Karel Vrána. Hrsg. v. Karel Mácha / 1985. VIII, 364 S. / Br (3-597-10545-9)
- 7: Aspekte kultureller Integration. Festschrift zu Ehren von Prof. Dr. Antonin Mestan. Hrsg. v. Peter Drews; Karel Mácha / 1991. XXXIV, 358 S. / Br (3-597-10698-6)

International Archival Bibliography 1958-59. Ed. by International Council on Archives (Bibliography No. 1) / 1964. 290 S. / Br (3-7940-3773-1) [1975 v. International Council on Archives übernommen]

International Association of Law Libraries - Bulletin (IALL Bulletin). Reprint Edition by Arrangement with the International Association of Law Libraries. No. 1 September 1960 - No. 30, December 1972 / 2 Vols. 1975. X, 972 S. (ISSN 0340-0611) (3-7940-3431-7)
- Pt. 1: No. 1 (1960) - No. 19 (1967) / 468 p. (3-7940-3435-X)
- Pt. 2: No. 20 (1967) - No. 30 (1972) / 487 p. (3-7940-3436-8)

International Association Statutes Series. Ed.: Union of International Associations (ISSN 0933-2588) [Erscheinen eingestellt]
- Vol. 1. International Association Statutes Series / 1988. 600 p. / Hard (3-598-21671-8)

International Bibliography of Biography 1970-1987. Ed. by The British Library / 12 vols. 1988. Cplt. 6,400 p. / Hard (3-598-32780-3)

International Bibliography of Historical Sciences. Internationale Bibliographie der Geschichtswissen-schaften / Bibliografia internacional de ciencias historicas / Bibliographie international des sciences historiques / Bibliografia internazionale delle scienze storiche. Hrsg.: International Committee of Historical Sciences / Comité International des Sciences Historiques (bis Vol. 61); Massimo Mastrogregori (ab Vol. 62)
- Vol. 39/40 (1970/1971). Ed. by Michel François and Nicolas Tolu / 1973. XXVIII, 567 p. / Soft (3-598-10340-9)
- Vol. 41 (1972). Ed. by Michel François and Nicolas Tolu / 1975. XXVII, 369 p. / Soft (3-598-10341-7)
- Vol. 42 (1973). Ed. by Michel François and Nicolas Tolu / 1976. XXVII, 361 p. / Soft (3-598-10342-5)
- Vol. 43/44 (1974/1975). Ed. by Michel François and Nicolas Tolu / 1979. XXVII, 469 p. / Soft (3-598-20401-9)
- Vol. 45/46 (1976/1977). Ed. by Michel François, Nicolas Tolu and Michael Keul / 1980. XXVII, 492 p. / Hard (3-598-20402-7)
- Vol. 47/48 (1978/1979). Ed. by Michel François and Michael Keul / 1982. XX, 458 p. / Hard (3-598-20403-5)
- Vol. 49 (1980). Ed. by Jean Glénisson and Michael Keul / 1984. XXVII, 406 p. / Hard (3-598-20404-3)
- Vol. 50 (1981). Ed. by Jean Glénisson and Michael Keul / 1985. XXIII, 331 p. / Hard (3-598-20405-1)
- Vol. 51 (1982). Ed. by Jean Glénisson and Michael Keul / 1986. XXV, 399 p. / Hard (3-598-20406-X)
- Vol. 52 (1983). Ed. by Jean Glénisson and Michael Keul / 1987. XXVI, 390 p. / Hard (3-598-20407-8)
- Vol. 53 (1984). Ed. by Jean Glénisson and Michael Keul / 1988. XXV, 398 p. / Hard (3-598-20408-6)
- Vol. 54 (1985). Ed. by Jean Glénisson and Michael Keul / 1989. XXVI, 423 p. / Hard (3-598-20409-4)
- Vol. 55 (1986). Ed. by Jean Glénisson and Michael Keul / 1989. XXV, 417 p. / Hard (3-598-20410-8)
- Vol. 56 (1987). Ed. by Jean Glénisson and Michael Keul / 1991. XXVI, 614 p. / Hard (3-598-20411-6)
- Vol. 57 (1988). Ed. by Jean Glénisson and Michael Keul / 1992. XX, 511 p. / Hard (3-598-20412-4)
- Vol. 58 (1989). Ed. by Jean Glénisson and Michael Keul / 1993. XXIV, 472 p. / Hard (3-598-20413-2)
- Vol. 59 (1990). Ed. by Jean Glénisson and Michael Keul / 1994. XXIV, 447 p. / Hard (3-598-20414-0)
- Vol. 60 (1991). Ed. by Jean Glénisson and Michael Keul / 1995. XXII, 477 p. / Hard (3-598-20415-9)
- Vol. 61 (1992). Ed. by Jean Glénisson and Michael Keul / 1996. XXIV, 498 p. / Hard (3-598-20416-7)
- Vol. 62 (1993). Ed. by Massimo Mastrogregori / 1999. XIV, 400 S. / Hard (3-598-20417-5)

International Bibliography of Jewish History and Thought. Ed. by Jonathan Kaplan, Rothberg School for Oversea Students, The Hebrew University / Dor Hemesch Institutes, The World Zionist Organisation / 1984. XVIII, 483 p. / Hard (3-598-07503-0)

International Bibliography of Maps and Atlases / Bibliographie der Karten und Atlanten / Bibliographie des cartes géographiques et atlas / Bibliografia delle Carte Geografiche e degli Atlanti / Bibliografía de Mapa. CD-ROM-Edition / 1998 (3-598-40399-2)

International Bibliography of Plant Protection 1965-1987 / Internationale Bibliographie der Pflanzenschutzliteratur 1965-1987 / Bibliographie Internationale de la Protection des Plantes 1965-1987 / Bibliografia International de la Protection de la Plantes 1965-1987. Comp. by Wolfrudolf Laux. In cooperation with Dedo Blumenbach, Dieter Jaskolla, Peter Kronowski, Wulf J. Pieritz, Michael Scholz and Wolfgang Sicker / 35 Vols. 1989/91. / Hard (3-598-21960-1)
- Vol. 1. 1989. XXV, 439 p. / Hard (3-598-21961-X)
- Vol. 2. 1989. XXV, 439 p. / Hard (3-598-21962-8)
- Vol. 3. 1989. XXV, 439 p. / Hard (3-598-21963-6)
- Vol. 4. 1989. XXV, 439 p. / Hard (3-598-21964-4)
- Vol. 5. 1989. XXV, 439 p. / Hard (3-598-21965-2)
- Vol. 6. 1989. XXV, 439 p. / Hard (3-598-21966-0)
- Vol. 7. 1989. XXV, 439 p. / Hard (3-598-21967-9)
- Vol. 8. 1989. XXV, 439 p. / Hard (3-598-21968-7)
- Vol. 9. 1989. XXV, 439 p. / Hard (3-598-21969-5)
- Vol. 10. 1989. XXV, 439 p. / Hard (3-598-21970-9)
- Vol. 11. 1989. XXV, 439 p. / Hard (3-598-21971-7)
- Vol. 12. 1989. XXV, 439 p. / Hard (3-598-21972-5)
- Vol. 13. 1989. XXV, 439 p. / Hard (3-598-21973-3)
- Vol. 14. 1989. XXV, 439 p. / Hard (3-598-21974-1)
- Vol. 15. 1989. XXV, 439 p. / Hard (3-598-21975-X)
- Vol. 16. 1989. XXV, 439 p. / Hard (3-598-21976-8)
- Vol. 17. 1989. XXV, 439 p. / Hard (3-598-21977-6)
- Vol. 18. 1989. XXV, 439 p. / Hard (3-598-21978-4)
- Vol. 19. 1989. XXV, 439 p. / Hard (3-598-21979-2)
- Vol. 20. 1990. XXV, 439 p. / Hard (3-598-21980-6)
- Vol. 21. 1990. XXV, 439 p. / Hard (3-598-21981-4)
- Vol. 22. 1990. XXV, 439 p. / Hard (3-598-21982-2)
- Vol. 23. 1990. XXV, 439 p. / Hard (3-598-21983-0)
- Vol. 24. 1990. XXV, 439 p. / Hard (3-598-21984-9)
- Vol. 25. 1990. XXV, 439 p. / Hard (3-598-21985-7)
- Vol. 26. 1990. XXV, 439 p. / Hard (3-598-21986-5)
- Vol. 27. 1990. XXV, 439 p. / Hard (3-598-21987-3)
- Vol. 28. 1990. XXV, 439 p. / Hard (3-598-21988-1)
- Vol. 29. 1990. XXV, 439 p. / Hard (3-598-21989-X)
- Vol. 30. 1990. XXV, 439 p. / Hard (3-598-21990-3)
- Vol. 31. 1990. XXV, 439 p. / Hard (3-598-21991-1)
- Vol. 32. Index of Authors A - Leo / 1991. IV, 408 p. / Hard (3-598-21992-X)
- Vol. 33. Index of Authors Leo - Zyu / 1991. IV, 410 p. / Hard (3-598-21993-8)
- Vol. 34. Index of Descriptors A - Lev / 1991. XIII, 376 p. / Hard (3-598-21994-6)
- Vol. 35. Index of Descriptors Lev - Zyz / 1991. XIII, 376 p. / Hard (3-598-21995-4)

International Biographical Dictionary of Central European Émigrés 1933-1945 → **Biographisches** Handbuch der deutschsprachigen Emigration nach 1933

International Biographical Dictionary of Religion. An Encyclopedia of more than 4,000 Leading Personalities. Ed. by Jon C. Jenkins assisted by Cécile Vanden Bloock; Union of International Organisations, Brussels / 1994. XVIII, 385 p. / Hard (3-598-11100-2)

International Book Trade Directory → **Handbuch** der Internationalen Dokumentation und Information **Bd. 15**

International Books in Print (ISSN 0170-9348)
- 1979. English-language titles publ. outside the U.S.A. and the United Kingdom / 2 vols. 1979. Cplt. XXII, 1.251 p. / Bd. (3-598-07070-5)
- 2nd ed. 1981-82. Ed. by Archie Rugh / 2 vols. 1981. Cplt. XV, 1.200 S. / Bd. (3-598-10201-1)
- 3rd ed. 1984 / 4 vols. 1984 / Hard
  - Pt. I (2 vols.). Author-Title List and Publishers / Cplt. XVIII, 1737 p. (3-598-20582-X)
  - Pt. II (2 vols.). Subject Guide / Cplt. XXII, 1449 p. (3-598-20584-8)
- 4th ed. 1985 / 4 vols. 1985 / Hard
  - Pt. I (2 vols.). Author-Title List and Publishers / Cplt. XXIII, 1808 p. (3-598-20586-4)
  - Pt. II (2 vols.). Subject Guide / Cplt. XXIV, 1482 p. (3-598-20587-2)
- 5th ed. 1986 / 4 vols. 1986 / Hard (3-598-20588-0)
  - Pt. I (2 vols.). Author-Title List and Publishers / Cplt. XXV, 2046 p. (3-598-20589-9)
  - Pt. II (2 vols.). Subject Guide / Cplt. XXIV, 1646 p. (3-598-20590-2)
- 6th ed. 1987 / 4 vols. 1987 / Hard
  - Pt. I (2 vols.). Author-Title List / Cplt. XX, 2,187 p. (3-598-20591-0)
  - Pt. II (2 vols.). Subject Guide / Cplt. XXIV, 1762 p. (3-598-20592-9)
- 7th ed. 1988. A Listing of English-Language Titles Published in Canada, Continental Europe, Latin America, Oceania, Africa, Asia, and the Republic of Ireland. Ed. by Archie Rugh and Barbara Hopkinson / 4 vols. 1988 / Hard
  - Part I (2 vols.). Author-Title List / Cplt. XXVI, 2202 p. (3-598-22110-X)
  - Part II (2 vols.). Subject Guide / Cplt. XXX, 1786 p. (3-598-22111-8)
- 1989 / 4 vols. 1989 / Hard
  - Part I (2 vols.). Author-Title List / Cplt. XXVI, 2257 p. (3-598-22112-6)
  - Part II (2 vols.). Subject Guide / Cplt. XXX, 1847 p. (3-598-22113-4)
- 1990. Ed. by Barbara Hopkinson / 4 vols. 1990 / Hard
  - Part I (2 vols.). Author-Title List / Cplt. XX, 2439 p. (3-598-22115-0)
  - Part II (2 vols.). Subject Guide / Cplt. XXIV, 2010 p. (3-598-22116-9)
- 1991 / 4 vols. 1991 / Hard
  - Part I (2 vols.). Author-Title List / Cplt. XX, 2,570 p. (3-598-22118-5)
  - Part II (2 vols.). Subject Guide / Cplt. XXIV, 2,130 p. (3-598-22119-3)
- 1992 / 4 vols. 1992 / Hard
  - Part I (2 vols.). Author-Title List / Cplt. XVI, 2,789 p. (3-598-22121-5)
  - Part II (2 vols.). Subject Guide / Cplt. XXVIII, 2,191 p. (3-598-22122-3)
- 1993 / 4 vols. 1993 / Hard
  - Part I (2 vols.). Author-Title List / Cplt. XVI, 2,982 p. (3-598-22124-X)
  - Part II (2 vols.). Subject Guide / Cplt. XXIII, 2,341 p. (3-598-22125-8)
- 1994 / 4 vols. 1994 / Hard
  - Part I (2 vols.). Author-Title List / Cplt. XXI, 3,124 p. (3-598-22127-4)
  - Part II (2 vols.). Subject Guide / Cplt. XXIII, 2,500 p. (3-598-22128-2)
- 1995 / 4 vols. 1995 / Hard
  - Part I (2 vols.). Author-Title List / Cplt. XXI, 3,198 p. (3-598-22130-4)
  - Part II (2 vols.). Subject Guide / Cplt. XXIII, 2,540 p. / Hard (3-598-22131-2)
- 1996. English-Language Titles. Published in Africa, Asia, Australia, Canada, Continental Europe, Latin America, New Zealand, Oceania, and The Republic of Ireland / 4 vols 1996 / Hard
  - Part I (2 vols.). Author-Title List / Cplt. XXI, 3,396 p. (3-598-22133-9)
  - Part II (2 vols.). Subject Guide / Cplt. XXVIII, 2,697 p. / Hard (3-598-22134-7)
- 1997. Ed. by Barbara Hopkinson and Irene Izod / 4 vols. 1997 / Hard
  - Part I (2 vols.). Author-Title List / Cplt. XXII, 3,319 p. (3-598-22285-8)
  - Part II (2 vols.). Subject Guide / Cplt. XXIX, 2,674 p. / Hard (3-598-22286-6)
- 1998. Ed. by Irene Izod / 4 vols. 1998 / Hard
  - Part I (2 vols.). Author-Title List / Cplt. XXI, 2,766 p. (3-598-22288-2)
  - Part II (2 vols.). Subject Guide / Cplt. XXVIII, 2,385 p. / Hard (3-598-22289-0)

International Books in Print PLUS. English-Language Titels published in Africa, Asia, Australia, Canada, Continental Europe, Latin America, New Zealand, Oceania and the Republic of Ireland. CD-ROM-Edition
- 1st ed. International Books in Print. CD-ROM-Edition. English-Language Titles. Published in Africa, Asia, Australia, Canada, Continental Europe, Latin America, New Zealand, Oceania, and the Republic of Ireland / 1990 (3-598-40210-4)
- 2nd ed. 1992 (3-598-40248-1)
- 3rd ed. 1993 (3-598-40261-9)
- 4th ed. 1994 (3-598-40274-0)
- 5th ed. International Books in Print PLUS 1995. English-Language Titels published in Africa, Asia, Australia, Canada, Continental Europe, Latin America, New Zealand, Oceania and the Republic of Ireland. CD-ROM-Edition / 1995 (3-598-40290-2)
- 6th ed. 1996 (3-598-40347-X)
- 7th. ed. 1997-1998 / 1997 (3-598-40349-6)
- 8th. ed. 1999 / 1999 (3-598-40395-X)
- International Books in Print Handbuch. Red.: Sylvia Nalbach / 1991. III, 82 S. / Br (3-598-40243-0)
- International Books in Print Manual. Ed. by Sylvia Nalbach / 1991. III, 72 S. / Soft (3-598-40245-7)

International Books in Print auf CD-ROM / siehe auch: → **Global** Books in Print

International Classification. Ed. by Ingetraut Dahlberg, Alwin Diemer, Jean M. Perreault and A. Neelameghan (ISSN 0340-0050)
- Vol. 1/1974 - 4/1977. A Journal on Theory and Practice of Universal and Special Classification Systems and Thesauri / Zeitschrift zur Theorie und Praxis universaler und spezieller Klassifikationssysteme und Thesauri (2 issues annually)
- Vol. 5/1978 - 7/1980. A Journal Devoted to Concept Theory, Organization of Knowledge and Data, and to Systematic Terminology (3 issues annually)

International Congress Calendar. Ed by the Union of International Organizations / Vol. 28/1988 - Vol. 34/1994 (4 issues annually) (ISSN 0538-6349)

International Directory of Arts / Internationales Kunstadressbuch / Annuaire International des Beaux-Arts / Annuario Internazionale delle Belle Arti / Anuario Internacional de las Artes 1993/94 / 2 vols. 21st ed. 1993. Cplt. 1816 p. / Hard (3-598-23070-2)
- 1995/96 / 2 vols. 22nd ed. 1994. Cplt. 1436 p. / Hard (3-598-23072-9)
- 1997/98 / 3 vols. 23rd ed. 1996. Cplt. 2152 p. / Hard (3-598-23075-3)

- 1998/99 / 3 vols. 24th ed. 1998. Cplt. 2398 p. / Hard (3-598-23078-8)

International Directory of Arts & Museums of the World. CD-ROM-Edition / 1999 (3-598-40377-1)

International Directory of Children's Literature Specialists. Ed.: International Board on Books for Young People / 1986. 263 p. / Hard (3-598-10623-8)

International Directory of Cinematographers Set- and Costume Designers in Film. Ed. by Alfred Krautz. Comp. by International Federation of Film Archives (FIAF).
- Vol. 1. German Democratic Republic (1946-1978). Poland (From the beginnings to 1978) / 1981. 280 p. / Hard (3-598-21431-6)
- Vol. 2. Bild- und Tonträger-Verzeichnisse France (From the beginnings to 1980) / 1983. 761 p. / Hard (3-598-21432-4)
- Vol. 3. Albania, Bulgaria, Greece, Rumania, Yugo-slavia (From the beginnings to 1980). Coordinator: Bujor T. Ripeanu / 1983. 297 p. / Hard (3-598-21433-2)
- Vol. 4. Internationaler Biographischer Index / World Biographical Index Germany (From the beginnings to 1945). Coordinator: Eberhard Spiess and Alfred Krautz / 1984. 605 p. / Hard (3-598-21434-0)
- Vol. 5. Denmark, Finland, Norway, Sweden (From the beginnings to 1984). Coordinator: Bujor T. Ripeanu / 1986. 588 p. / Hard (3-598-21435-9)
- Vol. 6. Bild- und Tonträger-Verzeichnisse. Supple-mentary Volume. New Entries, Additions, Correc-tions 1978-1984. Albania, Bulgaria, France, German Democratic Republic, Germany (until 1945), Poland, Romania, Sweden, Yugoslavia / 1986. 455 p. / Hard (3-598-21436-7)
- Vol. 7. Direttori di Fotografia Scenografi e Costumisti del Cinema Italiano Italy (From the beginnings to 1986). Coordinator: Eberhard Spiess / 1988. 666 p. / Hard (3-598-21437-5)
- Vol. 8. Portugal, Spain (From the beginnings to 1988). Coordinator: Rui Brito / 1988. 439 p. / Hard (3-598-21438-3)
- Vol. 9. Hungary (From the beginnings to 1988). Coordinator: Bujor T. Ripeanu / 1989. 215 p. / Hard (3-598-21439-1)
- Vol. 10. Czechoslovakia Comp. by Vladimir Opela / 1991. 281 p. / Hard (3-598-21440-5)
- Vol. 11. General Index Volume 1 to 10. Part A: „Film Titles", Part B: „Film Directors" Albania, Bulgaria, Czechoslovakia, Denmark, Finland, France, German Democratic Republic, Germany (to 1945), Greece, Hungary, Italy, Norway, Poland, Portugal, Romania, Spain / 2 parts 1993. Cplt. XII, 630 p. / Hard (3-598-21442-1)
- Vol. 12 Cuba (From the beginnings to 1990) / 1992. 64 p. / Hard (3-598-21443-X)
- Vol. 13 Soviet Union / 1994. 518 p. / Hard (3-598-21441-3)

International directory of social science research councils and analogous bodies. Publ. for and comp. by the Conference of Social Science Councils and Analogous Bodies (CNSSC) / 1978. VII, 159 S. (3-598-43048-4)

International Documents on Palestine 1970. Ed. by Walid Khadduri / 1970. 1056 S. (3-7940-5157-2)
- 1971. Ed. by Anne R. Zahlan / 1974. XLII, 746 S. (3-7940-5158-0)

International Encyclopedia of Abbreviations and Acronyms of Science and Technology / Internationa-le Enzyklopädie der Abkürzungen und Akronyme in Wissenschaft und Technik. Founded by Peter Wennrich. Comp. by Michael Peschke / 17 vols. 1995 ff. / Hard (3-598-22970-4)
- Vol. 1. 1995. VIII, 406 p. / Hard (3-598-22971-2)
- Vol. 2. 1996. VIII, 347 p. / Hard (3-598-22972-0)
- Vol. 3. 1996. VIII, 354 p. / Hard (3-598-22973-9)
- Vol. 4. 1996. VIII, 351 p. / Hard (3-598-22974-7)
- Vol. 5. 1996. VIII, 360 p. / Hard (3-598-22975-5)
- Vol. 6. 1997. VIII, 349 p. / Hard (3-598-22976-3)
- Vol. 7. 1997. IX, 370 p. / Hard (3-598-22977-1)
- Vol. 8. 1997. IX, 352 p. / Hard (3-598-22978-X)
- Vol. 9. 1998. VIII, 353 p. / Hard (3-598-22979-8)
- Vol. 10. 1998. VIII, 347 p. / Hard (3-598-22988-7)
- Vol. 11. 1998. VIII, 367 p. / Hard (3-598-22989-5)
- Vol. 12. 1999. VIII, 381 p. / Hard (3-598-22990-9)

International Encyclopedia of Systems and Cybernetics. Ed. by Charles François / 1997. 423 p. / Hard (3-598-11357-9)

The **International** Exchange of publications. Pro-ceedings of the European Conference Vienna 1972. International Federation of Library Associations. Ed. by Maria J. Schiltmann / 1973. 135 S. (3-7940-4311-1)

International Exposition of Rural Development. IERD series. Ed.: Institute of Cultural Affairs International
- 1: Directory of Rural Development Projects. Project descriptions prepared for international exposition of rural development / 1985. 514 p. / Hard (3-598-21041-8)
- 2: Voices of Rural Practioners. Self-analysis of local rural development initiatives worldwide / 1987. 474 p. / Hard (3-598-21047-7)
- 3: Approaches That Work in Plural Development. Emering trends, participatory methods and local initiatives. Ed. by John Burbridge / 1988. 414 p. / Hard (3-598-21043-4)

International Guide to Microform Masters CD-ROM-Edition / 2 CD-ROM 1997 (3-598-40282-1)

International Journal of Archives / Journal international des achives. Journal of the Internat. Council on Archives. Ed. by James O'Neill / Vol. 1/ 1980-2/1981 (ISSN 0194-1364)

International Journal of Law Libraries. Official publication of the Internat. Assoc. of Law Libraries. Ed.-in-chief: Klaus Menzinger / Jg. 1/1973 - Jg. 10/1982 / Ab Jg. 10, 1982 u. d. T. : International Journal of Legal Information

International Journal of Legal Information → *International* Journal of Law Libraries

International Librarianship Today and Tomorrow. A Festschrift for William J. Welsh. Intro. by Daniel S. Boorstin. Comp. by Joseph W. Price / 1985. 174 p. / Cloth (3-598-10586-X)

International Organization Participation. 1983/84. Country directory of secretariats and membership. Ed. by Union of International Associations / 1983.

International Publishing Today. A Festschrift in honour of Manuel Salvat. Ed. by O. P. Ghai; Narendra Kumar / 1983. XII, 237 p. / Cloth (3-598-07505-7) [Lizenzausg. m. Genehm. v. The Bookman's Club]

International Symposium Patent Information and Documentation. Munich, May 16 - 18, 1977. Lectures - Submitted Papers - Discussions. Organized by the German Society for Documentation (DGD / APD) and the German Patent Office (DPA) in cooperation with the World Intellectual Property Organization (WIPO) / 1978. 479 S. / Hard (3-598-07042-X) *German Ed.* → *Internationales* Symposium über Patentinformation

International Yearbook of Library Service for Blind and Physically Handicapped Individuals. Ed. by Barbara Freeze; Judith Dixon; Catherine O'Connor; Friends of Libraries for the Blind and Physically Handicapped Individuals in North America Inc. [Erscheinen eingestellt]
- Vol. 1. 1993 / 1993. 117 p. / Hard (3-598-23100-8)

Internationale Bibliographie der Bibliographien 1959-1988 / International Bibliography of Bibliog-raphies 1959-1988. (IBB). Hrsg. v. Hartmut Walra-vens. Bearb. v. Michael Peschke / 12 Bde. 1998 ff. / Ln (3-598-33734-5)
- Bd. 01. 1998. X,358 S. / Ln (3-598-33735-3)
- Bd. 02. 1998. X, 325 S. / Ln (3-598-33736-1)

Internationale Bibliographie der Reprints / International Bibliography of Reprints
- Bd. 1. Bücher und Reihen / Books and Serials. Bearb. v. Christa Gnirss / / 3 Tle. 1976. 1.768 S. / Lin (3-7940-3433-3)
- Bd. 2. Zeitschriften, Zeitungen, Jahrbücher, Konfe-renzberichte usw / Periodicals, Newspapers, Annuals, Conference Reports etc. Ein Produkt der Zeitschriften-Datenbank. Hrsg.: Deutsches Biblio-theksinstitut u. Staatsbibliothek Preussischer Kulturbesitz. Bearb. v. Hans Dettweiler / 1980. IX, 566 S. / Lin (3-598-03434-2)

Internationale Bibliographie des Atomenergierechts 1976-1987 → *Göttinger* Atomrechtskatalog

Internationale Bibliographie zur Deutschen Klassik 1750-1850. Bibliographien und Kataloge der Herzogin Anna Amalia Bibliothek zu Weimar. Hrsg. v. d. Stiftung Weimarer Klassik (ISSN 0323-5734)
- Flg. 29 / 30. 1982-83 / 2 Bde. 1987. Zus. 859 S. / Br (3-598-07240-6)
- Flg. 31 / 32. 1984-85 / 2 Bde. 1987. Zus. 906 S. / Br (3-598-07247-3)
- Flg. 39. 1992. Mit Nachträgen zu früheren Jahren von Heidi Zeilinger / 1994. 542 S. / Gb (3-598-23370-1)
- Flg. 40. 1993. Von Heidi Zeilinger / 1995. 585 S. / Gb (3-598-23371-X)
- Flg. 41. 1994. Von Heidi Zeilinger / 1996. 597 S. / Gb (3-598-23372-8)
- Flg. 42. 1995. Von Heidi Zeilinger / 1997. 580 S. / Gb (3-598-23373-6)
- Flg. 43. 1996. Von Heidi Zeilinger / 1997. 625 S. / Gb (3-598-23374-4)
- Flg. 44. 1997. Von Heidi Zeilinger / 1998 568 S. / Gb (3-598-23375-2)
- Gesamtregister zu den Folgen 1-41. Bearb. v. Lothar Wiesmeier / 3 Bde. 1998. Zus. 1199 S. / Gb (3-598-32885-0)

Internationale Bibliographie zur Deutschen Klassik 1750-1850 Folge 1. 1959 bis Folge 41. 1994. Mikrofiche-Edition. Hrsg.: Stiftung Weimarer Klassik / 1997. 91 Fiches, 3 Registerbde, Lesefaktor 24x / Silberfiche (3-598-32884-2) / Diazofiche (3-598-32883-4)

Internationale Bibliographie zur deutschsprachigen Presse der Arbeiter- und sozialen Bewegungen von 1830-1982. Hrsg. v. Alfred Eberlein. Bearb. v. Ursula Eberlein. Mit Unterstützung d. Deutschen Forschungsgemeinschaft,. Hrsg. v. d. Universitätsbibliothek Bochum / 8 Bde. 2. aktualis. u. wes. erw. Aufl. 1996. Zus. LII, 3.134 S. / Gb (3-598-23280-2)

Internationale Bibliographie zur Geschichte der deutschen Literatur. Bearb. v. deutschen, sowjetischen, bulgarischen, jugoslawischen, polnischen, rumänischen, tschechoslowakischen und ungarischen Wissenschaftlern, unter Leitung v. Günter Albrecht; Günther Dahlke / 4 Bde. / Lin (3-7940-4273-5)
- Tl. 1. Von den Anfängen bis 1789 / 1969. 1.045 S. / Lin (3-7940-4269-7)
- Tl. 2 / 1. Von 1789 bis zur Gegenwart / 1971. 1.031 S. / Lin (3-7940-4270-0)
- Tl. 2 / 2. Von 1789 bis zur Gegenwart / 1972. 1.126 S. / Lin (3-7940-4271-9)
- Tl. 3. Sachregister, Personen-Werk-Register / 1977. 378 S. / Lin (3-7940-4272-7)
- Tl. 4. Zehnjahres-Ergänzungsband. Berichtszeitraum 1965-1974. Nachträge zum Grundwerk / (2 Bde) 1983 / 1984. Zus. 1.881 S. / Lin (3-598-10448-0)

Internationale Bibliographien zur Kunstliteratur zwischen 1500 und 1850 → **Nachschlagewerke** *und Quellen zur Kunst*

Internationale Germanistische Bibliographie (IGB). Hrsg. v. Hans A. Koch; Uta Koch.
- 1980. 1981. XIV, 851 S. / Lin (3-598-10405-7)
- 1981. 1983. LI, 1.003 S. / Lin (3-598-21181-3)
- 1982. 1984. LXVIII, 1.398 S. / Lin (3-598-21182-1)

Internationale Künstlerdatenbank (IKD) → **Allgemeines** *Künstlerlexikon - Internationale Künstlerdatenbank*
Internationale Musik-Sachlexika vom 17. bis zum 19. Jahrhundert → **Nachschlagewerke** *zur Musik*

Internationale Zeitungsbestände in Deutschen Bibliotheken / International Newspaper Holdings in German Libraries Ein Verzeichnis von 18.000 Zeitungen, Amtsblättern und zeitungsähnlichen Periodika mit Besitznachweisen und geographischem Register / A Catalogue of 18,000 Newspapers, Gazettes and Related Periodicals with Locations and Geographical Index. Hrsg.: Staatsbibliothek zu Berlin - Preußischer Kulturbesitz; Hartmut Walravens / 2. Aufl. 1993. XXI, 801 S. / Gb (3-598-11154-1)

Internationaler betriebswirtschaftlicher Zeitschriftenreport (ibz). Hrsg. von Günter Sieben mit Gerhard Coenenberg; Karl-Heinz Grass; Eberhard Luckan / Jg. 11/1977 bis 13/1979 (monatlich mit Jahresregister) (ISSN 0340-871X) [Vorher im IdW-Vlg., Düsseldorf]

Internationaler Biographischer Index / World Biographical Index. CD-ROM-Edition / 1. Ausg. 1994 (3-598-40257-0)
- 2. Ausg. 1996 (3-598-40335-6)
- 3. Ausg. 1997 (3-598-40371-2)
- 4. Ausg. 1998 (3-598-40396-8)
- 5. Ausg. 1998. (3-598-40415-8)

Internationaler Biographischer Index der Bildung und Erziehung / World Biographical Index of Education Lehrer, Hochschullehrer, Erzieher und Studen-

ten / Teachers, Professors, Tutors and Students / 2 Bde. 1996. Zus. LIV, 1.018 S. / Ln (3-598-11301-3)

Internationaler Biographischer Index der Darstellenden Kunst / World Biographical Index of the Performing Arts Regisseure, Schauspieler und Tänzer / Directors, Actors and Dancers / 1998. XXX, 421 S. / Ln (3-598-11361-7)

Internationaler Biographischer Index der Geisteswissenschaften / World Biographical Index of the Humanities Gelehrte, Philosophen, Historiker, Philologen, Kunst- und Musikwissenschaftler / Scholars, Philosophers, Historians, Philologists, Art and Music Scholars / 2 Bde. 1996. Zus. L, 797 S. / Ln (3-598-11300-5)

Internationaler Biographischer Index der Medizin / World Biographical Index of Medicine Ärzte, Naturheilkundler, Veterinärmediziner und Apotheker / Physicians, Homeopaths, Veterinarians and Pharmacists / 3 Bde. 1995. Zus. LVIII, 1.200 S. / Ln (3-598-11289-0)

Internationaler Biographischer Index der Musik / World Biographical Index of Music. Komponisten, Dirigenten, Instrumentalisten und Sänger / Composers, Conductors, Instrumentalists and Singers. Geleitw. v. Kurt Dorfmüller / 2 Bde. 1995. Zus. XLVI, 792 S. / Ln (3-598-33810-4)

Internationaler Biographischer Index der Naturwissenschaften / World Biographical Index of the Natural Sciences / 1998. XLII, 439 S. / Ln (3-598-11360-9)

Internationaler Biographischer Index der Politik und der Sozial- und Wirtschaftswissenschaften / World Biographical Index of Politics, Social Sciences and Economics Politiker, Sozial-, Wirtschafts- und Politikwissenschaftler, Psychologen / Politicians, Sociologists, Economists, Political Scientists and Psychologists. Geleitw. v. Heinz D. Fischer / 3 Bde. 1996. Zus. LXI, 1.378 S. / Ln (3-598-11302-1)

Internationaler Biographischer Index der Publizistik und der Literatur / World Biographical Index of Journalism and Literature / ca. 5 Bde. 1998. Zus. ca. 2.450 S. / Ln (3-598-11362-5)

Internationaler Biographischer Index der Religion / World Biographical Index of Religion. Theologen, Prediger, Rabbiner und Ordensleute / Theologians, Preachers, Rabbis and Members of a Religious Order / 5 Bde. 1997. Zus. LXVIII, 2.504 S. / Ln (3-598-11299-8)

Internationaler Biographischer Index der Technik / World Biographical Index of Technology Techniker, Ingenieure, Erfinder und Architekten / Technicians, Engineers, Inventors and Architects / 2 Bde. 1997. Zus. XXXIX, 655 S. / Ln (3-598-11359-5)

Internationaler Biographischer Index des Militärwesens / World Biographical Index of Military Affairs / 3 Bde. 1996. Zus. LV, 1.535 S. / Ln (3-598-11304-8)

Internationaler Biographischer Index des Rechts und der Rechtswissenschaften / World Biographical Index of Law and Legal Science. Geleitw. v.

Friedrich Ebel / 3 Bde. 1996. Zus. LIX, 1.537 S. / Ln (3-598-11303-X)

Internationaler Generalkatalog der Zeitschriften für Technik und Wirtschaft 1960/61
- Ausg. A: 49 Fachkataloge. 49 Spezialgebiete mit den Periodika je Fachgebiet in alphabetischer Reihenfolge / 49 Hefte 1960 / Br
- Ausg. B: Alphabetische Gesamtausgabe / 6 Bde. 1960. Zus. 1.500 S. / Br
- Ausg. C: Länderkatalog. Nach den Erscheinungsländern der Periodika / Karteikarten in 2 Karteikästen 1960.
Die folgende Ausgabe erschien 1962 als Bd. 6 des Handbuchs der technischen Dokumentation und Bibliographie → **Handbuch** *der Internationalen Dokumentation und Information* **Bd. 6**

Internationaler Kongreß für Reprographie und Information (4.) 1975. Teil 1: Fachreferate (3-7940-3260-4). Teil 2: Plenarvorträge (3-7940-3261-6) / . Hrsg. von Günter Haase; Georg Thiele / 2 Teile / 1976. 484 S. (3-7940-3262-4)

Internationaler Literaturdienst. 59 Ausgaben, unterteilt nach 1.405 Spezialgebieten / 1951-1958.
- Fachausgaben:
  - Technik allgemein, Forschung, Ausbildung, Fachwörterbücher
  - Akustik, Optik, Feinmechanik, Kino, Photo, Röntgen
  - Spektroskopie, Phonetik, Uhren, Ultraschall
  - Atomtechnik, Atomphysik, Reaktortechnik, Radiologie
  - Chemie, Verfahrenstechnik, Labor- u. Apparatetechnik
  - Klimatechnik, Kälte, Wärme, Trocknung, Heizung, Lüftung, Staub
  - Mathematik, Fachrechnen
  - Medizinische Technik, Arbeitspsychologie, Hygiene, Strahlung, Therapie
  - Pharmazie, Medikamente, Vitamine
  - Physik, Mechanik, Statik, Dynamik, Strömungslehre
  - Unfall-, Feuer- u. Luftschutz, Betriebsmedizin, Gesundheitswesen
  - Anstrich, Farben, Lack
  - Glas, Keramik, Steine, Erden, Email
  - Gummi, Kautschuk
  - Holz
  - Kunststoff
  - Messtechnik, Messgeräte, Werkstoffprüfung, mechanische Technologie
  - Papier, Zellstoff
  - Textil, Bekleidung
  - Fertigung, Werkzeugmaschinen, Wehrtechnik, Materialwirtschaft
  - Maschinenbau, Maschinenelemente, Konstuktion, Antriebe
  - Motoren, Strahltriebwerke, Pumpen, Kompressoren
  - Regelungstechnik, Automatik, Cybernetik, Steuern, Hydraulik
  - Eisenbahn
  - KfZ, Traktoren, Fahrtechnik
  - Luftfahrt
  - Schiffe, Seefahrt, Häfen, Navigation
  - Transportanlagen, Hebezeuge, innerbetriebl. Transport
  - Verkehr: Straße, Schiene, Wasser, Luft
  - Energie: Atom-, Wärme-, Wasser-, Gaserzeugung u. -verteilung
  - Feuerungstechnik, feste und flüssige Brennstoffe
  - Lichttechnik, Starkstromtechnik
  - Nachrichtentechnik, Elektronik, Halbleiter, Radiologie
  - Meteorologie, elektr. Prüf- und Messwesen

- Bergbau, Markscheidekunde, Geologie
- Eisenhüttenwesen, Gießerei, Metallographie
- Metallhüttenwesen u. Metallkunde, Korrosion
- Bau allgemein, Statik, Vermessung
- Baustoffe, Baumaschinen,
  Baustelleneinrichtung, Fertigbau
- Hochbau, Städtebau, Architektur, Industriebau
- Tiefbau, Eisenbahn-, Brücken-, Stahl-,Straßen-
  u. Flughafenbau
- Wasserbau, Wasserversorgung, Abwasser,
  Badewesen
- Landwirtschaft u. Forsttechnik, Fischerei,
  Schädlingsbekämpfung
- Nahrungsmitteltechnik
- Betriebswirtschaft, Betriebsorganisation,
  betriebl. rechnungswesen, betriebl.
  Vorschlagswesen, Human Relation,
  Leistungslohn, Werbung
- Sozialwissenschaften, Ein- u. Verkauf, Statistik,
  Wirtschaftsprüfung
- Bürotechnik, Büromaschinen
- Dokumentation, Bibliothekswesen
- Druck u. graphische Technik, Papier, Farbe,
  Presse
- Gewerbl. Technik, Konsumgüter, Verpackung,
  Handwerk, Leder, Möbel, Schmuck, Tabak
- Industrie, Volkswirtschaft, Wirtschaftspolitik,
  Konjunkturforschung, Export, Import, Finanzen,
  Banken, Versicherung, Ausstellungstechnik
- Patentwesen, Urheberrecht, Erfinderberatung

Internationaler Nekrolog. Verzeichnis verstorbener
Personen aus Politik, Wirtschaft, Kultur, Wissen-
schaft und Gesellschaft. Mit einem ausführlichen
Pressespiegel von Nachrufen aus deutschsprachigen
Zeitungen und Zeitschriften. Hrsg. v. Willi Gorzny.
- Jg. 1993. 1995. VIII, 776 S. / Lin (3-598-23010-9)
- Jg. 1994. 1996. VIII, 796 S. / Lin (3-598-23011-7)
[anschließend Vlg. Willi Gorzny]

Internationaler Restauratorentag. Hrsg.: Arbeitsgemeinschaft der Archivrestauratoren, Bibliotheksrestauratoren und Graphikrestauratoren / 1967. 191
S. / Br (3-7940-5118-1)

Internationales Bibliotheksadreßbuch →
Handbuch *der Internationalen Dokumentation und*
Information Bd. 8

Internationales Handbuch der Spezialbibliotheken
→ Handbuch der Internationalen Dokumentation
und Information Bd. 17

Internationales ISBN-Verlagsregister / Internation-
al ISBN Publishers Index. Bearb. u. Hrsg. von d.
Internat. ISBN-Agentur, Staatsbibliothek Preuß. Kul-
turbesitz, Berlin / 1975. 213 S. / Br (3-7940-1273-9)
- 2nd. ed. International ISBN Publishers' Index /
  Internationales ISBN-Verlagsregister / 1977. 270 p.
  / Soft (3-7940-1373-5)
Weitere Ausgaben → Handbuch der
Internationalen Dokumentation und Information
Bd. 7 / 8th ed. ff.

Internationales Symposium über Patentinforma-
tion und -dokumentation. Vorträge, eingereichte
Beiträge, Diskussion. München, 16.-18. Mai 1977.
Veranstl. durch d. Deutsche Gesellschaft für Doku-
mentation, Ausschuß für Patentdokumentation
(DGD / APD) u.d. Deutsches Patentamt (DPA) in
Zusammenarbeit mit d. World Intellectual Property
Organization (WIPO) / 1978. 800 S. / Lin
(3-598-07041-1)
Engl. Ausg. → International Symposium Patent
Information

Internationales Verlagsadreßbuch → Handbuch
der Internationalen Dokumentation und
Information Bd. 7

Internationales Verlags- und Bezugsquellen-
Verzeichnis von Fachzeitschriften. Bearb. v. Karl-Otto
Saur / 1967. 509 S. / Br (Vlg.-Nr. 02027)

Internationales Verzeichnis der Wirtschaftsverbän-
de → Handbuch der Internationalen Dokumenta-
tion und Information Bd. 12

Internationales Verzeichnis wissenschaftlicher
Verbände und Gesellschaften → Handbuch der
Internationalen Dokumentation und Information
Bd. 13

Inventar archivalischer Quellen des NS-Staates. Die
Überlieferung von Behörden und Einrichtungen des
Reichs, der Länder und der NSDAP (Texte und
Materialien zur Zeitgeschichte Bd. 3. Red.: Werner
Röder u. Christoph Weisz)
- Tl. 1. Reichszentralbehörden, regionale Behörden
  und wissenschaftliche Hochschulen für die zehn
  westdeutschen Länder sowie Berlin. Im Auftr. d.
  Instituts f. Zeitgeschichte bearb. v. Heinz
  Boberach, unter Mitw. v. Dietrich Gessner; Kurt
  Metschies; Gustav H. Seebold u. Angehörigen d.
  Archive (Texte und Materialien zur Zeitgeschichte
  Bd. 3/1) / 1991. XXXV, 717 S. / Ln (3-598-10861-3)
- Tl. 2. Regionale Behörden und wissenschaftliche
  Hochschulen für die fünf ostdeutschen Länder, die
  ehemaligen preussischen Ostprovinzen und
  eingegliederte Gebiete in Polen, Österreich und
  der Tschechischen Republik. Mit Nachträgen zu
  Teil 1. Im Auftr. d. Instituts f. Zeitgeschichte bearb.
  v. Heinz Boberach, unter Mitw. v. Oldrich Sladek;
  Günter Weber; Wolfgang Weissleder sowie Angehörigen d. Archive (Texte und Materialien zur
  Zeitgeschichte Bd. 3/2) / 1994. XXII, 396 S. / Ln
  (3-598-11135-5)

Inventar der Befehle des Obersten Chefs der
Sowjetischen Militäradministration in Deutschland
(SMAD) 1945-1949. Offene Serie. Im Auftr. d.
Instituts f. Zeitgeschichte Zsgst. u. bearb. v. Jan
Foitzik. (Texte und Materialien zur Zeitgeschichte
Bd. 8. Red.: Werner Röder u. Christoph Weisz) /
1994. 229 S. / Ln (3-598-11261-0)

Inventar zu den Nachlässen der deutschen Arbeiter-
bewegung. Für die zehn westdeutschen Länder und
West-Berlin. Bearb. v. Hans H. Paul, im Auftr. d.
Archivs d. sozialen Demokratie d. Friedrich-Ebert-
Stiftung / 1992. X, 996 S. / Lin (3-598-11104-5)

Inventar zu den Nachlässen emigrierter deutsch-
sprachiger Wissenschaftler in Archiven und
Bibliotheken der Bundesrepublik Deutschland.
Hrsg.: Die Deutsche Bibliothek. Bearbeitet im
Deutschen Exilarchiv 1933-1945 der Deutschen
Bibliothek, Frankfurt am Main / 2 Bde. 1994. Zus.
XX, 1.327 S. / Gb (3-598-23130-X)

Inventar zur Geschichte der deutschen Arbeiterbewegung in den staatlichen Archiven der Bundesrepublik Deutschland. Hrsg. im Auftrag der Historischen Kommission zu Berlin v. Heinz Boberach;
Wolfram Fischer und Peter Lösche
- Reihe B: Überlieferungen aus den Flächenstaaten
  in den Staatsarchiven der alten Bundesländer

- Bd. 1. Hessisches Hauptstaatsarchiv Wiesbaden.
  Bearb. v. Volker Eichler / 1996. XXXI, 444 S. /
  Gb (3-598-23250-0)
- Bd. 2. Landeshauptarchiv Koblenz. Bearb. v.
  Marli Beck; Irma Löffler; Bernhard Simon;
  Manfred Simonis / Nordrhein-Westfälisches
  Hauptstaatsarchiv Düsseldorf. Bearb. v. Dieter
  Lück / 1998. XIX, 254 S. / Gb (3-598-23251-9)
- Bd. 3. Hessisches Staatsarchiv Darmstadt.
  Bearb. v. Martin Kukowski / 1998. XXVIII, 175
  S. / Gb (3-598-23252-7)
Bd. 4. Brandenburgisches Landeshauptarchiv
[i.Vb.]
- Reihe C: Überlieferungen der Stadtstaaten
  Bd. 2/2. Staatsarchiv Hamburg. Archiv der
  Hansestadt Lübeck [i.Vb.]
- Bd. 3. Landesarchiv Berlin. Bearb. v. Eckhardt
  Fuchs und Rosemarie Lewin / 1997. XXXVI,
  173 S. / Gb (3-598-23255-1)

Inventory of major research facilities in the Euro-
pean Community. Inventar der großen Forschungs-
einrichtungen in der Europäischen Gemeinschaft.
Comp. by the Comm. of the Europ. Communities,
Directorate-General for Research, Science and
Education / 2 Bde. / 1977. 1.561 S. (3-7940-3019)

ISBN Review. Internationale Standard-Buchnummer
/ Jg. 1/1977 - Jg. 2/1978 [Forts.: Internationale
ISBN-Agentur, Berlin (West)]

Italien-Index. Bilddokumentation zur Kunst in
Italien. Mikrofiche-Edition. Hrsg.: Bildarchiv Foto
Marburg; Deutsches Dokumentationszentrum f.
Kunstgeschichte Philipps-Universität Marburg /
1991/1992. ca. 602.713 Fotos auf 620 Silberfiches.
Lesefaktor 24x (3-598-33130-4)

Jablonska-Skinder, Hanna / Ulrich Teichler: Hand-
book of Higher Education Diplomas in Europe
Survey of Study Programmes and of Diplomas,
Degrees and other Certificatesgranted by Higher
Education Institutions in the Europe Region. Ed.:
UNESCO - European Centre for Higher Education
(CEPES), in collab. with Mathias Lanzendörfer /
1992. 304 p. / Hard (3-598-11073-1)

Jahrbuch der historischen Forschung in der Bundesrepublik Deutschland. Hrsg.: Arbeitsgemeinschaft
ausseruniversitärer historischer Forschungseinrichtungen der Bundesrepublik Deutschland.
- Berichtsjahr 1982. 1983. 632 S. / Lin
  (3-598-20082-X)
- Berichtsjahr 1983. 1984. 617 S. / Lin
  (3-598-20083-8)
- Berichtsjahr 1984. 1985. 660 S. / Lin
  (3-598-20084-6)
- Berichtsjahr 1985. 1986. 716 S. / Lin
  (3-598-20085-4)
- Berichtsjahr 1986. 1987. 511 S. / Lin
  (3-598-20086-2)
- Berichtsjahr 1987. 1988. 569 S. / Lin
  (3-598-20087-0)
- Berichtsjahr 1988. 1989. 527 S. / Lin
  (3-598-20088-9)

Jahrbuch der öffentlichen Meinung →Allensbacher
Jahrbuch der Demoskopie

Jahrbuch des Instituts für Deutsche Geschichte.
Hrsg. v. Walter Grab.
- Bd. IV. 1975. XXVII, 546 S. / Ln (3-7940-6287-6)
- Bd. V. 1976. 620, XVI S. / Ln (3-7940-6288-4)

- Bd. XII. 1983. 586, XV S. / Ln (3-598-21154-6)
- Bd. XIII. 1984. XIII, 491 S. / Ln (3-598-21155-4)
- Bd. XIV. 1986. XV, 507 S. / Ln (3-598-21159-7)
- Bd. XV. 1987. 697, XIV S. / Ln (3-598-21160-0)
- Beiheft 1. Germany and the Middle East 1835-1939. International Symposium April 1975 / 1975. 211 S. (3-7940-6251-4)
- Beiheft 2. Juden und jüdische Aspekte in der deutschen Arbeiterbewegung 1848-1918. Tagungsbericht eines internationalen Symposiums. Planung und Leitung von Na'aman, Shlomo / 1977. 260 S. (3-7940-6292-2)
- Beiheft 4. Gegenseitige Einflüsse deutscher und jüdischer Kultur von der Epoche der Aufklärung zur Weimarer Republik Internationales Symposium April 1982 / 1982. 341 S. / Br (3-598-21153-8)
- Beiheft 6. Jüdische Integration und Identität in Deutschland und Österreich 1848-1918 Internationales Symposium April 1983 / 1984. 368 S. / Br (3-598-21156-2)
- Beiheft 9. Juden in der Weimarer Republik Internationales Symposium Oktober 1984 / 1986. 386 S. / Br (3-598-21158-9)
- Beiheft 10. Juden in der Deutschen Wissenschaft / 1987. 348 S. / Br (3-598-21161-9)

Jahrbuch des Internationalen Literaturdienstes für alle Gebiete der Technik
- 1952 / 1. Ausg. 1952 / Br
- 1953 / 2. Ausg. 1953. 104 S. / Br
  3. Ausg. →*Taschenbuch* 1954. *Dokumentationen der Technik*

Jahrbuch des Vereins „Gegen Vergessen - Für Demokratie". Hrsg. im Auftr. d. Vereins „Gegen Vergessen - Für Demokratie"
- 1: Erinnerungsarbeit und demokratische Kultur. Hrsg.: Hans J. Vogel; Ernst Piper / 1996. 130 S. / Br (3-598-23760-X)
- 2: Mahnung und Erinnerung / . Hrsg.: Hans J. Vogel; Rita Süssmuth / 1998. 225 S. / Br (3-598-23761-8)

Jahrbuch für die Geschichte Mittel- und Ostdeutschlands. Hrsg. im Auftr. d. Historischen Kommission zu Berlin v. Otto Büsch und Klaus Zernack, in Verbindung mit den Sektionsleitern der Historischen Kommission zu Berlin [Vorher im Colloquium Vlg., Berlin]
- Bd. 41. Historische Landschaften des östlichen Mitteleuropa in der Forschung, Vierter Teil / 1993. X, 557 S. / Gb (3-598-23190-3)
- Bd. 42. Berliner Demokratie / 1994. VII, 533 S. / Gb (3-598-23191-1)
- Bd. 43. Adel und Ständewesen in Brandenburg und Preussen / 1994. XVI, 531 S. / Gb (3-598-23192-X)
- Bd. 44. Preussisch-Russische Beziehung seit Peter dem Grossen / 1996. VII, 470 S. / Gb (3-598-23193-8)

Jahresbericht 1970. Hrsg. von der Studiengruppe für Systemforschung, Heidelberg / 1971. 147 S. (3-7940-3502-X)

Jahresbibliographie der Universität München. Hrsg.: Universitätsbibliothek im Auftr. d. Präsidialkollegiums der Ludwig-Maximilians-Universität
- Bd. 11. Für das Jahr 1979. Zsgst. von Christa Willner / 1981. 354 S. (3-598-21520-7)
- Bd. 12. Für das Jahr 1980. Zsgst. v. Almut Tietze-Netolitzky / 1982. 381 S. (3-598-21521-5)

- Bd. 13. Für das Jahr 1981. Zsgst. v. Almut Tietze-Netolitzky / 1983. 388 S. (3-598-21522-3)
- Bd. 14. Für das Jahr 1982. Zsgst. v. Almut Tietze-Netolitzky / 1984. 364 S. / Lin (3-598-21523-1)
- Bd. 15. Für das Jahr 1983. Zsgst. v. Almut Tietze-Netolitzky / 1985. 523 S. / Lin (3-598-21524-X)
- Bd. 16. Für das Jahr 1984. Zsgst. v. Almut Tietze-Netolitzky / 1987. X, 518 S. / Lin (3-598-21525-8)
- Bd. 17. Für das Jahr 1985. Zsgst. v. Almut Tietze-Netolitzky / 1988. XI, 437 S. / Lin (3-598-21526-6)
- Bd. 18. Für das Jahr 1986. Zsgst. v. Tietze-Netolitzky, Almut / 1989. XI, 470 S. / Lin (3-598-21527-4)

Jahreskolloquium des Instituts für Systemtechnik und Innovationsforschung (ISI) der Fraunhofer-Gesellschaft, Karlsruhe. Systemtechnik und Innovationsforschung / 1973/1974. 83 S. (3-7940-4241-7)

James Hogg: Scottish Pastorals, Poems, Songs, ect. Mostly written in the dialect of South. Ed. by Elaine E. Petrie / 1988. 136 p. / Hard (3-598-10767-6)

Jan Tschichold. Leben und Werk des Typographen Jan Tschichold. Einl. v. Werner Klemke / 1988. 303 S., 224 Abb. / Ln (3-598-07224-4)

Janusz Korczak. Bibliographie. Hrsg. v. Rainer Pörzgen / 1982. 97 S. / Br (3-598-10447-2)

Japanese American World War II. Evacuation Oral History Project. Ed. by Arthur A. Hansen / 6 vols. in 5 parts 1991-1995. Cplt. 2,361 p. / Hard (3-598-41479-X) [1991 von Meckler Publishing, Westport, CT übernommen]
- Part I. Internees / 1991. XV, 227 p. / Hard (3-598-41480-3)
- Part II. Administrators / 1991. XV, 284 p. / Hard (3-598-41481-1)
- Part III. Analysts / 1994. XXXIV, 457 p. / Hard (3-598-41482-X)
- Part IV. Resisters / 1995. XXXV, 395 p. / Hard (3-598-41483-8)
- Part V. Guards and Townspeople. Ed. by Arthur A. Hansen; Nora M. Jesch / 2 vols. 1993. Clpt XXXIV, 808 p. / Hard (3-598-41704-7)

Jessen, Jens: Bibliographie der Autobiographien.
- Bd. 1. Selbstzeugnisse, Erinnerungen, Tagebücher und Briefe deutscher Schriftsteller und Künstler / 1987. 229 S. / Gb (3-598-10673-4)
- Bd. 2. Selbstzeugnisse, Erinnerungen, Tagebücher und Briefe deutscher Geisteswissenschaftler / 1987. 256 S. / Gb (3-598-10674-2)
- Bd. 3. Selbstzeugnisse, Erinnerungen, Tagebücher und Briefe deutscher Mathematiker, Naturwissenschaftler und Techniker / 1989. 371 S. / Gb (3-598-10675-0)
- Bd. 4. Jessen, Jens / Reiner Voigt: Selbstzeugnisse, Erinnerungen, Tagebücher und Briefe deutschsprachiger Ärzte / 1995. X, 630 S. / Gb (3-598-10862-1)

Jewish Immigrants of the Nazi Period in the USA. Ed. by Herbert A. Strauss (3-598-08005-0)
- Vol. 1. Archival Resources. Comp. by W. Siegel / 1979. 300 p. / Hard (3-598-08006-9)
- Vol. 2. Classified and Annotated Bibliography of Books and Articles on the Immigration and Acculteration of Jews From Europe to the U.S.A. Since 1933. Comp. by H. Friedlaender; A. Gardner; K. Schwerin; Herbert A. Strauss; S. Wassermann / 1981. XXIV, 286 p. / Hard (3-598-08007-7)

- Vol. 3/1. Guide to the Oral History Collections of the Research Foundation for Jewish Immigration. Comp. by Joan Lessing / 1982. XXXVI, 152 p. / Hard (3-598-08008-5)
- Vol. 3/2. Classified List of Articles Concerning Emigration in Germany. Jewish Periodicals January 30, 1933 to November 9, 1938. Comp. by Daniel R. Schwart; Daniel Niederland / 1982. XXII, 177 p. / Hard (3-598-08013-1)
- Vol. 4. Jewish Emigration from Germany 1933-1942 Documentary History: Part 1: Programs and Policies until 1937. Part 2: Restrictions on Emigration and Deportation to Eastern Europe. Comp. by Norbert Kampe / 2 parts 1993. Cplt. 740 p. / Hard (3-598-08009-3)
- Vol. 5. The Individual and Collective Ex-perience of German-Jewish Immigrants 1933-1984. An oral History Record. Comp. by Dennis Rohrbough / Research Foundation of Jewish Immigration, with the assist. of Antje Schubert / 1986. 308 p. / Hard (3-598-08010-7)
- Vol. 6. Essays on the History, Persecution and Emigration of the German Jews. Comp. by Herbert A. Strauss,, with the assist. of Antje Schubert / 1987. 411 p. / Hard (3-598-08011-5)

Jewish Life in Modern Britain. 1962 - 1977. Papers and proceedings of a reference held at Hillel House, London on March 13, 1977 by the Board of Deputies of British Jews and the Inst. of Jewish Affairs. Ed. by Sonja L. Lipman; Vivian D. Lipman / 1981. XVII, 203 S. (3-598-40005-5)

Jokusch, Peter / Manfred Hegger: Betriebsanalyse und Nutzungsmessung als Instrument der Bedarfsplanung. Dargrstellt an Beispielen aus den Naturwissenschaften und der nichtklinischen-theoretischen Medizin (NATHMED). Hrsg.: Zentralarchiv für Hochschulbau, Stuttgart / 1974. Getr. Pag. (3-598-10019-1)

Jost, Dominik: Deutsche Klassik. Goethes „Römische Elegien". / 1974. 196 S. (3-7940-3032-X)
- 2. Aufl. 1978. 196 S. / Kst. (UTB 851) (3-7940-2663-2)

Journal of Commonwealth literature. Annual bibliography of Commonwealth literature. Ed. by Andrew Gurr; Alastair Niven; John Thieme; Shirley Chew; Alan Bower / Vol. 14/1979 - Vol. 29/1994 (2 issues annually) (ISSN 0021-9894) [Hans Zell Publishers] [Anschließend bei Bowker-Saur, London]

Jüdisches Biographisches Archiv / Jewish Biographical Archive (JBA). Eine alphabetische Kumulation von biographischen Einträgen aus 102 biographischen Nachschlagewerken und Enzyklopädien von der zweiten Hälfte des 18. Jahrhunderts bis zur Gründung des Staates Israel. Mikrofiche-Edition Bearb. v. Hilmar Schmuck. Beratender. Hrsg. u. Vorw.: Pinchas Lapide / 1994-1996. 690 Fiches. Lesefaktor 24x / Silberfiche (3-598-33603-9) / Diazofiche (3-598-33590-3)
- Supplement 1998, 127 Fiches / Silberfiche (3-598-33516-4) / Diazofiche (3-598-33513-X)
- Jüdischer Biographischer Index / Jewish Biographical Index (JBI). Bearb. v. Hilmar Schmuck / 4 Bde. 1998. Zus. LII, 1.491 S. / Ln (3-598-33616-0)

Jüttner, Gerald / Ulrich Güntzer: Methoden der Künstlichen Intelligenz für Information Retrieval / 1988. 254 S. / Lin (3-598-10784-6)

Kalkulationstabelle für Mehrwertsteuer
- 5,5% / 1968. 63 S. / Kt
- 11% / 1968. 63 S. / Kt

Kapp, Friedrich / Johann Goldfriedrich: Geschichte des Deutschen Buchhandels. Im Auftrage des Börsenvereins der Deutschen Buchhändler. Hrsg. von der Historischen Kommission. Bd1: Geschichte des Deutschen Buchhandels bis in das siebzehnte Jahrhundert. Bd2: Vom Westfälischen Frieden bis zum Beginn der klassischen Literaturperiode (1648 - 1740). Von Friedrich Kapp. Bd. 3: Geschichte des Deutschen Buchhandels vom Beginn der klassischen Literaturperiode bis zum Beginn der Fremdherrschaft. Von Johann Goldfriedrich. Bd4: Vom Beginn der Fremdherrschaft bis zur Reform des Börsenvereins im neuen Deutschen Reiche (1805 - 1889). Von Johann Goldfriedrich. Register. - Nachdruck der Ausgaben Leipzig 1913 und 1923 / 4 Bde. 1971. 2.820 S. (3-7940-5018-5)

Kapr, Albert: Art of Lettering. The History, Anatomy, and Aesthetics of the Roman Letter Forms. Transl. by Ida M. Kimber / 1983. 470 p., 485 ills. / Hard (3-598-10464-2)

Kapr, Albert: Schriftkunst Geschichte, Anatomie und Schönheit der lateinischen Buchstaben / 3. Aufl. 1983. 473 S., 485 Abb. / Ln (3-598-10463-4) [Lizenzausg. d. VEB Vlg. d. Kunst, Dresden]

Katalog der Bibliothek des schiitischen Schrifttums im Orientalischen Seminar der Universität zu Köln / Catalogue of the Schi'ite Collection in the Oriental Department of the University of Cologne. Hrsg. v. Abdoldjavad Falaturi
- 1. Aufl. u. d. T.: Katalog des schiitischen Schrifttums im Orientalischen Seminar der Universität zu Köln / Catalogue of the Schi'ite Collection in the Oriental Department of the University of Cologne / 1988. XXVII, 517 S. / Ln (3-598-10761-7)
- 2. erheb. erw. Aufl. 6 Bde. 1996. Zus. LIV, 2.927 S. / Ln (3-598-23350-7)

Katalog der Graphischen Porträts in der Herzog August Bibliothek Wolfenbüttel: 1500-1850 Reihe A: Die Porträtsammlung der Herzog August Bibliothek Wolfenbüttel. Hrsg.: Herzog August Bibliothek Wolfenbüttel. Vorw. v. Paul Raabe. Bearb. v. Peter Mortzfeld / 38 Bde. 1986 - 2005 / Ln (3-598-31480-9)
- Bd. 01. 1986. X, 429 S. / Gb (3-598-31481-7)
- Bd. 02. 1987. X, 473 S. / Gb (3-598-31482-5)
- Bd. 03. 1987. IV, 387 S. / Gb (3-598-31483-3)
- Bd. 04. 1987. IV, 417 S. / Gb (3-598-31484-1)
- Bd. 05. 1988. IV, 476 S. / Gb (3-598-31485-X)
- Bd. 06. 1988. IV, 537 S. / Gb (3-598-31486-8)
- Bd. 07. 1988. IV, 401 S. / Gb (3-598-31487-6)
- Bd. 08. 1989. IV, 419 S. / Gb (3-598-31488-4)
- Bd. 09. 1989. IV, 381 S. / Gb (3-598-31489-2)
- Bd. 10. 1989. IV, 406 S. / Gb (3-598-31490-6)
- Bd. 11. 1989. IV, 419 S. / Gb (3-598-31491-4)
- Bd. 12. 1990. IV, 424 S. / Gb (3-598-31492-2)
- Bd. 13. 1990. IV, 423 S. / Gb (3-598-31493-0)
- Bd. 14. 1990. IV, 396 S. / Gb (3-598-31494-9)
- Bd. 15. 1990. IV, 416 S. / Gb (3-598-31495-7)
- Bd. 16. 1991. IV, 415 S. / Gb (3-598-31496-5)
- Bd. 17. 1991. IV, 433 S. / Gb (3-598-31497-3)
- Bd. 18. 1991. IV, 452 S. / Gb (3-598-31498-1)
- Bd. 19. 1991. IV, 446 S. / Gb (3-598-31499-X)
- Bd. 20. 1992. IV, 518 S. / Gb (3-598-31500-7)
- Bd. 21. 1992. IV, 432 S. / Gb (3-598-31501-5)
- Bd. 22. 1993. IV, 361 S. / Gb (3-598-31502-3)
- Bd. 23. 1993. IV, 377 S. / Gb (3-598-31503-1)
- Bd. 24. 1993. IV, 387 S. / Gb (3-598-31504-X)
- Bd. 25. 1993. IV, 371 S. / Gb (3-598-31505-8)
- Bd. 26. 1994. IV, 388 S. / Gb (3-598-31506-6)
- Bd. 27. 1994. IV, 388 S. / Gb (3-598-31507-4)
- Bd. 28. 1995. VI, 465 S. / Gb (3-598-31508-2)
- Bd. 29. (Textband) 1996. XVI, 359 S. / Gb (3-598-31509-0)
- Bd. 30. (Textband) 1997. VIII, 385 S. / Gb (3-598-31510-4)
- Bd. 31. (Textband) 1998. VI, 408 S. / Gb (3-598-31511-2)
- Bd. 32. (Textband) 1999. VI, 407 S. / Gb (3-598-31512-0)

Katalog der Thomas-Mann-Sammlung der Universitätsbibliothek Düsseldorf. Hrsg. v. Günter Gattermann und Elisabeth Niggemann / 9 Bde. 1991. Zus. 3977 S. / Ln (3-598-22270-X)

Katalog der Universitätsbibliothek Hannover und Technischen Informationsbibliothek → TIB - Katalog der Technischen Informationsbibliothek Hannover

Katalog der Zentralbibliothek der Landbauwissenschaft, Bonn. Alphabetischer Gesamtkatalog 1847-April 1986. Schlagwortkatalog 1960-April1986. Mikrofiche-Edition / 1986. ca. 375.000 Katalogktn. auf 275 Fiches. Lesefaktor 42x (3-598-30288-6)

Katalog der Zentralbibliothek für Medizin, Köln. Mikrofiche-Edition / 2 Tle. 1984 / 1985 (3-598-30352-1)
- Autoren-Sachtitel-Katalog. Schlagwort Katalog / 1984. ca. 3.302.808 Karten auf 176 Fiches. Lesefaktor 42x / Silberfiche (3-598-30353-X)
- Hochschulschriften bis ES 1976 / 1985. 232.500 Karten auf 170 Fiches. Lesefaktor 42x / Silberfiche (3-598-30354-8)

Katalog zur Volkserzählung. Spezialbestände des Seminars für Volkskunde und der Enzyklopädie des Märchens, Göttingen, des Instituts für europäische Ethnologie, Marburg und des Instituts für Volkskunde, Freiburg im Breisgau. Bearb. v. Hans J. Uther; Elisabeth Fritzsching; Elvira Scheide und Anne Willenbrock / 2 Bde. 1987. 1135 S. / Ln (3-598-10669-6)

Kataloge der Bibliothek des Zentralinstituts für Kunstgeschichte in München. Mikrofiche-Edition / 4 Tle. u. Begleitbd. / 1982-1991. 1,2 Mill. Karten auf 1.025 Fiches. Lesefaktor 42x / Silberfiche (3-598-30348-3) / Diazofiche (3-598-30346-7)
- Alphabetischer Katalog. Stand 30. 6. 1984. Hauptteil, Museums-, Ausstellungskataloge, Private Sammlungen / 1982-1985. 295.020 Karten auf 220 Fiches / Silberfiche (3-598-30349-1) / Diazofiche (3-598-30343-2)
- Katalog der unselbständigen Schriften. (Aufsatzkatalog). Stand 30. 6. 1984 / 1983-1985. 315.000 Karten auf 231 Fiches / Silberfiche (3-598-30350-5) / Diazofiche (3-598-30344-0)
- Sachkatalog. Stand: 30. 6. 1984. Systematischer Teil - Topographischer-, Künstler-, Portraitkatalog / 1984. 6.502.818 Karten auf 480 Fiches / Silberfiche (3-598-30351-3) / Diazofiche (3-598-30345-9)
- Begleitheft zum Sachkatalog / 1985. 280 S. / Br (3-598-30347-5)
- Supplement, Stand 31. 7. 1989 / 1991. 94 Fiches / Silberfiche (3-598-30391-2)

Katalog der Dresdener Hasse-Musikhandschriften. CD-ROM-Edition mit Begleitband. Die handschriftlich überlieferten Kompositionen von Johann Adolf Hasse in der Sächsischen Landesbibliothek - Staats- und Universitätsbibliothek Dresden. Beschr. u. Komm. v. Ortrun Landmann / 1999. 1 CD, 96 S. / Br (3-598-40435-2)

Kataloge der Internationalen Jugendbibliothek. Hrsg.: Internationale Jugendbibliothek, München.
- 2: Preisgekrönte Kinderbücher-Children's Prize Books. Ein Katalog der Internationalen Jugendbibliothek über 67 Preise. Hrsg. u. mit einer Einführung v. Walter Scherf / 1969. 238 S. / Br (Vlg.-Nr. 03359) / Ln (Vlg.-Nr. 03249)
- 2nd rev. and enl. ed. Children's Prize Books / Preisgekrönte Kinderbücher / Ouvrages Pour Enfants Dotés d'un Prix Littéraire. An international listing of 193 children's literature prices / Ein internationales Verzeichnis von 193 Kinderbuchpreisen / Un repertoire international comprenant 193 prix. Ed. by Jess R. Moransee. Introd. by Walter Scherf / 1983. XXII, 620 p., 50 ills. / Hard (3-598-03250-1)
- 3: Die Besten der Besten. Bilder-, Kinder- u. Jugendbücher aus 57 Ländern od. Sprachen. Hrsg.: Walter Scherf / 1971. 189 S.
- 2., erw. Aufl. Die Besten der Besten / The Best of the Best. Bilder-, Kinder- u. Jugendbücher aus 110 Ländern oder Sprachen. Hrsg.: Walter Scherf / 1976. 324 S. (3-7940-3253-5)
- 4: Bewältigung der Gegenwart? Emanzipatorische und gesellschaftspolitische Tendenzen in der Kinder- und Jugendliteratur. Hrsg. v. Elisabeth Scherf / 1974. 128 S., Abb. / Lin (3-7940-3252-7)

Kataloge der Technischen Informationsbibliothek (TIB) Hannover. Alphabetischer Katalog bis 1982. Konferenzkatalog bis 1983 / Mikrofiche-Edition 1983. ca. 1.254.960 Eintragungen auf 664 Fiches in 4 Schuppentaf. Lesefaktor 42x / Diazofiche (3-598-30445-5)
Katalog der Technischen Informationsbibliothek. CD-ROM → TIB-Kataloge der Technischen Informationsbibliothek

Kayser, Christian Gottlob: Vollständiges Verzeichnis der von 1750 bis zu Ende des Jahres 1832 in Deutschland und in den angrenzenden Ländern gedruckten Romane und Schauspiele. Tl. 1: Romane. Tl. 2: Schauspiele. - Nachdr. d.Ausg. Leipzig 1836 / 2 Tle. in 1 Bd. 1972. 155, 114 S. / Ln (3-7940-5026-6)

Kecskeméti, Károly: La Hongrie et le Reformisme Liberal Problems Politiques et Sociaux (1790-1848). (Fonti e Studi di Storia moderna e contemporanea I). Ed.: Centro di Ricerca, Roma / 1989. 413 S. / Gb (3-598-07560-X)

Keller, Ulrike: Otto Zoffs dramatische Werke. Vom Theater zum Hörspiel. Vorw. v. Wolfgang Greisenegger / 1988. X, 490 S. / Lin (3-598-10762-5)

Kempkes, Wolfgang: Bibliographie der internationalen Literatur über Comics / International Bibliography of Comics Literature / 1971. 213 S. / Lin (3-7940-3394-9)
- 2., verb. Aufl. 1974. 293 S. / Lin (3-7940-3251-9)

Kerner, Immo Ottomar / Gerhard Zielke: Einführung in die algorithmische Sprache Algol. Mit e. Geleitw. von H. Zemanek u. e. Anh. v. R. Strobel / 3., berichtigte Aufl. 1969. 283 S. / Ln (Vlg.-Nr. 04239)

Keudell, Elisev / Karl Bulling: Goethe als Benutzer der Weimarer Bibliothek. Ein Verzeichnis der von ihm entliehenen Werken. Beigefügt: Goethe als Erneurer und Benutzer der Jenaischen Bibliotheken. Gedenkgabe der Universitätsbibliothek Jena zu Goethe's hundertstem Geburtstag. - Nachdr. d. Ausg. Berlin 1931 u. 32 / 1982. XII, 391 S., 1 Faks. XI, 67 S. / Ln (3-598-10326-3)

Kimminich, Otto: Einführung in das Völkerrecht / 1975. 326 S. (UTB 469) (3-7940-2640-3)
• 2. vollst. überarb. Aufl. 1983. 548 S., Reg. (UTB 469) / Kst. (3-598-02673-0)
• 2. vollst. überarb. Ausg. 1984. 548 S. (Parallelausgabe zur 2. Aufl. 1983) / Ln (3-598-10536-3)
• 3. erg. u. verb. Aufl. 1987. 548 S., Reg. (UTB 469) / Kst. (3-598-02677-3)
• 4. erg. u. verb. Aufl. 1990. 548 S. (UTB 469) / Br (3-598-11003-0)

Kind, Friedbert: Bausysteme im Hochschulbau. Konkordanz,. Zusammenstellung der Worte, Begriffe und Bezeichnungen zur Bausystem-Datenbank des Zentralarchivs für Hochschulbau / Bearb. von Reinhard Bouché; Robert K. Jopp / 1975. 302 S. (3-598-20028-5)

Kinderbücher aus Italien - Spanien - Portugal / Libri per Raggazi dall Italia / Libros Infantiles de Espa¤a / Livros Infantis de Portugal Annotationen in Deutsch, Italienisch, Spanisch, Portugiesisch. Hrsg.: Staatsinstitut für Frühpädagogik; Wassilios E. Fthenakis; Michaela Ulich, unter Mitw. v. Pamela Oberhuemer; Dörte Thorbeck-Hess / 1984. VIII, 226 S., 171 Abb. / Lin (3-598-10558-4)

Kinderbücher aus Türkei - Yugoslawien - Griechenland Annotationen in Deutsch, Türkisch, Serbokroatisch, Slovenisch, Griechisch. Hrsg.: Staatsinstitut für Frühpädagogik; Wassilios E. Fthenakis; Michaela Ulich, unter Mitw. v. Pamela Oberhuemer; Dörte Thorbeck-Hess / 1984. VIII, 234 S. / Lin (3-598-10566-5)

Der **Kirchenkampf.** The Gutteridge-Micklem collection held in the Bodleian Library. Microfiche edition / 1988. Intro. and index, 515 fiches, Reader factor 24x / Silverfiches (3-598-32599-1)

Kirchlicher Zentralkatalog beim Evangelischen Zentralarchiv in Berlin. Mikrofiche-Edition. Hrsg. im Auftr. d. Evangelischen Zentralarchivs Berlin, v. Uwe Czubatynski / 1997. ca. 40.2842 Katalogkarten auf 216 Fiches. Lesefaktor 42x / Silberfiche (3-598-32014-0) / Diazofiche (3-598-32013-2)
• Begleitband zur Mikrofiche-Edition / 1997. 32 S. / Br (3-598-32015-9)

Kirchner, Dietrich: Topographie umweltrelevanter Literaturbestände in Bibliotheken der Bundesrepublik Deutschland einschließlich Berlin (West). Hrsg. v. Umweltbundesamt / 1983. XXXIII, 182 S. / Lin (3-598-10482-0)

Klassifikation für Allgemeinbibliotheken.
• Bd. 1. Teil Wissenschaftliche und Fachliteratur. Belletristik (KAB / E). Gliederung und Alphabetisches Schlagwortregister. Hrsg.: Zentralinstitut f. Bibliothekswesen / 1990. 320 S. / Lin (3-598-07039-X)

Kleine philosophische Bibliographien
• 1: Henrichs, Norbert: Bibliographie der Hermeneutik und ihrer Anwendung seit Schleiermacher. - 2. unveränd. Aufl. d. Ausg. 1968 / 1972. XX S., 491 Spalten / Lin (3-7940-3148-2)
• 2: Hogrebe, Wolfram / Rudolf Kamp; Gert König: Periodica Philosophica. Eine internationale Bibliographie philosophischer Zeitschriften von den Anfängen bis zur Gegenwart. / 1972. XII S., 728 Spalten / Lin (3-7940-3149-0)
• 3: Sass, Hans M.: Inedita Philosophica. Ein Verzeichnis von Nachässen deutschsprachiger Philosoophen des 19. und 20. Jahrhunderts / 1974. VIII S., 86 Spalten / Kst. (3-7940-3150-4)
• 4: Geldsetzer, Lutz: In Honorem. Eine Bibliographie philosophischer Festschriften und ihrer Beiträge / 1975. VI S., 226 Spalten / Gb (3-7940-3151-2)
• 5: Geldsetzer, Lutz: Bibliography of the international congresses of philosophy. Bibliographie der internationalen Philosophie-Kongresse. Proceedings / 1900-78. / 1981. 207 S.

Klose, Albrecht: Sprachen der Welt / Languages of the World. Ein weltweiter Index der Sprachfamilien, Einzelsprachen und Dialekte, mit Angabe der Synonyma und fremdsprachigen Äquivalente / A Multi-lingual Concordance of Languages, Dialects and Language-Families / 1987. XLVIII, 410 S. / Lin (3-598-10443-X)

Knaus, Hermann: Studien zur Handschriftenkunde. Ausgewählte Aufsätze. Hrsg. v. Gerard Achten; Thomas Knaus; Kurt H. Staub. Mit einer Bibliogr. d. Schriften Hermann Knaus' v. Christa Staub / 1992. 297 S., 42 Abb. / Gb (3-598-10975-X)

Kneipp, Sebastian: Meine Wasser-Kur, durch mehr als 35 Jahre erprobt und geschrieben zur Heilung der Krankheiten und Erhaltung der Gesundheit. - Ein originalgetreuer Nachdr. d. 50. Jubiläumsausgabe, Kempten, 1894 / 1963. 2, VIII, 376 S. / Ln (3-7940-3013-3)

Knigge, Adolph von: Sämtliche Werke. Hrsg. v. Paul Raabe, in Zus.-Arb. mit Ernst O. Fehn; Manfred Grätz; Gisela von Hanstein und Claus Ritterhoff / 24 Bde. 1978-1993. 12.315 S. / Gb (3-598-22870-8)

Knowledge for Europe: Librarians and Publishers Working Together. European Conference, 11-13 November 1992, Brussels. Ed. by Hans-Peter Geh; John Davies; Marc Walckiers; European Foundation for Library Cooperation / 1993. 235 p. / Hard (3-598-11164-9)

Koch, Nikolaus / Peter Rath: Das pädagogische Medienwesen in der Diskussion. Empfehlungen zur Sanierung d. päpagog. Medienwesens / 1977. 95 S. / Br (3-7940-7019-4)

Körperschaftsnamendatei Index of corporate bodies. Stand: 1. Juni 1973. Hrsg. v. d. Bayerischen Staatsbibliothek, München / 1973. XI, 969, 21 S. (3-7940-3040-0)

Kössler, Franz: Verzeichnis von Programm-Abhandlungen deutscher, österreichischer und schweizerischer Schulen der Jahre 1825-1918. Alphabetisch geordnet nach Verfassern / 4 Bde. 1987. Zus. 2.134 S. / Ln (3-598-10665-3)

• Bd. 5. Ergänzungsband. Alphabetisch geordnet nach Verfassern / 1991. XI, 351 S. / Ln (3-598-10684-X)

Köster, Hermann Leopold: Geschichte der deutschen Jugendliteratur in Monographien. - Nachdruck der 4. Aufl. Braunschweig 1927. - Hrsg. mit einem Nachwort u. einer annotierten Bibliographie v. Walter Scherf, in Zus.-Arb. mit d. Internationalen Jugendbibliothek, München / 1970. VI, 571 S. / Gb (3-7940-3188-1)
• Studienausgabe. 1971. VI, 571 S. / Br (3-7940-3190-3)
• Unveränd. Nachdruck. 1972. 572 S. / Kst. (UTB 125) (3-7940-2606-3)

Die **Kolonne.** Zeitung der jungen Gruppe Dresden. (Ab Nr. 7 / 8 1930: Zeitschrift für Dichtung). Jg. 1 (1929-30) - Jg. 3 (1932). (Alles Erschienene). Hrsg. v. Artur A. Kuhnert und Martin Raschke. Nachw. v. Wolfgang Stein / 1990. IV, 206 S., zahlr. Abb. u. faks. Briefe / Gb (3-598-07269-4) [Lizenzausg. m. Genehm. v. Zentralantiquariat der DDR, Leipzig]

Kommunale Sozialpolitik. Hrsg. v. Walter H. Asam; Michael Heck [Minerva Publikation]
• 1: Subsidiarität und Selbsthilfe. Hrsg. v. Walter H. Asam; Michael Heck / 1985. 240 S. / Br (3-597-10575-0)
• 2: Schafft, Sabine: Psychische und soziale Probleme krebskranker Frauen Über die Bewältigung einer Krebserkrankung im sozialen Umfeld, in der medizinischen Versorgung und in Selbsthilfegruppen / 1987. 300 S. / Br (3-597-10576-9)
• 3: Hilfe zur Selbsthilfe Ein Konzept zur Unterstützung von Selbsthilfegruppen Von Walter H. Asam; Michael Heck; Iris Knerr; Michael Krings / 1989. 246 S. / Br (3-597-10577-7)
• 4: Frauen - Alltag - Politik Eine Zwischenbilanz. Hrsg. v. Rita Heck; Annette Keinhorst / 1986. 216 S. / Br (3-597-10578-5)
• 5: Schulz-Nieswandt, Frank: Wirkungen von Selbsthilfe und freiwilliger Fremdhilfe auf öffentliche Leistungssysteme. Auswertung deutsch- und englischsprachiger Literatur Vorw. v. Klaus Mackscheidt / 1989. 94 S., 12 Abb. / Br (3-597-10579-3)
• 6: Abele, Petra: Organisations- und Teamentwicklung in der Sozialverwaltung. Theoretische Konzepte und Anwendungsmöglichkeiten / 1989. 179 S. / Br (3-597-10673-0)
• 7: Asam, Walter H. / Uwe Altmann; Wolfgang Vogt: Altsein im ländlichen Raum. Ein Datenreport / 1990. 202 S. / Br (3-597-10674-9)
• 8: Altern und Altenhilfe auf dem Lande Zukunftsperspektiven. Hrsg. v. Ingeborg Langen; Ruth Schlichting / 1992. 308 S. / Br (3-597-10675-7)

Kommunikation und Politik. Hrsg. v. Jörg Aufermann; Hans Bohrmann; Otfried Jarren; Winfried B. Lerg; Elisabeth Löckenhoff.
• 1: Hollstein, Dorothea: Antisemitische Filmpropaganda. Die Darstellung der Juden im nationalsozialist. Spielfilm / 1971. 367 S. (3-7940-4017-1)
• 2: Traumann, Gudrun: Journalistik in der DDR. Sozialist. Journalistik u. Journalistenausbildung an d. Karl-Marx-Universität: Leipzig / 1971. 232 S. (3-7940-4018-X)
• 3: Aufermann, Jörg: Kommunikation und Modernisierung. Meinungsführer und Gemeinschaftsempfang im Kommunikationsprozess / 1971. 239 S., 20 Abb., 3 Tab. / Lin (3-7940-4019-8)
• 4: Bohrmann, Hans: Strukturwandel der deutschen Studentenpresse. Studentenpolitik und Studentenzeitschriften 1848-1974 / 1975. 337 S. / Lin (3-7940-4020-1)

- 5: Hoffmann, Gabriele: NS-Propaganda in den Niederlanden. Organisation und Lenkung der Publizistik unter deutscher Besatzung 1940-1945 / 1972. 296 S. / Lin (3-7940-4021-X)
- 6: Rudolf, Günther: Presseanalyse und zeitgeschichtliche Forschung. Telegraf und WAZ zur Berlin-Krise 1948/49 / 1972. 342 S. / Lin (3-7940-4022-8)
- 7: Schneider, Beate: Konflikt, Krise und Kommunikation. Eine quantitative Analyse innerdeutscher Politik / 1975. 276 S. / Lin (3-7940-4023-6)
- 8: Ciolek-Kümper, Jutta: Wahlkampf in Lippe. Die Wahlkampfpropaganda der NSDAP zur Landtagswahl am 15. Januar 1933 / 1976. 406 S., zahlr. Tab., Abb., Ktn. / Lin (3-7940-4024-4)
- 9: Eurich, Claus: Politische Meinungsführer. Theoretische Konzeptionen und empirische Analysen der Bedingungen persönlicher Einflussnahme im Kommunikationsprozess / 1976. 320 S. / Lin (3-7940-4025-2)
- 10: Lorenzen, Ebba: Presse unter Franco. Zur Entwicklung publizistischer Institutionen und Prozesse im politischen Kräftespiel. Nachw. v. Horst Hano / 1978. XI, 408 S. / Lin (3-7940-4028-7)
- 11: Vaillant, Jérôme: Der Ruf. Unabhängige Blätter der jungen Generation (1945-1949). Eine Zeitschrift zwischen Illusion und Anpassung. Vorw. v. H. Hurwitz / 1978. XIII, 250 S. / Lin (3-598-04029-6)
- 12: Petrick, Birgit: Freies Deutschland - die Zeitung des National-Komitees „Freies Deutschland" (1943-1945). Eine kommunikationsgeschichtliche Untersuchung / 1979. 392 S. / Lin (3-598-20542-2)
- 13: Schneider, Sigrid: Das Ende Weimars im Exilroman. Literarische Strategien zur Vermittlung von Faschismustheorien / 1980. XI, 575 S. (3-5978-20543-1)
- 14: Hilchenbach, Maria: Kino im Exil. Die Emigration deutscher Filmkünstler / 1982. 286 S. / Lin (3-598-20544-9)
- 15: Crone, Michael: Hilversum unter dem Hakenkreuz. Die Rundfunkpolitik der Nationalsozialisten in den besetzten Niederlanden 1940-1945 / 1983. 350 S. / Br (3-598-20545-7)
- 16: Schönbach, Klaus: Das unterschätzte Medium. Politische Wirkungen von Presse und Fernsehen im Vergleich. Vorw. v. Winfried Schulz / 1983. 159 S. / Br (3-598-20546-5)
- 17: Stoop, Paul: Niederländische Presse unter Druck. Deutsche auswärtige Pressepolitik und die Niederlande 1933-1940 / 1987. 453 S. / Br (3-598-20547-3)
- 18: Ravenstein, Marianne: Modellversuch Kabelkommunikation. Problemanalyse zum ersten Kabelpilotprojekt Ludwigshafen / Vorderpfalz / 1988. 488 S., 20 Tab., 4 Abb. / Br (3-598-20548-1)
- 19: Schulte, Franz G.: Der Publizist Hellmut von Gerlach (1866-1935) Welt und Werk eines Demokraten und Pazifisten Vorw. v. Dirk Sager / 1988. XVI, 388 S. / Br (3-598-20549-X)
- 20: Publizistik und Journalismus in der DDR. Acht Beiträge zum Gedenken an Elisabeth Löckenhoff. Hrsg. v. Rolf Geserick; Arnulf Kutsch / 1988. 256 S. / Br (3-598-20550-3)
- 21: Basse, Dieter: Wolff's Telegraphisches Bureau 1849 bis 1933. Agenturpublizistik zwischen Politik und Wirtschaft / 1991. VIII, 346 S. / Br (3-598-20551-1)
- 22: Macias, José: Die Entwicklung des Bildjournalismus / 1990. XIV, 302 S. / Br (3-598-20552-X)
- 23: Kohlmann-Viand, Doris H.: NS-Pressepolitik im Zweiten Weltkrieg. Die „Vertraulichen Informationen" als Mittel der Presselenkung / 1991. 199 S. / Br (3-598-20553-8)

- 24: Telegraphenbüros und Nachrichtenagenturen in Deutschland Untersuchungen zu ihrer Geschichte bis 1949. Hrsg. v. Wilke, Jürgen / 1991. 360 S. / Br (3-598-20554-6)
- 25: Kopper, Gerd G.: Anzeigenblätter als Wettbewerbsmedien. Eine Studie zu Typologie, publizistischem Leistungsbeitrag, Entwicklung von Wettbewerbsrecht und Wettbewerbsstrukturen auf der Grundlage einer Gesamterhebung im Werbemarkt Nielsen II, Nordrhein-Westfalen / 1991. 264 S. / Br (3-598-20555-4)
- 26: Holler, Regina: 20. Juli 1944 - Vermächtnis oder Alibi? / 1994. 359 S. / Br (3-598-20556-2)
- 27: Minholz, Michael / Uwe Stirnberg: Der Allgemeine Deutsche Nachrichtendienst (ADN). Gute Nachrichten für die SED / 1995. XVI, 486 S. / Br (3-598-20557-0)

Kommunikations- und Verhaltenstraining für Erziehung, Unterricht und Ausbildung. Hrsg. v. Bernd Fittkau; Hans M. Müller-Wolf; Friedemann Schulz von Thun / 2. überarb. u. erg. Aufl. 1977. 238 S., 2 Abb. / Kst. (UTB 350) (3-7940-2625-X)

Kommunikative Gesellschaft. Beitr. e. Interkulturellen Tagung zwischen Japanern u. Europäern vom 4. bis 8. Sept. 1977 im Haus der Völker und Kulturen, St: Augustin bei Bonn; Ostasien-Institut e.V.; Gesellschaft für Mathematik und Datenverarbeitung / 1979. 232 S. (3-598-10022-1)

Komorowski, Manfred: Bibliographien zur Sportwissenschaft. E. Überblick über ihre Entwicklung im internat. Rahmen. / 1978. VIII, 81 S.
- Promotionen an der Universität Königsberg 1548-1799. Bibliographie der pro-gradu-Dissertationen in den oberen Fakultäten und Verzeichnis der Magisterpromotionen in der philosophischen Fakultät / 1988. XX, 98 S. / Lin (3-598-10760-9)

Konfrontation und Koexistenz. Realität und Utopie im Verhältnis der Juden und Nichtjuden. Hrsg.: Renate Heuer, im Auftr. des Archiv Bibliographia Judaica e.V. / 1994. 504 S. (3-598-11132-0)

Konrad Mellerowicz. Bibliographie seiner Veröffentlichungen und Aufsatzsammlung. Einf. u. Hrsg. v. Aribert Peeckel / 2 Bde. 1990. Zus. 1.577 S. / Ln (3-598-10770-6)

Kopper, Gerd G.: Medien- und Kommunikationspolitik der Bundesrepublik Deutschland. Ein chronologisches Handbuch 1944 bis 1988 / 1992. 518 S. / Lin (3-598-10799-4)

Koppitz, Hans J.: Grundzüge der Bibliographie / 1977. 327 S. / Br (3-7940-3182-2)

Korte, Werner B.: Neue Medien und Kommunikationsformen - Auswirkungen auf Kunst und Kultur. Untersuchungen zum künftigen Verhältnis von neuen Techniken und traditionellen Kunst- und Kulturbereichen. Unter Mitarb. v. Klaus Haller; Sabine Mertsch / 1985. IX, 393 S. / Br (3-598-10579-7)

Korwitz, Ulrich: Zur Geschichte der fachlichen Literaturerschließung. Die medizinischen Referateblätter des J. Springer Verlages / 1985. XI, 204 S., 44 Abb., 22 Tab. / Lin (3-598-10580-0)

Koschnick, Wolfgang J.: Enzyklopädisches Wörterbuch Marketing / Encyclopedic Dictionary Marketing. Englisch-Deutsch - Deutsch-Englisch / 2 Bde. in 4 Tlbdn / Gb (3-598-23175-X)
- Vol. 1 / Bd. 1: English-German A-Z / 2 vols. 1994. Cplt. X, 1715 p. / Hard (3-598-23176-8)
- Bd. 2 / Vol. 2: Deutsch-Englisch A-Z / 2 Bde. 1995. Zus. IX, 1647 S. / Gb (3-598-23177-6)

Koschnick, Wolfgang J.: Kompaktwörterbuch der Sozialwissenschaften / Compact Dictionary of the Social Sciences (3-598-11281-5)
- Bd. 1 / Vol. 1: Deutsch-Englisch / German-English / 1995. VIII, 620 S. / Gb (3-598-11282-3)
- Vol. 2 / Bd.2: Englisch-German / Englisch-Deutsch / 1995. VIII, 555 p. / Gb (3-598-11283-1)

Koschnick, Wolfgang J.: Media-Lexikon Österreich / 1995. IX, 759 S., 160 Abb., ca. 90 Formeln / Gb (3-598-11246-7)

Koschnick, Wolfgang J.: Media-Lexikon Schweiz / 1995. IX, 760 S., 160 Abb., ca. 90 Formeln / Gb (3-598-11245-9)

Koschnick, Wolfgang J.: Standard Dictionary of the Social Sciences / Standard-Wörterbuch für die Sozialwissenschaften. Englisch-Deutsch / English-German. Deutsch-Englisch / German-English / 2 vols. in 3 parts 1984-1993 / Hard (3-598-10866-4)
- Vol. 1: English-German / Englisch-Deutsch / 1984. X, 664 p. / Hard (3-598-10526-6)
- Vol. 2: German-English / Deutsch-Englisch / 2 parts 1992 / 1993. Cplt. XIV, 2,291 p. / Hard (3-598-10527-4)

Koschnick, Wolfgang J.: Standard-Lexikon für Marketing, Marktkommunikation, Markt- und Mediaforschung / 1987. VI, 940 S. / Lin (3-598-10583-5)
- 2. aktualis. u. erw. Aufl. Standard-Lexikon für Marketing / 2 Bde. 1996. Zus. XI, 3071 S. / Ln (3-598-11185-1)

Koschnick, Wolfgang J.: Standard-Lexikon für Markt- und Konsumforschung / 2 Bde. 1995. Zus. IX, 1.052 S., 400 Abb., 220 Formeln / Gb (3-598-11247-5)

Koschnick, Wolfgang J.: Standard-Lexikon für Mediaplanung und Mediaforschung / 1988. 600 S., ca. 220 Abb. / Gb (3-598-10699-8)
- 2. überarb. u. erw. Aufl. / 2 Bde. 1995. Zus. XIII, 1.977 S., 500 Abb., 220 Formeln / Gb (3-598-11170-3)

Koschnick, Wolfgang J.: Standardlexikon Werbung-Verkaufsförderung-Öffentlichkeitsarbeit / 2 Bde. 1996. Zus. IX, 1.235 S. / Gb (3-598-11280-7)

Koszyk, Kurt / Karl Hugo Pruys: Wörterbuch zur Publizistik. / 1970. 539 S.

Krämer, Louise: Die Fachbücherei in Wirtschaft, Verwaltung und Wissenschaft. Eine Einf. für d. Praxis / 1968. 180 S. / Br.

Kratzsch, Konrad: Verzeichnis der Luther-Drucke 1517-1546. Aus den Beständen der Zentralbibliothek der Deutschen Klassik / 1986. 213 S., 7 Abb. / Br (3-598-07243-0)

Kreuter, Alma: Deutschsprachige Neurologen und Psychiater. Ein biographisch-bibliographisches Lexikon von den Vorläufern bis zur Mitte des 20. Jahrhunderts. Vorw. v. Hanns Hippius; Paul Hoff / 3 Bde. 1995. Zus. XXV, 1.629 S. / Gb (3-598-11196-7)

Krüger, Henning Anders: Betriebsabrechnung für die Druckindustrie. 2 Bde
• Bd. 1.. Leistungserfassung u. Betriebsabrechnungs-bogen / 5., völlig neubearb. Aufl. 1973. 131 S. / Lin (3-7940-8715-1)
• Bd. 2.. Kostenträgerrechnung, Kostenkatalog, Auswertung d. Betriebsabrechnung / 4., völlig neubearb. Aufl. 1972. 124 S. / Lin (3-7940-8716-X)

Krüger, Henning Anders: Kalkulationsleitfaden Buchdruck. Unter Mitarb. von Manfred Bluthardt / 12., völlig neubearb. Aufl. 1973. 184 S. / Lin (3-7940-8717-8)

Krüger, Karl-Heinz: Wörterbuch der Datenverarbeitung. Englisch - Deutsch / Data Processing Dictionary. English-German / 1968. 296 S. / Lin (Vlg.-Nr. 03098)
2. Aufl.: → Oppermann, Alfred: Wörterbuch der Datenverarbeitung

Krumholz, Walter: Die politische Dokumentation in der Bundesrepublik Deutschland. Untersuchung über den heutigen Stand. / 1968. 304 S. / Br (Vlg.-Nr. 03218)
• 2. Aufl. 1969. 304 S. / Br (Vlg.-Nr. 03218)
• 3. Aufl. 1971. 412 S. / Lin (3-7940-3218-7))

Kruse, Lenelis / Reiner Arlt: Environment and Behavior. An International Multidisciplinary Bibliography / Part I u. II.
• Part I: Kruse, Lenelis / Reiner Arlt: Environment and Behavior 1970-1981. Volume 1: Alphabetical Listing by Authors, Keyword Index. Volume 2: Abstracts / 2 vols. 1984. Cplt. XVI, 1424 p. / Hard (3-598-10494-4)
• Part II: Schwarz, Volker / Lenelis Kruse; Reiner Arlt: Environment and Behavior 1982-1987. Volume 1: Alphabetical Listing by Authors, Keyword Index. Volume 2: Abstracts / 2 vols. 1988. Cplt. 964 p. / Hard (3-598-10783-8)

Kruszynski, Gisela: Die komischen Volkskalender Adolf Glaßbrenners 1846 bis 1854. Untersuchungen zur satirischen Illustration in Deutschland / 1979. 183 S. (3-597-00012-6) [Minerva Publikation]

Künstler der jungen Generation / Artists of the „Young Generation". Literaturverzeichnis zur Gegenwartskunst in der Amerika-Gedenkbibliothek - Berliner Zentralbibliothek / Literature on Contemporary Art in the American Memorial Library - Berlin Central Library. (Berichtszeitraum: 1960-1985). Hrsg.: Amerika-Gedenkbibliothek - Berliner Zentralbibliothek. Bearb. v. Gritta Hesse; Marie A. Bingel / 2. erw. u. überarb. Aufl. 1987. VI, 353 S. / Gb (3-598-10693-9)

Kürschners Deutscher Literatur-Kalender [Seit 1998. Vorher bei Walter de Gruyter, Berlin]
• 61. Jg. 1998 / 2 Tlbde. 1998. XXVIII, 1.696 S. / Ln (3-598-23581-X)

Kürschners Deutscher Literatur-Kalender 1922-1988. 40. Jg. (1922) - 60. Jg. (1988), Nekrolog 1901-1935 (1935), Nekrolog 1936-1970 (1973). Mikrofiche-Edition / 1998. ca. 22.740 S. auf 240 Fiches, Lesefaktor 24x / Silberfiche (3-598-33754-X) / Diazofiche (3-598-33755-8)

Kuhles, Doris: Deutsche literarische Zeitschriften von der Aufklärung bis zur Romantik. Bibliographie der kritischen Literatur von den Anfängen bis 1990. Hrsg.: Stiftung Weimarer Klassik, Herzogin Anna Amalia Bibliothek / 2 Bde. 1994. Zus. LVIII, 554 S. / Ln (3-598-11159-2)

Kulturaustausch zwischen Orient und Okzident. Über die Beziehungen zwischen islamisch-arabischer Kultur und Europa (12.-16. Jahrhundert). Hrsg.: Deutsche UNESCO-Kommission. Red.: Hans-Dieter Dyroff / 1985. 263 S. / Br (3-598-10585-1)

Kulturförderung und Kulturpflege in der Bundesrepublik Deutschland. Hrsg.: Deutsche. UNESCO-Kommission, Köln / 1974. 86 S. (3-7940-3059-1)

Kunstadressbuch Deutschland, Österreich, Schweiz. 8. Ausg. 1992/93 Redakt. Leitung: Michael Zils / 1991. 580 S. / Soft (3-598-23068-0)
• 9. Ausg. 1994/95 Redakt. Leitung: Michael Zils / 1994. XII, 674 S. / Br (3-598-23071-0)
• 10. Ausg. 1996/97 Redakt. Leitung: Michael Zils / 1996. 528 S. / Br (3-598-23074-5)
• 11. Ausg. 1998/99 Redakt. Leitung: Michael Zils / 1998. 6, 506 S. / Br (3-598-23079-6)

Kunstadressbuch und Kalendarium. Mit e. Verz. d. Austellungskataloge, Kunstzeitschriften u. Kunstpreise d. Bundesrepublik Deutschland, Österreich, Schweiz / 1975. 469 S. (3-7940-3380-9)

Kunz, Werner / Wolf Reuter; Horst W. J. Rittel: UMPLIS, Entwicklung eines Umwelt-Planungs-Informationssystems. Fallstudie. Unter Mitarb. von Peter Menke-Glückert u.e. Schlußbetrachtung von Jürgen Seggelke / 1980. 300 S. (3-598-21108-2)

Kunz, Werner / Horst W. J. Rittel; Werner Schwuchow: Methods of analysis and evaluation of information needs. A critical review / 1977. 84 S. (3-7940-3450-3)

Kunze, Horst: Grundzüge der Bibliothekslehre. (Nachdruck der 3. völlig veränderten Aufl.) / 1969. 530 S.
• 4. neu bearb. Aufl. 1977. 602 S. / Ln (3-7940-4088-0)

Kunze, Horst: Über das Registermachen / 1966. 68 S.
• 4. erw. u. verb. Aufl. 1992. 83 S. / Br (3-598-11090-1)

Kunze, Horst: Vom Bild im Buch. Hrsg. v. Friedhilde Krause; Renate Gollmitz / 1988. 288 S., 20 Abb. / Ln (3-598-07245-7)

Kutsch, Karl J. / Leo Riemens: Grosses Sängerlexikon. Mit einem Anhang: Verzeichnis von Opern und Operetten / 2 Bde. 1987. Zus. XII S., 3.452 Spalten / Ln (3-907820-67-3) [1991 vom Francke Vlg, Bern übernommen]
  - Bd. 3. Ergänzungsband / 1991. VIII S., 2.002 Spalten / Ln (3-907820-68-1)
  - Bd. 4. Ergänzungsband II. Unter Mitarb. v. Hansjörg Rost / 1994. VIII S., 1.598 Spalten (3-907820-69-X)
• 2. Aufl. 3 Bde. 1993. XVI, 5.454 Spalten / Br iKass. (3-907820-70-3)

• 3., erw. Aufl. Insgesamt ca. 14.800 Bibliographien zu Sängerinnen und Sängern weltweit / 5 Bde. 1997. Zus. XXXIII, 3.980 S. / Gb (3-598-11250-5)
• Unveränd. Broschurausgabe / 5 Bde. 1999. Zus. XXXIII, 3.980 S / Br (3-598-11419-2)

Kutsch, Karl J. / Leo Riemens: Unvergängliche Stimmen. Sängerlexikon. / 2., neubearb. u. erw. Aufl. 1982. 782 S. (3-317-01555-1) [1991 vom Francke Vlg, Bern übernommen]

Kybernetische Analysen geistiger Prozesse. Neue Ergebnisse kybernetisch-psychologischer Forschungen. Hrsg. Friedhart Klix / 1968. 234 S., 66 Abb. 4 Tab. / Br (Vlg.-Nr. 04188)
• 2. Aufl. 1970. / Br (3-7940-4188-7)

Ladenpreistabelle für die Mehrwertsteuer / 1967 / Kt
• 2. Aufl. / 1967 / Kt
• 3. Aufl. / 1967 / Kt
• 4. Aufl. 5 und 10% / 1968. 32 S. / Kt
• 5. Aufl. 5,5 und 11% / 1968. 32 S. / Kt

Länderkatalog Afrika der Übersee-Dokumentation Hamburg 1971-1984. Mikrofiche-Edition. Hrsg.: Deutsches Übersee-Institut Hamburg / 1986 ca. 250.500 Katalogktn. auf 231 Fiches. Lesefaktor 42x. Begleitbr. Deutsch / Engl. / Franz. / Span. / Silberfiche (3-598-30982-1) / Diazofiche (3-598-30981-3)

Lageberichte (1920-1929) und Meldungen (1930-1933). Reichskommissar für die Überwachung der öffentlichen Ordnung und Nachrichtensammelstelle im Reichsministerium des Innern. Bestand R 134 des Bundesarchivs Koblenz. Microfiche-Edition. Hrsg. v. Ernst Ritter / 1979. 399 Fiches u. Registerbd. Lesefaktor 24x. Silberfiche (3-598-10004-3)

Lager, Front oder Heimat. Deutsche Kriegsgefangene in Sowjetrussland 1917 bis 1920. Hrsg.: Institut für Geschichte der Arbeiterbewegung, unter d. Leitung v. Inge Pardon / 2 Bde. 1994. Zus. X, 792 S., 16 Abb. / Gb (3-598-11106-1)

Laminski, Adolf: Die Kirchenbibliotheken zu St. Nicolai und St. Marien. Ein Beitrag zur Berliner Bibliotheksgeschichte. (Beiheft 98 zum Zentralblatt für Bibliothekswesen) / 1990. 104 S., 16 Taf / Br (3-598-07032-2)

Lamm, Hans: Hans Lamm. Deutsch-Jüdischer Publizist. Ausgewählte Aufsätze 1933-1983 mit ausführlicher Bibliographie. Vorw. v. Werner Nachmann / 1984. VIII, 279 S. / Lin (3-598-10557-6)

Langridge, Derek W.: Inhaltsanalyse. Grundlagen und Methoden. Übers. v. Ute Reimer-Böhner / 1994. 159 S. / Br (3-598-11071-5)

Large Libraries and New Technological Developments. Proceedings of a Symposium held on the occasion of the inauguration of the new building of the Royal Library, The Hague, September 30 - October 2, 1982. Ed. by C. Reedijk; W. R. H. Koops and Carol K. Henry / 1985. 194 p. / Hard (3-598-10508-8)

LaRoche, Sophie von: Pomona für Teutschlands Töchter. - Reprint der Ausgabe Speier (Selbstverlag) 1783-1784. - Gedruckt mit Enderesischen (Johann Paul Enderes) Schriften. Vorw. v. Jürgen Vorderstemann / 4 Bde. 1987. Zus. 2.381 S. / Hl iKass. (3-598-10736-6)

Laubinger, Hans-Dieter: Materialien über die Bibliotheksplanung Schwedens / 1973. 351 S. (3-7940-3206-3)

Leesch, Wolfgang: Die deutschen Archivare 1500-1945 / Lin (3-598-10606-8)
• Bd. 1. Verzeichnis nach ihren Wirkungsstätten / 1985. 268 S. / Lin (3-598-10530-4)
• Bd. 2. Biographisches Lexikon / 1992. 737 S. / Lin (3-598-10605-X)

Lehrer und Lernprozess. Der Unterricht und seine Voraussetzungen. Hrsg. v. Robert D. Strom; Georg E. Becker. Aus d. Amerik. v. Georg E. Becker.
• Bd. 1. 1976. 246 S. / Br (UTB 567) (3-7940-2651-9)
• Bd. 2. 1976. VIII, 247-416 S. / Br (UTB 568) (3-7940-2652-7)

Leiningen-Westerburg, K. E. von: Deutsche und österreichische Bibliothekszeichen und Exlibris. Ein Handbuch für Sammler, Bücher und Kunstfreunde. (Nachdr. d. Ausg. 1901 Stuttgart) / 1980. XVIII, 610 S., 254 Abb. / Ln (3-598-10023-X)

Leistner, Georg: Abbreviations' guide to French forms in justice and administration. Abkürzungsverzeichnis zur französischen Rechts- und Verwaltungssprache / 2. ed. 1975. 191 S. (3-7940-3016-8)

Leistungsmessung in wissenschaftlichen Bibliotheken. Internationale Richtlinien. Hrsg.: IFLA Section of University Libraries & other General Research Libraires; Roswitha Poll; Peter Boekhorst, in Zus.-Arb. mit Ramon A. Hiraldo; Aase Lindahl; Rolf Schuursma; Gwenda Thomas; John Willemse / 1998. 172 S. / Br (3-598-11387-0)

Leleu-Rouvray, Geneviève / Gladys Langevin: Bibliographie Internationale de la Marionette. Ouvrages en Anglais 1945-1990 / International Bibliography on Puppetry. English Books 1945-1990. Ed.: Institut International de la Marionette avec la collaboration de l'Association Marionette et Thérapie / 1993. XXXII, 281 p., 17 illus. noires et blanches / Hard (3-598-22651-9)

Lengrand, Paul: Permanente Erziehung. Eine Einführung. Einl. v. Gottfried Hausmann / 1972. 120 S. / Kst. (UTB 149) (3-7940-2611-X)

Leonov, Valerii: The Library Syndrome / 1999. IV, 302 p. / Hard (3-598-11407-9)

Lernsysteme. Theorie und Praxis der Unterrichtstechnologie. Hrsg.: CERI. Nach Arbeitsunterlagen des Centre for Educational Research and Innovation (CERI). Dt. Ausg. v. Klaus Hinst / 1973. 157 S., 1 Tab., 4 Übersichten / Kst. (UTB 156) (3-7940-2609-8)
• 4. erg. u. überarb. Aufl. 1985. 380 S., Reg / Kst. (3-598-02675-7)

Leser und Lektüre vom 17. zum 19. Jahrhundert. Die Ausleihbücher der Herzog August Bibliothek Wolfenbüttel 1664-1806. Bearb. v. Mechthild Raabe. Vorw v. Paul Raabe / 8 Bde. in 3 Tln. 1989-1997 / Lin (3-598-10700-5)
• Tl. A. Leser und Lektüre im 17. Jahrhundert. Die Ausleihbücher der Herzog August Bibliothek Wolfenbüttel 1664-1713. Bd. 1: Leser und Lektüre - Lesergruppen und Lektüre. Bd. 2: Alphabetisches und systematisches Verzeichnis der entliehenen Bücher / 2 Bde. 1997. Zus. 89, 1.150 S. / Lin (3-598-10701-3)

• Tl. B. Leser und Lektüre im 18. Jahrhundert. Die Ausleihbücher der Herzog August Bibliothek Wolfenbüttel 1714-1799. Band 1: Leser und Lektüre. Band 2: Lesergruppen und Lektüre. Band 3: Alphabetisches Verzeichnis der Bücher. Band 4: Systematisches Verzeichnis der Bücher / 4 Bde. 1989. Zus. 2.300 S. / Lin (3-598-10651-3)
• Tl. C. Ergänzungen und Zusammenfassungen. Bd. 1: Leser und Lektüre 1800-1806 - Chronologisches Verzeichnis 1664-1750. Bd. 2: Chronologisches Verzeichnis 1751-1806. Gesamtstatistik / 2 Bde. 1997. Zus. 26, 1.191 S. / Lin (3-598-10702-1)

Lesky, Albin: Geschichte der griechischen Literatur / 3. neubearb. u. erw. Aufl. 1971. 1.023 S. / Gb (3-907820-82-7) [1991 vom Francke Vlg, Bern übernommen]
• Unveränderter Nachdruck (Jubiläumsausgabe). 1999. 1.023 S. / Gb (3-598-11423-0)

Lessing-Bibliographie 1971-1985. Bearb. v. Doris Kuhles, unter Mitarb. v. Erdmann von Wilamowitz-Moellendorf / 1988. X, 473 S. / Ln (3-598-07253-8)

Leuchtmann, Horst: Wörterbuch Musik / Dictionary of Terms in Music. Deutsch-Englisch / Englisch-Deutsch / 2., ergänzte Aufl. 1976 (3-7940-3186-5) [Vorher im Langenscheidt Vlg, München]
• 3. Aufl. 1977. XVI, 493 S. (3-7940-3186-5)
• 4th rev. and enl. ed. Dictionary of Terms in Music / Wörterbuch Musik English-German, German-English / Englisch-Deutsch, Deutsch-Englisch / 1992. XVII, 411 p. / Hard (3-598-10913-X)

Lewanski, Richard: Eastern Europe and Russia: A Handbook of Western European Archival and Library Resources / 1981. XV. 317 p. / Hard (3-598-08012-3)

Lexikon der Buchkunst und Bibliophilie. Hrsg. v. Karl K. Walther / 1988. 386 S., 180 farb. u. 220 schw.-w. Abb. / Ln (3-598-07236-8) [Lizenzausg. m. Genehm. v. VEB Bibliographisches Institut, Leipzig]

Lexikon der graphischen Technik. Bearb. im Inst. für graph. Technik, Leipzig, Hugo Weschke / 2., verb. u. erw. Aufl. 1967. 608 S. (Vlg-Nr. 04077)
• 3. Aufl. 1970. 608 S.
• 4., neubearb Aufl. 1977. 656 S. (3-7940-4078-3)

Lexikon des Bibliothekswesens. Bearb. v. Horst Kunze; Gotthard Rückl. - Nachdr. d. 2. Aufl. 1975 / 2 Bde. 1986. Zus. XV, 2.112 Spalten / Ln (3-598-07020-9) [Lizenzausg. m. Genehm. v. VEB Bibliographisches Institut, Leipzig]

Lexikon der Kinder- und Jugendliteratur. Personen-, Länder- u. Sachartikel zu Geschichte u. Gegenwart d. Kinder- u. Jugendliteratur. Erarbeitet im Institut für Jugendbuchforschung d. Johann-Wolfgang-Goethe Universität Frankfurt a. M. Hrsg. v. Klaus Doderer. Red.: Peter Aley / Bd. 1: A - H / 1975. XIV, 579 S. (3-7940-3414-7)
Lexikon deutsch-jüdischer Autoren. Hrsg.: Archiv Bibliographia Judaica / ca. 16 Bde, davon 2 Reg.-Bde. 1992 ff. (3-598-22680-2)
• Bd. 1. A-Benc / 1992. XXXIV, 488 S. / Gb (3-598-22681-0)
• Bd. 2. Bend-Bins / 1993. XLIII, 474 S. / Gb (3-598-22682-9)
• Bd. 3. Birk-Braun / 1994. XLVI, 457 S. / Gb (3-598-22683-7)

• Bd. 4. Brech-Carle / 1995. LII, 452 S. / Gb (3-598-22684-5)
• Bd. 5. Carmo-Donat / 1996. LVII, 498 S. / Gb (3-598-22685-3)
• Bd. 6. Dore-Fein / 1998. XLVIII, 560 S. / Gb (3-598-22686-1)
• Bd. 7. 1998. XLIV, 463 S. / Gb (3-598-22687-X)
• Bd. 8. 1999. ca. 550 S. / Gb (3-598-22688-8)

Lexikon fremdsprachiger Schriftsteller. Von den Anfängen bis zur Gegenwart. Hrsg. v. Gerhard Steiner; Herbert Greiner-Mai; Wolfgang Lehmann / 3 Bde. 1981. Zus. 1.820 S. / Ln (3-598-10391-3)

Li Heng: Dictionary of Library and Information Sciences. English-Chinese / 1984. XX, 328 p. / Hard (3-598-10532-0)

LIBER QUARTERLY. The European Journal of Research Libraries. Ed. on behalf of the Ligue des Bibliothèques Européennes de Recherche (LIBER) by Esko Häkli / since Vol. 8/1998 (4 issues annually) (ISSN 1435-5205)

Libraries, Information Centers and Databases for Science and Technology / Bibliotheken, Informationszentren und Datenbasen für Wissenschaft und Technik. A World Guide / Ein internationales Verzeichnis. Ed. by Helga Lengenfelder,. Comp. by Gertrud Conrad-Bergweiler; Barbara Fischer; Angelika Riedel / 1984. XXV, 561 p. / Hard (3-598-10533-9)
• 2nd ed. Ed.: Helga Lengenfelder. Ed. by Barbara Fischer; Ruth Lochar; Bettina Pautler / 1988. XXXII, 665 p. / Hard (3-598-10757-9)

Libraries, Information Centers, and Databases in Biomedical Sciences. A World Guide. Ed. by Bettina Bartz; Helmut Opitz; Elisabeth Richter / 1991 XXIV, 392 p. / Hard (3-598-11001-4)

Library Automation and Networking - New Tools for a New Identity European Conference, 9-11 May 1990, Brussels. Ed. by Herman Liebaers; Marc Walckiers; European Foundation for Library Cooperation / Groupe de Lausanne / 1991. 370 p. / Hard (3-598-10935-0)

Library Problems in the Humanities. Case, studies in reference services, coll. building, and management. Ed. by Thomas P. Slavens / 1981. IX, 113 S. (3-598-40026-8)

Libretti in deutschen Bibliotheken. Katalog der gedruckten Texte zu Opern, Oratorien, Kantaten, Schuldramen, Balletten, Gelegenheitskompositionen von den Anfängen bis zur Mitte des 19. Jahrhunderts. Mikrofiche-Edition. Hrsg.: Répertoire International des Sources Musicales (RISM); Arbeitsgruppe Deutschland / 1992. 107 Fiches, Lesefaktor 24x (3-598-30660-1)

Libri. International Journal of Libraries and Information Services. Ed. by Nancy R. John; Svend Larsen; Irene Wormell / since Vol. 47/1997 (4 issues annually) (ISSN 0024-2667) [Seit 1997. Vorher im Vlg. Munksgaard, Kopenhagen]

Liebich, Ferdinand Karl: Die Kennedy-Runde. Eine Analyse des weltweiten Genfer Zollsenkungsabkommens / 1968. 237 S. (3-7940-5169-6)

Liebich, Ferdinand Karl: Kultur ohne Handels-schranken. Empfehlungen zur Liberalisierung d. Welthandels mit erzieher., wissenschaftl. u. kulturellen Erzeugnissen / 1972. 119 S. (3-7940-3388-4)

Lingenberg, Jörg: Das Fernsehspiel in der DDR. Ein Beitr. zur Erforschung künstlerischer Formen marxist.-leninist. Publizistik / 1968. 340 S. / Br (Vlg.-Nr. 03268)

Linschmann, Theodor: Ludwig Bechsteins Schriften. - Nachdruck aus „Neue Beiträge zur Geschichte des Altertums", Meiningen 1907 / 1972. 152 S. (3-7940-5030-4)

Lion Feuchtwanger: A Bibliographic Handbook / Ein bibliographisches Handbuch. Ed. by John M. Spalek; Sandra H. Hawrylchak / 4 vols. 1998 ff. / Hard (3-598-11377-3)
• Vol. I. German Editions / Deutschsprachige Ausgaben / 1998. XXVIII, 392 p. / Hard (3-598-11378-1)
  Vol. II - IV [i.Vb.]

Literatur des gesamten Eisenbahnwesens. Zusammengestellt von August von Kaven 1868. Nach d. Orig. Hrsg. v. Günter Breuer / 1988. XVIII, 546 S., 8 Abb. / Lin (3-598-10790-0)

Literatur und Archiv. Hrsg. v. Thomas Feitknecht; Georg Jäger; Christoph König; Walter Methlagl; Siegfried Seifert
• 1: König, Christoph: Verwaltung und wissenschaftliche Erschliessung von Nachlässen in Literaturarchiven. Österreichische Richtlinien als Modell. Hrsg.: Forschungsinstitut „Brenner Archiv" (Innsbruck) / 1988. 175 S., 20 Abb. / Ln (3-598-10741-2)
• 2: Schenkel, Martin: Dokumentation literarischer Quellen in Bibliotheken. Drei Modellprojekte zur Zeitschrifteninhaltserschliessung in Göttingen, Frankfurt und Marbach / 1988. 179 S. / Ln (3-598-22081-2)
• 3: Wimmer-Webhofer, Erika: Die Konservierung von Archivalien in Literaturarchiven. Empfehlungen zur Lagerung, Benützung und Sicherung von Nachlässen / 1989. 144 S. / Ln (3-598-22082-0)
• 4: Heusinger, Lutz: Marburger Informations-, Dokumentations- und Administrations System (MIDAS). Handbuch. Vorw. v. Christoph König. Hrsg.: Bildarchiv Foto Marburg. Deutsches Dokumentationszentrum f. Kunstgeschichte; Philipps-Universität Marburg / 1989 XX, 560 S. / Ln (3-598-22084-7)
  - 2. überarb. u. erw. Aufl. 1992. XXVI, 577 S. / Ln (3-598-22086-3)
  - 3. überarb. Aufl. 1994. XXVIII, 593 S. / Ln (3-598-22087-1)
• 5: Literarische Ausstellungen von 1949 bis 1985. Diskussion, Dokumentation, Bibliographie. Hrsg. v. Susanne Ebeling; Otto Hügel; Ralf Lubnow / 1991. 484 S. / Ln (3-598-22083-9)
• 6: Historische Bestände der Herzogin Anna Amalia Bibliothek zu Weimar. Beiträge zu ihrer Geschichte und Erschliessung. Mit Bibliographie. Zsgst. u. wiss. Redakt.: Konrad Kratzsch; Siegfried Seifert / 1992. IX, 355 S. / Ln (3-598-22085-5)
• 7: Bestandserschliessung im Literaturarchiv. Arbeitsgrundsätze des Goethe- und Schillerarchivs in Weimar. Hrsg. v. Gerhard Schmid / 1996. 277 S. / Ln (3-598-22088-X)

• 8: Literaturarchiv und Literaturforschung. Hrsg. v. Christoph König; Siegfried Seifert / 1996. 255 S. / Ln (3-598-22089-8)

Literaturinformationen zur Berufsbildungsforschung (BBF). Hrsg.: Bundesinstitut für Berufsbildungsforschung, Berlin in Zusammenarb. mit der Universitätsbibliothek der Technischen Universität Berlin / Jg. 2/1975 - Jg.5/1978 (ISSN 0340-2614)
• 4. Jahrg. 1977. (ISSN 0340-2614)

Literaturzusammenstellung Wasserbau. 1.480 Literaturnachweise / 1949. 370 S. / Br

Lith, Ulrich van: Die Kosten der akademischen Selbstverwaltung. Eine vergleichende Untersuchung über den Zeitaufwand und die Kosten der Gremientätigkeit an vier Universitäten / 1979. 168 S. (3-598-10002-7)

Le **livre**, les bibliothèques et la documentation. Une bibliographie sélective. Préparée sous la direction de Jacques Breton; Catherine Gaillard; Daniel Renoult (Bibliothèques et organismes documentaires) / 1985. XVIII, 192 S. / Br (3-598-20456-6)

Le **livre** scientifique et le livre de vulgarisation scienifique en France. Actes du colloque par l'Assoc. des Bibiothécaires Francais dans le cadre du Festival Internat. du Livre de Nice le sammmedi 13 mai 1978 / 1980. 136 S. (3-598-10335-2)

Löw, Konrad: Die Grundrechte. Verständnis und Wirklichkeit in beiden Teilen Deutschlands / 1977. 419 S. (UTB 735) (3-7940-2662-4)
• 2. überarb. Aufl. 1982. 471 S., Reg / Kst. (3-598-02668-4)

Löw, Konrad: Warum fasziniert der Kommunismus? Eine systematische Untersuchung / 1985. 380 S. / Kst. (UTB 1345) (3-598-02675-7)

Loh, Gerhard: Geschichte der Universitätsbibliothek Leipzig von 1543-1832. Ein Abriss / 1989. 174 S. / Br (3-598-07266-X) [Lizenzausg. m. Genehm. v. VEB Bibliographisches Institut, Leipzig]

Longerich, Peter: Hitlers Stellvertreter. Führung der Partei und Kontrolle des Staatsapparates durch den Stab Hess und die Parteikanzlei Bormanns. Hrsg.: Institut f. Zeitgeschichte, München / 1992. V, 283 S. / Gb (3-598-11081-2)

Lorck, Carl B.: Handbuch der Geschichte der Buchdruckerkunst. - Reprint d. Orig.-Ausg. 1882 / 83 / 1989. 828 S. / Ln (3-598-07239-2)

Ludes, Peter: Bibliographie zur Entwicklung des Fernsehens / Bibliography on the Development of Television Fernsehsysteme und Programmgeschichte in den USA, Grossbritannien und der Bundesrepublik Deutschland / Television Systems and Programme Development inthe USA, Great Britain and the Federal Republic of Germany Unter Mitarb / in collab. with Ulrich Brandt; Leonore Georg-Shirley; Thomas Krusche; Frauke Piwodda; Irmela Schneider; Mario, / 1990. 241 S. / Gb (3-598-10973-3)

Lübeck-Schrifttum. Hrsg.: Stadtbibliothek Lübeck. Bearb. v. Gerhard Meyer; Antjekathrin Grassmann / 1976. 413 S. / Ln (3-7940-3038-9)

Lülfing, Hans: Johannes Gutenberg und das Buchwesen des 14. und 15. Jahrhunderts / 1969. 165 S. / Lin (Vlg.-Nr. 04249)

Lugert, Alfred C.: Auslandskorrespondenten im internationalen Kommunikationssystem. E. Kommunikator-Studie / 1974. 230 S. (3-7940-3057)

Mácha, Karel: Glaube und Vernunft. Die böhmische Philosophie in geschichtlicher Übersicht.
• Tl. I. 863-1800 / 1985. 166 S. / Lin (3-598-20130-3)
• Tl. II. 1800-1900 / 1987. 215 S. / Lin (3-598-20131-1)
• Tl. III. 1900-1945 / 1989. 273 S. / Lin (3-598-20132-X)

MacKenzie, Norman / Michael Eraut; Hywel C. Jones: Lehren und Lernen. Eine Einführung in neue Methoden und Lernmittel in der Hochschuldidaktik. Eine UNESCO-Studie. Aus d. Engl. v. Helga Lengenfelder; Armgard de Vivanco-Luyken / 1973. 256 S., 4 Diagr., 2 Tab. / Kst. (UTB 157) (3-7940-2613-6)

Magnetband-Austauschformat für Dokumentationszwecke. MADOK. Hrsg. von d. Zentralstelle für Maschinelle Dokumentation (ZMD), Frankfurt am Main / 1977. 131 S. (3-7940-7004-6)

Maher, Mustafa / Wolfgang Ule: Deutsche Autoren in arabischer Sprache, arabische Autoren in deutscher Sprache, Bücher über Deutsche und Deutschland in arabischer Sprache. E. Bibliogr. / Mu'allaft likuttb almn mutargama ila 'l-lua 'l-'arabya. In Zus.-Arb. mit Inter Nationes e.V., Bonn-Bad Godesberg / 3., erw. Aufl. 1979. 224 S. / Br (3-598-10027-2)

Main Catalog of the Library of Congress 1898-1980 (MCLC). Microfiche edition. The complete main dictionary catalog of the Library / 1984-1989. approx. 7,5 million titles, 26 million entries in a single alphabetical sequence on 8926 fiches. Reader factor 48x / Silverfiche (3-598-41061-1) / Diazofiche (3-598-41041-7)

Makowski, Ilse: Emanzipation oder „Harmonie" - zur Geschichte der gleichnamigen Mannheimer Lesegesellschaft in der ersten Hälfte des 19. Jahrhunderts / 1988. 166 S. / Lin (3-598-10763-3)

Der **Malik** Verlag 1916-1947. Ausstellungskatalog. Zsgst. u. Hrsg. v. Wieland Herzfelde, mit Verwendung d. Bibliographie des Malik Verlages v. Heinz Gittig. - Nachdr. d. Ausg. 1967 / 1985. 160 S. / Br (3-598-07214-7)

Management technologischer Innovationen. Beiträge des Innovationsseminars für die Wirtschaft vom 6. bis 10. Nov. 1972. Hrsg. im Auftrag des Deutschen Komitees d. AIESEC von Hans Günther Meissner und Herfried A. Kroll / 1974. 222 S. (3-7940-4240-9)

Management of Recorded Information / La gestion de l'information consignée Converging Disciplines. Proceedings of the International Council on Archives' Symposium of Current Records. National Archives of Canada, Ottawa. May 15-17, 1989 / Les Disciplines Convergentes. Compte rendu du colloque sur les archives courantes du son. Comp. by / par Cynthia J. Durance / 1990. 218 p. / Soft (3-598-10897-4)

Manecke, Hans J. / Elfriede Dietrich: Information International. Die Informationstätigkeit internationaler Organisationen / Internationale Magnetbanddienste für Wissenschaft und Technik / 1984. 187 S. / Br (3-598-10476-6)

Manecke, Hans J. / Steffen Rückl; Karl H. Tänzer: Informationsbedarf und Informationsnutzer / 1986. 144 S. / Gb (3-598-07235-X)

Manuscrits musicaux après 1600 / Music Manuscripts after 1600 / Musikhandschriften nach 1600. Catalogue Thématique sur CD-ROM / Thematic Catalogue on CD-ROM / Thematischer Katalog auf CD-ROM. Ed.: Répertoire International des Sources Musicales; Zentralredakt. an d. Stadt- und Universitätsbibliothek Frankfurt am Main / Série / Series / Reihe: A / II 3rd cumulative ed. 1996 (3-598-40293-7)
- 4th cumulative ed. Série / Series / Reihe: A / II 1996 (3-598-40355-0)
- 5th cumulative ed. Série / Series / Reihe: A / II 1997 (3-598-40372-0)
- 6th cumulative ed. Série / Series / Reihe: A / II 1998 (3-598-40390-9)

The **Map** Librarian in the Modern World. Essays in honour of Walter W. Ristow. Ed. by Helen Wallis and Lothar Zögner. Presented by the Geography and Map Libraries Section (IFLA) / 1979. 295 p. / Hard (3-598-10063-9)

Marburger Beiträge zur Vergleichenden Erziehungswissenschaft und Bildungsforschung. Hrsg.: Marburger Forschungsstelle für Vergleichende Erziehungswissenschaft [Minerva Publikation]
- 7: Stübig, Heinz: Bildungswesen, Chancengleichheit und Beschäftigungssystem. Vergleichende Daten und Analysen zur Bildungspolitik in England 1980 XII, 334 S. / Br (3-597-10202-6)
- 8: Nieser, Bruno: Bildungswesen, Chancengleichheit und Beschäftigungssystem. Vergleichende Daten und Analysen zur Bildungpolitik in Frankreich 1980 XII, 256 S. / Br (3-597-10203-4)
- 9: Blumenthal, Viktor von: Bildungswesen, Chancengleichheit und Beschäftigungssystem. Vergleichende Daten und Analysen zur Bildungspolitik in Italien. Vorw. v. Leonhard Froese. Hrsg. v. Viktor von Blumenthal; Bruno Nieser; Heinz Stübig; Bodo Willmann in Verb. mit Leonhard Froese / 1980. XII, 275 S. / Br (3-597-10204-2)
- 10: Willmann, Bodo: Bildungswesen, Chancengleichheit und Beschäftigungssystem. Vergleichende Daten und Analysen zur Bildungspolitik in Schweden Vorw. v. Leonhard Froese,. Hrsg v. Viktor von Blumenthal; Bruno Nieser; Heinz Stübig; Bodo Willmann in Verb. mit Leonhard Froese / 1980. XII, 378 S. / Br (3-597-10205-0)
- 11: Bode, Herbert F.: Bildungswesen, Chancengleichheit und Beschäftigungssystem. Vergleichende Daten und Analysen zur Bildungspolitik in den USA. Vorwort v. Leonhard Froese. Hrsg. v. Viktor von Blumenthal; Bruno Nieser; Heinz Stübig; Bodo Willmann in Verb. mit Leonhard Froese / 1980. XXIV, 395 S. / Br (3-597-10206-9)
- 12: Blumenthal, Viktor von: Die Reform der Sekundarstufe II in Italien. Zum Legitimationsproblem bildungspolitischer Entscheidungen / 1980. X, 244 S. / Br (3-597-10207-7)
- 13: Staatliche Massnahmen zur Realisierung des Rechts auf Bildung in Schule und Hochschule. Zur Situation in England, Frankreich, Italien, Schweden und den USA / 1981. X, 532 S. / Br (3-597-10281-6)
- 14: Krauter, Armin: Abhängige Entwicklung und Veränderungen im Bildungswesen. Das nonformale Bildungswesen in Kenya / 1981. XII, 459 S. / Br (3-597-10282-4)
- 15: Buttlar, Annemarie: Rassisch getrennte Schulen im Süden der USA, 1890-1950. Politische, rechtliche und ökonomische Faktoren / 1981. XII, 305 S. / Br (3-597-10283-2)
- 16: Programm und Realität der Gesamtschule im Ausland. Zum Stand der strukturell-organisatorischen Massnahmen in England, Frankreich, Italien, Schweden und den USA Beitr. v. Heinz Stübig; Bruno Nieser; Viktor von Blumenthal; Bodo Willmann und Annemarie Buttlar / 1981. VIII, 531 S. / Br (3-597-10284-0)
- 17: Behinderte in ausländischen Schulen - Wege zur Integration Von Heinz Stübig; Bruno Nieser; Viktor von Blumenthal; Bodo Willmann; Annemarie Buttlar und Leonhard Froese / 1982. XX, 516 S. / Br (3-597-10285-9)
- 18: Brämer, Rainer: Anspruch und Wirklichkeit sozialistischer Bildung. Beiträge zur Soziologie des DDR-Bildungswesens / 1983. IV, 176 S. / Br (3-597-10286-7)
- 19: Froese, Leonhard: Ausgewählte Studien zur Vergleichenden Erziehungswissenschaft. Positionen und Probleme / 1983. VIII, 158 S. / Br (3-597-10471-1)
- 20: Jacobi-Dittrich, Juliane: 'Deutsche' Schulen in den Vereinigten Staaten von Amerika. Historisch-vergleichende Studie zum Unterrichtswesen im mittleren Westen (Wisconsin 1840-1900) / 1988. 261 S., 2 Abb. / Br (3-597-10552-1)
- 21: Makarenko-Diskussionen international. Protokoll des 2. Marburger Gesprächs (1.- 4. Mai 1986). Hrsg. v. Götz Hillig und Siegfried Weitz / 1989. X, 284 S. / Br (3-597-10563-7)
- 22: Stübig, Heinz: Bildungspolitik in England (1975-1985). Vergleichende Daten und Analysen / 1990. XVI, 276 S. / Br (3-597-10564-5)
- 23: Blumenthal, Viktor von: Bildungspolitik in Italien (1975-1985). Vergleichende Daten und Analysen / 1990. XXII, 334 S. / Br (3-597-10565-3)
- 24: Nieser, Bruno: Bildungspolitik in Frankreich (1975-1985). Vergleichende Daten und Analysen / 1990. XIV, 226 S. / Br (3-597-10569-6)
- 25: Blumenthal, Viktor von: Bildungspolitik in der Schweiz. Vergleichende Daten und Analysen zur Entwicklung in den 80er Jahren / 1991. XXIV, 358 S. / Br (3-597-10696-X)
- 26: Willmann, Bodo: Bildungspolitik in Österreich. Vergleichende Daten und Analysen zur Entwicklung in den 80er Jahren / 1991. XXIV, 431 S. / Br (3-597-10697-8)
- 27: Stand und Perspektiven der Makarenko-Forschung. Materialien des 6. internationalen Symposiums (28. April 2. Mai 1989). Hrsg. v. Götz Hillig und Siegfried Weitz / 1994. IV, 402 S. / Br (3-597-10706-0)

Marburger Index. Inventar der Kunst in Deutschland. Mikrofiche-Edition. Hrsg.: Bildarchiv Foto Marburg im Kunstgeschichtlichen Institut d. Philipps Universität Marburg / 1977 ff. ca. 1, 3 Mio. Fotos auf ca. 14.427 Fiches. Lesefaktor 24x / Silberfiches (3-598-30700-4)
- Lfg. 1-20. Foto-Inventar der Kunst in Deutschland / 1977-1982. 4.803.115 Fotos auf 4.800 Silberfiches (3-7940-6200-0)
- Lfg. 21-30. Foto-Inventar der Kunst in Deutschland / 1983-1987. 3.031.170 Fotos auf 3000 Silberfiches (3-598-30711-X)
- Lfg. 31-38. Foto-Inventar der Kunst in Deutschland / 1988-1990. 1.503.119 Fotos auf 1350 Silberfiches (3-598-30727-6)
- Lfg. 39-63. Foto-Inventar der Kunst in Deutschland / Fotografien der „Deutschen Fotothek" der Sächsischen Landesbibliothek Dresden / 1991 ff. Ca. 470.000 Fotos auf ca. 4.400 Silberfiches (3-598-33200-9)
- Lfg. 1-5. Text-Inventar und Register zur Kunst in Deutschland. Objekt-Inventar mit 1031240 Kunstwerken. 30 Register (Kataloge) zu 1031250 Kunstwerken / 1983-1987. 600 Silberfiches (3-598-30712-8)
- Lfg. 6-8. Text-Inventar und Register zur Kunst in Deutschland / 1988-1991. 877 Silberfiches (3-598-30726-8)
- Heusinger, Lutz: DV-Anleitung. Vorw. v. Christoph König,. Hrsg.: Bildarchiv Foto Marburg. Deutsches Dokumentationszentrum f. Kunstgeschichte, Philipps-Universität Marburg / 2. überarb. u. erw. Aufl. 1992. XI, 272 S. / Ln (3-598-11109-6)
- Heusinger, Lutz: Gebrauchsanleitung / 2. Aufl. 1985. 132 S. / Br (3-598-30722-5)
- Heusinger, Lutz: Handbuch des Marburger Index / 1988 (3-598-30718-7)
- Heusinger, Lutz: Handbuch. Vorw. v. Christoph König. Hrsg.: Bildarchiv Foto Marburg - Deutsches Dokumentationszentrum f. Kunstgeschichte; Philipps-Universität Marburg / 3. überarb. Aufl. 1994. XXVIII, 593 S. / Ln
- Heusinger, Lutz: Marburger Informations-, Dokumentations- und Administrations-System (MIDAS). DV-Anleitung. Vorw. v. Christoph König. Hrsg.: Bildarchiv Foto Marburg; Deutsches Dokumentationszentrum f. Kunstgeschichte; Philipps-Universität Marburg / 1989. VIII, 246 S. / Ln (3-598-10888-5)
- Heusinger, Lutz: Users' Manual Transl. by Christopher Moos / 2nd ed. 1985. 127 p. / Soft (3-598-30721-7)

Marburger Index. Österreich-Index. Bilddokumentation zur Kunst in Österreich / 1996. Ca. 24.000 Fotos auf 244 Silberfiches (3-598-33199-1)

Marburger Index. Schweiz-Index. Bilddokumentation zur Kunst in der Schweiz / 1995. 15.000 Fotos auf 148 Silberfiches. Lesefaktor 24x (3-598-33219-X)

Marburger Index. Wegweiser zur Kunst in Deutschland. CD-ROM Datenbank. Hrsg.: Bildarchiv Foto Marburg im Kunstgeschichtlichen Institut d. Philips-Universität Marburg / 1. Ausg. 1995 (3-598-40280-5)
- 2. Ausg. 1996 (3-598-40340-2)
- 3. Ausg. 1997. 2 discs (3-598-40367-4)
- 4. Ausg. 1998. 3 discs (3-598-40386-0)
- 5. Ausg. 1999. 3 discs. (3-598-40424-7)

Market Economy and Planned Economy / Marktwirtschaft und Planwirtschaft. An Encyclopedic Dictionary: German-English-Russian / Ein enzyklopädisches Wörterbuch: Deutsch-Englisch-Russisch. On behalf of the Fédération Internationale d'Information et de Documentation (FID) by T. Földi; P. Knirsch; L. Szamuely; N. Watts, in association with W. Demmler; S. Krasavtchenko; F. Seton / 2 vols. 1992. Cplt. XLI, 1045 p. / Hard (3-598-11070-7)

Marks, Erwin: Die Entwicklung des Bibliothekswesens der DDR. (Beiheft 94 zum Zentralblatt für Bibliothekswesen) / 1987. 201 S. / Br (3-598-07030-6)

Marti, Hanspeter: Philosophische Dissertationen deutscher Universitäten 1660-1750. Eine Auswahlbibliographie. Unter Mitarb. v. Karin Marti / 1982. 705 S. / Lin (3-598-10445-6)

Martin Buber. A Bibliography of his Writings / Eine Bibliographie seiner Schriften. 1897-1978. Compiled by Margot Cohn and Rafael Buber / 1980. 160 p. / Hard (3-598-10146-5) [Coproduktion mit Magnes Press, The Hebrew University]

Marwinski, Felicitas: Die Freie öffentliche Bibliothek Dresden-Plauen und Walter Hofmann. Ein Beitrag zur Geschichte des Volksbüchereiwesens zu Beginn des 20. Jahrhunderts. (Beiheft 6 zur Zeitschrift „der bibliothekar"). Hrsg.: Zentralinstitut für Bibliothekswesen, Berlin / 1983. 134 S., 19 Abb. / Br (3-598-07213-9) [Lizenzausg. m. Genehm. v. VEB Bibliographisches Institut Leipzig]

Mass Communication Culture and Society in West Africa. Ed. by Frank O. Ugboajah / 3., unveränd. Nachdr. 1985. 329 p. / Hard (3-598-10569-X) [Hans Zell Publishers]

Die **Matrikel** der Universität Jena. Veröffentlichungen der Universitätsbibliothek Jena. Hrsg. v. Konrad Marwinski.
- Bd. III. Die Matrikel der Universität Jena 1723-1764
  - Lfg. 8. Ortsregister. Erster Teil: A-Hi. Bearb. v. Otto Köhler / 1986, S. 1.133-1.276 / Br (3-598-07049-7)
  - Lfg. 9. Ortsregister. Zweiter Teil: Hi-Zy. Bearb. v. Otto Köhler / 1990, S. 1.277-1.472 / Br (3-598-07050-0)
  - Lfg. 10. Einleitung von Heinz Wiessner sowie Berichtigungen. Bearb. v. Otto Köhler / 1993. LXII S. / Br (3-598-11096-0)

Mayer, Rudolf A.: Probleme der Planung und des Aufbaus von Informations- und Dokumentationssystemen im Bildungsbereich / 1982. XIII, 405 S. / Br (3-598-10409-X)

Mayer, Rudolf A. / Diethelm Prauss: Fachinformation Bildung / Information System in Education Ein Topographisches Verzeichnis der Institutionen / Register of InformationServices and Agencies. Hrsg.: Dokumentationsring Pädagogik und Ges. f. soz. u. erziehungswiss. Forschung und Dokumentation (DOPAED und GFD). Stand: Oktober 1986. / 1987. 236 S. / Br (3-598-10734-X)

Medien und Archive. Beitr. zur Rolle moderner Archive in Information u. Dokumentation. Hrsg. von Gerhard Mantwill / 1974. 348 S. (3-7940-3229-2)

Mediendramaturgie und Zuschauerverhalten. (Fernsehen und Bildung Jg. 16/1982, 1-3). Hrsg. v. Helmut Oeller; Hertha Sturm. Redakt.: Marianne Grewe-Partsch; Käthe Nowacek / 1983. 245 S. / Br (3-598-00156-8)

Meier, Karin / Helga Schirmeister; Sigfried Zeimer: Vertrauensbildung in der internationalen Politik / 1991. 208 S. / Gb (3-598-11035-9)

Memorial Book. The Gypsies at Auschwitz-Birkenau / Ksiega Pamieci. Cyganie w obozie Koncentracyjnym Auschwitz-Birkenau / Gedenkbuch. Die Sinti und Roma im Konzentrationslager Auschwitz-Birkenau. Ed.: State Museum of Auschwitz-Birkenau in cooperation with Documentary and Cultural Center of German Sintis and Roms. Preface by Romani Rose. Introd. v. Jan Parcer / 2 vols. 1993 Cplt. 1800 p. / Hard (3-598-11162-2)

Mensch und Medien. Zum Stand von Wissenschaft und Praxis in nationaler und internationaler Perspektive. Zu Ehren von Hertha Sturm. Hrsg. v. Marianne Grewe-Partsch; Jo Groebel / 1987. 349 S. / Gb (3-598-10629-7)

Die **Menschenrechte**. Erklärungen, Verfassungsartikel, internat. Abkommen. Hrsg. u mit e. Einf.von Wolfgang Heidelmeyer / 2., vollst. überarb. Aufl. 1977. 276 S. / Kst. (UTB 123) (3-506-99160-4)

Metropolitan libraries on their way into the eighties. Festschrift Jürgen Eyssen zum 60. Geburtstag. Hrsg. von Marion Beaujean / 1982. 188 S. (3-598-10429-4)

Mette, Günther / Eva Schöppl: Fachinformation Wirtschaftswissenschaften. Literaturhinweise, Informationsbeschaffung und Informationsverarbeitung. Hrsg. von Reinhard Horn; Wolfram Neubauer / (3-598-10899-0)

Meyer, Gerhard: Wege zur Fachliteratur. Geschichtswissenschaft / 1980. 144 S. / Kst. (UTB 1001)

Microform & Imaging Review. Ed. by Thomas A. Bourke (Vol.19); Susan Szasz Palmer (Vol. 20 - 22 / Vol. 27/2); Wendy Thomas (Vol. 23 - 27/1); Julie Arnott (since Vol. 27/3) / since Vol. 19/1990 (4 issues annually) [Seit 1990. Vorher bei Meckler Publishing, Westport, CT]
- Vol. 19/1990 – Vol. 25/1996. Microform Review (ISSN 0002-6530)
  - Parallel ed.: Vol. 20/1991 – 25/1996. Microfiche edition / (4 fiches annually) Reader factor 24x
- Vol. 26/1997 - Vol. 28/1999. Microform & Imaging Review (ISSN 0949-5770)

Microform Market Place 1990-1991. An International Directory of Micropublishing. Ed. by Barbara Hopkinson / 1991. VIII, 234 p. / Soft (3-598-10933-4)
- 1992-1993 / 1993. VIII, 235 p. / Soft (3-598-11094-4)
- 1994-1995 / 1994. VIII, 227 p. / Soft (3-598-11182-7)
- 1996-1997 / 1996. VIII, 223 p. / Soft (3-598-11311-0)

Microform Review →*Microform & Imaging Review*

Microform Review. Cumulative Index, 1972-1981 Volumes 1-10 / 1982. 57 p. / Hard (3-598-10944-X) [1990 von Meckler Publishing, Westport, CT übernommen]

Microform Review Series in Library Micrographics Management [1990 von Meckler Publishing, Westport, CT übernommen]
- 1: Microforms in Libraries: A Reader. Ed. by Albert Diaz / 1975. 443 p. / Hard (3-598-10948-2)
- 2: Studies in Micropublishing, 1853-1976: Documentary Sources. Ed. by Allen B. Veaner / 1976. 489 p. / Hard (3-598-10951-2)
- 3: Microforms and Library Catalogs: A Reader. Ed. by Albert Diaz / 1978. 320 p. / Hard (3-598-10947-4)
- 4: Serials Management and Microforms: A Reader. Ed. by Patricia A. Walsh / 1979. 302 p. / Hard (3-598-10950-4)
- 5: Microforms Management in Special Libraries: A Reader. Ed. by Judy H. Fair / 1979. 272 p. / Hard (3-598-10945-8)
- 6: Hernon, Peter: Microforms and Government Information / 1981. 287 p. / Hard (3-598-10946-6)

- 7: Boss, Richard W. / Deborah Raikes: Developing Microforms Reading Facilities / 1981. 198 p. / Hard (3-598-10939-3)
- 9: Microform Research Collections: A Guide. Ed. by Suzanne Cates Dodson / 2nd ed. 1984. 670 p. / Hard (3-598-10943-1)
- 10: Michaels, Georg H. / Mindy S. Kerber; Hal W. Hall: Microform Reader Maintenance Manual / 1984. 194 p. / Hard (3-598-10942-3)
- 11: An Index to Microform Collections. Ed. by Ann Niles / 1984. XVIII, 891 p. / Hard (3-598-11030-8)
- 12: Cumulative Microform Reviews, 1977-1984 / 1986. 1047 p. / Hard (3-598-10938-5)
- 13: An Index to Microform Collections Volume 2. Ed. by Ann Niles / 1988. 1002 p. / Hard (3-598-10940-7)

Mikrostandorte für eine Gesamthochschule in Osnabrück im Rahmen der Stadtentwicklung. Planungsgutachten. Hrsg.: Zentralarchiv für Hochschulbau, Stuttgart / 1974. 111 S. (3-598-10020-5)

Milano, Ernesto / Renzo Margonari; Mauro Bini: Xilografia dal Quattrocento al Novecento Percorso Storico-Artistico sui Fondi della Biblioteca Estense / 1993. 237 p. / Hard (3-598-07618-5)

Mikroformen und Bibliothek. Hrsg. von Gert Hagelweide / 1977. 471 S. (3-7940-3053-2)

Minerva-Fachserie Geisteswissenschaften
- Andersen, Arne: Lieber im Feuer der Revolution sterben, als auf dem Misthaufen der Demokratie verrecken! Die KPD in Bremen von 1928-1933. Ein Beitrag zur Bremer Sozialgeschichte / 1987. VI, 526 S. / Br (3-597-10263-8)
- Barloewen, Constantin von: Gleichheit und Freiheit: Alexis de Tocqueville in Amerika. Seine Darstellung des Verhältnisses von zentralstaatlicher Lenkung und lokaler Eigenverantwortung, untersucht am historischen Beispiel Pennsylvanias / 1978. XII, 336 S. (3-597-10025-2)
- Baumann, Heidrun: Der Geschichtsschreiber Philippe de Commynes und die Wirkung seiner politischen Vorstellungen in Frankreich um die Mitte des 16. Jahrhunderts / 1981. XII, 279 S. (3-597-10334-0)
- Blumenthal, Bert: Der Weg Arno Schmidts. Vom Prosaprotest zur Privatprosa. / 1980. XIV, 267 S. (3-597-10238-7)
- Bohne, Werner / Felix Semmelroth: Die künstlerische Objektivation des psychisch deformierten Individuums bei George Meredith. / 1978. VIII, 430 S. (3-597-10047-3)
- Carls, Hans G.: Alt-Hormoz - ein historischer Hafen an der Strasse von Hormoz (Iran). Retrospekt und Prospekt zu einem ungelösten archäologischen, geographischen und orientalischen Problem / 1982. XXXII, 191 S. / Br (3-597-10409-6)
- Dickerhof-Fröhlich, Hedwig: Das historische Studium an der Universität München im 19. Jahrhundert. Vom Bildungsfach zum Berufsstudium / 1979. II, 229 S. (3-597-10068-6)
- Eberan, Barbro: Luther? Friedrich 'der Grosse'? Wagner? Nietzsche? Wer war an Hitler schuld? Die Debatte um die Schuldfrage 1945-1949 / 1983. IV, 281 S. (3-597-10434-7)
  - 2. erw. Aufl. 1985. VIII, 290 S. / Br (3-597-10533-5)
- Freudenreich, Carla: Zwischen Loen und Gellert. Der deutsche Roman 1740-1747 / 1979. 325 S. (3-597-10172-0)

- Grossmann, Anton J.: Irische Nationalbewegungen, 1884-1915 / 1979. II, 373 S. (3-597-10057-0)
- Groth, Michael: The Road to New York: The Emigration of Berlin Journalists 1933-1945 / 1984. X, 383 S. / Br (3-597-10518-1)
- Grube-Hell, Ingrid: Landeskunde als Teilbereich des Anglistik-Studiums. Systematische empirische Untersuchungen zum Stellenwert der Landeskunde innerhalb verschiedener Bezugsfelder des Anglistikstudiums / 1980. IV, 289 S. (3-597-10110-0)
- Güldenpfennig, Sven: Sport als Gegenstand gewerkschaftlicher Politik / 1980. XII, 550 S. (3-597-10112-7)
- Hemje-Oltmanns, Dirk: Materielle Bedingungen der Entwicklung des Verhältnisses von Sozialreform und Revolution in Deutschland (1890-1924). Unter besonderer Berücksichtigung der Bremer Werftarbeiterbewegung / 1983. II, 393 S. / Br (3-597-10145-3)
- Hetzer, Friedrich: Les débuts narratifs de Julien Gracq (1938 - 1945) / 1980. XII, 166 S. (3-597-10179-8)
- Huber, Rudolf: Redaktionelles Marketing für den Lokalteil. Die Zeitungsregion als Bezugspunkt journalistischer Themenplanung und -recherche / 1986. XX, 318 S., 122 S. Anh / Br (3-597-10262-X)
- Ignatow, Assen: Aporien der marxistischen Ideologielehre. Zur Kritik der Auffassung der Kultur als „Ideologie in letzter Instanz" / 1984. 176 S. / Br (3-597-10515-7)
- Karlson, Ilse: Die Funktion der Lieder in den Dublin Plays von Sean O'Casey / 1984. XII, 428 S. / Br (3-597-10513-0)
- Kawashima, Takeyoshi: Die japanische Gesellschaft Familismus als Organisationsprinzip. Aus d. Jap. v. Kunihiro Kamiya; Reinhold Gerd / 1985. 245 S. / Br (3-597-10571-8)
- Keil, Werner: Zwischen „Vorbildlichkeit" und „Adaptabilität". Studien zum vorpädagogischen Wortgebrauch von „bildsam": - Heinrich von Meissen - Johann von Neumarkt / 1985. IV, 78 S. / Br (3-597-10149-6)
- Kleine, Peter: Zur Figurencharakteristik in Shakespeares 'Henry VI'. Ein Vergleich mit den Quellen unter Berücksichtigung der Textüberlieferung und der Konzeption moderner Historik / 1980. X, 282 S. (3-597-10187-9)
- Kronsbein, Joachim: Autobiographisches Erzählen. Die narrativen Strukturen der Autobiographie / 1984. 197 S. / Br (3-597-10494-0)
- Lange, Bernd P.: Die Theorie literarischer Gattungen in der englischen Aufklärung. Poetische Regeln und bürgerliche Gesellschaft / 1979. VIII, 558 S. (3-597-10171-2)
- Marckwardt, Wilhelm: Die Illustrierten der Weimarer Zeit. Publizistische Funktion, ökonomische Entwicklung und inhaltliche Tendenzen (unter Einschluss einer Bibliographie dieses Pressetypus 1918-1932) / 1982. XIV, 186 S. / Br (3-597-10133-X)
- Mees, Imke: Die Hui - Eine moslemische Minderheit in China. Assimilationsprozesse und politische Rolle vor 1949 / 1984. XII, 131 S. / Br (3-597-10530-0)
- Mehnert, Volker: Protestantismus und radikale Spätaufklärung. Die Beurteilung Luthers und der Reformation durch aufgeklärte deutsche Schriftsteller zur Zeit der Französischen Revolution / 1982. 240 S. / Br (3-597-10142-9)
- Meiners, Reinhard: Methodenprobleme bei Marx und ihr Bezug zur Hegelschen Philosophie / 1980. X, 367 S. (3-597-10223-9)
- Moerchel, Joachim: Die Wirtschaftspolitik Maria Theresias und Josephs II. in der Zeit von 1740 bis 1780 / 1979. IV, 67 S. (3-597-10099-6)

- Paus-Haase, Ingrid: Soziales Lernen in der Sendung „Sesamstrasse". Versuch einer Standortbestimmung / 1986. XVIII, 550 S. / Br (3-597-10260-3)
- Peltz-Dreckmann, Ute: Nationalsozialistischer Siedlungsbau. Versuch einer Analyse der die Siedlungspolitik bestimmenden Faktoren am Beispiel des Nationalsozialismus / 1978. II, 472 S. (3-597-10004-X)
- Philippoff, Eva: Alfred Polgar. Ein moralischer Chronist seiner Zeit. 1980. X, 414 S. (3-597-10250-6)
- Philippoff, Eva: Kurt Tucholskys Frankreichbild / 1978. XVI, 155 S. (3-597-10018-X)
- Pitschl, Florian: Das Verhältnis vom Ding an sich und den Ideen des Übersinnlichen in Kants kritischer Philosophie. Eine Auseinandersetzung mit T. I. Oisermann / 1979. X, 328 S. (3-597-10160-7)
- Pöschl, Rainer: Vom Neutralismus zur Blockpolitik. Hintergründe der Wende in der türkischen Aussenpolitik nach Kemal Atatürk / 1985. XX, 395 S. / Br (3-597-10215-8)
- Rogalla von Bieberstein, Johannes: Preussen als Deutschlands Schicksal. Ein dokumentarischer Essay über Preussen, Preussentum, Militarismus, Junkertum und Preussenfeindschaft / 1981 VIII, 194 S. / Br (3-597-10336-7)
- Roggausch, Werner: Das Exilwerk Anna Seghers 1933 - 1939. Volksfront und antifaschistische Literatur / 1979. IV, 420 S. (3-597-10104-6)
- Rottstock, Felicitas: Studien zu den Nuntiaturberichten aus dem Reich in der zweiten Hälfte des sechzehnten Jahrhunderts. Nuntien und Legaten in ihrem Verhältnis zu Kurie, Kaiser und Reichsfürsten / 1980. ILII, 312 S. (3-597-10185-2)
- Sander-Schauber, Irma: Zur Entwicklung des Minne-Begriffes vor Walther von der Vogelweide / 1978. VI, 247 S. (3-597-10039-2)
- Schubert, Barbara: Karl Stedmann (1804-1878). Kindheit, Jugend und die Zeit seines politischen Wirkens / 1985. VIII, 468 S. / Br (3-597-10532-7)
- Schulte am Hülse, Heinrich: Die verbindliche sechsjährige Grundschule in Bremen als Politikum (1949-1957) / 1982. X, 399 S. (3-597-10367-7)
- Schwarzenbeck, Engelbert: Nationalsozialistische Pressepolitik und die Sudentenkrise 1938 / 1979. 630 S. (3-597-10095-3)
- Sinti in der Grafschaft Lippe Studien zur Geschichte der Zigeuner im 18. Jahrhundert. Hrsg. v. Karin Bott-Bodenhausen / 1988. 247 S. / Br (3-597-10546-7)
- Smets, Franz: Rijeka-Triest: Die Verlagerung eines italienisch-jugoslavischen Konflikts / 1979. II, 258 S. (3-597-10168-2)
- Thöne, Albrecht W.: Das Licht der Arier. Licht-Feuer- und Dunkelsymbolik des Nationalsozialismus. 1979. IV, 106 S. (3-597-10085-6)
- Wersich, Rüdiger B.: Zeitgenössischer Rechtsextremismus in den Vereinigten Staaten. Organisation, Ideologie, Methoden und Einfluss dargestellt unter besonderer Berücksichtigung der John Birch Society / 1978. VIII, 478 S. (3-597-10029-5)
- Wolf, Eva: Der Schriftsteller im Querschnitt: Aussenseiter der Gesellschaft um 1900? Ein systematischer Vergleich von Prosatexten / 1978. VI, 244 S. (3-597-10027-9)

Minerva-Fachserie Kunst
- Burgmer, Brigitte: Ausdrucksformen. Eine Studie zu den Szondi-Test-Personen / 1983. 111 S., davon 56 S. Fotos / Br (3-597-10452-5)
- Corleis, Gisela: Die Bildergeschichten des Genfer Zeichners Rodolphe Töpffer (1799-1846). Ein Beitrag zur Entstehung der Bildergeschichte im 19. Jahrhundert / 2. Aufl. 1979. VI, 274 S. (3-597-10064-3)

- Düchting, Hajo: Robert Delaunays „Fenêtres": peinture pure et simultané. Paradigma einer modernen Wahrnehmungsform 1982. IV, 667 S. / Br (3-597-10293-X)
- Engemann, Alwin: Prüfen als Handlung Eine handlungspsychologische Analyse des Urteilens in mündlichen Prüfungen / 1983. 274 S. / Br (3-597-10428-2)
- Mikorey, Stefan: Klangfarbe und Komposition / 1982. 300 S. / Br (3-597-10373-1)
- Sievers, Gudrun B.: Bauernstuben im Museum und historische Wirklichkeit. Ländliches Wohnen im Dithmarschen des 19.Jahrhunderts und seine Präsentation in kulturhistorischen Museen Norddeutschlands / 1980. X, 241 S. (3-597-10237-9)
- Straten, Adelheid: Das Judith-Thema in Deutschland im 16. Jahrhundert Studien zur Ikonographie - Materialien und Beiträge / 1983. II, 200 S., 7 schw.-w. Abb. / Br (3-597-10486-X)

Minerva-Fachserie Medizin
- Behr, Barbara: Zum Stoffwechselverhalten einiger Radionuklide bei Säugetieren. Eine Literaturstudie unter besonderer Berücksichtigung haustierartlicher Unterschiede am Beispiel von Jod, Cäsium, Strontium sowie von Transuranen / 1978. VIII, 336 S. (3-597-10023-6)
- Birau, Nikolaus: Melatonin. Endokrinologische und humangenetische Untersuchungen. / 1982. VIII, 153 S. (3-597-10412-6)
- Droste, Conrad: Schmerz bei koronarer Herzerkrankung. Eine Literaturübersicht und Psychophysiologische Untersuchung an symptomatischen und asymptomatischen Patienten mit Myokardischämie / 1983. 427 S. / Br (3-597-10467-3)
- Hirsch, Rolf D.: Das Basaliom. Vorwort von O. Braun-Falco / 2., überarb. Aaufl. 1978. XIV, 222 S. (3-597-10013-9)
- Joraschky, Peter: Das Körperschema und das Körper-Selbst als Regulationsprinzipien der Organismus-Umwelt-Interaktion / 1983. 539 S. / Br (3-597-10485-1)
- Kolb, Adam: Grundlagen und logische Analyse des Krebsproblems / 1982. 147 S. (3-597-10295-6)
- Prävention im Betrieb. Workshop auf der „24. Wissenschaftlichen Jahrestagung der Deutschen Gesellschaft für Sozialmedizin und Prävention". Hrsg. v. Uwe Brandenburg; Bodo Marschall; Thomas Schmidt; Friedrich W. Schwartz / 1990. IV, 215 S. / Br (3-597-10692-7)
- Schlottke, Peter F. / Werther Schneider: Motivation zur Blutspende. Eine empirische Studie zur Gewinnung und Erhaltung von Stammspendern / 1983. 92 S. / Br (3-597-10435-5)
- Schmitt, Rainer M.: Plazentadurchblutungsmessung. Untersuchungen mit der Gasaustauschmethode / 1978. VIII, 87 S. (3-597-10054-6)
- Siegrist, Karin: Sozialer Rückhalt und kardiovaskuläres Risiko. Ein medizinsoziologischer Beitrag zum Verständnis menschlicher Adaptation / 1986. II, 323 S. / Br (3-597-10473-8)
- Weber, Ingbert: Berufstätigkeit, Belastungserfahrung und koronares Risiko. Ein Beitrag zum Krankheitsverständnis aus medizinsoziologischer Sicht / 1984. XII, 302 S. / Br (3-597-10516-5)
- Wolff, Johannes: Die funktionelle Gestalt der menschlichen Unterkiefersymphyse / 1982. 240 S. (3-597-10380-4)

Minerva-Fachserie Naturwissenschaften
- Aguirre, Raul H.: Immobilization as a tool to investigate the function and stability of oligomeric enzymes. Studies on rat and beef liver arginase / 1982. VI, 92 S. (3-597-10114-3)

- Balci, Dervis: Zur Theorie der topologischen n-Gruppen / 1981. XII, 74 S. (3-597-10341-3)
- Bienhold, Holger: Beteiligung cholinerger Synapsen an den zentralen Kompensationsprozessen nach unilateraler Labyrinthektomanie bei Rana temporaria L / 1981. IV, 105 S. (3-597-10340-5)
- Boetticher, Heiner von: Gamma-Gamma-Koinzidenzszintigraphie mit der Szintillationskamera und einem schichtisolierenden Detektorsystem / 1982. X, 65 S. (3-597-10126-7)
- Cornelius, Gerd: Zur Frage circarianer Rhythmen bei menschlichen Erythrocyten / 1981. IV, 11105 S. (3-597-10343-X)
- Fichtner, Wolfgang: Synthesen, Charakterisierung und Reaktion von Übergangsmetallcarbonylkomplexen mit Tetraalkyl- und Tetraphenyldistiban-Liganden / 1981. VI, 125 S. (3-597-10338-3)
- Grubert, Meinhard: Mucilage or gum in seeds and fruits of angiosperms. A review / 1981. VIII, 397 S. (3-597-10279-4)
- Hildebrandt, Jan P.: Kreislaufuntersuchungen am medizinischen Blutegel Hirudo medicinalis L., insbesondere unter Belastungen des Salz / Wasserhaushaltes / 1987. 96 S. / Br (3-597-10555-6)
- Kattner, Evelyn: Charakterisierung und vergleichende Untersuchungen zur Organspezifik verschiedener N-Acetyltransferasen der Ratte / 1981. VIII, 142 S. (3-597-10344-8)
- Krüger, Michael: Ordnungsmasse auf Binärsequenzen / 1982. II, 92 S. / Br (3-597-10129-1)
- Luthardt, Gisela: Verhaltensbiologische Untersuchungen zum visuell gesteuerten Beutefangverhalten von Salamandra salamandra L / 1981. VIII, 91 S. (3-597-10363-4)
- Maass, Wolfgang: Contribution to alpha- and beta--Recursion Theory / 1978. IV, 107 S. (3-597-10015-3)
- Pfeifer, Tilo / Ulrich Schwerhoff; Hans J. Held: Struktur und Aufbau eines Programmiersystems zur flexiblen Prüfablauferstellung für automatisierte Mess- und Prüfsysteme / 1985. II, 93 S., 23 Abb. / Br (3-597-10147-X)
- Rappenglück, Michael: „Mensch und Kosmos". Eine Geschichte der Astronomie vom Mythos zum Nomos / 1982. 300 S. (3-597-10376-6)
- Rump, Wolfgang: Beiträge zur Darstellungstheorie der Zahlringe / 1978. XII, 68 S. (3-597-10041-4)
- Schnittker, Peter: Beiträge zur Theorie chemischer und biochemischer Oszillationen / 1979. XII, 113 S. (3-597-10077-5)
- Schönfeld, Wolfgang: Gleichungen in der Algebra der binären Relationen / 1981. VI, 74 S. (3-597-10350-2)
- Tausch, Michael: Theoretische und experimentelle Untersuchungen von Valenzisomerisierungen / 1981. VIII, 146 S. (3-597-10346-4)
- Ulbricht, Kurt: Wahrscheinlichkeitsfunktionen im neunzehnten Jahrhundert. Die verschiedenen Ausprägungen des Begriffs der Wahrscheinlichkeitsfunktion im neunzehnten Jahrhundert / 1980. XX, 189 S. (3-597-10224-7)
- Weigel, Helmut: Eigenschaften der oszillierenden Briggs-Rauscher-Reaktion. Spektralphotometrische Untersuchungen mit mikrocomputergesteuerter Photonenzählung / 1981. XII, 171 S. (3-597-10339-1)
- Wölfel, Wolf: Kernrelaxationsuntersuchungen der Molekülbewegung im Flüssigkristall PAA / 1978. IV, 199 S. (3-597-10030-9)

Minerva-Fachserie Pädagogik

- Asmus, Sieghard: Zum Verhältnis von technischem und ökonomischem Lernen / 1978. VI, 234 S. (3-597-10019-8)

- Bauer, Wolfgang: Lehrerarbeit und Lehrerbewusstsein. Empirische Untersuchung zum Verhältnis von Arbeits- und Berufssituation der Lehrer und Lehrerbewusstsein / 1980. 240 S. (3-597-10234-4)
- Bennack, Jürgen: Schule und Wirtschaft. Meinungen Beteiligter / 1979. 155 S. (3-597-10087-2)
- Bickel, Horst: Diagnose der Studierfähigkeit für die Philologien / 1980. XII, 308 S. (3-597-10183-6)
- Bittlinger, Ludwig: Elemente einer Theorie des Bildungsprozesses und der Curriculare Lehrplan / 1978. XII, 905 S. (3-597-10050-3)
- Blandow-Wechsung, Steffi: Heimerziehung und Devianz. Eine Untersuchung pragmatischer Devianztheorie und Sympathie von Erziehern in ihren Auswirkungen auf Heimkinder / 1978. XX, 419 S. (3-597-10045-7)
- Blum, Franz: Studieneignung für die Ingenieurwissenschaften / 1980. XXII, 417 S. (3-597-10184-4)
- Bonnemann, Arwed: Studenten an der Hochschule der Bundeswehr Hamburg / 1982. IV, 179 S. (3-597-10414-2)
- Brammerts, Hermann: Die Bildungsarbeit der Gewerkschaften im Kontext der Erwachsenenbildung / 1982. 259 S. / Br (3-597-10418-5)
- Breitbach, Irmgard: Identitätsentwicklung im Kindergartenalter unter den Bedingungen kollektiver Erziehung. Die Analyse dreier Kindergärten in israelischen Kibbutzim / 1979. VI, 355 S. (3-597-10071-6)
- Christian, Hatto: Studierfähigkeit für das Fach Rechtswissenschaft / 1980. XII, 244 S. (3-597-10182-8)
- Daugs, Reinhard: Programmierte Instruktion und Lerntechnologie im Sportunterricht / 1979. XII, 483 S. (3-597-10078-3)
- Ebener, Manfred: Praxisbezogene Ausbildung für Jugendarbeit. Evaluation einer Ausbildung für evangelische Jugendarbeit und Materialien zur Curriculum-Entwicklung / 1979. VI, 433 S. (3-597-10093-7)
- Eickhorst, Annegret: Innovation im Unterricht. Bestandsaufnahme und Klassifikation empirischer Forschungsergebnisse zur Lernorganisation / 1981. XII, 442 S. (3-597-10303-0)
- Forytta, Claus / Jürgen Linke: Ist Unterricht „gestörte" Kommunikation?. Eine Untersuchung zum sprachlichen Handeln im Unterricht der Primarstufe / 2 Bde. 1981. Bd. 1: XII, 430 S.; Bd. 2: VI, S. 431-815. (3-597-10273-5)
- Funiok, Rüdiger: Fernsehen lernen - eine Herausforderung an die Pädagogik. Theorie und Praxis der Fernseherziehung / 1981. XVIII, 473 S. (3-597-10337-5)
- Göldner, Hans D.: Elternmeinung, Elternwille, und ihr Einfluss auf die Schule. Ergebnisse einer schriftlichen Befragung von Eltern in Bayern / 1978. II, 315 S. (3-597-10038-4)
- Gonschorek, Gernot: Erziehung und Sozialisation im Internat. Ziele, Funktionen, Strukturen und Prozesse komplexer Sozialisationsorganisationen / 1979. XII, 515 S. / Br (3-597-10163-1)
- Gries, Jürgen: Umdenken in der Jugendhilfe / 1980. VI, 146 S. (3-597-10220-4)
- Groteloh, Elke: Kommunikation und Lernerfolg. Eine Vergleichsuntersuchung zur Lehreffektivität eines Einzel- und Gruppenlehrprogrammes mit dem gleichen Programmtext / 1981. XII, 342 S. (3-597-10278-6)
- Hörmann, Karl / Rüdiger Pfeiffer-Rupp: Musiklehrerausbildung und Kapazitätsermittlung. Ein Leitfaden für Studienreform und Selbstverwaltung an wissenschaftlichen Hochschulen. In Rede und Gegenrede dargestellt / Vorwort von Helmuth Hopf / 1982. 96 S. (3-597-10405-3)

- Kehlbacher, Monika: Volkshochschulen in Hessen Eine organisationssoziologische Untersuchung unter besonderer Berücksichtigung der Programmplanung / 1982. XII, 365 S. / Br (3-597-10123-2)
- Keller, Gustav / Rolf D. Thiel; Annette Binder: Studienerfolg durch studienvorbereitende Beratung. Die wesentlichen Ergebnisse des Ulmer Modellversuchs zur Studienberatung / 1982. X, 199 S. (3-597-10368-5)
- Kolem, Ute: Intelligenz, Sprache und Leistungsbewertung. Eine empirische Untersuchung der Zusammenhänge schulrelevanter Variablen in der Grundschule / 1981. XIV, 361 S. (3-597-10345-6)
- Krawitz, Rudi: Pädagogik als Handlungsorientierung. Die Bedeutung des transzendental-kritischen Aspekts der Pädagogik / 1980. 322 S. (3-597-10218-2)
- Lackmann, Jürgen: Sozialisation von Lehrern er Wirtschaftslehre. Eine empirische Untersuchung über die Implementation eines neuen Studienbereichs / 1980. 354 S. (3-597-10190-9)
- Langland, Elisabeth: Evaluierung eines Auswahlverfahrens für die Ermittlung der Studierbefähigung. Nacherhebung zur Oberprimanerauswahl der Studienstiftung des deutschen Volkes / 1978. VIII, 184 S. (3-597-10051-1)
- Mackert, Norbert: Inhalte in schulischen Interaktionen. Ein Beitrag zu einer Methode inhaltsspezifischer Unterrichtsanalyse anhand von Unterrichtsdokumenten (aufgezeigt an Beispielen des Deutsch-, Sozialkunde- und Sachunterrichts) / 1983. 483 S. / Br (3-597-10470-3)
- Maiwurm, Eberhard: Situative Bedingungen des Chemieunterrichts der Hauptschule in einer ländlichen Region. Eine empirische Untersuchung für curriculare Innovationen / 1982. 235 S. (3-597-10370-7)
- Mayer, Joachim: Interaktionsverläufe und unterrichtlicher Kontext. Eine Methode zur handlungsorientierten Analyse von Unterrichtsdokumenten / 1982. 230 S. / Br (3-597-10408-8)
- Michl, Werner: Der Beitrag der Kinderspielgruppe zu Erziehung und Sozialisation in afrikanischen Stammesgesellschaften / 1986. 271 S. / Br (3-597-10472-X)
- Polzin, Manfred: Kinderspieltheorien und Spiel- und Bewegungserziehung. Theorien des Kindlichen Spiels und ihre Bedeutung für die Spiel- und Bewegungserziehung in Kindergarten und Anfangsunterricht / 1979. VIII, 277 S. (3-597-10055-4)
- Pragst, Hans: Curriculare Reformen und institutionalisierte Lehrerfortbildung. Dargestellt an der Einführung der Arbeitslehre in ausgewählten Ländern der Bundesrepublik Deutschland / 1981. IV, 301 S. (3-597-10304-9)
- Räuchle, Luise: Geisteswissenschaft als Realwissenschaft. Zum Problem des dialektischen Prinzips im Frühwerk Theodor Litts / 1982. XXII, 267 S. / Br (3-597-10378-2)
- Reichmann, Erwin: Pädagogisch-psychologische Probleme im allgemeinbildenden Schul- und Berufsausbildungswesen der DDR / 2 Bde. 1982. Bd. 1: VI, 382 S., Bd. 2: XII, S. 383-723 / Br (3-597-10289-1)
- Scheffer, Walther: Aufarbeitung von Problemen aus der Unterrichtspraxis einer Lehreeeergruppe. Fallstudie zu einem Kooperationsprojekt im Rahmen des Modellversuchs Regionale Lehrerfortbildung in Hessen / 1980. X, 318 S. (3-597-10180-1)
- Schmid-Schönbein, Gisela: Möglichkeiten und Grenzen des gesteuerten Fremdsprachenerwerbs im Vorschulalter / 1978. XIV, 289 S. (3-597-10034-3)

- Seel, Norbert M.: Lernerleben im Geschichtsunterricht der Sekundarstufe I. Eine experimentelle Analyse / 1980. X, 200 S. (3-597-10217-4)
- Seel, Norbert M.: Wertungen im Geschichtsunterricht. Experimentelle Analyse der Effekte wertender Stellungnahmen zu geschichtlichen Sachverhalten und Schülerantworten in programmierter Unterweisung auf Kenntnisse und Einstellung der Lernenden / 1979. XII, 288 S. (3-597-10072-4)
- Späth, Franz: Anthropologische Aspekte berufspädagogischer Theorien. Eine problemgeschichtliche Untersuchung / 1981. 180 S. (3-597-10369-3)
- Steinforth, Harm: Schulfunkverwendung im Unterricht. Vergleichende Analyse struktureller Merkmale / 1980. V, 356 S. (3-597-10108-9)
- Stelljes, Helmut: Evaluation eones Curriculums der Vorschulmathematik / 1981. 467 S. (3-597-10352-9)
- Stübig, Frauke: Erziehung zur Gleichheit. Die Konzepte der „éducation commune" in der Französischen Revolution / 2., unveränd. Aufl. 1979. XVI, 519 S. (3-597-10086-4)
- Studien zur Pädagogik Tolstojs. Hrsg. v. Horst E. Wittig und Ulrich Klemm / 1988. 80 S. / Br (3-597-10591-2)
- Sturzebecher, Klaus / Wichard Klein: Sozialarbeit als Ausbildungsförderung. Differentialpsychologische Wirkungsanalysen sozialpädagogischer Hilfsmassnahmen bei der beruflichen Erstausbildung benachteiligter Jugendlicher / 1987. 197 S. / Br (3-597-10480-0)
- Swillims, Wolfgang: Ökologische Faktoren und Hemmnisse des Lernens. Empirische Studie an Auszubildenden in betrieblichen Lernsituationen / 1982. XX, 408 S. (3-597-10128-3)
- Trost, Günter / Horst Bickel: Studierfähigkeit und Studienerfolg / 1979. II, 211 S. (3-597-10024-4)
- Vierke, Wolfgang: Verlaufsformen methodologischer Prüfungstheorien / 1979. VI, 132 S. (3-597-10088-0)
- Weck, Manfred: Zum Verhältnis administrativer, parteipolitischer und wissenschaftlicher Einflüsse auf curriculare Planungen und Entscheidungen. Untersuchung der „pragmatischen Lehrplanentwicklung" bei der Einführung der Orientierungsstufe in Bremen - am Beispiel des Schulfachs Deutsch / 1980. IV, 467 S. (3-597-10106-2)
- Wenzel, Hartmut: Lernziele im Physikunterricht. Die Lernzieldiskussion in der Physikdidaktik der Bundesrepublik unter besonderer Berücksichtigung gesellschaftsorientierter Lernziele / 1978. XIII, 232 S. (3-597-10046-5)
- Zaiss, Konrad: Entwicklungsfaktoren und Erscheinungsformen von Kinderbedürfnissen / 1982. XVI, 419 S. (3-597-10292-1)

Minerva-Fachserie Philosophie
- Stock, Wolfgang: Wissenschaftliche Informationen - metawissenschaftlich betrachtet. Eine Theorie der wissenschaftlichen Information / 1980. XVIII, 147 S. (3-597-10188-7)
- Zittlau, Dieter: Die Philosophie von Hans Reichenbach / 1981. VI, 328 S. (3-597-10315-4)

Minerva-Fachserie Politik
- Riescher, Gisela / Raimund Gabriel: Die Politikwissenschaft und der Systemwandel in Osteuropa / 1993. 188 S. / Br (3-597-10700-1)
- Verstehen und Erklären von Konflikten Beiträge zur nationalen und internationalen Politik. Hrsg. v. Oscar Gabriel / 1993. 183 S. / Br (3-597-10635-8)

Minerva-Fachserie Psychologie
- Aktivierungsforschung im Labor-Feld-Vergleich. Zur Vorhersage von Intensität und Mustern psychophysischer Aktivierungsprozesse während wiederholter psychischer und körperlicher Belastung. Von Jochen Fahrenberg; Friedrich Foerster; H.-J. Schneider; Wolfgang Müller; Michael Myrtek / 1984. VIII, 390 S. / Br (3-597-10531-9)
- Below, Elke: Metatheoretische Analysen psychologischer Handlungstheorien / 1981. XIV, 250 S. (3-597-10366-9)
- Bischoff, Dietmar: Grundlagen der Interaktionnsanalyse / 1982. 120 S. (3-597-10216-6)
- Bongartz, Walter: Selektive Verarbeitung tachistokopisch dargebotener Reizmuster / 1978. VI, 132 S. (3-597-10026-0)
- Born, Jan: Effekte eines ACTH 4-9 Analogs auf Aufmerksamkeit. Eine experimentelle Analyse mit akustisch evozierten Potentialen / 1985. IV, 231 S. / Br (3-597-10148-8)
- Brodner, Gerhard: Beiträge zur Prognose des Rehabilitationserfolges bei „jugendlichen Herzinfarktpatienten". Multivariate Analyse psychophysiologischer Befunde / 1983. XII, 571 S. / Br (3-597-10484-3)
- Day, Peter: Bewegungsverhalten geistig Behinderter im Sportunterricht. Eine psychologische Analyse / 1980. VIII, 203 S. (3-597-10221-2)
- Elbert, Thomas: Biofeedback langsamer kortikaler Potentiale / 1978. X, 272 S. (3-597-10031-7)
- Engel, Rolf: Aktivierung und Emotion Psychophysiologische Experimente zur Struktur physiologischer Reaktionsmuster unter psychischer Belastung / 1986. VIII, 239 S. / Br (3-597-10376-6)
- Fiedler, Klaus: Urteilsbildung als kognitiver Vorgang / 1981. VIII, 257 S. (3-597-10287-5)
- Fink, Norbert: Lehrbuch der Schlaf- und Traumforschung. Eine kritische Gegenüberstellung aller psychologischen und physiologischen Ansätze für eine Verwendung des Traumes als Hilfsmittel in Diagnostik, Beratung und Therapie / 2., überarb. Aufl. 1979. X, 331 S. (3-597-10170-4)
- Fischer, Christoph: Der Traum in der Psychotherapie. Ein Vergleich Freud'scher und Jung'scher Patiententräume / 1978. VIII, 257 S. (3-597-10052-X)
- Foerster, Friedrich / Hermann J. Schneider; Peter Walschburger: Psychophysiologische Reaktionsmuster. Zur Theorie und Methodik der Analyse individualspezifischer, stimulusspezifischer und motivationsspezifischer Reaktionsmuster in Aktivierungsprozessen / 1983. IV, 223 S. / Br (3-597-10468-1)
- Friedrich, Brigitte: Emotionen im Alltag. Versuch einer deskriptiven und funktionalen Analyse / 1982. XII, 330 S. / Br (3-597-10372-3)
- Gattig, Ekkehard: Identifizierung und primäre Sozialisation / 1982. VIII, 180 S. / Br (3-597-10140-2)
- Gessulat, Siegfried: Selbstmordverhalten. Stadien der Suizidalität / 1983. VI, 325 S. (3-597-10450-9)
- Gessulat, Siegfried: Therapie einer depressiven Patientin. Stadien der Ermutigung / 1984. VIII, 83 S. / Br (3-597-10529-7)
- Hübner, Helmut: Biosignalverarbeitung. Als Mittel zur Verbesserung der motorischen Regulationsfähigkeit von bewegungsgestörten Personen / 1982. VIII, 240 S. / Br (3-597-10138-0)
- Huber, Dorothea: Psychophysiologie des Migränekopfschmerzes. Eine klinisch experimentelle Studie unter besonderer Berücksichtigung der Habituation / 1984. VIII, 233 S. / Br (3-597-10498-3)
- Jacoby, Klaus: Schulangst. Zur Frage der Entwicklung von Personenwahrnehmungsfertigkeiten im Jugendalter / 1979. X, 214 S. (3-597-10100-3)
- Kieselbach, Thomas: Politische Sozialisation in der Hochschule. Theoretische Überlegungen zur politischen Sozialisationsfunktion der Hochschule und eine empirische Untersuchung über die Studiengangsphase Psychologie / 1979. XVI, 392, XVII S. (3-597-10107-0)
- Krämer, Hans J.: Zu Konzept und Diagnose der Originalität / 1979. VIII, 206 S. (3-597-10156-9)
- Kruse, Peter: Bewegungsgestalt und Frequenzanalyse. Ein psychophysiologischer Modellentwurf und seine erste empirische Überprüfung / 1984. II, 136 S. / Br (3-597-10500-9)
- Landry, Carolyn: Ein Vergleich der Theorien von Freud und Jung anhand der Traumdeutung / 1984. XIV, 169 S. / Br (3-597-10501-7)
- Müsseler, Jochen: Aufmerksamkeitsverlagerungen und Relativität. Ein experimenteller Beitrag zur Raum-Zeit-Wahrnehmung anhand eines Kontraktionsphänomens (Tandem-Effekt) / 1987. 257 S. / Br (3-597-10475-4)
- Myrtek, Michael: Typ-A-Verhalten Untersuchungen und Literaturanalysen unter besonderer Berücksichtigung der psychophysiologischen Grundlagen / 1983. VIII, 196 S. / Br (3-597-10491-6)
- Nell, Verena: Interaktives Verhalten in Situationen der Kontaktaufnahme / 1982. X, 390 S. / Br (3-597-10377-4)
- Otto, Jürgen: Regulationsmuster in Warte- und Vollzugssituationen / 1981. II, 215 S. (3-597-10364-2)
- Pietzcker, Frank: Wilhelm Busch - Schuld und Strafe in Werk und Leben / 1984. 104 S., zahlr. Abb. / Br (3-597-10511-4)
- Psychophysiologische Aktivierungsforschung. Ein Beitrag zu den Grundlagen der multivariaten Emotions- und Stress-Theorie. Von Jochen Fahrenberg; Peter Walschburger; Friedrich Foerster; Michael Myrtek und Wolfgang Müller / 1979. X, 436 S. / Br (3-597-10082-1)
- Psychosoziale Probleme und psychotherapeutische Interventionsmöglichkeiten bei Herzinfarktpatienten. Vorwort von Roskamm;. Hrsg. von Wolfgang Langosch / 1980. X, 307 S. (3-597-10249-2)
- Reuter, Helmut: Betrachtungen zur Sozialpsychologie der bildenden Kunst. Dargestellt am Beispiel Emil Noldes und seiner Zeit / 1984. X, 353 S. / Br (3-597-10130-5)
- Rist, Fred: Reaktionen chronisch schizophrener Patienten auf Reize mit affektiver Bedeutung / 1981. XIV, 211 S. (3-597-10333-2)
- Rohr, Gabriele: Objektbedeutung, Objektpräsentation und Scanningstrategien / 1983. 384 S. / Br (3-597-10424-X)
- Ruoff, Bernd A.: Psychologische Analysen zum Alltag jugendlicher Leistungssportler. Eine empirische Untersuchung (kognitiver Repräsentationen) von Tagesabläufen / 1981. XIV, 383 S. (3-597-10365-0)
- Schandry, Rainer: Habituation psychophysiologischer Grössen in Abhängigkeit von der Reizintensität / 1978. VI, 190 S. (3-597-10009-0)
- Schlottke, Peter F.: Selbstinstruktion und Bewältigung von Belastung. Eine empirische Studie mit 9- bis 11jährigen Kindern / 1980. VIII, 320 S. (3-597-10175-5)
- Schwartz, Richard L.: Der Begriff des Begriffes in der philosophischen Lexikographie Ein Beitrag zur Begriffsgeschichte / 1983. X, 171 S. / Br (3-597-10482-7)
- Steinmetz, Patricia: Psychophysiologische Untersuchungen. EMG, EKG, EEG und subjektive Variabeln in Theorie und Empirie / 1979. VIII, 245 S. (3-597-10059-7)

- Utz, Hans E.: Empirische Untersuchungen zum Belohnungsaufschub. Ein Beitrag zur Konstruktvalidierung / 1979. XII, 257 S. (3-597-10081-3)
- Walch, Sylvester: Subjekt, Realität und Realitätsbewältigung. Struktur und Genese eines subjektbezogenen Realitätsentwurfs als Voraussetzung für die vorläufige Deskription von Realitätsbewältigung / 2., überarb. Aufl. 1981. XII, 174 S. (3-597-10307-3)
- Weitkunat, Rolf: Psychologische und elektrokortikale Korrelate kardiozeptiver Prozesse / 1987. X, 759 S. / Br (3-597-10554-8)
- Die Welt unserer Kinder im Krankenhaus von heute. Eine Diskussion von Fachleuten und Laien. Hrsg. v. Hans L. Schmidt; Erhard Hischer; Peter Sachtleben / 1983. VIII, 122 S. / Br (3-597-10489-4)

Minerva-Fachserie Rechts- und Staatswissenschaften

- Access to Higher Education and its Restrictions under Constitution. Decision of the Federal Constitutional Court of Federal Republic of Germany. Hrsg. von Ulrich Karpen / 1980. X, 66 S. (3-597-10264-6)
- Ahlers, Jürgen: Die Sozialisierung von Grund und Boden. Historische Wurzeln, Zulässigkeit nach dem Grundgesetz sowie Vor- und Nachteile der Durchführung in der Bundesrepublik Deutschland / 1982. XIV, 290 S. (3-597-10120-8)
- Barnert, Harald: Das illegale Staatsgeheimnis / 1978. X, 202 S. (3-597-10043-0)
- Berberich, Volker: Rechtsschutz bei Verkehrslärmimmissionen? / 1983. XVIII, 182 S. / Br (3-597-10488-6)
- Böve, Bernd: Unfallforschung und strafrechtliche Sozialkontrolle des Strassenverkehrs. Eine Untersuchung über Unfallursachen und ihre Ermittlung bei fahrlässigen Erfolgsdelikten im Strassenverkehr - dargestellt anhand der fahrlässigen Körperverletzung gemäss § 230 StGB / 1982. XIV, 260 S. / Br (3-597-10132-1)
- Braune-Steiniger, Franz: Die Bedeutung des XX. Parteitages der KPdSU für die sowjetische Aussenpolitik. Vorwort von Gottfried Erb / 1981. 131 S. (3-597-10351-0)
- Cadmus, Manuel: Zivilrechtliche Haftung des nachgeordneten ärztlichen Dienstes / 1982. 165 S. / Br (3-597-10375-8)
- Chon, Tuk Chu: Die Beziehungen zwischen der DDR und der Koreanischen Demokratischen Volksrepublik 1949-1978. Unter besonderer Berücksichtigung der Teilungsproblematik in Deutschland und Korea sowie der Beziehungsstruktur zwischen einem sozialistischen Mitgliedsstaat des Rates für gegenseitige Wirtschaftshilfe sowie des Warschauer Paktes und einem sozialist / 1982. VI, 183 S. / Br (3-597-10419-3)
- Fuhrkop, Andreas: Der Steuervorteilsbegriff im Sinne des § 370 AO. Eine Abgrenzung zu dem Begriff des Vermögensvorteils im Sinne des § 263 StGB und dem Begriff der Subvention im Sinne des § 264 des StGB / 1979. VIII, 124 S. (3-597-10084-8)
- Die Funktion und Struktur der Freien Wählergemeinschaften am Beispiel der Christlichen Wählereinheit des Landkreises Fulda. Eine empirische Studie / 1982. XX, 281 S. / Br (3-597-10417-7)
- Galtung, Johan: Self-Reliance. Beiträge zu einer alternativen Entwicklungsstrategie. Hrsg. v. Mir A. Ferdowsi. Aus d. Engl. v. Ingrid Langer / 1983. II, 193 S. / Br (3-597-10371-5)
- Gleichenstein, Hans von: Die Allgemeinheit des Rechts. Zum fragwürdigen Gerechtigkeitspathos sozialstaatlicher Rechtsreformen / 1979. VI, 498 S. (3-597-10103-8)

- Gromoll, Bernhard: Rechtliche Grenze der Privatisierung öffentlicher Aufgaben. Untersucht am Beispiel kommunaler Dienstleistungen / 1982. XLVIII, 309 S. (3-597-10116-X)
- Günther, Uwe: Zum Verhältnis von Arbeitsrecht, Arbeitsrechtswissenschaft und gewerkschaftlicher Rechtspolitik am Beispiel von Tarifvertragsklauseln / 1983. XII, 193 S. 20, 9 x 14, 7 cm / Br (3-597-10146-1)
- Kaiser, Rolf: Die künstliche Unfruchtbarmachung von sexuellen Triebtätern / 1981. XXII, 229 S. (3-597-10308-1)
- Kim, Byung-ung: Nationalismus und Grossmachtpolitik. Das Dilemma des Nationalismus in Korea unter der US-Militärbesetzung 1945-1948. Vorwort von Kindermann / 1981. XXII, 499 S. (3-597-10326-X)
- Le-Mong, Chung: Der Untergang Südvietnams. Eine Untersuchung der primären endogenen und exogenen Ursachen des Zusammenbruchs der Republik Vietnam / 1982. 311 S. / Br (3-597-10415-0)
- Lennartz, Hans A.: Zur Rechtsprechung des Bundesverfassungsgerichts zu den politischen Parteien / 1982. X, 224 S. / Br (3-597-10119-4)
- Marbach, Johannes: Strafrechtspflege in den hessischen Städten an der Werra am Ausgang des Mittelalters / 1980. XII, 231 S. (3-597-10186-0)
- Möller, Thomas: Ethisch relevante Äusserungen von Max Weber zu den von ihm geprägten Begriffen der Gesinnungs- und Verantwortungsethik / 1983. VI, 58 S. / Br (3-597-10455-X)
- Möller, Thomas: Die kommunalen Wählergemeinschaften in der Bundesrepublik Deutschland / 1981. XIV, 275 S. (3-597-10348-0)
  - 2. erw. Aufl. 1985 XXX, 275 S. / Br (3-597-10574-2)
- Müller-Boysen, Ulrike: Die Rechtsstellung des Betroffenen vor dem parlamentarischen Untersuchungsausschuss. Ein Beitrag zu den Mitwirkungsrechten und zum Schweigerecht des Betroffenen / 1980. 191 S. (3-597-10252-2)
- Ochi, Hisashi: Der aussenpolitische Entscheidungsprozess Japans. Der zur Normalisierung der Beziehungen zwischen Japan und der Volksrepublik China führende politische Entscheidungsprozess in Japan / 1982. XIV, 314 S. (3-597-10288-3)
- Polke, Heinrich: Der Zahlungsverkehr der Banken im In- und mit dem Ausland. Risikoverteilung bei Schliessung eines Kreditinstituts / 1978. XXXVI, 223 S. (3-597-10053-8)
- Reinhard, Volker: Die AGB-Reform. Darstellung und kritische Würdigung / 1979. 168 S. (3-597-10056-2)
- Reuss, Andreas: Grenzen steuerlicher Mitwirkungspflichten. Dargestellt am Beispiel der erhöhten Mitwirkungspflicht des Steuerpflichtigen bei Auslandsbeziehungen / 1979. XXXIV, 169 S. (3-597-10090-2)
- Schrimpf, Henning: Herrschaft, Individualinteresse und Richtermacht im Übergang zur bürgerlichen Gesellschaft. Studien zum Rechtsschutz gegenüber der Ausübung öffentlicher Gewalt in Preussen 1782 - 1821 / 1979. XVI, 498 S. (3-597-10079-1)
- Seneca, Clara: Begriffsreduktion. Zur Theoriereduktion in der analytischen Wissenschaftstheorie / 1982. X, 200 S. (3-597-10411-8)
- Sommer, Karl L.: Gustav Heinemann und die SPD in den sechziger Jahren. Die Entwicklung politischer Zielsetzungen in der SPD in den Jahren 1960 bis 1969, dargestellt am Beispiel der politischen Vorstellungen Gustav Heinemanns / 1980. 299 S. (3-597-10222-0)
- Sonne, Rolf: Die Anpassung betrieblicher Ruhegelder. Eine Untersuchung über die Dynamisie-

rung von Betriebsrenten unter besonderer Beachtung von § 16 des Gesetzes zur Verbesserung der betrieblichen Altersversorgung / 1978. X, 446 S. (3-597-10037-6)
- Spiekermann, Klaus: Bürgerliche Herrschaft und Bildungsfreiheitsrechte. Eine historisch materialistische Untersuchung anhand der Grundrechte der Ausbildungsfreiheit und der Wissenschaftsfreiheit / 1982. XXX, 135 S. / Br (3-597-10134-8)
- Strzyz, Wolfgang: Die Abgrenzung von Strafverteidigung und Strafvereitelung / 1983. IV, 325 S. / Br (3-597-10436-3)
- Traeger, Burkhardt: Die Reichweite des arbeitsrechtlichen Sozialschutzes. Zur Abgrenzungsproblematik des durch Arbeitsgesetze zu schützenden Personenkreises / 1981. XXXVIII, 217 S. (3-597-10349-9)

Minerva-Fachserie Technik

- Arbeitsschutz und Umweltschutz bei der Verarbeitung von Farben und Lacken. Hrsg.: Gesellschaft f. Technologiefolgenforschung und Dt. Lackinstitut GmbH / 1983. VIII, 237 S., 61 Abb. / Br (3-597-10487-8)
- Bremer, Joachim: Binäre stochastische Pulsfolgen. Aspekte der Erzeugung, Anwendung und Filterung / 1982. XIV, 135 S. (3-597-10413-4)
- Brücker-Steinkuhl, Kurt: Spatial Guidance and Proportional Navigation / 1984. II, 44 S. 29, 5 x 21 cm / Br (3-597-10492-4)
- Fehn, Heinz G.: Untersuchungen von Mehrwegesuchverfahren zur Codierung von Modell- und Sprachquellen / 1982. VIII, 134 S. (3-597-10131-3)
- Felten, Klaus: Optimalbildbestimmung bei elektromikroskopischen Fokussieren. Entwicklung eines rechnergesteuerten Schnellverfahrens / 1981. XII, 112 S. (3-597-10347-2)
- Haase, Volkmar: Notizen zur Konstruktion von Realzeit-Software / 1978. IV, 130 S. (3-597-10042-2)
- Larsen, Günter R.: Nachbarimpulsbeeinflussung bei der Übertragung binärer Signale / 1981. VI, 183 S. (3-597-10301-4)
- Morell, Frank: Das Einsetzen unterkühlten Siedens bei Zwangskonvektion in Ringkanalgeometrie / 1978. VIII, 106 S. (3-597-10040-6)
- Pauls, Werner: Verfahren zum schnellen Abschalten dynamisch streuender Flüssigkeitskristall-Anzeigen / 1978. X, 96 S. (3-597-10033-3)
- Saulich, Gisbert: Filter aus ortsabhängig gekoppelten TEM-Wellenleitungen / 1978. VIII, 84 S. (3-597-10048-1)
- Traenkle, Carl A.: Flugmechanik I. Mechanik der Flugleistungen / 2., durchges. Aufl. 1978. II, 146 S. (3-597-10005-8)
- Traenkle, Carl A.: Flugmechanik II. Stabilität und Steuerung / 1977. II, 188 S. (3-597-37672-9)
- Wessels, Günter: Digitale Sprachübertragung in paketvermittelnden Netzen / 1982. VIII, 137 S. / Br (3-597-10143-7)

Minerva-Fachserie Theologie

- Fellermayr, Josef: Tradition und Sukzession im Lichte des römisch-antiken Erbdenkens. Untersuchungen zu den lateinischen Vätern bis zu Leo dem Grossen / 1979. X, 468 S. (3-597-10158-5)
- Weclawski, Tomasz: Zwischen Sprache und Schweigen. Eine Erörterung der theologischen Apophase im Gespräch mit Vladimir N. Lossky und Martin Heidegger / 1985. XII, 238 S. / Br (3-597-10572-6)
- Wimmer, Walter: Eschatologie der Rechtfertigung. Paul Althaus' Vermittlungsversuch zwischen uneschatologischer und nureschatalogischer Theologie / 1979. XII, 527 S. (3-597-10058-9)

Minerva-Fachserie Wirtschafts- und
Sozialwissenschaften

- Backes, Wieland: Planung und Raumentwicklung im Mittleren Neckarraum. Sozialökonomische Determinanten der Lebensbedingungen in einer verdichteten Region, dargestellt unter besonderer Berücksichtigung der Waiblinger Bucht / 1978. XVI, 426 S. (3-597-10021-X)
- Bäuerle, Gerhard: Der Freiraum als räumliches Wertobjekt. Ein Ansatz zu einer räumlichen Werttheorie, konkretisiert am Beispiel des freizeitorientierten Freiraums unter besonderer Berücksichtigung von Wohnergänzungsflächen / 1984. XVI, 311 S. / Br (3-597-10499-1)
- Bartol, Gerda: Ideologie und studentischer Protest. Untersuchungen zur Entstehung deutscher Studentenbewegungen im 19. und 20. Jahrhundert / 1977. IX, 271 S. (3-7940-7015-1)
- Becker, Barbara von: Berufssituation der Journalistin. Eine Untersuchung der Arbeitsbedingungen und Handlungsorientierungen von Redakteurinnen bei einer Tageszeitung / 1980. 281 S. (3-597-10189-5)
- Beckmann, Michael: Theorie der sozialen Bewegung. Anwendung sozialpsychologischer Hypothesen zur Erklärung der Entstehungsbedingungen sozialer Bewegungen. IV, 370 S. / (3-597-10061-9)
- Beiträge des Instituts für empirische Soziologie
  - 1: Grieswelle, Detlef: Studenten aus Entwicklungsländern. Eine Pilot-Studie / 1978. 130 S. (3-597-10001-5)
  - 2: Grieswelle, Detlef: Rhetorik und Politik. Kulturwissenschaftliche Studien / 1978. 155 S. (3-597-10002-3)
  - 3. Grieswelle, Detlef: Jugend und Freizeit. Bedingungen ausserschulischer Jugendarbeit / 1978. 183 S. (3-597-10003-1)
  - 4: Grieswelle, Detlef: Jugendliche Arbeitslose und Jungarbeiter. Berufsbezogene Einstellungen und Verhaltensweisen berufsschulpflichtiger Jugendliche in sogenannten Jungarbeiterklassen im Saarland / 1978. 168 S. (3-597-10011-2)
- Bierkandt, Rudolf: Der soziale Alltag von Kindern in Fernsehserien / 1978. VI, 206 S. (3-597-10016-3)
- Blaes, Ruth: Qualifikationsstruktur in der journalistischen Ausbildung. Dargestellt am Beispiel der Entwicklung eines berufsbezogenen Studienganges / 1981. VI, 160 S. (3-597-10277-8)
- Blaum, Verena: Marxismus-Leninismus, Massenkommunikation und Journalismus. Zum Gegenstand der Journalistikwissenschaft in der DDR / 1980. 242 S. (3-597-10191-7)
- Blome, Gerhard: Gesellschaftliche Lage, Bewußtsein und Handlungspotential der Arbeiter der „grossen Industrie". Zur Konstitutionsproblematik basaler Solidarisierungs- und Selbstorganisationsprozesse / 1980. VIII, 389 S. (3-597-10232-8)
- Bode, Herbert F.: Arbeit und Qualifikation. Studien zum Verhältnis von Arbeitsplatz- und Qualifikationsstruktur. / 2., durchges. Aufl. 1978. 405, XIII S. (3-597-10007-4)
- Bossow, Manfred: Öffentlicher Nahverkehr und Politik. Fallstudien über Interessenverpflichtungen und -kollisionen am Beispiel der Städte Stuttgart und Freiburg / 1980. 358 S. (3-597-10272-7)
- Bura, Josef: Obdachlosigkeit in der Bundesrepublik: Ursachen und Entwicklung. Ein Beitrag zur Theoriebildung / 1979. VI, 227 S. (3-597-10080-5)
- Canibol, Hans P.: Zur Erklärung und Prognose der kurzfristigen Beschäftigtenentwicklung / 1982. XX, 478 S. / Br (3-597-10410-X)
- Conen, Marie L.: Mädchen flüchten aus der Familie. „Abweichendes" Verhalten als Ausdruck gesellschaftlicher und psychischer Konflikte bei

weiblichen Jugendlichen / 1983. 174 S. / Br (3-597-10454-1)
- Corsten, Hans / Klaus O. Junginger-Dittel: Zur Bedeutung von Forschung und Entwicklung / 1983. XIV, 172 S. / Br (3-597-10453-3)
- Corsten, Hans / Klaus O. Junginger-Dittel: Zur Berufs- und Motivationsstruktur von Diplom-Wirtschaftsingenieuren und Diplom-Ingenieuren. Ein vergleichender empirischer Beitrag zur Berufsökonomik / 1982. VI, 133 S. (3-597-10379-0)
- Dahlgaard, Knut: Rationalisierung und Personalplanung im Krankenpflegebereich. Eine Untersuchung über die Problematik von Rationalisierungsbestrebungen in öffentlichen Krankenhäusern / 1982. XXII, 292 S. / Br (3-597-10122-4)
- Degkwitz, Peter: Gesellschaftliche Bedeutung der Bürorationalisierung und gewerkschaftliche Angestelltenpolitik / 1982. VIII, 644 S. / Br (3-597-10125-9)
- Deisenberg, Anna M.: Die Schweigespirale: Die Rezeption des Modells im In- und Ausland / 1986. XVI, 375 S. / Br (3-597-10362-6)
- Dey, Günther: Arbeitsorganisation als Instrument zur flexiblen Marktanpassung. Eine Untersuchung unter betriebswirtschaftlichen und industrie-soziologischen Aspekten / 1985. 302 S. / Br (3-597-10176-3)
- Dogas, Elias: Abhängige Entwicklung in einer Bauernkultur des Mittelmeergebietes. Dargestellt am Beispiel das Dorfes Pera Melana in Griechenland / 1982. X, 200 S. / Br (3-597-10139-9)
- Epple, Karl: Theorie und Praxis der Systemanalyse. Eine empirische Studie zur Überprüfung der Relevanz und Praktikabilität des Systemansatzes / 1979. X, 268 S. (3-597-10162-3)
- Erfolgschancen ausserschulischer Jugendarbeit / Tl. II. 1989. 245 S. / Br (3-597-10582-3)
- Fahling, Ernst: Die logische Struktur der Krisentheorie bei Karl Marx / 1978. VIII, 364 S. (3-597-10049-X)
- Fellmeth, Rainer: Staatsaufgaben im Spiegel politischer Ökonomie. Zum Verhältnis von Wirtschaft und Staatstätigkeit in Werken von Adam Smith und Adolph Wagner / 1981. 366 S. (3-597-10113-5)
- Ferdowsi, Mir A.: Der positive Frieden. Johan Galtungs Ansätze und Theorien des Friedens / 1981. VIII, 222 S. (3-597-10276-X)
- Filzen, Reinhard: Infrastrukturpolitik, Marktmechanismus und Kapitalverwertung. Grundlagen zu einer Kritik der Infrastrukturtheorien / 1982. VIII, 218 S. (3-597-10118-6)
- Fink, Matthias: Nationales Interesse und Entwicklungshilfe. John F. Kennedy's „Alliance for Progress" / 1978. VIII, 286 S. (3-597-10020-1)
- Fritz, Angela: Die Familie in der Rezeptionssituation. Grundlage zu einem Situationskonzept für die Fernseh- und Familienforschung / 1984. 276 S. / Br (3-597-10514-9)
- Funke, Dörte: Die Rolle von Jugendlichen im Jugendhilfeprozess / 1981. X, 279 S. (3-597-10309-X)
- Geserick, Rolf: 40 Jahre Presse, Rundfunk und Kommunikationspolitik in der DDR / 1989. VI, 519 S. / Br (3-597-10568-8)
- Giesen Frank: Hochschulunterricht im Medienverbund. Evaluation des Reformprojekts „Einführung in die Kommunikationswissenschaft" / 1978. 162 S. (3-597-10008-2)
- Gikas, Michael: Funktion und Wirkungsweise der Sozialtechnologie. Materialien zur kritischen Einschätzung der Aufgabe der Soziologie im System der Arbeitswissenschaft / 1980. X, 398 S. (3-597-10111-9)
- Gikas, Michael / Wolfgang Vierke: Methodologische Probleme des soziologischen Funktionalismus / 1981. VI, 113 S. (3-597-10342-1)

- Greca, Rainer: Die Grenzen rationalen Handelns in sozialen Organisationen. Sozialhistorische und empirische Untersuchungen / 1990. II, 141 S. / Br (3-597-10691-9)
- Greca, Rainer: Handlungsmuster in der Sozialarbeit / 2 Tle. / Br (3-597-10580-7)
- Griepenkerl, Heiko: Führungsschulung und Persönlichkeitsformung Kritische Analyse der heutigen Managementausbildung / 1983. 296 S. / Br (3-597-10483-5)
- Gröning, Gert: Dauercamping. Analyse und planerische Einschätzung einer modernen Freizeitwohnform / 1979. VIII, 230 S. (3-597-10098-8)
- Grube, Lutz: Hafenentwicklung und Industrieansiedlungspolitik am Beispiel Bremerhavens / 1978. XIV, 209, XXI S. (3-597-10101-1)
- Grunsky-Peper, Konrad: Deutsche Volkskunde im Film. Gesellschaftliche Leitbilder im Unterrichtsfilm des Dritten Reiches / 1978. II, 353 S. (3-597-10035-X)
- Hass, Egmont: Zum Verhältnis von Arbeit, Qualifikation und Ausbildung. Eine Kritik der Bildungsökonomie / 1980. VIII, 222 S. (3-597-10178-X)
- Hauff, Hanns J.: Aufstieg zur Macht Zur Kader- und Personalpolitik im Leitungsbereich / 1983. 87 S. / Br (3-597-10422-3)
- Hausarbeit und geschlechtsspezifischer Arbeitsmarkt im deutschen Industrialisierungsprozess Frauenpolitik und proletarischer Frauenalltag zwischen 1800 und 1933 / Bd. 2 / 2. unveränd. Aufl. 1986. XX, 704 S. / Br (3-597-10497-5)
- Heimeran, Silvia: Der Einsatz neuer Telekommunikationssysteme. Ein Handbuch für kommunale Entscheidungsträger / 1980. XVI, 270 S. (3-597-10253-0)
- Hennicke, Peter: Die entwicklungstheoretischen Konzeptionen Mao Tse-tungs. Historische Grundlagen und sozialökonomische Bedingungen der Entwicklungspolitik der Volksrepublik China (1927 - 1957) / 1978. L, 609 S. (3-597-10017-1)
- Heyn, Jürgen: Partizipation und Lokalkommunikation in Grossbritannien. Video, Fernsehen, Hörfunk und das Problem der Demokratisierung kommunaler Kommunikation / 1979. XVI, 492 S. (3-597-10076-7)
- Hösel, Konrad: Marktstruktur und Wettbewerb im Bereich komplexer Spitzentechnologien. Dargestellt am Weltmarkt für Verkehrsflugzeuge / 1981. IV, 357 S. (3-3597- 10361-8)
- Hübner, Heinz: Integration und Informationstechnologie im Unternehmen / 1979 XXIV, 380 S. 20, 9 x 14, 6 cm / Br (3-597-10157-7)
- Kleinhaus, Ursula: Theoriegeschichtlicher Kontext und erfahrungswissenschaftliche Rationalität der Neo-Quantitätstheorie. Ein Beitrag zur wissenschaftstheoretischen Diskussion des Monetarismus / 1979. 300 S. (3-597-10169-0)
- Knapp, Ulla: Frauenarbeit in Deutschland / 2 Bde. / 2. unveränd. Aufl. 1986 / 2 (3-597-10495-9)
- Körnig, Helga: Bildungsexpansion und Fernstudium als bildungs- und gesellschaftpolitische Aufgaben / 1979. XII, 389 S. (3-597-10062-7)
- Koll, Robert: Regionales Wachstum. Eine empirische Untersuchung seiner Bestimmungsfaktoren unter besonderer Berücksichtigung der Raumstruktur / 1979. X, 223 S. (3-597-10167-4)
- Komar, Gerhard: Ansatz zu einer zeichentheoretisch orientierten Wirtschaftsdidaktik / 1985. IV, 225 S. / Br (3-597-10150-X)
- Koopmann, Klaus: Gewerkschaftliche Vertrauensleute. Darstellung und kritische Analyse ihrer Entwicklung und Bedeutung von den Anfängen bis zur Gegenwart unter besonderer Berücksichtigung des Deutschen Metallarbeiter-Verbandes

(DMV) und der Industriegewerkschaft Metall (IGM) / 2 Bde. 1979. Bd. 1: XIV, 610 S., Bd. 2: VIII, S. 611-1136 / Br (3-597-10164-X)

- Kreder, Martina: Situation - Struktur - Erfolg. Eine Analyse des Erfolgsbeitrages situationsadäquater Strukturformen / 1983. XIV, 267 S. / Br (3-597-10421-5)
- Kremzow, Heinz F.: Theorie und Praxis der DKP im Lichte des DKP-Verbots durch das Bundesverfassungsgericht / 1982. XII, 213 S. (3-597-10416-9)
- Krüer-Buchholz, Wiebke: Steuerpolitik und Steuerreformen in der Bundesrepublik Deutschland unter besonderer Berücksichtigung der „Grossen Steuerreform" 1975. Eine Untersuchung der Steuerreformmassnahmen in ihrem historischen Zusammenhang, ihren wirtschaftlichen und politischen Grenzen sowie ihren sozialen Auswirkungen / 1982. XVI, 410 S. (3-597-10127-5)
- Lang, Norbert: Lehre und Fernsehen. Überlegungen und Untersuchungen zur Rolle der öffentlichen Erziehung im Prozess der Massenkommunikation. Dargestellt am Beispiel Fernsehen / 1978. VI, 327 S. (3-597-10012-0)
- Lawrence, Peter A.: Technische Intelligenz und Soziale Theorie. Aus d. Engl. von Rudolf Bierkandt / 1981. 143 S. (3-597-10306-5)
- Leisentritt, Gudrun: Das eindimensionale Theater. Beitrag zur Soziologie des Boulevardtheaters / 1979. 343 S. (3-597-10097-X)
- Li Han lin: Ausdifferenzierung der Wissenschaftsbewertung in der Wissensproduktion / 1984. XII, 282 S. / Br (3-597-10493-2)
- Liebsch, Dieter: Führung und Leistung in Managementgruppen. Kritische Auswertung sozialwisenschaftlicher Untersuchungen zum Verhältnis von Führung und Leistung aus betriebswirtschaftlicher Sicht / 1982. IV, 192 S. (3-597-10115-1)
- Löhmann, Reinhard: Bildung und soziale Emanzipation. Studien zum Verhältnis von sozialistischer Demokratie und Massenbildung in der frühen Sowjetunion (1917 - 1927) / 1979. VIII, 557 S. (3-597-10096-1)
- Lösel, Jörg: Die politische Funktion des Spielfilms in der Volksrepublik China zwischen 1949 und 1965 / 1980. X, 503 S. (3-597-10177-1)
- Malchau, Joachim: Technologie und Emanzipation. Zum Bestimmungszusammenhang der Rationalitätskriterien wissenschaftlich-technischer Entwicklung und deren Auswirkungen auf industrielle Arbeitssituationen / 1983. VIII, 185 S. / Br (3-597-10425-8)
- Manns, Walter: Führungskräfte der brasilianischen Wirtschaft / 1986. II, 149 S. / Br (3-597-10474-6)
- Matthes-Nagel, Ulrike: Latente Sinnstrukturen und objektive Hermeneutik. Zur Begründung einer Theorie der Bildungsprozesse / 1982. XVI, 337 S. (3-597-10137-2)
- Meier-Braun, Karl H.: Freiwillige Rotation - Ausländerpolitik am Beispiel der baden-württembergischen Landesregierung / 1979. VIII, 280 S. (3-597-10089-9)
- Meudt, Volker: Drogen und Öffentlichkeit. Soziale Probleme, gesellschaftliche Konflikte und öffentliche Kommunikation dargestellt am Beispiel der Drogenberichterstattung der Presse / 1979. VIII, 364 S. (3-597-10065-1)
- Möller, Wolf P.: Der Erfolg von Unternehmenszusammenschlüssen. Eine empirische Untersuchung / 1983. II, 403 S. / Br (3-597-10456-8)
- Möller, Wolfgang / Heidrun Wimmersberg: Public Broadcasting in den USA. Nichtkommerzielle Hörfunk- und Fernsehstrukturen in einem kom-

merziell geprägten Rundfunksystem Vorw. v. Hofmann-Riem / 1988. XVI, 286 S. / Br (3-597-10547-5)

- Müller, Egon: Erziehung zum verantwortungsbewußten Verbraucher. Plädoyer für eine neue Konsum-Ethik. Nachwort von Steffens, TU Berlin / 1980. 256 S. (3-597-10239-5)
- Müller-Vogg, Hugo: Public Relations für die Soziale Marktwirtschaft. Die Öffentlichkeitsarbeit der Bundesvereinigung der Deutschen Arbeitgeberverbände, des Bundesverbandes der Deutschen Industrie und des Instituts der Deutschen Wirtschaft zwischen 1966 und 1974 / 1979. II, 242 S. (3-597-10066-X)
- Nagel, Gerhard: Grundbedürfniskonzepte zur Entwicklungspolitik. Wurzeln - Umriss - Chancen / 1985. 90 S. / Br (3-597-10573-4)
- Naroska, Hans J.: Neue Formen sozialer Infrastrukturleistungen. Struktur und Wirkungen am Beispiel selbstverwalteter Jugendfreizeiteinrichtungen / 1985. 288 S. / Br (3-597-10538-6)
- Neumaier, Herbert: Die Steuerung des städtischen Strassenverkehrs unter dem Gesichtspunkt des Umweltschutzes,. Luftreinhaltung, und Lärmschutz an Hauptverkehrsadern durch ein anwendungsreifes, quantitatives Verfahren / 1978. VI, 198 S. (3-597-10010-4)
- Neumann-Bechstein, Wolfgang: Altensendungen im Fernsehen als Lebenshilfe. Eine Inhaltsanalyse der Altensendungen Mosaik und Schaukelstuhl mit einem Exkurs zu den Serien „Unternehmen Rentnerkommune" und „UnRuhestand" / 1982. XII, 386 S. / Br (3-597-10407-X)
- Neuss, Beate: Europa mit der linken Hand? Die deutschen Personalentscheidungen für die Kommission der Europäischen Gemeinschaften / 1988. VI, 481 S. / Br (3-597-10548-3)
- Neverla, Irene: Arbeitszufriedenheit von Journalisten / 1979. XII, 313 S. (3-597-10094-5)
- Oesterdickhoff, Peter: Hemmnisse und Widersprüche in der Entwicklung armer Länder. Darstellung am Beispiel Afghanistans / 1978. IV, 695 S. (3-597-10102-X)
- Petit, Helmut: Die Spezifikation von Modellen mit verteilten Verzögerungen. Ökonomische Theorie, ökonometrische Ansätze, empirische Analysen / 1979. XII, 313 S. (3-597-10161-5)
- Politikwissenschaft und gesellschaftliche Praxis. Normative Selbststeuerung oder Arbeitsmarktorientierung? Vorwort von Heinrich Scholler. Hrsg. von Alfred Büllesbach; Mir A. Ferdowsi / 1979. VI, 237 S. (3-597-10159-3)
- Pott, Philipp: Direktinvestitionen im Ausland. Investitionsmotive, Standortfaktoren und Hilfsmittel bei der Entscheidung für die optimale Auslandsinvestition / 1983. XIV, 278 S. / Br (3-597-10451-7)
- Regionalismus '90. Zur Dialektik des westeuropäischen Einigungsprozesses. Hrsg. v. Gisela Riescher / 1991. 279 S. / Br (3-597-10693-5)
- Regionalpolitik des bürgerlichen Staates. Eine exemplarische Studie anhand der Ansprüche, Planungsvorhaben und Ergebnisse der staatlichen Arbeitsmarktpolitik im bayerischen Regierungsbezirk Oberpfalz. Von Thomas Stahl; Peter Höhmann; Rainer Schreiber; Edgar Straub; Heinz Wieferink; Bärbel Kolb; Walter Simböck / 1980. VIII, 268 S. (3-597-10270-0)
- Riechey, Gunter: Möglichkeiten und Grenzen der Beeinflussung der Preissetzungsmöglichkeiten der Unternehmen / 1982. 286 S. / Br (3-597-10144-5)
- Rockenbach, Helga G.: Komponenten der gesellschaftlichen und wirtschaftlichen Integration der Bundeswehr / 1983. 143 S. / Br (3-597-10449-5)

- Röske, Volker: Der lautlose Zwang der Computer. Eine Untersuchung über den Einzug computergestützter Verwaltungsarbeit in den Arbeitsalltag öffentlicher Verwaltungen / 1987. II, 258 S., 23 Abb. / Br (3-597-10553-X)
- Rota, Franco P.: Menschen - Staaten - Umwelt. Ethologisch-sozialwissenschaftliche Grundlagen und Skizzen internationaler Umweltpolitik / 1986. XII, 355 S. / Br (3-597-10261-1)
- Schertler, Walter / Wolfgang Popp: Attraktivitätsanalyse von Dienstleistungen Ein empirischer Ansatz zur Entwicklung von Ausbaustrategien für Unternehmen und Staat / 1983. 183 S. / Br (3-597-10420-7)
- Schmieder, Arnold: Bewußtsein im Widerspruch. Theorieversuche über den Zusammenhang von industrieller Arbeit und Bewusstsein / 1980. 175 S. (3-597-10236-0)
- Schönecker, Horst G.: Bedienerakzeptanz und technische Innovationen. Akzeptanzrelevante Aspekte bei der Einführung neuer Bürotechniksysteme. Vorwort von Ralf Reichwald / 1980. 312 S. (3-597-10251-4)
- Schor, Ambros: Politische Bildung an Hauptschulen. Eine kritische Bestandsaufnahme verschiedener Konzeptionen in den Bundesländern Bayern, Hessen, Nordrhein-Westfalen und Rheinland-Pfalz / 1982. IV, 295 S. / Pr (3-597-10406-1)
- Schulze, Peter M.: Region und Informationssystem / 1980. 198 S. (3-597-10219-0)
- Schwartz, Wolf: Professionelle Kommunikation und Standesideologie. Zum Verhältnis von beruflicher Praxis, Wissenschaft und Verbandspolitik / 1981. II, 260 S. (3-597-10302-2)
- Schwarz, Michael: Grenzen und Möglichkeiten einer beschäftigungsorientierten betrieblichen Interessenvertretung. Ergebnisse und Schlussfolgerungen aus betrieblichen Fallstudien in der Automobilindustrie / 1987. VIII, 257 S. / Br (3-597-10481-9)
- Sörgel, Angelina: Investitionsdeterminanten und Wirtschaftspolitik. Eine Abhandlung unter besonderer Berücksichtigung des monopolistischen Investitionsverhaltens / 1982. XX, 284 S. (3-597-10117-8)
- Ständischer und bürgerlicher Patriarchalismus Frauenarbeit und Frauenrolle im Mittelalter und im Bürgertum des 19. Jahrhunderts / Bd. 1 / 2. unveränd. Aufl. 1986. XVI, 294 S. / Br (3-597-10496-7)
- Steinle, Holger: Partizipative Planung neuer Formen der Produktionsorganisation / 1982. X, 159 S. / Br (3-597-10404-5)
- Stroetmann, Karl A.: Neue Medien und Informationstechnologien. Eine Darstellung der Auswirkungen auf die Presse am Beispiel der USA. Unter Mitarb. von Günter Knieps / 1980. XXXII, 152 S. (3-597-10271-9)
- Taubner, Friedrich W.: Die Bereitschaft, vom Gegner zu lernen. Zur Ideenpolitik des Sozialliberalismus / 1979. IV, 132 S. (3-597-10063-5)
- Tholen-Struthoff, Karin: Sisyphus. Über die Politik einer Reproduktionskultur / 1982. X, 124 S. (3-597-10141-0)
- Trieba, Volker / Ulrich Mentrup: Entwicklung der Arbeitswissenschaft in Deutschland. Rationalisierungspolitik der deutschen Wirtschaft bis zum Faschismus / 1983. 230 S. / Br (3-597-10427-4)
- Valdueza, Rafael: Die spanische Gewerkschaftsbewegung unter Franco / 1982. X, 277 S. (3-597-10124-0)
- Volkmann, Rainer: Preisverhalten und Beschäftigungsentwicklung unter monopolisierten Konkurrenzverhältnissen / 1982. XIV, 265 S. / Br (3-597-10121-6)

- Volland, Joachim: Der Prozess der Unternehmensentwicklung. Eine branchenbezogene Analyse / 1987. X, 446 S. / Br (3-597-10549-1)
- Vossberg, Henning: Studentenrevolte und Marxismus. Zur Marxrezeption in der Studentenbewegung auf Grundlage ihrer politischen Sozialisationsgeschichte / 1979. XXVIII, 648 S. (3-597-10105-4)
- Wagner, Dieter: Konfliktsituationen in alternativen Organisationsmodellen. Analyse und Handhabung von organisationsbedingten Konflikten in Komplexen, mehrdimensionalen Strukturen / 1988. 270 S. / Br (3-597-10567-X)
- Wallner, Ernst M.: Fischerwesen und Fischerbevölkerung in Sizilien. Bestand - Besonderheiten - Bedeutung heute / 1981. IV, 268 S. (3-597-10332-4)
- Weber, Harald: Statistische Prognoseverfahren / 1983. 164 S. / Br (3-597-10490-8)
- Weber, Werner: Strategien zur Verbesserung des Managements in kleinen und mittleren Unternehmen / 1981. XVI, 332 S. (3-597-10335-9)
- Weingarten, Richard: Eine kapitaltheoretische Untersuchung zum Gesetz vom tendenziellen Fall der Profitrate einschliesslich der Bestimmung seiner empirischen Relevanz / 1982. X, 342 S. / Br (3-597-10136-4)
- Wersebe, Irene von: Qualifizierte Spezialisierung und Autoritätsstruktur / 1978. VIII, 172 S. (3-597-10014-7)
- Weyer, Helmut: Die Gewinn- und Kapitalbeteiligung der Mitarbeiter. Funktionselemente, finanzwirtschaftliche Wirkungen und Determinanten aus der Sicht der Unternehmungen / 1978. IV, 175 S. (3-597-10032-5)
- Winkler, Helmut: Zur Theorie und Praxis der Gesamthochschulplanung. Unter besonderer Berücksichtigung der Studiengangmodelle, Entscheidungsplanung und -organisation. / 2. Aufl. 1979. IV, 474 S. (3-597-10060-0)
- Wischnewsky, Barbara: Kommunale Jugendplanung / 1982. X, 301 S. (3-597-10135-6)
- Woll, Helmut: Die Untauglichkeit des Indikators Sozialprodukt als Wohlfahrtsmass / 1981. VI, 85 S. (3-597-10305-7)
- Woll, Helmut: Monopol und technischer Fortschritt. Probleme der Realanalyse / 1980. XII, 355 S. (3-597-10109-7)
- Zeller, Maria: Struktur- und Verhaltensformen. Eine empirische Analyse des Verhaltens von Führungskräften in grossen Industrieunternehmungen unter Berücksichtigung der Organisationsstruktur / 1983. XVI, 371 S. / Br (3-597-10423-1)
- Ziegeler, Werner: Die Bilanzierung von Preisänderungen in der harmonisierten EG-Rechnungslegung. Ein Beitrag zu Art. 32 / 33 der Vierten EG-Richtlinie unter besonderer Berücksichtigung der Ausschüttungsbemessungsfunktion / 1980. X, 267 S. (3-597-10233-6)
- Zur Verberuflichung sozialer Hilfe. Eine berufssoziologische Untersuchung generativer Handlungsschemata in der Sozialarbeit / Tl. I. 1988. 145 S. / Br (3-597-10581-5)

Mitteilungsblatt. Verband der Bibliotheken des Landes Nordrhein-Westfalen / Jg. 28/1978 bis Jg. 37/1987 (4 Hefte jährlich) (ISSN 0042-3629)
- Sonderh. 1: Leihverkehrsordnung für die deutschen Bibliotheken Ausg. für Nordrhein-Westfalen. Stand: 1. Okt. 1983. Richtlinien für den Regionalen Leihverkehr Nordrhein-Westfalen; Mit d. amtl. Leihverkehrsliste d. Landes Nordrhein-Westfalen, Liste d. zum Regionalen Leihverkehr zugelassenen bibliothekar. Einrichtungen. Mit Erläuterung Hochschulbibliothekszentrum d. Landes Nordrhein-Westfalen. Zsgst. u. bearb. von Jürgen Heydrich / 1983. 58 S

Modern American Protestantism and its World. Historical Articles on Protestantism in American Religious Life. Ed. by Martin E. Marty, University of Chicago / 14 vols. 1992 / 1993. Cplt. CLXXXV, 4,671 p. / Hard (3-598-41530-3)
- Vol. 1. The Writings of American Religious History / 1992. XIII, 338 p. / Hard (3-598-41531-1)
- Vol. 2. Trends in American Religion and the Protestant / 1992. XIV, 280 p. / Hard (3-598-41532-X)
- Vol. 3. Civil Religion, Church and State / 1992. XIV, 485 p. / Hard (3-598-41533-8)
- Vol. 4. Theological Themes in the American Protestant / 1992. XIV, 452 p. / Hard (3-598-41535-4)
- Vol. 5. Varieties of Protestantism / 1992 XIII, 258 p. / Hard (3-598-41536-2)
- Vol. 6. Protestantism and Social Christianity / 1992. XIII, 409 p. / Hard (3-598-41537-0)
- Vol. 7. Protestantism and Regionalism / 1992. XIII, 234 p. / Hard (3-598-41538-9)
- Vol. 8. Ethnic and Non-Protestant Themes / 1993. XIII, 297 p. / Hard (3-598-41539-7)
- Vol. 9. Native American and Black Religion Protestantism / 1993 XIII, 330 p. / Hard (3-598-41540-0)
- Vol. 10. Fundamentalism and Evangelicalism / 1993. XIII, 342 p. / Hard (3-598-41541-9)
- Vol. 11. New and Intense Movements / 1993. XIII, 390 p. / Hard (3-598-41542-7)
- Vol. 12. Women and Women's Issues / 1993. XIII, 366 p. / Hard (3-598-41543-5)
- Vol. 13. Missions and Ecumenical Expressions / 1993. XIII, 208 p. / Hard (3-598-41544-3)
- Vol. 14. Varieties of Religious Expression / 1993. XIII, 282 p. / Hard (3-598-41545-1)

Moderne Systeme und Informationsnetze und die Zuverlässigkeit der Information. Modern systems and networks and the reliability of information. Kongreßbericht, Hamburg 17. - 22. April 1978. Eurarians und Dokumentalirop. Regionalkongreß d. Bibliothekswesens u. d. Dokumentation d. Agrarwiss. Internat. Assoc. of Agricultur Librarians and Dokumentalists (IAALD); Gesellschaft für Bibliothekswesen und Dokumentation des Landbaus / 1978. 347 S. (3-598-07096-9)

Moderne Unterrichtstechnologie. Situationsanalyse u. Basisinformationen zur Forschung u. Anwendung in d. BRD. Von Hans Bertram u.a / 1972. 117 S. / Kst. (UTB 158) (3-7940-2612-8)

Mostar, Gerhart Herrmann. In diesem Sinn - Verlag und Buch. Eine Rede in Versen / 1968. 40 S. / Br (Vlg.-Nr. 03178)

Mühlner, Manfred: Julius Petzholdt. Wegbereiter für Bibliographie und Bibliothekswesen im 19. Jahrhundert. (Zentralblatt für Bibliothekswesen: Beih., 95) / 1987. 127 S. / Br (3-598-07267-8)

Müller-Wolf, Hans M.: Lehrverhalten an der Hochschule Dimensionen, Zusammenhänge, Trainingsmöglichkeiten / 1977. 241 S., Tab., Reg / Kst. (UTB 687) (3-7940-2661-6)

Münchner Ethnologische Abhandlungen. Hrsg: Institut f. Völkerkunde u. Afrikanistik [Minerva Publikation].
- 1: Müller, Claudius C.: Untersuchungen zum „Erdaltar" she im China der Chou- und Han-Zeit. Hrsg.: Inst. f. Völkerkunde und Afrikanistik der Ludwig-Maximilians-Universität München / 1980. XVI, 424 S. / Br (3-597-10173-9)
- 2: Hildebrand, Hartmut K.: Die Wildbeutergruppen Borneos / 1982. XXII, 374 S. / Br (3-597-10294-8)
- 3: Forkl, Hermann: Die Beziehungen der zentralsudanischen Reiche Bornu, Mandara und Bagirmi sowie der Kotoko-Staaten zu ihren südlichen Nachbarn Unter besonderer Berücksichtigung des Sao-Problems / 1983. 638 S., 20 S. Abb., 3 S. Summary / Br (3-597-10426-6)
- 4: Gundert, Sibylle: Der historische Rahmen der wirtschaftlichen und politischen Entwicklung von Vanuatu / 1984. 275 S. / Br (3-597-10512-2)
- 5: Forkl, Hermann: Der Einfluss Bornus, Mandaras, Bagirmis, der Kotoko-Staaten und der Jukun-Konföderation auf die Kulturentwicklung ihrer Nachbarn südlich des Tschadsees / 1985. 505 S. / Br (3-597-10570-X)
- 6: Link, Hilde: Der Olofat-Zyklus in der Erzähltradition Mikronesiens / 1986. XII, 457 S. / Br (3-597-10605-6)
- 7: Gundert-Hock, Sibylle: Mission und Wanderarbeit in Vanuatu. Eine Studie zum sozialen Wandel in Vanuatu 1863-1915 / 1986. 370 S. / Br (3-597-10606-4)
- 8: Öhrig, Bruno: Bestattungsriten alttürkischer Aristokratie im Lichte der Inschriften / 1988. VIII, 370 S., 4 Abb. / Br (3-597-10649-8)

Münchner Theaterzettel. 1807-1982. Altes Residenztheater, Nationaltheater, Prinzregenten-Theater, Odeon. Ausgew. von Klaus Schultz. Vorwort von August Everding / 1982. 400 S. (3-598-10462-6)

Münnich, Monika: PC-Katalogisierung mit RAK. Nach dem Format des DBI-Pflichtenheftes / 1992. 362 S. / Br (3-598-11068-5)

Multi-script, Multi-lingual, Multi-character. Issues for the Online Environment. Proceedings of a Workshop. Sponsored by the IFLA Section on Cataloguing, Istanbul, Turkey, August 24, 1995. 1267 1 / . Ed. by John D. Byrum jr; Olivia Madison / 1998. IV,123 p. / Hard (3-598-21814-1)

Multilingual Lexicon of Higher Education. Ed.: UNESCO. Prepared by the European Centre for Higher Education (CEPES) / 2 vols. Cplt. XXXIV, 661 p. / Hard (3-598-108834)
- Vol. 1. Western Europe and North America / 1993. XIX, 346 p. / Hard (3-598-11058-8)
- Vol. 2. Central and Eastern Europe also including Greece, Israel and Turkey / 1996. 268 p. / Hard (3-598-11059-6)

Musculus, Carl Theodor: Inhalts- und Namen-Verzeichnisse über sämtliche Goethe'sche Werke. Nach der Ausgabe letzter Hand und dem Nachlasse verfertigt. Unter Mitwirkung des Hofraths und Bibliothekars Dr. Riemer: - Nachdr. der Ausgabe Stuttgart und Tübingen 1835 / 1977. VIII, 304 S. (3-7940-4284-0)

Museen der Deutschen Demokratischen Republik / 1974. X, 200 S. (3-7940-34309)

Museologie. Neue Wege - Neue Ziele Bericht über ein internationales Symposium, veranstaltet von den ICOM-Nationalkomitees der Bundesrepublik Deutschland, Österreichs und der Schweiz vom 11. bis 14. Mai 1988 am Bodensee. Hrsg. v. Hermann Auer; Deutsches Nationalkomitee d. Internationalen

Museumsrates ICOM / 1989. 289 S., 28 Abb. / Br
(3-598-10809-5)

Das **Museum** im technischen und sozialen Wandel.
Bericht über ein internationales Symposium,
veranstaltet von den ICOM-Nationalkomitees der
Bundesrepublik Deutschland, Österreichs und der
Schweiz vom 13.-19. Mai 1973 am Bodensee. Hrsg.
v. Hermann Auer / 1975. 188 S. / Br (3-7940-3427-9)

Museum und Denkmalpflege. Bericht über ein
internationales Symposium, veranstaltet von den
Nationalkomitees von ICOM und ICOMOS der
Bundesrepublik Deutschland, Österreichs und der
Schweiz vom 30. Mai bis 1. Juni 1991 am Bodensee.
Hrsg. v. Hermann Auer; Deutsches Nationalkomitee
d. Internationalen Museumsrates ICOM / 1992. 257
S., 39 Abb. / Br (3-598-11107-X)

Das **Museum** und die Dritte Welt. Bericht über ein
internationales Symposium, veranstaltet von den
ICOM Nationalkomitees der Bundesrepublik
Deutschland, Österreichs und der Schweiz vom
7.-10. Mai am Bodensee. Hrsg. v. Hermann Auer /
1981. 357 S. / Br (3-598-10346-8)

Museums of the World. Museen der Welt. Ein
Handbuch über 17.000 Museen in 148 Ländern
einschließlich eines Registers der Sammelgebiete /
1973. XIV, 762 S. (3-7940-3426-0)
• 2., erw. Aufl. Ein Handbuch über 17.500 Museen
in 150 Ländern mit einem Register der Sammelgebiete / 1975. 808 S. (3-7940-3429-5)
weitere Auflagen: → *Handbuch der*
internationalen Dokumentation und Information
Bd. 16

Music Publishers' International ISMN Directory.
Ed.: International ISMN Agency 1998/99 / 2nd ed.
1998. XXXII, 381 p. / Hard (3-598-22248-3)
• 1995/96 ed. 1996. XXVIII, 339 p. / (ISSN
0948-5678) / Hard (3-598-22232-7)

Musikhandschriften der Staatsbibliothek zu Berlin - Preussischer Kulturbesitz. Hrsg.: Staatsbibliothek zu Berlin - Preussischer Kulturbesitz
• 1: Die Bach-Sammlung. Mikrofiche-Edition /
1998/99. Ca. 115.000 S. auf ca. 1.500 Fiches in 5
Lfgn. Lesefaktor 24x / Silberfiche (3-598-34420-1)

Muziol, Roman: Pressedokumentation. Anleitung
für die Arbeit in Pressearchiven. Mit e. Geleitwort
und e. Bibliographie von Roland Seeberg-Elverfeldt /
2. unveränd. Aufl. 1968. 152 S. / Br (Vlg.-Nr. 03228)
• 3., erw. u. erg. Aufl. 1971. 196 S. (3-7940-3228-4)
• unveränd. Nachdr. 1988. VII, 151 S.

Nachlässe und Sammlungen zur deutschen Kunst
und Literatur des 20. Jahrhunderts Die Bestände der
Stiftung Archiv der Akademie der Künste Berlin.
Hrsg.: Stiftung Archiv d. Akademie d. Künste Berlin
/ 1995. XV, 231 S. / Gb (3-598-11293-9)

Nachrichten für Dokumentation (NfD). Zeitschrift
für Information und Dokumentation mit Schrifttum
zu den Informationswissenschaften. Hrsg.: Deutsche
Gesellschaft für Dokumentation e. V. / Jg 25 / 1974 -
Jg. 37/1986 (6 Hefte jährl.) (ISSN 0027-7436)

Nachschlagewerke und Quellen zur Kunst
• Internationale Bibliographien zur Kunstliteratur
zwischen 1500 und 1850. Mikrofiche-Edition.

Hrsg. v. Ulrich Schütte / 1998/1999. Ca. 44 Titel
mit ca. 26.000 S. auf ca. 165 Fiches, Lesefaktor
24x / Silberfiche (3-598-34541-0) / Diazofiche
(3-598-34540-2)

Nachschlagewerke zur Musik. Hrsg. v. Harald
Heckmann
• Internationale Musik-Sachlexika vom 17. bis zum
frühen 19. Jahrhundert. Mikrofiche-Edition /
1998/99. Ca. 203.374 S. auf ca. 120 Fiches,
Lesefaktor 24x / Silberfiche (3-598-33863-5) /
Diazofiche (3-598-33862-7)

Näther, Günter: Bibliothekswesen in Italien. Eine
Einführung In Zus.-Arb. mit Leonore Näther / 1991.
IX, 93 S. / Br (3-598-10759-5)

Napier, Paul: Index to Micrographics Equipment
Evaluations / 2nd ed. 1984. 28 p. / Hard
(3-598-10941-5)

The **Nazi** Holocaust Historical Articles on the
Destruction of European Jews. Ed. by Michael R.
Marrus, University of Toronto / 9 parts in 15 vols.
Cplt. 6957 p. (3-598-21550-9) [Vorher bei Meckler
Publishing, Westport, CT]
• Part 1. Perspectives on the Holocaust / 1989. IX,
449 p. / Hard (3-598-21551-7)
• Part 2. The Origins of the Holocaust / 1989. IX,
737 p. / Hard (3-598-21552-5)
• Part 3. The Final-Solution The Implementation of
Mass Murder / 2 vols. 1989. Cplt. XI, 847 p. / Hard
(3-598-21553-3)
• Part 4. The Final Solution Outside Germany / 2
vols. 1989. Cplt. XI, 793 p. / Hard (3-598-21555-X)
• Part 5. Public Opinion and Relations to the Jews
in Nazi Europe / 2 vols. 1989. Cplt. XI, 786 p. /
Hard (3-598-21557-6)
• Part 6. The Victims of the Holocaust / 2 vols.
1989. Cplt. X, 1259 p. / Hard (3-598-21559-2)
• Part 7. Jewish Resistance to the Holocaust / 1989.
IX, 563 p. / Hard (3-598-21561-4)
• Part 8. Bystanders to the Holocaust / 3 vols. 1989
Cplt. XII, 1497 p. / Hard (3-598-21562-2)
• Part 9. The End of the Holocaust / 1989. IX, 734 p.
/ Hard (3-598-21565-7)

Nelde, Peter H.: Wortatlas der deutschen Umgangssprachen in Belgien / 1987. 31 S., 60 Ktn. / Kt
(3-907820-84-3) [1991 vom Francke Vlg., Bern
übernommen]

Nestler, Friedrich: Bibliographie. Einführung in die
Theorie, Methoden und Geschichte der bibliogra-
phischen Literaturinformation und in die allgemei-
nen bibliographischen Verzeichnisse / 2., neubearb.
Aufl. 1989 / Gb (3-598-07029-2) [Lizenzausg. m.
Genehm. v. VEB Bilbiographisches Institut Leipzig]

Neubauer, K. W. / G. Schwarz; W. Schwuchow:
Kommunikation in Forschung und Entwicklung.
Übersicht über die Problematik. Ergebnisse einer
empirischen Studie / . Hrsg. von U. Möller / 1972.
200 S. (3-7940-3291-8)

Die **neue** Bibliothek. Festschrift für Harro Heim
zum 65. Geburtstag. Hrsg. v. Günther Pflug; Hans-
jochen Hancke / 1984. 415 S. / Lin (3-598-10529-0)

Neue Informations- und Kommunikationstechnik.
Ein Kolloquium zur fachlichen Meinungssbildung in
Bereich Bibliothek, Information und Dokumenta-
tion. Hrsg. v. Brigitte Endres-Niggemeyer / 1984. 175
S. / Br (3-598-10535-5)

Neue Informations- und Kommunikationstechnologien für wissenschaftliche Bibliotheken. Bericht der
IKB-Kommission. Hrsg.: Bayerisches Staatsministerium f. Unterricht, Kultus, Wissenschaft u. Kunst /
1996. 74 S. / Br (3-598-11350-1)

Neue Preussische Zeitung Kreuzzeitung Berlin,
1848-1939. Microfilm-Edition / 1984. 2.403.395 S.
auf 201 35 mm Filmen / Diazofilm (3-598-30673-3)

Neue Rheinische Zeitung. Politisch-ökonomische
Revue (1850). Redigiert von Karl Marx und Friedrich
Engels. Einl. u. Verz. v. Druck u. Sachfehlern v.
Martin Hundt (Nachdr. d. Ausg. 1850) / 1983. XLIII,
180 S. / Ln (3-598-10318-2)

Neue Verfahren für die Dateneingabe und Daten-
ausgabe in Bibliotheken. Berichte e. Symposiums
März 1973. Arbeitsstelle f. Bibliothekstechnik bei der
Staatsbibliothek Preuß. Kulturbesitz / 1973. 158 S.
(3-7940-3213-6)

Die **neue** Weltbühne 1933-1939 (Exil-Weltbühne).
Vorw. v. Thomas A. Eckert / 14 Bde. 1992. Zus.
12.000 S. / Ln (3-598-22320-X)

Neuere Formate für Verarbeitung und Austausch
bibliographischer Daten. Bericht e. Symposiums,
veranst. von d. Arbeitsstelle für Bibliothekstechnik
am 30. April 1974. Hrsg. Arbeitsstelle für Bibliothekstechnik bei d. Staatsbibliothek Preuß. Kulturbesitz / 1975. 101 S. (3-7940-3051-6)

Neuheuser, Hanns P.: Internationale Bibliographie
Liturgische Bücher / International Bibliography on
Liturgical Books. Eine Auswahl kunsthistorischer
und liturgiewissenschaftlicher Literatur zu liturgi-
schen Handschriften und Drucken / A Selection of
Art - Historical and Liturgical Science Literature on
Liturgical Manuscripts and Printed Books / 1991.
XXII, 147 S. / Lin (3-598-10798-6)

Neumann, Gerda: Das Portrait der Frau in der
zeitgenössischen Jugendliteratur (Untersuchungen
der Internationalen Jugendbibliothek Bd. 1. Hrsg. v.
Walter Scherf) / 1977. 389 S., 50 Abb. / Lin
(3-7940-3371-X)

New Contents Librarianship. Contents of current
periodicals in the fields of librarianship, archive
admin. and documentation. Ed. by Niedersächsische
Staats- u. Universitäts-Bibliothek, Göttingen / Vol. 1/
1979 (4 issues annually) (ISSN 0171-9122)

New Contents Slavistics. Contents of current
periodicals in the field of Slavic linguistics, literature
and folklore / Inhaltsverzeichnisse slavistischer Zeitschriften (ISZ). Ed. by Bayerische Staatsbibliothek
München / Jg. 1/1980 (4 issues) (ISSN 0173-6388)
[Erscheinen eingestellt]

New Horizons for Academic Libraries. Papers
presented at the First National Conference of the
Association of College and Research Libraries.
Boston, Massachusetts, November 8-11, 1978. Ed. by
Robert D. Stueart and Richard D. Johnson / 1979.
582 p. / Hard (3-598-40002-0)

The **New** ICC World Directory of Chambers of
Commerce 1990 / Nouvel Annuaire Mondial CCI
des Chambres du Commerce 1990. Ed. by International Chamber of Commerce / 2nd ed. 1990. XII,
551 p. / Soft (3-598-23067-2)

A **New** International Economic Order. Selected Documents 1945 - 1975. Zsgst. von Alfred G. Moss and Harry N. Winton / 2 Teile 1976. XIII, 964 S. / (3-7940-7011-9)

The **New** Library Legacy: Essays in Honor of Richard De Gennaro. Ed. by Susan A. Lee / 1998. XVIII, 117 p. / Hard (3-598-11389-7)

New, Peter G.: Education for Librarianship. Decisions in Organising a System of Professional Education / 1978. 174 S. (0-85157-250-2) [Clive Bingley]

Nicholas, David / Maureen Ritchie: Literature and Bibliometrics / 1978. 184 S. (0-85157-228-6) [Clive Bingley]

Nietzscheana. Hrsg. von Karel Màcha [Minerva Publikation]
• Kopf, Albert: Der Weg des Nihilismus von Friedrich Nietzsche bis zur Atombombe / 1988. X, 316 S., 1 Abb. / Br (3-597-10551-3)
• Zur Genealogie einer Moral. Beiträge zur Nietzsche Forschung. Hrsg. v. Karel Mácha / 1985. VI, 136 S. / Br (3-597-10550-5)

Normenlogik. Grundprobleme der deontischen Logik. Mit Beitr. v. I. Berkemann u.a. Hrsg. v. Hans Lenk / 1974. 256 S. / Kst. (UTB 414) (3-7940-2637-3)

NS-Presseanweisungen der Vorkriegszeit. Edition und Dokumentation. Hrsg.: Institut für Zeitungsforschung der Stadt Dortmund / Hans Bohrmann. Bearb. v. Gabriele Toepser-Ziegert / 7 Bde.
• Bd. 1. 1933 Vorw. v. Fritz Sänger / 1984. 197, XIV, 351 S. / Lin (3-598-10552-5)
• Bd. 2. 1934 / 1985. 36, XXVI, 694 S. / Lin (3-598-10553-3)
• Bd. 3. 1935 / Tl. I u. II
  - Tl. I / 1987. 556 S. / Lin (3-598-10554-1)
  - Tl. II / 1987. 556 S. / Lin (3-598-10678-5)
• Bd. 4. 1936 / 4 Tlbde. 1993. Zus. 1.858 S. / Lin (3-598-11004-9)
• Bd. 5. 1937 / in 3 Tlbdn. 1998. Zus. 25*, 1.046 S. / Lin (3-598-11256-4)
  - Registerband / 1998. 223 S. / Lin (3-598-11353-6)
  Bde. 6 & 7 [i.Vb.]

NTWZ. Verzeichnis Neuer Technisch-Wissenschaftlicher Zeitschriften. Berichtsstand: 1.April 1969. Im Auftr. f. Arbeitsgemeinschaft der Spezialbibliotheken e.V. (ASpB) bearb. u. hrsg. v.. Günther Reichardt / 1970. 143 S. / Lin (3-7940-3379-5)

Nyéki-Körösy, Maria: Les documents sonores Précis de discothéconomie / 1987. 379 p., 117 illus. / Br (3-598-10435-9)

Das **Oberkommando** der sowjetischen Streitkräfte 1985. Tafelband und Index. Hrsg.: Institut für Sowjet-Studien / 2 Tle. 1985. Tafelbd: 18 Taf. iM, Index: 100 S. / Br (3-598-07518-9)

Oelrichs, Johann C.: Entwurf einer Geschichte der Königlichen Bibliothek zu Berlin / 1986. 164 S., 20 Abb. / Gb (3-598-07238-4)

Österreich-Index →Marburger Index.
Österreich-Index.

Oettingen-Wallerstein'sche-Musiksammlung. Mikrofiche-Edition / 1985. 3.819 Diazofiches. Lesefaktor 24x (3-598-30730-6)

ÖVK-WB Österreichischer Verbund-Katalog Wissenschaftliche Bibliotheken. CD-ROM-Edition. Hrsg.: Bundesministerium f. Wissenschaft, Forschung u. Kunst / 2 Ausgaben jährl.
• 1. Ausg. 1994 (3-598-40277-5)
• 2. Ausg. 1995 (3-598-40291-0)
• 3. Ausg. 1995 (3-598-40292-9)
• 4. Ausg. 1996 (3-598-40333-X)
• 5. Ausg. 1996 (3-598-40334-8)
• 6. Ausg. 1997 (3-598-40358-5)
• 7. Ausg. 1997 (3-598-40359-3)
• 8. Ausg. 1998 (3-598-40384-4)
• 9. Ausg. 1998 (3-598-40385-2)
• 10. Ausg. 1999 (3-598-40422-0)

Ohne Seil und Haken. Frauen auf dem Weg nach oben. Hrsg. v. Claudia Bernardoni und Vera Werner, Deutsche UNESCO-Kommission, Bonn. Red.: Hans-Dieter Dyroff / 1987 451 S. / Br (3-598-10717-X)

Ollé, James G.: Library History / 1979. 116 p. (0-85157-271-5) [Clive Bingley]

Online Information. Organized by Online Review. The International Journal of Online & Teletext Information Systems
• 1st Meeting, London December 1977 / 1977. 286 p. / Soft (3-598-10073-6)
• 2nd Meeting, London December 1978 / 1978. 286 p. / Soft (3-598-10074-4)

Oppermann, Alfred: Wörterbuch der Datenverarbeitung / Data processing Dictionary. Englisch - Deutsch / 2., wesentl. erw. u. neu bearb. Aufl. 1973. 343 S. / Lin (3-7940-3099-0)
*1. Aufl.: → **Krüger**, Karl-Heinz: Wörterbuch der Datenverarbeitung*

Oppermann, Alfred: Wörterbuch der Elektronik / Dictionary of Electronics. Englisch - Deutsch / English - German / 1980. 690 S. / Lin (3-598-10312-3)

Oppermann, Alfred: Wörterbuch Kybernetik / Cybernetics Dictionary. Deutsch - Englisch / English - German / 1969. 241 S. / Br (Vlg.-Nr. 03258)

Oppermann, Alfred: Wörterbuch der modernen Technik. Deutsch-Englisch / Englisch-Deutsch. / 2., erw. Neuaufl. 5 Bde. 1965-1968. 4.038 S. / Kst. (Vlg.-Nr. 06068) [Die 1. Aufl. erschien u.d.T.: "Aeronautical English" im Alfred Oppermann, Aeronautischer Vlg., München]
• 3., völlig neubearb. Aufl. Wörterbuch der modernen Technik / Dictionary of Modern Engineering.
  - Bd. 1 Englisch-Deutsch / English- German / 1972. 912 S. / Lin (3-7940-6001-6)
  - Bd. 2 Deutsch - Englisch / German - English / 1974. 952 S. / Lin (3-7940-6002-4)
• 4. Aufl. Bd. 1: Englisch - Deutsch. Bd. 2: Deutsch - Englisch / 1982. 912 S.; 952 S. / Lin (3-598-10471-5)

Organisations- und Geräteprobleme bei der Datenerfassung. Bericht e. Symposiums, veranst. von d. Arbeitsstelle für Bibliothekstechnik am 27. / 28. Februar 1978 / 1978. 115 S. (3-7940-7068-2)

Orgel-Köhne, Armin / Liselotte Orgel-Köhne: Staatsbibliothek Berlin. Mit Textbeitr. von Ulrich Conrads u.a. / 1980. 152 S., 120 S. Abb. / Lin (3-598-10135-X) [Koprod. m. arani-Vlg., Berlin]

Ostdeutsche Funkschriften - Rückblick und Ausblick. Hrsg. im Auftr. d. Mediengesprächskreises d. Ost- u. Westpreussenstiftung in Bayern u. d. Redakt. f. Ostfragen d. Bayerischen Rundfunks v. Heinz Radke; Hans Ulrich Engel.
• 1: Geschichtsbewußtsein, groß geschrieben. E. gesamtdt. Programm. Red. Bearb. Dorothee Radke / 1984. 147 S. (3-598-21031-0)

Ostdeutsche und Südostdeutsche Heimatbücher und Ortsmonographien nach 1945. Eine Bibliographie zur historischen Landeskunde der Vertreibungsgebiete. Hrsg.: Stiftung Ostdeutscher Kulturrat. Bearb. v. Wolfgang Kessler / 1979. 300 S., 15 Ktn. / Lin (3-598-10039-6)

Ostdeutsches Kulturgut in der Bundesrepublik Deutschland. Ein Handbuch der Sammlungen, Vereinigungen und Einrichtungen mit ihren Beständen. Hrsg.: Stiftung Ostdeutscher Kulturrat OKR. Bearb. v. Wolfgang Kessler / 1989. 739 S. / Lin (3-598-10577-0)

Outstanding International Press Reporting. Pulitzer Prize Winning Articles in Foreign Correspondence. Ed. by Heinz D. Fischer
• Vol. 1: 1928-1945 / From the Consequences of World War I to the End of World War II / 1984. LIII, 368 p. / Hard (3-598-23020-6)
• Vol. 2: 1946-1962 / From the End of World War II to the Various Stations of the Cold War / 1985. LXVII, 304 p. / Hard (3-598-23021-4)
• Vol. 3: 1963-1977 / From the Escalation of the Vietnam War to the East Asian Refugee Problems / 1986. LXXIII, 309 p. / Hard (3-598-23022-2)
• Vol. 4: 1978-1989 / From Roarings in the Middle East to the Destroying of the Democratic Movement in China / 1991. XLVII, 332 p. / Hard (3-598-23023-0)

Over, Albert: Die deutschsprachige Forschung über Hochschulen in der Bundesrepublik Deutschland. Eine kommentierte Bibliographie 1965-1985. Vorw. v. Dietrich Goldschmidt; Christoph Oehler; Ulrich Teichler. Hrsg.: Gesamthochschule Kassel - Universität / 1988. 716 S. / Lin (3-598-10682-3)

Paenson, Isaac: Handbuch der Terminologie des Völkerrechts. (Friedensrecht) und der Internationalen Organisationen. Deutsche Teilausgabe / 1993. XIII, 179 S. / Gb (3-598-10917-2)

Paenson, Isaac: Manual of the Terminology of Public International Law. (Law of Peace) andof International Organizations. English / French / Spanish / Russian Edition / 1983. XLVIII, 846 p. / Hard (3-598-07558-8)

Paenson, Isaac: Manual of the Terminology of the Law of Armed Conflicts and of International Humanitarian Organizations. English / French / Spanish / Russian Edition / 1989. XXXVIII, 844 p. / Hard (3-598-07559-6)

Palestine and the Arabic-Israeli Conflict. An Annotated Bibliography. Ed. by Walid Khalidi; Jill Khadduri / 1974. 736 S. (3-7940-5156-4)

Panskus, Hartmut: 40 Jahre K. - G.- Saur - Verlag. 1949 - 1989 / 1989. 31 S.

Panzer, Georg W.: Literarische Nachricht von den allerältesten gedruckten deutschen Bibeln aus dem fünfzehnten Jahrhundert, welche in der öffentlichen Bibliothek der freien Reichsstadt Nürnberg aufbewahrt werden . - Geschichte der Nürnbergischen Ausgaben der Bibel von Erfindung der Buchdruckerkunst an bis auf unsere Zeit. Ausführliche Beschreibung der ältesten Augspurgischen Ausgaben der Bibel mit litterarischen Anmerkungen. Versuch einer kurzen Geschichte der römisch-catholischen deutschen Bibelübersetzung / 4 Teile in 1 Bd. / Nachdr. der Ausg. Nürnberg 1777 - 1781 / 1971. 720 S. (307940-5013-4)

Papritz, Johannes: Archivwissenschaft. Band 1, Teil I: Einführung, Grundbegriffe, Terminologie / Teil II/1: Organisationsformen des Schriftgutes in Kanzlei und Registratur. Erster Teil / Band2, Teil II/2: Organisationsformen des Schriftgutes. Zweiter Teil / Band 3, Teil III/1: Archivische / 4 Bde. / 2. durchges. Aufl. 1983. Bd. 1: XL, 357 S., Bd. 2: 503 S., 39 Abb., Bd. 3: XXII, 303 S., 3 Abb, Bd. 4: 345 S. / Br (3-598-07521-9)

Papyri Bodmer. Papyri Bodmer XXX-XXXVII. Codex des Visions - Poèmes divers. Hrsg. v. André Hurst und Jean Rudhardt / 1999. VI, 226 S. / Gb (3-598-22554-7) [Seit 1999. Die vorherigen Bände sind seit 1954 beim Herausgeber im Selbstverlag erschienen]

Parlament und Bibliothek / Parliament and Library / Parlement et Bibliothèque. Internationale Festschrift für Wolfgang Dietz zum 65. Geburtstag. Hrsg. v. Gerhard Hahn; Hildebert Kirchner / 1986. 452 S. / Gb (3-598-10634-3)

Partizipation als Lernziel. Curriculare Modellvorstellungen u. Materialien zum Hochschulstudium. Das Problem erziehungs- und gesellschaftswissenschaftlich dimensionierter Fachstudiengänge an den Hochschulen der Bundeswehr. Hrsg. von d. Arbeitsgruppe Anleitstudium. Mit Beitr. von Katrin Fließ u.a / 1975. 270 S. / Br (3-7940-3187-3)

Patentinformation and Documentation in Western Europe. An Inventory of Services available to the Public in the European Community. Ed.: The Commission of the European Communities / 1976. 173 p. / Bound (3-7940-5167-X)
- 2nd rev. and enl. ed. Ed. by H. Bank; M. Fenat-Haessig und M. Roland; The Commission of the European Communities / 1980. 268 p. / Bound (3-598-10158-9)
- 3rd rev. and enl. ed. Ed. by Brenda Rimmer; Commission of the European Communities / 1988. 219 p. / Hard (3-598-10744-7)

French ed. ➤ *Information et documentation en matière de brevets en Europe occidentale*
Germ. ed ➤ **Patentinformation** *und Dokumentation in Westeuropa*

Patentinformation und Patentdokumentation in Westeuropa. Ein Bestandsverzeichnis von öffentlich zugänglichen Diensten in den Mitgliedsstaaten der Europäischen Gemeinschaft. Hrsg. v. d. Kommission der Europäischen Gemeinschaften, Generaldirektion Wissenschaftliche und Technische Information und Informationsmanagement, Sekretariat der Arbeitsgruppe „Patentdokumentation", Europazentrum Kirchberg, Luxemburg. / 1976. 180 S. / Lin (3-7940-5166-1)

- 2. neu bearb. Aufl. Hrsg. v. H. Bank; M. Fenat-Haessig und M. Roland; Kommission der Europäischen Gemeinschaften / 1980. 283 S. / Lin (3-598-10156-2)
- 3. überarb. u. erw. Aufl. Hrsg. v. Brenda Rimmer; Kommission der Europäischen Gemeinschaften / 1988. 219 S. / Lin (3-598-10156-2)
- 4. überarb. u. erw. Aufl. 1990. 231 S. / Lin (3-598-10768-4)

Engl. Ausg. ➤ **Patentinformation** *and Documentation in Western Europe*
Franz. Ausg. ➤ **Information** *et documentation en matiere de brevets en Europe occidentale*

Paul, Hans-Holger: Inventar zu den Nachlässen der deutschen Arbeiterbewegung für die zehn westdeutschen Länder un West-Berlin. Hrsg. im Auftr. d. Archivs der Sozialen Demokratie der Friedrich-Ebert-Stiftung; Projektmitarb. Karl Kollmann u.a. / 1993. X, 996 S. (3-598-11104)

Paulys Realencyclopädie Der Classischen Altertumswissenschaft Mikrofiche-Edition / 1991. ca. 55.714 S., 37 Abb. auf 598 Fiches, davon 5 Fiches mit Abb., Lesefaktor 24x (3-598-33221-1)

Periodicals Administration in Libraries. A Collection of Essays. Hrsg. von Paul Mayer / 1978. 158 S. (0-85157-262-6) [Clive Bingley]

Persönlichkeiten der Weinkultur deutscher Sprache und Herkunft Kurz-Biographien aus 16 Jahrhunderten. Hrsg. v. Paul Claus; Gesellschaft für Geschichte des Weines e.V. / 1991. XII, 132 S. / Br (3-598-11093-6)

Peschel, Manfred: Anwendung algebraischer Methoden / 2. Aufl. 1971. 422 S. / Br (3-7940-4288-3)

Petermann, Kurt: Tanzbibliographie. Verzeichnis der in deutscher Sprache veröffentlichten Schriften und Aufsätze zum Bühnen-, Gesellschafts-, Kinder-, Volks- und Turniertanz sowie zur Tanzwissenschaft, Tanzmusik und zum Jazz. Hrsg.: Akademie der Künste, Tanzarchiv der DDR, Leipzig / 6 Bde.
- Bd. 1 u. 2 / 1978. Zus. 1952 S. / Lin (3-598-07024-1)
  - 2. unveränd. Aufl. 1981. Zus. 1952 S. / Lin (3-598-07025-X)
- Bd. 3 / 2. unveränd. Aufl. 1982. 549 S. / Lin (3-598-07026-8)
- Abschlussband / 2 Tle. 1981. Zus. 814 S. / Br (3-598-10398-0)
- Registerband. Verzeichnis des deutschsprachigen Schrifttums über den Volks-, Gesellschafts- und Bühnentanz / 1987. VIII, 592 S. / Lin (3-598-07241-4)

Peters, Klaus: Fachinformation Rechtswissenschaften. Literaturhinweise, Informationsbeschaffung und Informationsverarbeitung. Hrsg. von Reinhard Horn; Wolfram Neubauer / 1992. 200 S. (3-598-10898-2)

Pfeiffer-Rupp, Rüdiger: Studien zur Kapazitätsermittlung. Untersuchungen zur Kapazitätsvorordnung (KapVo) mit einer Analyse der Ausbildungskapazität der Anglistik / Amerikanistik an Hochschulen in der Bundesrepublik und in der Republik Österreich / 1977. XX, 448 S. (3-7940-7047-X)

Phaedrus. An International Journal of Children's Literature Research. Ed. by James H. Fraser / Vol. 6/ 1979 - Vol. 7/1980 (2 (1979) and 3 (1980) issues annually) (ISSN 0098-3365)

Phaedrus Bibliographic Series.
- No. 1: Fraser, James H.: Children's authors and illustrators. A guide to ms. collections in United States research libraries. With the assistance of Renée I Weber / 1980. XI, 119 S. (3-598-40504-9)

Pharmazie. Jahresfachbibliographie 1964. Bearb. von Klaus Gerhard Saur / 1965. 96 S. / Kst.

Pichert, Dietrich: Kostenprobleme der Filmproduktion / 1975. 224 S. / Lin (3-7940-3273-X)

PIK - Praxis der Informationsverarbeitung und Kommunikation. Fachzeitschrift für den Einsatz von Informationssystemen. Hrsg. v. Hans W Meuer, Universität Mannheim. Redakt.: Hans G Kruse, Rechenzentrum der Universität Mannheim / seit Jg. 14/1991 (4 Hefte jährlich) (ISSN 0930-5157)

Plakatsammlung des Instituts für Zeitungsforschung der Stadt Dortmund. Teil A: Deutsche Plakate. Teil B: Deutsche Plakate. Teil C: Ausländische Plakate. Mikrofiche-Edition. Bearb. v. Posthoff, Barbara / 1988-1992 Begleitbde, 4.731 Plakate auf 111 farb. Fiches. Lesefaktor 24x (3-598-32563-0)
- Begleitbd. 1. Teil A: Deutsche Plakate / 1992. 174 S. / Gb (3-598-21311-5)
- Begleitbd. 2. Teil B und C: Ausländische und deutschsprachige Plakate / 1992. 139 S. / Gb (3-598-21312-3)

Planen und Bauen. Hrsg.: Zentralarchiv für Hochschulbau, Stuttgart [1979 übernommen]
- 2: Löwenhauser, Paul / W. Oswald Grube; Michael Sepp: Ideenwettbewerb Zentrum der Technischen Universität München in Garching / 1972. 54 S. (3-598-20024-2)
- 4: Büttner, Otto: Kostenplanung von Gebäuden. Aspekt einer umfassenden Baukostenplanung mit Entwicklung und Anwendung eines Simulationsmodells / 1972. 143 S. (3-598-20026-9)
- 7: Wichmann, Heinrich W.: Neue Hochschulen in Skandinavien und Finnland / 1975. 193 S. (3-598-20027-7)
- 8: Kind, Friedbert: Bausysteme im Hochschulbau / Br (3-598-20028-5)
- 9: Bayer, Werner / Ulrich Hempel: Betriebssteuerung in den Hochschulen. Auswirkungen auf Flächenbedarf und Flächenkapazität / 1976. 85 S. (3-598-20029-3)
- 10: Keller, Klaus / A. G. Tillyart: Verfahren zur Kostenbewertung. Dargestellt am Beispiel der Kostenplanung beim Ausbau der TU Berlin / 1976. (3-598-20030-7)
- 11: Grosse Universitätskliniken. Colloquium in Mannheim, am 4. und 5. November 1976. Bearb. von Ludwig Heck; Hans U. Schmidt / 1977. 137 S. (3-598-20031-5)
- 12: Nicolai, Manfred / Jochen Schaaf: Bautechnische Flächenarten. Ergebnisbericht Geisteswissenschaften und Geowissenschaften. Zusammenfassung und vergleichende Darstellung der Natur-, Ingenieur- und Geisteswissenschaften / 1977. 289 S. (3-598-20032-3)
- 13: Kind, Friedbert: Automatische Klassifikation und Bewertung von Bausystemen. Dargestellt an Beispielen aus dem Universitäts- und Hochschulbau / 1977. 313 S. (3-598-20033-1)
- 14: Hochschulplanung in Belgien und in den Niederlanden. 6. Internationales Symposion Hochschulplanung. Bearb. von Robert K. Jopp / 1977. 208 S. (3-598-20034-X)

- 15: J. Dietrich Besch / A. G. Tillyart: Verfahren zur Kostenplanung am Beispiel der Länderbauverwaltungen Baden-Württemberg, Berlin, Hessen und Nordrhein-Westfalen. Unter Mitarb. von Klaus Keller / 1977. 136 S. (3-598-20035-8)
- 16: Bedarfsermittlung und Bemessungsgrundlagen für die Planung von Kunsthochschulen. Auszug aus Arbeiten des Sonderforschungsbereichs 63 - Hochschulbau. Bearb. von Hans Groh / 1978. 236 S. (3-598-20036-6)
- 17: Nicolai, Manfred / Jochen Schaaf: Probleme der Umnutzung. Verfahren zur Bewertung d. Nutzungseignung von Gebäuden / 1979. 215 S. (3-598-20037-4)
- 18: Heck, Ludwig: Kriterienkatalog zur Erhebung des Baubestandes von Hochschulkliniken. Verabschiedet als Empfehlung d. Arbeitskreises Med. Forschungs- und Ausbildungsstätten d. Zentralarchivs für Hochschulbau / 1979. 40 S. (3-598-20038-2)
- 19: Raumkapazität von Hochschulen. Erfahrungen mit d. Nutzungsplanung in Hochschuleinrichtungen 8. Internat. Symposium Hochschulplanung vom 10. bis 12. Mai 1979 in Konstanz / 1979. 235 S. (3-598-20039-0)

Plasger, Uwe: Wörterbuch zur Musik / Dictionnaire de la Terminologie Musicale Deutsch-Französisch / Französisch-Deutsch / Allemand-Français / Français-Allemand. Unter Mitw. v. Jean J. Legrand; Hermann Rudolph / 1995. XVI, 289 S. / Gb (3-598-11242-4)

Plaul, Hainer: Bibliographie deutschsprachiger Veröffentlichungen über Unterhaltungs- und Trivialliteratur vom letzten Drittel des 18. Jahrhunderts bis zur Gegenwart / 1980. XI, 298 S. / Ln (3-598-10025-6)

Plaul, Hainer: Illustrierte Karl-May-Bibliographie. Unter Mitw. v. Gerhard Klussmeier / 1989. 436 S. / Gb (3-598-07258-9)

Plutokraten und Sozialisten / Plutocrats and Socialists. Berichte deutscher Diplomaten und Agenten über die amerikanische Arbeiterbewegung, 1878-1917 / Reports by German Diplomats and Agents on the American Labor Movement, 1878-1917. Hrsg. v. Dirk Hoerder / 1981. XXXV, 422 S. / Lin (3-598-10347-6)

Pocztar, Jerry: The Theory and Practice of Programmed Instruction. A guide for teachers (Monographs on Education VII) / 1972. 179 p., 4 p. illus. / Soft (3-7940-5123-8) [Koprod. m. UNESCO, Paris]

Poliakov, Leon / Joseph Wulf: Das Dritte Reich und die Juden. - Nachdr. d. Ausg. 1955 / 1978. X, 457 S. / Lin (3-598-04602-2)
- Das Dritte Reich und seine Denker. Dokumente. - Nachdr. d. Ausg. 1959 / 1978. XI, 560 S. / Lin (3-598-04601-4)
- Das dritte Reich und seine Diener. Enthüllungen über Auswärtiges Amt, Justiz und Wehrmacht. - Nachdr. d. 2., durchges. Aufl. 1956 / 1978. XVI, 540 S. (3-598-04600-6)

Political Manifestos in the Post War Era 1945 - 1988. Microfiche edition. Ed.: European Consortium for Political Research Manifestos Group / 1991. 150 fiches. (0-86291-800-6)

The **Political** Risk Yearbook 1992. Ed. by William D. Coplin and Michael O'Leary / 7 vols. 4th ed. 1992. Cplt. 4,000 p. / Hard (3-598-11105-3)

Politik - Recht - Gesellschaft. Interdisziplinäre Reihe. Hrsg.: Vereinigung der Politologen an d. Hochschule f. Politik München [Minerva Publikation]
- 1: Scholler, Heinrich: Die bayerische Gemeindegebietsreform als Konflikt zwischen grundrechtlich verstandener Selbstverwaltung und staatlicher Reformpolitik / 1980. VIII, 163 S. (3-597-10254-9)
- 2: Schissler, Jakob: Symbolische Sicherheitspolitik. Die Bedeutung der KSZE-Schlussakte für die Sicherheitspolitik der Bundesrepublik Deutschland / 1980. VI, 233 S. (3-597-10255-7)
- 3: Busse-Steffens, Meggy: Systemtheorie und Weltpolitik. Eine Untersuchung systemtheoretischer Ansätze im Bereich internationaler Beziehungen / 1980. VIII, 152 S. (3-597-10256-5)
- 4: Afrika im Geschichtsunterricht europäischer Länder. Von der Kolonialgeschichte zur Geschichte der Dritten Welt. Hrsg. v. Walter Fürnrohr / 1982. IV, 233 S. / Br (3-597-10257-3)
- 5: Die Verträge von Lomé zwischen Modell und Mythos. Zur Entwicklungspolitik der EG in der Dritten Welt. Hrsg. von Mir A. Ferdowsi / 1983. 160 S. (3-597-10258-1)

Politikwissenschaftliche Forschung. (fodok). Hrsg. von der Leitstelle Politische Dokumentation an der Freien Universität Berlin (ISSN 0340-9244)
- Bd. 1969. Einf. v. Nils Diederich / 1971. XVI, 474 S. / Lin (3-7940-3611-5)
- Bd. 1970. 1972. XVIII, 218, 54 S. Reg. u. 4 S. Anlage / Lin (3-7940-3612-3)
- Bd. 1971. Red. Hildegard Kellmann u.a / 1973. XV, 311, 95 S. / Lin (3-7940-3613-1)
- Bd. 1972. 1975. 395, 139 S., 2 Anl / Lin (3-7940-3614-X)
- Bd. 1973 1976. 580, 286 S. Reg / Lin (3-7940-3615-8)
- Bd. 1974. 1976. IX, 525, 391 S. Reg / Lin (3-7940-3616-6)
- Bd. 1975 / 1976. 1977. 948 S. / Lin (3-7940-3617-4)
- Bd. 1977. 1977. XVI, 1.218 S. / Lin (3-7940-3618-2)
- Bd. 1978/79: Referatedienst / Forschungsarbeiten Zsgst. u. hrsg. v. Monika Grefkes-Heinz und Walter Krumholz / 1981. 925 S. (3-598-20988-6)

Politique de Préservation du Patrimoine Archivistique / Policies for the Preservation of the Archival Heritage Actes de la vingt-cinquième conférence internationale de la table ronde des archives Gardone Riviera 1987 / Proceedings of the twenty-fifth international archival round table conference Gardone Riviera 1987. Ed.: Conseil International des Archives / International Council on Archives / 1989. 271 p. / Br (3-598-11033-2)

Politische Dokumentation poldok. Hrsg. von der Leitstelle Politische Dokumentation, in Verbindung mit dem Otto-Suhr-Institut an der Freien Universität Berlin. Red.: Monika Grefkes-Heinz; Walter Krumholz; Oskar Kruss / Jg. 1/1965 bis Jg. 24/1988 (bis Jg. 14/1978 monatlich / ab Jg. 15/1979: 6 Doppelhefte u. Reg jährlich / ab Jg. 17/1981: 3 Hefte jährlich) (ISSN 0032-3438)
- Poldok Referatedienst, deutschsprachige Zeitschr. Zweijahresreg. 1978/79. Hrsg.: Freie Univ. Berlin, Leitstelle Polit. Dokumentation;. Hrsg. u. bearb. von Monika Grefkes-Heinz u. Walter Krumholz / 1980. 704 S. in getr. Zählung (3-598-00136-3)

Politische Information. Zeitschrift der deutschen Antifaschisten in Schweden. (Stockholm 1943-1945). Nachw. v. Jan Peters. Hrsg. v. Lennart Brick / 1988. 602 S. / Ln (3-598-07265-1)

Polnische Drucke und Polonica 1501-1700 / Druki Polskie i Polonica 1501-1700. Katalog der Herzog August Bibliothek Wolfenbüttel / Katalog zbióraw Herzog August Bibliothek Wolfenbüttel / 2 Bde. in 5 Tlbdn. 1992/1994 (3-598-32809-5)
- Band 1: 1501-1600 / Tom 1: 1501-1600. Bearb. v. Malgorzata Goluszka; Marian Malicki / 2 Bde: Tl. 1 u. 2 / Czesch 1 u. 2 / 1992. Zus. XXVII, 547 S. / Ln (3-598-32810-9)
- Band 2: 1601-1700 / Tom 2: 1601-1700 / 3 Bde: Tl. 1, 2 u. 3 / Czesch 1, 2 u. 3 / 1994. Zus. XLVI, 879 S. / Ln (3-598-32811-7)

Polskie Archiwum Biograficzne / Polnisches Biographisches Archiv / Polish Biographical Archive (PAB). Eine Kumulation von Einträgen aus 246 biographischen Nachschlagewerken von ca. 1700 bis 1945. Mikrofiche-Edition. Bearb. v. Gabriele Baumgartner und Dieter Hebig / 1992-1995. 625 Fiches, Lesefaktor 24x / Silberfiche (3-598-32701-3) / Diazofiche (3-598-32700-5)
- Supplement. Bearb. v. Gabriele Baumgartner / 1997. 93 Fiches / Silberfiche (3-598-32735-8) / Diazofiche (3-598-32734-X)
- Polski Indeks Biograficzny / Polnischer Biographischer Index / Polish Biographical Index (SOFTI). Bearb. v. Gabriele Baumgartner / 4 Bde. 1998. Zus. LXXXIV, 1.457 S. / Ln (3-598-32728-5)

Polskie Archiwum Biograficzne. Seria Nowa / Polnisches Biographisches Archiv. Neue Folge / Polish Biographical Archive. Series II (PAB II). Mikrofiche-Edition / 1999 ff. Ca. 430 Fiches in 12 Lfgn, Lesefaktor 24x / Silberfiche (3-598-34481-3) / Diazofiche (3-598-34480-5)

Popkin, J. M.: Musical monuments / 1986. X, 250 S. (3-598-10628-9)

Portrayal of the Child in Children's Literature / La Représentation de l'enfant dans la littérature d'enfance et de jeunesse. Proceedings of the 6th Conference of the IRSCL / Actes du 6ème Congrès de l'IRSCL Bordeaux, 1983. Ed.: International Research Society on Children's Literature by Denise Escarpit / 1985. XII, 392 p., numerous illus. / Hard (3-598-10581-9)

La **pratique** du catalogage. Receuil d'exercices. Tome 1: Exercice. Tome 2: Corrigés. Choisis et presentés par une réunion de professeurs sous la responsabilité de Jacques Breton (Bibliothèques et organismes documentaires) / 2ᵉ éd. entièrement revue et corr / 1981. XX, 193 S.; VIII, 111 S. (3-598-20453-1)
- 3éme ed. entièrement revue et corrigée / 1983. Tome 1: XV, 193 S., Tome 2: VI, 111 S. / Br (3-598-20455-8)

Pre-school Children and Television. Two studies carried out in three countries. Bearb. von Erentraud Hömberg. Hrsg.: Stiftung Prix Jeunesse, München / 1978. 78 S. (3-7940-7048-8)

Presse-, Rundfunk- und Filmarchive - Mediendokumentation. Von d. Fachgruppe Presse-, Rundfunk- und Filmarchivare im Verein deutscher Archivare. Zsgst. v. Marianne Englert
- 1: Protokoll der Tagung der Fachgruppe Presse-, Rundfunk- und Filmarchivare im Verein Deutscher Archivare (35.) Baden-Baden / Offenburg 8.-11. Mai 1979 / 1979. 160 S. / Br (3-598-10200-3)
- 2: Protokoll der Tagung der Fachgruppe Presse-, Rundfunk- und Filmarchivare im Verein Deut-

scher Archivare in Nürnberg 5.-8. Mai 1980 / 1980. 214 S. / Br (3-598-10333-6)
- 3: Elektronische Datenverarbeitung in Medienarchiven und Fachinformationssystemen. Protokoll der 38. Tagung der Fachgruppe Presse-, Rundfunk- und Filmarchivare im Verein deutscher Archivare in Kassel 4.-7. Mai 1981 / 1981. 192 S. / Br (3-598-10358-1)
- 4: Medienarchive in Gegenwart und Zukunft. Protokolle der 39. und 40. Tagung der Fachgruppe Presse-, Rundfunk- und Filmarchivare im Verein deutscher Archivare. Heilbronn, Oktober 1981, Köln, Mai 1982 / 1983. 279 S. / Br (3-598-10480-4)
- 5: Entwicklungsperspektiven zukünftiger Informationssysteme. Protokoll der 41. und 42. Tagung der Fachgruppe Presse-, Rundfunk und Filmarchivare im Verein deutscher Archivare. Münster Oktober 1982; München, Mai 1983 / 1983. 240 S. / Br (3-598-20335-7)
- 6: Dokumentation in Presse und Rundfunk. Fünfundzwanzig Jahre Fachgruppe Presse-, Rundfunk- und Filmarchivare im Verein deutscher Archivare. Protokoll der 43. und 44. Tagung der Fachgruppe, Saarbrücken, Oktober 1983 und Frankfurt am Main, Mai 1984. Zsgst. v. Marianne Englert; Gerhard Mantwill / 1985. 251 S. / Br (3-598-20336-5)
- 7: Informations- und Dokumentationsleistungen in Presse und Rundfunk. Protokoll der 45. Tagung der Fachgruppe Presse-, Rundfunk- und Filmarchivare im Verein deutscher Archivare, Oldenburg / Bremen vom 7. bis 9. Mai 1985 Zsgst. v. Marianne Englert; Gerhard Mantwill / 1986. 186 S. / Br (3-598-20337-3)
- 8: Presse- und Rundfunkarchive: Partner von Redaktion, Produktion und Programm. Protokoll der 47. Tagung der Fachgruppe der Archivare an Presse-, Rundfunk- und Filmarchiven im Verein deutscher Archivare vom 12. bis 15. Mai 1986 in Mainz Bearb. v. Marianne Englert; Eckhard Lange / 1987. 207 S. / Br (3-598-20338-1)
- 9: Die neuen Entwicklungen im EDV-Bereich und ihre Bedeutung für die Mediendokumentation. Protokoll der 49. Tagung der Fachgruppe der Archivare an Presse-, Rundfunk-, und Filmarchiven im Verein deutscher Archivare in Zusammenarbeit mit der Schweizerischen Vereinigung für Dokumentation (SVD) und der Fachgruppe Mediendokumentation der SVD. Bearb. v. Marianne Englert / 1988. 248 S. / Br (3-598-20339-X)
- 10: Das Medienarchiv als Dienstleistungsbereich und betriebswirtschaftlicher Faktor. Protokoll der 50. und 51. Tagung der Fachgruppe der Archivare an Presse-, Rundfunk- und Filmarchiven im Verein deutscher Archivare Frankfurt am Main, Oktober 1987 und Stuttgart, Mai 1988. Bearb. v. Marianne Englert; Eckhard Lange / 1989. 257 S. / Br (3-598-20340-3)
- 11: Die Informationsvermittler und die Informationsgesellschaft. Protokoll der 52. Tagung der Fachgruppe der Archivare an Presse-, Rundfunk- und Filmarchiven im Verein deutscher Archivare vom 24. bis 27. April 1989 in Hamburg / 1990. 182 S. / Br (3-598-20341-1)

Pressearchiv zur Geschichte Deutschlands sowie zur internationalen Politik von 1949-60. Hrsg.: Fachbereich Politische Wissenschaft d. Freien Universität Berlin - Pressearchiv d. Bibliotheks- u. Informationssystems / Mikrofiche-Edition 1995. 203.561 S. auf 4.681 Fiches, Lesefaktor 24x / Incl. Findbuch / Diazofiche (3-598-33850-3)
- Findbuch zur Mikrofiche-Edition. Bearb. v. Erdmute Horn-Sauder; Michael Peschke / 1996. XVI, 252 S. / Br (3-598-33853-8)

Pressekonzentration. Eine kritische Materialsichtung und -systematisierung. Hrsg. von Jörg Aufermann; Peter Heilmann; Hubertus Hüppauf; C. Wolfgang Müller; Ulrich Neveling; Gernot Wersig / 1970. 389 S. / Br (3-7940-3289-6) / Lin (3-7940-3239-X)
- unveränd. Ausg. 1971. 389 S. / Kst. (UTB 69) (3-7940-2605-5)

Pretzell, Klaus A.: Topographie asienkundlicher Schrifttumssammlungen in der BRD und Berlin (West) „TOPAS". Hrsg.: Institut für Asienkunde im Verbund der Stiftung Deutsches Übersee-Institut Dokumentations-Leitstelle Asien / 1979. X, 358 S. (5-598-10036-1)

Printed for children. [This catalogue was ed. by the Ausstellungs- u. Messe-GmbH d. Börsenvereins d. Dt. Buchhandels, on the occasion of the focus on „The child and the book" at the Frankfurt Book Fair 1978 and of the International year of the child 1979. Comp. by Forew. by Amadou-Mahtar M'Bow. World Children's Book Exhibition] / 1978. 448 S. / Br

Prinz-Albert-Studien / Prince Albert Studies. Hrsg.: Prinz Albert Gesellschaft.
- 1: Viktorianisches England in deutscher Perspektive. Hrsg. v. Adolf M. Birke; Kurt Kluxen / 1983. 154 S., 16 Abb. / Ln (3-598-21401-4)
- 2: Kirche, Staat und Gesellschaft im 19. Jahrhundert / Church, State, and Society in the 19th Century. Ein deutsch-englischer Vergleich / An Anglo-German Comparison. Einl. v. Adolf M. Birke. Hrsg. v. Adolf M. Birke; Kurt Kluxen / 1984. 164 S. / Ln (3-598-21402-2)
- 3: Deutscher und Britischer Parlamentarismus / British and German Parliamentarism. Hrsg. v. Adolf M. Birke; Kurt Kluxen / 1985. 192 S. / Ln (3-598-21403-0)
- 4: England und Hannover / England and Hanover. Hrsg. v. Adolf M. Birke; Kurt Kluxen / 1986. 194 S. / Ln (3-598-21404-9)
- 5: Die europäische Herausforderung / The European Challenge England und Deutschland in Europa / Britain and Germany in Europe. Hrsg. v. Adolf M. Birke; Kurt Kluxen, unter Mitarb. v. Manfred Hanisch / 1987. 160 S. / Ln (3-598-21405-7)
- 6: Wettlauf in die Moderne. England und Deutschland seit der industriellen Revolution / The Race for Modernisation Britain and Germany since the Industrial Revolution. By Adolf M. Birke; Lothar Kettenacker, unter Mitarb. v. Manfred Hanisch / 1988. 151 S. / Ln (3-598-21406-5)
- 7: Bürgertum, Adel und Monarchie / Middle Classes, Aristocracy and Monarchy. Wandel der Lebensformen im Zeitalter des bürgerlichen Nationalismus / Patterns of Change and Adaptation in the Age of Modern Nationalism. Hrsg. v. Adolf M. Birke; Lothar Kettenacker, unter Mitarb. v. Helmut Reifeld / 1989. 159 S. / Ln (3-598-21407-3)
- 8: Das gestörte Gleichgewicht / Upsetting the balance. Deutschland als Problem britischer Sicherheit im neunzehnten und zwanzigsten Jahrhundert / German and British Security Interests in the nineteenth andtwentieth Century. Hrsg. v. Adolf M. Birke; Marie L. Recker / 1990. 189 S. / Ln (3-598-21408-1)
- 9: Grossbritannien und Ostdeutschland seit 1918 / Britain and East Germany since 1918. Hrsg. v. Adolf M. Birke; Günther Heydemann, unter Mitarb. v. Hermann Wentker / 1992 151 S. / Ln (3-598-21409-X)
- 10: Föderalismus im deutsch-britischen Meinungsstreit. Historische Dimensionund politische Aktualität / The Federalism Debate in Britain and Germany. A Historical and Political Controversy. Hrsg. v. Adolf M. Birke; Hermann Wentker / 1993. 177 S. / Ln (3-598-21410-3)
- 11: Deutschland und Russland in der britischen Kontinentalpolitik seit 1815 / Germany and Russia in British Policy towards Europe since 1815. Hrsg. v. Adolf M. Birke; Hermann Wentker / 1994. 217 S. / Ln (3-598-21411-1)
- 12: Politikverdrossenheit. Der Parteienstaat in der historischen und gegenwärtigen Diskussion / Disillusioned with Politics. Party Government in the Past and Present Discussion Ein deutsch-britischer Vergleich / An Anglo-German Comparison. Hrsg. v. Adolf M. Birke; Magnus Brechtken / 1995. 136 S. / Ln (3-598-21412-X)
- 13: Kommunale Selbstverwaltung / Local Self-Government. Geschichte und Gegenwart im deutsch-britischen Vergleich. Hrsg. v. Adolf M. Birke; Magnus Brechtken / 1996. 164 p. / Ln (3-598-21413-8)
- 14: Das Kreditwesen in der Neuzeit / Banking Systems in Modern History. Ein deutsch-britischer Vergleich. Hrsg. v. Franz Bosbach; Hans Pohl, in Zus.-Arb. mit Gabriele Jachmich; Christoph Kampmann / 1997. 183 S. / Ln (3-598-21414-6)
- 15: Künstlerische Beziehungen zwischen England und Deutschland in der viktorianischen Epoche / Art in Britain and Germany in the Age of Queen Victoria and Prince Albert. Hrsg. v. Franz Bosbach; Frank Büttner, in Zus.-Arb. mit Michaela Braesel; Christoph Kampmann / 1998. 230 S., 76 Abb. / Ln (3-598-21415-4)

Prismata. Dank an Bernhard Hanssler. Hrsg. von Dieter Grimm / 1974. 512 S. (3-7940-3030-3)

Protest, Direct Action, Repression. Dissent in American Society from Colonial Times to the Present. A Bibliography. Including the call numbers of books and journals available from the Library of the John-F.-Kennedy-Institute at the Free University Berlin to facilitate (international) interlibrary loan procedures. Zsgst. u. Hrsg. v. Dirk Hoerder / 1977. XXVI, 434 S. / Lin (3-7940-7009-7)

Protokoll der „Brüsseler Konferenz" der KPD 1935. Reden, Diskussionen und Beschlüsse, Moskau von 3.-15. Oktober 1935. Hrsg. v. Erwin Lewin; Elke Reuter; Stefan Weber, unter Mitarb. v. Marlies Coburger; Günther Fuchs; Marianne Jentsch; Rosemarie Lewin / 2 Bde. 1997. Zus. 897 S. / Gb (3-598-11228-9)

Protokolle des Vermittlungsausschusses des Deutschen Bundestages und des Bundesrates 1.-11. Wahlperiode 1949-1990. Loseblattausgabe. Vorw. v. Gebhard Zilk / Mikrofiche-Edition / 1986-1991. 32 Fiches, 306 S. Reg / Silberfiche (3-598-30245-2) / Diazofiche (3-598-30244-4)
siehe auch: ➤ *Verhandlungen des deutschen Bundestages und des Bundesrates*

Proverbien-Kodex. Faksimileausgabe des Codex Ms. or. oct. 987 der Deutschen Staatsbibliothek zu Berlin. M. e. Nachwort v. A. Böhlig. - Faksimilereprod. im Lichtdruck / 1969. 166 S. / Ganzleder i. Kass. (3-7940-5198-X)

Public Library Purpose. A Reader. Hrsg. von Barry Totterdell / 1978. 160 S. (0-85157-244-8) [Clive Bingley]

Publikation der Vereinigung linksgerichteter Verleger (1925-1926). Hrsg. v. Wolfgang U. Schuette / 1989. 298 S. / Br (3-598-07260-0)

Publisher's Practical Dictionary in 20 Languages.
- 1. Ausg. u.d.T.: Wörterbuch des Verlagswesens in zwanzig Sprachen / The Publisher's Practical Dictionary in 20 Languages. Hrsg v. Imre Móra. Mitarb.: Sándor Skripecz / 1974. 389 S. / Lin (3-7940-4112-7)
- 2. Aufl. 1977. 389 S. (3-7940-4112-7) *(963-05-0996-2)*
- 3rd rev. ed. Publisher's Practical Dictionary in 20 Languages / Dictionnaire pratique de l'Edition en 20 Langues / Wörterbuch des Verlagswesens in 20 Sprachen. Hrsg. v. Móra, Imre / 1983. XI, 418 p. / Hard (3-598-10449-9)

Publishers' *International Directory with ISBN Index* → ***Handbuch*** *der Internationalen Dokumentation und Information Bd. 7*

Publishers' *International ISBN-Directory (PIID)* → ***Handbuch*** *der Internationalen Dokumentation und Information Bd. 7*

Publishers' International ISBN-Directory PLUS (PIID PLUS). CD-ROM-Edition. Machine-readable database. Ed.: International ISBN Agency, Berlin (Annual Update) (ISSN 4213-1716)
- 1996 (0-8352-3883-0)
- 2nd ed. 1997 (0-8352-4001-0)
- 3rd ed. 1998 (3-598-40402-6)
Buch-Ausgabe: → ***Handbuch*** *der Internationalen Dokumentation und Information Bd. 7*

Publizistik-historische Beiträge. Hrsg. von Heinz-Dietrich Fischer
- Bd. 1: Deutsche Publizisten des 15. bis 20. Jahrhunderts / Hrsg. v. Heinz-Dietrich Fischer / 1971. 419 S. / Lin (3-7940-3601-8)
- Bd. 2: Deutsche Zeitungen des 17. bis 20. Jahrhunderts. Hrsg.: Heinz-Dietrich Fischer / 1972. 415 S. / Lin (3-7940-3602)
- Bd. 3: Deutsche Zeitschriften des 17. bis 20. Jahrhunderts. Hrsg. v. Heinz-Dietrich Fischer / 1973. 445 S. / Lin (3-7940-3603-4)
- Bd. 4: Deutsche Presseverleger des 18. bis 20. Jahrhunderts / Hrsg. v. Heinz-Dietrich Fischer / 1975. 375 S. / Lin (3-7940-3604-2)
- Bd. 5: Deutsche Kommunikationskontrolle des 15. bis 20. Jahrhunderts. Hrsg. v. Heinz-Dietrich Fischer / 1982. 359 S. / Lin (3-598-21345-X)
- Bd. 6: Deutsche Staatspropaganda. Hrsg. von Fischer, Heinz-Dietrich / 1978. 400 S. / Lin (3-7940-3606-9)

Publizistikwissenschaftlicher Referatedienst (prd). Hrsg.: Institut für Publizistik und Dokumentationswissenschaft der Freien Universität Berlin. Redakt.: Dagmar Yü-Dembski [Früher im Westdeutschen Vlg., Wiesbaden. Nach Bd. 19 wurde das Erscheinen eingestellt]
- Bd. 9. (1974) / 1976. 184 S. / Br (3-7940-4054-6)
- Bd. 10. (1975) / 1977. 278 S. / Br (3-7940-4055-4)
- Bd. 11. (1976) / 1978. 328 S. / Br (3-7940-4056-2)
- Bd. 12. (1977) / 1979. 384 S. / Br
- Bd. 13. (1978) / 1980. 373 S. / Br (3-598-20473-6)
- Bd. 14. (1979) / 1982. XVI, 422 S. / Br (3-598-20474-4)
- Bd. 15. (1980) / 1983. XVI, 500 S. / Br (3-598-20475-2)
- Bd. 16. (1981) / 1984. XVIII, 572 S. / Br (3-598-20476-0)

- Bd. 17. (1982) / 1985. XVIII, 604 S. / Br (3-598-20477-9)
- Bd. 18. (1983) / 1988. XVIII, 607 S. / Br (3-598-20478-7)
- Bd. 19. (1984) / 1988. XVII, 636 S. / Br (3-598-20479-5)

The **Pulitzer** Prize Archive. A History and Anthology of Award-winning Materials in Journalism, Letters and Arts. Ed. by Heinz Dietrich Fischer; Erika J. Fischer / 16 vols. 1987 ff. (3-598-30170-7)
- Part A. Reportage Journalism / 3 vols.
  - Vol. 1. International Reporting 1928-1985. From the Activities of the League of Nations to present-day Global Problems. Ed. by Heinz D. Fischer; Erika J. Fischer / 1987. LXXXVI, 352 p. / Hard (3-598-30171-5)
  - Vol. 2. National Reporting 1941-1986. From Labor Conflicts to the Challenger Disaster / 1988. 388, LXII p. / Hard (3-598-30172-3)
  - Vol. 3. Local Reporting 1947-1987. From a County Vote Fraud to a Corrupt City Council / 1989. LIV, 388 p. / Hard (3-598-30173-1)
- Part B. Opinion Journalism / 3 vols.
  - Vol. 4. Political Editorial 1916-1988. From War-related Conflicts to Metropolitan Disputes / 1990. LXXIV, 376 p. / Hard (3-598-30174-X)
  - Vol. 5. Social Commentary 1969-1989. From University Troubles to a California Earthquake / 1991. XLVI, 398 p. / Hard (3-598-30175-8)
  - Vol. 6. Cultural Criticism 1969-1990. From Architectural Damages to Press Imperfections / 1992. LII, 420 p. / Hard (3-598-30176-6)
- Part C. Nonfiction Literature / 3 vols.
  - Vol. 7. American History Awards 1917-91. From Colonial Settlement to the Civil Rights Movements / 1993. LXVIII, 366 p. (3-598-30177-4)
  - Vol. 8. Biography / Autobiography Awards 1917-92 / 1993. LXX, 406 p. (3-598-30178-2)
  - Vol. 9. General Nonfiction Award 1962-93 / 1995. LII, 362 p. / Hard (3-598-30179-0)
- Part D. Belles-Lettres / 3 vols.
  - Vol. 10. Novel / Fiction Awards 1917-94 / 1996. LXXXII, 305 p. / Hard (3-598-30180-4)
  - Vol. 11. Poetry / Verse Awards 1918-1995 / 1997. XXII, 302 p. / Hard (3-598-30181-2)
  - Vol. 12. Drama / Comedy Awards / 1998. LXXXIV, 366 p. (3-598-30182-0)
  Part E [i.Vb.]

Qualifizierung und wissenschaftlich-technischer Fortschritt am Beispiel der Sekundarschulreform in ausgewählten Industriestaaten [Minerva Publikation]
- Bd. 1: Blumenthal, Viktor von: Bildungsplanung und Sekundarschulreform in Italien. - Bodo Willmann: Qualifikationsbedarf und Sekundarschulreform in Schweden. Mit einer grundlegenden Einführung von Leonhard Froese / 1975. 248 S. (3-597-42549-6)
- Bd. 2: Hein, R.: Qualifikationsplanung und Sekundarschulreform in Frankreich. - Stübig, H.: Bildungspolitik in England / 1975. 224 S. (3-597-00010-X)
- Bd. 3: Bode,Herbert F.: Qualifikationsstruktur und Sekundarschulreform in den USA. - Zänker, U.: Qualifikationsanforderungen und Sekundrschulreform in der UdSSR / 1975. 288 S. (3-597-00011-8)

Quellen zur Geschichte der Juden in den Archiven der neuen Bundesländer. Hrsg.: Leo-Baeck-Institut, Wissenschaftliche Arbeitsgemeinschaft Berlin; Historische Kommission zu Berlin; Stefi Jersch-Wenzel; Reinhard Rürup.

- 1: Eine Bestandsübersicht. Bearb. v. Andreas Reinke; Barbara Strenge, unter Mitarb. v. Bernd Braun; Nathan Sznaider / 1995. XXVI, 602 S. / Gb (3-598-22441-9)
- 2: Geheimes Staatsarchiv Preussischer Kulturbesitz. Teil 1: Ältere Zentralbehörden bis 1808 / 10 und Brandenburg-Preussisches Hausarchiv. Bearb. v. Meta Kohnke. Redakt.: Bernd Braun; Manfred Jehle; Andreas Reinke / 1999. XVII, 777 S. / Gb (3-598-22442-7)

Rabattabelle für die Mehrwertsteuer.
- 5,5% / 1968. 63 S. / Kt
- 11% / 1968. 63 S. / Kt

Rahmen deutscher Buchtitel im 16. Jahrhundert. Hrsg. von Julius von Pflugk-Harttung. Mit ergänzendem Kommentar von Helmuth Claus. - Nachdr. d. Ausg. Stuttgart 1906 / 1980. XXXVII, 32 S., 102 Tafeln / Ln (3-598-10024-8)

Rajec, Elizabeth M.: The study of names in literature. A bibliography / 1978. XII, 261 S. (3-7940-8002-5)
- Supplement. 1981. IX, 298 p. / Hard (3-598-10408-1)

RAK-Anwendung in der Deutschen Bibliothek. Präzisierungen und Erläuterungen zu den einzelnen Paragraphen. Hrsg.: Deutsche Bibliothek Frankfurt am Main / 2 Lieferungen. 1979. 308 S.; 288 S. Loseblattausg. (3-598-03024-X)
- 2. neubearb. Aufl. 1982. XVI, 571 S. (3-598-10433-2)

Rassler, Gerda: Pariser Tageblatt - Pariser Tageszeitung 1933-1940. Eine Auswahlbibliographie / 1989. 433 S. / Ln (3-598-07596-0)

Rassmann, Friedrich: Übersicht der aus der Bibel geschöpften Dichtungen älterer und neuerer deutscher Dichter mit Einschluß derartiger Übersetzungen. Ein Wegweiser für Literatoren, Freunde der Dichtkunst, Geistliche und Schullehrer. - Nachdruck der Ausg. Essen 1829 / 1971. 103 S. (3-7940-5029-0)

Raum, Objekt und Sicherheit im Museum. Bericht über ein internationales Symposium, veranstaltet von ICOM-Nationalkomitee der Bundesrepublik Deutschland, Österreichs und der Schweiz vom 9. bis 15. Mai 1976 am Bodensee. Hrsg. v. Hermann Auer / Deutsches Nationalkomitee des Internationalen Museumrates ICOM / 1978. 208 S., Abb. / Br (3-598-07074-8)

Rautenberg, Klaus U. / Oldrich Sova: Dokumentation computergestützter Informationssysteme. Ein praxisorientierter Leitfaden für die Gestaltung, Erstellung und Wartung einer DV-Dokumentation / 1983. 232 S., 98 Abb. / Lin (3-598-10438-3)

Die **rechtliche** Regelung der internationalen Energiebeziehungen der RWG-Länder. Studie u. Dokumentation. Hrsg.: Inst. für Völkerrecht d. Univ. Göttingen. Unter Mitarb. von Dieter Mentz u. Gottfried Pfeffer. Hrsg. von Gottfried Zieger / 1982. XXXVI, 765 S. (3-598-10441-3)

Reden für die Deutsche Nation 1848 / 1849. Stenographischer Bericht über die Verhandlungen der Deutschen Constituirenden Nationalversammlung zu Frankfurt am Main. Einf. v. Christoph Stoll. Hrsg. auf Beschluss d. Nationalversammlung durch d. Redactions-Commission im Auftr. v. Franz Wigard / 9 Bde. 1988. Zus. 7.146 S. / Gb (3-598-33040-5)

Redfern, Brian: Organising Music in Libraries / 2 vols. [Clive Bingley]
- Vol. 1: Erw. Aufl. 1978. (0-85157-231-6)
- Vol. 2: 2. Aufl. 1979. (0-85157-261-8)

Regeln für den Schlagwortkatalog. „Erlanger Regelwerk". Im Auftr. d. Universitätsbibliotheken Erlangen-Nürnberg bearb. v. Agnes Stählin, unter Mitw. v. Roswitha Poll / 4. neu bearb. Aufl. 1977. 128 S. / Kst. (UTB 686) (3-7940-2660-8)

Die „**Regeln** für die alphabetische Katalogisierung (RAK)" im nationalen und internationalen Rahmen. Vorträge einer Fortbildungsveranstaltung des Bibliothekar-Lehrinstituts des Landes Nordrhein-Westfalen am 27.-29. November 1978. Bearb. v. Rudolf Jung / 1980. 190 S. / Br (3-598-10122-8)

Rehbinder, Manfred: Entwicklung und gegenwärtiger Stand der Rechtstatsachenforschung in den USA. Ein bibliographischer Bericht / 1970. 47 S.

Rehm, Margarete: Lexikon Buch - Bibliothek - Neue Medien / 1991. VII, 294 S. / Gb (3-598-10889-3) / Br (3-598-10851-6)

Reichshandbuch der Deutschen Gesellschaft. Das Handbuch der Persönlichkeiten in Wort und Bild. Herausgeber: Deutscher Wirtschaftsverlag. Berlin 1931. Eingeleitet von Ferdinand Tönnies. Mikrofiche-Edition / 1995. 2.144 S. auf 36 Fiches, Lesefaktor 24x / Silberfiche (3-598-30664-4)

Der **Reichstagsbrand**. Die Provokation des 20. Jahrhunderts. Forschungsergebnis. Hrsg.: Internationales Komitee zur Erforschung der Ursachen und Folgen des Zweiten Weltkrieges / 1978. 444 S. / Br (3-598-10003-X)

Der **Reichstagsbrand**. Eine wissenschaftliche Dokumentation (Veröffentlichungen des Internationalen Komitees zur Wissenschaftlichen Erforschung der Ursachen und Folgen des Zweiten Weltkrieges) [Vorher im arani - Vlg, Berlin; ab 1978 in Koprod. mit d. arani -Vlg.]
- Bd. 1. Hrsg. v. Walther Hofer; Edouard Calic; Karl Stephan und Friedrich Zipfel / 1972. 293 S. / Br (3-598-04605-7)
- Bd. 2. Hrsg. v. Walther Hofer; Edouard Calic; Christoph Graf und Friedrich Zipfel. Mit Sachverständigen-Äußerungen von Karl Stephan und Heinz Leferenz / 1978. 487 S. / Lin (3-598-04604-9)

Reimers, J.: Handbuch für die Denkmalpflege. Hrsg.: Provinzial-Kommission zur Erforschung und Erhaltung der Denkmäler in der Provinz Hannover. (Nachdr. d. 2. umgearb. u. verm. Aufl. Hannover 1911) / 1986. V, 498 S. / Ln (3-598-07215-5)

Reinitzer, Sigrid / Marcus Gossler: Nachschlagetechniken in der Wissenschaft. Eine praktische Anleitung zur Benutzung von Index- und Abstractswerken und deren Struktur / 1987. 230 S. / Br (3-598-10751-X)

Répertoire automatisé des livres du XVIe siècle conservés à la bibliothèque municipale de Rouen I (1501-1550). Ed.: Centre National de la Recherche Scientifique / Institut de Recherche et d'Histoire des Textes. Etabli à partir du catalogage de Paul Subert; Edith Bayle; Marie J Beaud; Agnes Guillaumont,

Mitchiko Iagolnitzer; Jean F. Maillard / 1983. XVII, 516 S. / Gb (3-598-10479-0)

Répertoire des Bibliothèques en France. Ed. par Helga Lengenfelder / 1984. 150 S. / Gb (3-598-10524-X)

Répertoire International des Médiévistes / International Directory of Medievalists. Ed.: Centre National de la Recherche Scientifique. Institut de Recherche et d'Histoire des Textes / 2 vols. 5. ed. rev. et enl. 1979. XI, 811 S. (3-598-07023-3)
- 6. ed. rev. et enl / 2 vols. 1987. Cplt. IX, 1.259 S. / Gb (3-598-10683-1)

Research Libraries and Collections in the United Kingdom. A Selective Inventory and Guide. Hrsg. von Stephan Roberts; Alan Cooper; Lesley Gilder / 1978. 285 S. (0-85157-258-8) [Clive Bingley]

Research Priorities in African Literatures. Ed. by Bernth Lindfors / 1984. 222 p. / Hard (3-598-10570-3) [Hans Zell Publishers]

Responses to children's literature. Proceedings of the 4. symposium of the International Research Society for Children's Literature, held at the Univ.of Exeter, Sept 2-12, 1978. Ed. by Geoff Fox and Graham Hammond with Stuart Amor / 1980. IX, 141 S. / Hard (3-598-40025-X)
- 2nd ed. 1983. VII, 141 S. / Hard (3-598-10528-2)

Restaurator. International Journal for the Preservation of Library and Archival Material. Ed. by Helmut Bansa / since Vol. 18/1997 (4 issues annually) (ISSN 0034-5806) [Seit 1997. Vorher im Vlg. Munksgaard, Kopenhagen]

Richard A. Bermann alias Arnold Höllriegel. Österreicher - Demokrat - Weltbürger. Hrsg.: Deutsche Bibliothek / 1995. XII, 431 S. / Gb (3-598-11298-X) / Br (3-598-11297-1)

Richter, Brigitte: Precis de Bibliothéconomie. Avec la collaboration par Marcelle Ménil; Noë Richter / 3ème éd. corr. et augmentée 1980. XI, 233 S. (3-598-10419-7)
- 4ème revue et mise à jour 1987. IX, 298 p. / Br (3-598-10560-6)
- 5ème ed. revue et mise à jour 1992. IX, 298 S. / Gb (3-598-11077-4)

Richter, Gottfried Lebrecht: Allgemeines biographisches Lexikon alter und neuer geistlicher Liederdichter. -Nachdr. der Ausg. Leipzig, Martini 1804 / 1971. VIII, 487 S. (3-7940-5014-2)

Ritter, Paul: Frans Masereel. Eine annotierte Bibliographie des druckgraphischen Werkes / 1992. XXIV, 743 S., ca. 580 Abb., davon 3 farb / Gb (3-598-11103-7)

Roeber, Georg / Gerhard Jacoby: Handbuch der filmwirtschaftlichen Medienbereiche / 1973. 984 S., 70 Tab. u. Ktn. / Lin (3-7940-3272-1)

Rönsch, Hermann: Itala und Vulgata. Das Sprachidom der urchristlichen Itala und der katholischen Vulgata. Vorwort von Hans W. Klein. - Nachdruck der 2., berichtigten und vermehrten Ausgabe 1965 / 1978. XVI, 526 S. (3-597-43067-X) [Minerva Publikation]

Rösch-Sondermann, Hermann: Bibliographie der lokalen Alternativpresse. Vom Volksblatt zum Stadtmagazin / 1988. 156 S. / Lin (3-598-10816-8)

Roller, Franz A.: Systematisches Lehrbuch der bildenden Tanzkunst und körperlichen Ausbildung. Von der Geburt an bis zum vollendeten Wachstume des Menschen. Weimar 1843. Nachw. u. Reg. v. Marion Kant / 1989. XVI, 321, XXIX S., 15 Taf., 4 Tab. / Gb (3-598-07251-1)

Roloff, Heinrich: Lehrbuch der Sachkatalogisierung / 3., überarb. und erw. Aufl. 1968. 204 S. / Ln (3-7940-4148-8)
- Unveränd. Nachdr. d. 3. Aufl. 1968 / 1973. XX, 196 S. / Br (3-7940-4148-8)
- 4., überarb. Aufl. 1976. 236 S. / Lin (3-7940-4148-8)

Roloff, Heinrich: Sachkatalogisierung auf neuen Wegen. Von der Facettenklassifikation zu Descriptor und Thesaurus / 3., erw.und überarb. Aufl. 1972. 48 S. (3-7940-4199-2)

Rost, Maritta: Bibliographie Arnold Zweig. Unter Mitarb. v. Jörg Armer; Rosemarie Geist; Ilse Lange / 2 Bde. 1987. Bd. 1: 512 S., Bd. 2: 668 S. / Ln (3-598-07249-X)

Roth, Karl H.: Intelligenz und Sozialpolitik im „Dritten Reich". Eine methodisch-historische Studie am Beispiel des Arbeitswissenschaftlichen Instituts der Deutschen Arbeitsfront / 1993. 394 S., 3 Abb. / Ln (3-598-11166-5)

Rovelstad, Mathilde: Bibliotheken in den Vereinigten Staaten / 1974. 92 S. (3-7940-3050)

Rowley, Jennifer E.: Mechanised in-house information systems / 1979. 208 S. (0-85157-259 6) [Clive Bingley]

Der **Ruf**. Unabhängige Blätter der jungen Generation. Hrsg. v. Alfred Andersch; Hans Werner Richter; Erich Kuby; Walter von Cube - Reprint der Jahrgänge 1-3, 1946-1948 / 1987. 332 S. / Ln (3-598-07524-3)

Der **Ruf**. Zeitung der deutschen Kriegsgefangenen in USA. Faksimile-Ausgabe der Nummern 1 (1. März 1945) bis 26 (1. April 1946), Sonderbeilagen zu den Nummern 6 und 26 sowie der „Auslese" / 1986. 232 S., zahlr. Abb. / Ln (3-598-10663-7)

Rund um die Jugendliteratur. Auskunft aus d. Praxis für d. Praxis. Hrsg. v. Helge Adler / 2., neubearb. Aufl. 1974. 249 S. / Lin (3-7940-3248-9)

Rundfunk Bibliographien 1926-1942. Mikrofiche-Edition. Einf. v. Ansgar Diller / 1984. 25 Diazofiches, 568 S. Registerbd. Lesefaktor 24x (3-598-10545-2)

Rundfunkstudien.
- 1: Schiller-Lerg, Sabine: Walter Benjamin und der Rundfunk. Programmarbeit zwischen Theorie und Praxis / 1984. XVIII, 548 S. / Br (3-598-21571-1)
- 2: Kutsch, Arnulf: Rundfunkwissenschaft im Dritten Reich. Geschichte des Instituts für Rundfunkwissenschaft der Universität Freiburg / 1985. 600 S., 10 Tab. u. Abb. / Br (3-598-21572-X)
- 3: Pütter, Conrad: Rundfunk gegen das Dritte Reich. Deutschsprachige Rundfunkaktivitäten im Exil 1933-1945. Ein Handbuch. Unter Mitw. v.

Ernst Loewy. Ein Beitr. v. Elke Hilscher. Erarb. im Auftr. d. Deutschen Rundfunkarchives / 1986. 388 S. / Br (3-598-10470-7)
- 4: Feldinger, Norbert P.: Nachkriegsrundfunk in Österreich. Zwischen Föderalismus und Zentralismus von 1945 bis 1957 / 1990. 206 S. / Br (3-598-21573-8)
- 5: Soppe, August: Rundfunk in Frankfurt am Main 1923-1926. Zur Organisations-, Programm- und Rezeptionsgeschichte eines neuen Mediums. Mit einem Nachw. Hrsg. v. Jörg J. Berns / 1993. 500 S. / Br (3-598-21574-6)
- 6: Bauer, Thomas: Deutsche Programmpresse 1923 bis 1941. Entstehung, Entwicklung und Kontinuität der Rundfunkzeitschriften. Hrsg. v. Winfried B. Lerg; Ansgar Diller; Walter Klingler / 1993. 484 S. / Gb (3-598-21575-4)
- 7: Capellan, Frank: Für Deutschland und Europa: Der Deutschlandfunk Rundfunkanstalt mit besonderem Auftrag 1961-1989 / 1993 492 S. / Gb (3-598-21576-2)

Rupesinghe, Kumar / Berth Verstappen: Ethnic Conflict and Human Rights in Sri Lanka. An annotated Bibliography / 1988. 450 p. (3-598-10794-3)

Ruppert, Fritz: Initials. Abkürzungen von Namen internationaler Organisationen / 1966. 220 S. / (3-7940-4026-0) [Koproduktion mit Vulkan-Verl. Essen]

Rusch, Gerhard: Einführung in die Titelaufnahme [Lizenzausg. m. Genehm. v. VEB Vlg. für Buch u. Bibliothekswesen, Leipzig] / 2 Bde. in 1 Bd. 1967. 424, XVI S. u. ein Lösungsheft m. 70 S. / Ln
- 2., unveränd. Aufl. 1969 (Vlg.-Nr. 04118)
- 3., unveränd. Aufl. 1971 (3-7940-4118-6)
- 4., unveränd. Aufl. 1973 (3-7940-4118-6)
- 5., unveränd. Aufl. Einführung in die Titelaufnahme. Nach den „Regeln für die alphabetische Katalogisierung in wissenschaftlichen Bibliotheken" / 1974 (3-7940-4118-6)
- 6. unveränd. Aufl. Einführung in die Titelaufnahme. Nach den „Regeln für die alphabetische Katalogisierung in Wissenschaftlichen Bibliotheken" („Preussische Instruktionen"). Mit zahlreichen Beispielen und Aufgaben / 1975 (3-598-10413-8)

Russian National Bibliography PLUS. CD-ROM-Edition. Ed. in cooperation with Mir-Dialogue and the Russian Book Chamber (3-598-40339-9)
- 3rd ed. 1997 (3-598-40366-6)
- 4th ed. 1998 (3-598-40403-4)

Russisches Biographisches Archiv / Russian Biographical Archive (RBA). Mikrofiche-Edition. Bearb. v. Axel Frey / 1997 ff. Ca. 500 Fiches in 12 Lfgn, Lesefaktor 24x / Silberfiche (3-598-34061-3) / Diazofiche (3-598-34060-5)

Russisches Staatsarchiv für Literatur und Kunst - Vollständiger Archivführer / Russian State Archive of Literature and Art - The complete Archive Guide / Rossijskij Gosudarstvennyj Archiv literatury i iskusstva - Putevoditel' po archivu. CD-ROM-Edition. Hrsg. v. Klaus W. Waschik, Lotman-Institut f. russische u. sowjetische Kultur (Deutschland); Natalia B. Volkova, Russisches Staatsarchiv f. Literatur und Kunst (Russland) / 1996 (3-598-40341-0)

Sachkatalog der Deutschen Bücherei Leipzig 1913 - (Mai) 1945. Monographien - Zeitschriften - Karten. Mikrofiche-Edition / 1994. ca. 1, 4 Mio. Katalogktn.

auf 603 Fiches. incl. Indexband. Lesefaktor 24x Silberfiche (3-598-33679-9) / Diazofiche (3-598-33678-0)
- Supplement Karten Juni 1945 ff. / 1994. 27 Fiches / Silberfiche (3-598-33698-5) / Diazofiche (3-598-33686-1)
- Supplement Zeitschriften Juni 1945 ff. / 1995. 108 Fiches / Silberfiche (3-598-33697-7) / Diazofiche (3-598-33685-3)
- Karten 1913 ff. / 1994. 38 Fiches / Silberfiche (3-598-33701-9) / Diazofiche (3-598-33689-6)
- Zeitschriften 1913 ff. / 1994. 169 Fiches / Silberfiche (3-598-33700-0) / Diazofiche (3-598-33688-8)
- Index und alphabetisches Register der Hauptschlagwörter zur Mikrofiche-Edition. Bearb. v. Axel Frey / 1995. XVI, 347 S. / Br (3-598-33691-8)

Sachsse, Rolf: Photographie als Medium der Architekturinterpretation. Studien zur Geschichte der deutschen Architekturphotographie im 20. Jahrhundert / 1984. 469 S., 138 Abb. / Lin (3-598-10564-9)

Die **Sammlung** Hobrecker der Universitätsbibliothek Braunschweig. Katalog der Kinder- und Jugendliteratur 1565-1945. Hrsg.: Universitätsbibliothek d. Technischen Universität Braunschweig. Bearb. v. Peter Düsterdieck unter Mitarb. v. Ingrid Bernin-Israel; Ingrid Czarnecki; Beate Eschenburg und Christian Gärtner / 2 Bde. 1985. XII, 1157 S. / Lin (3-598-10559-2)

Samson, Benvenuto: Urheberrecht. Ein kommentierendes Lehrbuch / 1973. 259 S. / Kst. (UTB 24) (3-7940-2601-2)

Samurin, Evgenij I.: Geschichte der bibliothekarisch - bibliographischen Klassifikation. Bd. I: Vom Altertum bis zum Beginn des 19. Jahrhunderts. Bd. II Das 19. und 20. Jahrhundert / Bd. I: 1967. XVI, 405 S., 12 Abb. Bd. II: 1968, 796 S. / Br (3-7940-4068-6)
- 1. unveränderter Nachdr. der Ausg. 1969 / 2 Bde. in 1 Bd. 1977. XXXII, 1186 S., 12 Abb. (3-7940-7012-7)
- 2. unveränderter Nachdr. der Ausg. 1969 / 2 Bde. 1992. Zus XXXII, 1186 S., 12 Abb. / Lin (3-7940-7012-7)

Sanna, Salvatore A.: Sardinien - Bibliographie / Sardegna - Una Bibliografia. Deutsche Beitr. zur Erforschung d. Insel. Mit e. Einf. von Gerhard Rohlfs. Hrsg. von d. Deutsch-Italienischen Vereinigung e.V. Frankfurt a. M. / 1974. 136 S. / Br (3-7940-3180-6)

Sandeau, Georges: Internationale Marketing-Bibliographie / Bibliographie internationale du marketing et de la distibution. International bibliography of marketing and distribution / 1971. 322 Bl. (3-7940-6115-2)

Sauppe, Eberhard: Dictionary of Librarianship including a selection from the terminology of information science, bibliography, reprography and data processing / Wörterbuch des Bibliothekswesens unter Berücksichtigung der bibliothekarisch wichtigen Terminologie des Informations- und Dokumentationswesens, des Buchwesens, der Reprographie und der Datenverarbeitung. German-English, English-German / Deutsch-Englisch, Englisch-Deutsch / 1988. XX, 428 p. / Hard (3-598-10618-1)
- 2nd rev. and enl. ed. 1996. XXIV, 388 p. / Hard (3-598-11316-1)

Sawilla, Erwin: Thesaurus Feinwerktechnik. Ein alphabetisches und systematisches Verzeichnis von

Fachbegriffen in der Feinwerktechnik / 1978. XIX, 564, 195 S. (3-7940-7064-X)

Sawoniak, Henryk / Maria Witt: International Bibliography of Bibliographies in Library and Information Sciences and Related Fields / 2 vols. 1998 ff. (3-598-11143-6)
Vol. 1 [i.Vb.]
- Vol. 2. 1979-1990 / 3 parts 1998. Cplt. LXXIII, 1208 p. / Hard (3-598-11145-2)

Sawoniak, Henryk / Maria Witt: New International Dictionary of Acronyms in Library and Information Science and Related Fields / 2nd ed. 1988. X, 449 p. / Hard (3-598-10697-1)
- 2nd rev. and enl. ed. 1992. X, 497 p. / Hard (3-598-10972-5)
- 3rd rev. and enl. ed. 1994. XI, 522 p. / Hard (3-598-11171-1)

Scandinavian Biographical Archive / Skandinavisches Biographisches Archiv (SBA). A major work of biographical reference, covering around 150.000 individuals from the beginnings of Scandinavian history to the early 20th century-amounting to 576 printed volumes-have been used as sources. Microfiche edition. Ed. by David Metherell and Paul Guthrie / 1989-1991. 828 fiches. Reader factor 24x / Silverfiche (3-598-32670-X) / Diazofiche (3-598-32650-5)
- Scandinavian Biographical Index / Skandinavischer Biographischer Index (SBI). Ed. by Georgina Clark-Mazo and Helena Koivunen-Henderson / 4 vols. 1994. Cplt. 2580 p. / Gb (0-86291-829-4)

Schaumkell, Ernst: Geschichte der deutschen Kulturgeschichtsschreibung von der Mitte des 18. Jahrhunderts bis zur Romantik im Zusammenhang mit der allgemeinen geistigen Entwicklung. - Nachdruck der Ausg. Leipzig 1905 / 1971. 320 S. (3-7940-5060-6)

Schauspieltexte im Theatermuseum der Universität zu Köln / The Playbook-Collection of the Theatre Museum of the University of Cologne Ein Bestandskatalog mit theaterhistorischen Anmerkungen und Registern / A comprehensive catalogue with historical annotations and indexes Bearb. v. Roswitha Flatz / 8 Bde. u. 1 Reg.-Bd. 1990 / 1991. Zus. 3.900 S. / Ln (3-598-32890-7)

Scheele, Martin: Wissenschaftliche Dokumentation. Grundzüge, Probleme, Notwendigkeiten. Dargestellt an eigenen Arbeiten und Beispielen / 1967. 391 S. (3-7940-5119-X)

Scheibe, Johann A.: Abhandlungen vom Ursprunge und Alter der Musik in Sonderheit der Vokalmusik (Altona und Flensburg 1754) / 1987. 216 S. / Ln (3-598-07261-9)

Scheider, Gesine: Die Fachhochschulbibliotheken. Ergebnisse einer Umfrage. Mit einem Verzeichnis der Anschriften / 1984. 185 S., 22 tabellar. Übersichten / Br (3-598-10541-X)

Schenk, Elisabeth M.: Zwanzig Jahre demokratische Erwachsenenbildung in den böhmischen Ländern 1918 - 1938. Vorgeschichte, Einrichtungen, Gesetze u. Leistungen d. dt. öffentl. Bildungspflege / 1972. 184 S. (3-7940-3288)

Scherf, Walter: Die Herausforderung des Dämons. Form und Funktion grausiger Kindermärchen. Eine Volkskundliche und tiefenpsychologische Darstellung der Struktur, Motivik und Rezeption von 27 untereinander verwandten Erzählungen / 1987. 394 S., 30 Abb. / Ln (3-598-10664-5)

Schertel, Albrecht: Abkürzungen in der Medizin / Abbreviations in Medicine / Abréviations en Médecine / 1974. 198 S. / Br (3-7940-3014-1)
• 2., erw. u. verb. Aufl. 1977. 204 S. (3-7940-7071-8)

Scheven, Yvette: Bibliographies for African Studies 1980-1983 / 1984. XIII, 300 p. / Hard (3-598-10487-1) [Hans Zell Publishers]

Schiek, Gudrun: Emanzipation in der Erziehung. Von der Fremderziehung zur Selbsterziehung. Theorien, Ziele, Praxis / 1975. 203 S. / Kst. (UTB 470) (3-7940-2638-1)

Schilfert, Sabine: Die Simon-Protokolle (1906 / 07) Zur Situation der technischen Hochschulbibliotheken in Preussen nach der Jahrhundertwende. (Beiheft 97 zum Zentralblatt für Bibliothekswesen) / 1989. 152 S., 16 Taf / Br (3-598-07031-4)

Schiller, Friedrich von: Geschichte der merkwürdigsten Rebellionen und Verschwörungen aus den mittlern und neuern Zeiten. - Nachdr. der Ausg. Leipzig 1788 / 1971. 274 S. (3-7940-3470-8)

Schirmer, Wilhelm G. / Gerhard Meyer-Wöbse: Internationale Rohstoffabkommen. Vertragstexte mit einer Einführung und Bibliographie. Unter Mitarb. d. Institut für Völkerrecht der Universität Göttingen. Hrsg. v. Gottfried Zieger / 1980. XX, 907 S. / Ln (3-598-10120-1)

Schlieder, Wolfgang: Riesaufdrucke. Volkstümliche Grafik im alten Papiermachergewerbe / 1989. 184 S., 81 Abb. / Ln i. Sch. (3-598-07255-4) [Lizenzausg. m. Genehm. v. Fachbuchverlag Leipzig]

Schmidbauer, Michael: Kabelkommerz oder Kommunikationsgesellschaft? Ein amerikanisches System im Schlaglicht / 1984. 89 S. / Br (3-598-10509-6)

Schmidt, Ralph: Informationsagenturen im Bereich der Raumforschung, Geowissenschaften und Umweltdisziplinen / 3 Bde. / Lin (3-598-10607-6)
• Bd. 1. Raumrelevante Informations- und Dokumentationsstellen / 1986. 297 S. / Lin (3-598-10608-4)
• Bd. 2. Raumrelevante Datenbanken / 1989. 273 S. / Lin (3-598-10609-2)
• Bd. 3. Raumwissenschaftliche Informationsstrategien / 1990. 441 S. / Lin (3-598-10879-6)

Schneider, Klaus: Computer aided subject index system for the life sciences / 1976 XIII, 205 S. (3-7940-2810-4)

Schöttgen, Christian: Historie derer Buchhändler, wie solche in alten und mittleren Zeiten gewesen. Aus tüchtigen Nachrichten zusammengetragen. (Nachdr. d. Aufl. Nürnberg u. Altdorff 1722) / 1985. 60 S., 4 Abb. / Pp (3-598-07223-6)

Scholl, Hanns Karl: Wegmarken der Entwicklung der Schreib- und Drucktechnik / Neuaufl. 1967. 74 S. / engl. Br (3-7940-3137-7) / Ln (3-7940-3037-0)

Schreiber, Erhard: Kritik der marxistischen Kommunikationstheorie / 1984. 239 S. / Br (3-598-10531-2)

Schriften der Herbert und Elsbeth Weichmann Stiftung.
• Quellen zur deutschen politischen Emigration 1933-1945. Inventar von Nachlässen, nichtstaatlichen Akten und Sammlungen in Archiven und Bibliotheken der Bundesrepublik Deutschland. Hrsg. im Auftr. d. Herbert und Elsbeth Weichmann Stiftung v. Heinz Boberach; Patrik von zur Mühlen; Werner Röder; Peter Steinbach, unter Mitw. v. Ursula Adam; Volkmar Elstner u. Mitarb. in d. Archiven. Bearb. v. Ingrid Schulze-Bidlingmaier / 1994. XXII, 368 S. / Ln (3-598-23040-0)
• Schicksale deutscher Emigranten. Auf der Suche nach den Quellen - Arbeitsergebnisse der Herbert und Elsbeth Weichmann Stiftung. Einführungsband. Hrsg.: Herbert und Elsbeth Weichmann Stiftung / 1993. 70 S., 3 Fotos, 12 Abb. / Ln (3-598-23041-9)

Die **Schriften** der Mainzer Jakobiner und ihrer Gegner (1792-1802). Revolutionäre und gegenrevolutionäre Proklamationen und Verordnungen, Reden, Flugschriften, Abhandlungen, Schauspiele, Gedichte und Lieder, Reiseberichte, Zeitungen und Zeitschriften aus der Zeit der Mainzer Republik (1792 / 93) und der linksrheinischen Revolutionsbewegung bis zur frühen Franzosenzeit in Rheinhessen und der Pfalz (1797-1802). Mikrofiche-Edition. Hrsg.: Stadtbibliothek Mainz / 1993. 357 Fiches. Lesefaktor 24x / Silberfiche (3-598-33033-2) / Diazofiche (3-598-33034-0)
• Bibliographie und Begleitheft zur Mikrofiche-Edition / 1993. XI, 111 S. / Br (3-598-33035-9)

Schriftenreihe der Deutschen Gesellschaft für Photographie e.V.
• 1: Heidtmann, Frank / Hans J. Bresemann; Rolf H. Krauss: Die deutsche Photoliteratur 1839-1978 / German Photographic Literature 1839-1978. Theorie - Technik - Bildleistungen. Eine systematische Bibliographie der selbständigen deutschsprachigen Photoliteratur / 1980. XXX, 690 S. / Lin (3-598-10026-4)
• 3: Heidtmann, Frank: Bibliographie der Photographie Deutschsprachige Publikationen der Jahre 1839-1984 / Bibliography of German-language Photographic Publications 1839-1984. Technik - Theorie - Bild / Technology - Theory - Visual. (Veröffentlichung des Instituts für Bibliothekswissenschaft und Biblioth.) Hrsg.: Deutsche Gesellschaft f. Photographie e.V. / 2 Bde. 2. verb. u. erw. Ausg / 2nd rev. and enl. ed. 1989. Zus. XVI, 886 S. / Ln (3-598-10829-X)

Schriftenreihe der Georg-von-Vollmar-Akademie. Hrsg.: Georg-von-Vollmar-Akademie, München.
• 1: Kronawitter, Hildegard: Wirtschaftskonzeptionen und Wirtschaftspolitik der Sozialdemokratie in Bayern 1945-1949 / 1988. XX, 296 S., 21 Abb. / Ln (3-598-22020-0)
• 2: Mehringer, Hartmut: Waldemar von Knoeringen. Teil 1 - Eine politische Biographie Der Weg vom revolutionären Sozialismus zur sozialen Demokratie. Hrsg.: Forschungsinstitut d. Friedrich-Ebert-Stiftung; Institut für Zeitgeschichte / 1989. XVII, 529 S. / Ln (3-598-22021-9)
• 3: Hennig, Diethard: Johannes Hoffmann. Sozialdemokrat und Bayerischer Ministerpräsident. Biographie / 1990. XII, 634 S. / Ln (3-598-22022-7)

• 4: Pohl, Karl H.: Die Münchener Arbeiterbewegung. Sozialdemokratische Partei, Freie Gewerkschaften, Staat und Gesellschaft in München 1890-1914 / 1992. 580 S. / Gb (3-598-22023-5)
• 5: Von der Klassenbewegung zur Volkspartei. Wegmarken der bayerischen Sozialdemokratie 1892-1992. Hrsg. v. Hartmut Mehringer, im Auftr. d. Georg-von-Vollmar-Akademie, in Zus.-Arb. mit Marita Krauss; Rainer Ostermann; Wolf D. Krämer u. d. Historischen Arbeitskreis d. bayerischen SPD. Bildredakt.: Marita Krauss, in Verb. mit Wolf Krämer / 1992. XI, 355 S., 57 Abb. / Ln (3-598-22024-3)

Schriftenreihe des Landesinstituts für Arabische, Chinesische und Japanische Sprache
• Deutsche Fernostbibliographie / German Far East Bibliography / Te-kuo-han-yü-chung-hsin. Deutschsprach. Veröff. über Ost-, Zentral- u. Südostasien / German Language Publications on East-, Central- and Southeast Asia. Hrsg. von Helmuth Martin u. Günther Pflug, in Zus.-Arb. mit Rainer Krempien u. Hartmut Walravens (ISSN 0721-944X)
• 1: 1979 / 1981. XIX, 99 S. / Gb (3-598-20141-9)
• 6: 1980 / 1982. XVI, 183 S. / Gb (3-598-20142-7)
• 7: 1981 / 1982. XIV, 247 S. / Gb (3-598-20143-5)

Schriftenreihe der UFITA. Archiv für Urheber-, Film-, Funk- und Theaterrecht. Hrsg. v. Georg Roeber [Ab 1971 im J. Schweitzer Vlg., Berlin]
• H. 33: Hartlieb, Horst von: Die Freiheit der Kunst und das Sittengesetz / 1969. 67 S. / Br (Vlg.-Nr. 07033)
• H. 34: Asprogerakas-Grivas, Constantinos: Das Urheberrecht in Griechenland. Geleitw. von Eugen Ulmer / 1969 83 S. / Br (Vlg.-Nr. 07034)
• H. 35: Das Stockholmer Vertragswerk zum internationalen Urheberrecht. Hrsg. von Georg Roeber. Mit e. Vorwort von Gerhard Schneider. Dokumentarischer Teil bearb. von Kurt Schiefler / 1969. 516 S / Lin (Vlg.-Nr. 07035)
• H. 36: Riedel, Hermann: Die musikalische Bearbeitung. Studie über Musik und Musikurheberrecht / 1969. 240 S.
• H. 37: Rie, Robert: Zur Reform des amerikanischen Copyrights. Eine Studie de lege ferenda / 1969. 37 S. / Kt (Vlg.-Nr. 07037)
• H. 38: Möllering, Jürgen: Die internationale Coproduktion von Filmen. Eine vergleichende Darstellung nach deutschem, französischem und italienischem Zivil-, Urheber- und Internationalem Privatrecht / 1970. 261 S:. / Lin (3-7940-7038-0)
• H. 40: Brugger, Gustav: Die neuen audio-visuellen Systeme. Begriffsbestimmung u. recht. Beurteilg., insbes. d. sog. „Kassettenfernsehens" u. d. „Bildplatte". Vorwort von Helmut Jedele / 1970. 57 S. / Br (3-7940-77040-2)

Schriftenreihe Internationales Zentralinstitut für das Jugend- und Bildungsfernsehen. Hrsg.: Internationales Zentralinstitut für das Jugend- und Bildungsfernsehen, München.
• 9: Methoden der Medienwirkungsforschung. Einführung für Medienpraktiker. Ein Seminarbericht Von Jürgen Friedrich u.a. / 1976. 81 S. / Br (3-7940-3559-3)
• 10: Sturm, Hertha / Katharina Holzheuer; Reinhard Helmreich: Emotionale Wirkungen des Fernsehens - Jugendliche als Rezipienten / 1977. 128 S. / Br (3-7940-3560-7)
• 11: Schulfunk in Europa. Eine Dokumentation mit Beitr. zur Europ. Schulfunkkonferenz 1977 / 1978. 198 S. (3-598-03561-6)

- 12: Sturm, Hertha / Sabine Jörg: Informationsverarbeitung durch Kinder. Piagets Entwicklungstheorie auf Hörfunk und Fernsehen angewandt / 1980. 95 S. / Br (3-598-20752-2)
- 13: Schmidbauer, Michael: Kabelfernsehen in der Bundesrepublik Deutschland. Die Interessen von Wirtschaft, Politik und Publikum / 1982. 118 S. / Br (3-598-20755-7)
- 14: Frauenrollen, Kommunikation und Beruf. Beitr. v. Lois W. Hoffmann; Aimee Dorr; Gerald S. Lesser; Ellen Wartella; Gertrud Robinson; Mariann Jelinik; Joyce Hocker-Frost / 1983. 180 S. / Br (3-598-20754-9)
- 15: Gesundheitserziehung in Fernsehen und Hörfunk. Beiträge zu einer internationalen Konferenz mit einer annotierten Auswahlbibliographie. Hrsg. v. Manfred Meyer / 1982. 437 S., 25 Ill / Br (3-598-20753-0)
- 16: Schmidbauer, Michael / Paul Löhr: Die Kabelpilotprojekte in der Bundesrepublik Deutschland. Ein Handbuch / 1983. 175 S. / Br (3-598-20756-5)
- 17: Wie verstehen Kinder Fernsehprogramme. Forschungsergebnisse zur Wirkung formaler Gestaltungselemente des Fernsehens. Hrsg. v. Manfred Meyer / 1984. 315 S. / Br (3-598-20757-3)
- 18: Jörg, Sabine: Unterhaltung im Fernsehen. Show-Master im Urteil der Zuschauer / 1984. 170 S. / Br (3-598-20758-1)
- 19: Arbeitslose Jugendliche - eine Zielgruppe für Fernsehen und Hörfunk. Hrsg. v. Paul Löhr / 1985. 157 S. / Br (3-598-20759-X)
- 20: Schmidbauer, Michael / Paul Löhr: Der Markt der kommerziellen Kindermedien. Eine Dokumentation / 1985. 136 S., 68 Tab., 10 Graf / Br (3-598-20760-3)
- 21: Schmidbauer, Michael: Die Geschichte des Kinderfernsehens in der Bundesrepublik Deutschland. Eine Dokumentation / 1987. 187 S. / Br (3-598-20761-1)
- 22: Schmidbauer, Michael / Paul Löhr: Kinderfernsehen in der Bundesrepublik Deutschland. Eine Dokumentation von Forschungsergebnissen 1959-1988 / 1988. 178 S. / Br (3-598-20762-X)
- 23: Schmidbauer, Michael / Paul Löhr: Jugend und Fernsehen Plädoyer für ein jugendgeeignetes Programm / 1989. 100 S. / Br (3-598-20763-8)
- 24: Medienkommunikation im Alltag Interpretative Studien zum Medienhandeln von Kindern und Jugendlichen. Hrsg. v. Michael Charlton; Ben Bachmair / 1990. 283 S. / Br (3-598-20764-6)
- 25: Schmidbauer, Michael / Paul Löhr: Fernsehpädagogik. Eine Literaturanalyse / 1991. 126 S. / Br (3-598-20765-4)
- 26: Aufenanger, Stefan: Kinder im Fernsehen - Familien beim Fernsehen / 1993. 125 S. / Br (3-598-20766-2)
- 27: Kultur- und Bildungsprogramme im Fernsehen - Defizite, Unterstützung, Chancen. Beiträge zu einem Internationalen Symposium. Hrsg. v. Manfred Meyer. Veranstaltet v. Internationalen Zentralinstitut. d. Jugend- und Bildungsfernsehen (IZI), München und der Bayerischen Landeszentrale für neue Medien (BLM) / 1994. 367 S. / Br (3-598-20767-0)

Schritte zur Neuen Bibliothek. Rudolf Frankenberger zum Abschied aus dem Dienst. Hrsg. v. Otto Weippert / 1998. VIII, 266 S. / Ln (3-598-11348-X)

Schulte-Hillen, Jürgen / Beatrix von Wietersheim: IuD-online-Datenbankbenutzung in der Bundesrepublik Deutschland. Gefahren der Abhängigkeit von ausländischen Datenbanken für Wissenschaft und Forschung, Wirtschaft und Industrie / 1984. 146 S. / Br (3-598-10542-8)

Schulz-Torge, Ulrich J.: Who was Who in the Soviet Union. A Biographical Dictionary of more than 4600 leading Officials from the Central Apparatus and the Republics to 1991 / 1992 / Hard (3-598-10810-9)
- Who's Who in Russia Today. A Biographical Dictionary of more than 2,100 Individuals from the Russian Federation including the Other Fourteen Former USSR Republics / 1993. IX, 478 p. / Hard (3-598-11184-3)

Schuster-Schmah, Sigrid: Der Leserattenfänger. Mit Illustrationen v. Ulrike Bethge / 1979. 22 S. / Gb (3-598-10136-8)

Schwanhäußer, Gerhard / Friedbert Kind: Thesaurus Hochschschulforschung, Hochschulbau. Ein automat. erstellter Thesaurus / 1974. 461 S. (3-7940-3215-2)

Schwarz, Werner: Guitar Bibliography / Gitarre Bibliographie. An International Listing of Theoretical Literature on Classical Guitar from the Beginning to the Present / Internationales Verzeichnis der theoretischen Literatur der klassischen Gitarre von den Anfängen bis zur Gegenwart With ass. of Monika Haringer. Frwd / Vorw. v. Siegfried Behrend / 1984. XXXII, 257 p. / Hard (3-598-10518-5)

Schweers, Hans: Gemälde in deutschen Museen / Paintings in German Museums. Katalog der in der Bundesrepublik ausgestellten Werke / Catalogue of Works on Exhibition in the Federal Republic of Germany / 2 Bde. 1981 / 1982. Zus. XLVII, 1.268 S., 85 Abb. / Lin (3-598-10308-5)
- 2. aktualis., erhebl. erw. u. verb. Ausg. Katalog der ausgestellten und depotgelagerten Werke / Catalogue of Exhibited Works and Depository / 3 Tle. in 10 Bdn 1994. Zus. CCCXXIV, 4.871 S. / Lin (3-598-10927-X)
- Teil I / Part I: Künstler und ihre Werke / Artists and Their Works
  - Bd. 1 / Vol. 1. 1994. LXXII, 599 S. / Lin (3-598-11061-8)
  - Bd. 2 / Vol. 2. 1994. VI, 532 S. / Lin (3-598-11062-6)
  - Bd. 3 / Vol. 3. 1994. VI, 469 S. / Lin (3-598-11063-4)
  - Bd. 4 / Vol. 4. 1994. VI, 567 S. / Lin (3-598-11064-2)
- Teil II / Part II: Ikonographisches Verzeichnis / Iconographic Directory
  - Bd. 5 / Vol. 5. 1994. XLIII, 320 S. / Lin (3-598-11065-0)
  - Bd. 6 / Vol. 6. 1994. XLIII, 723 S. / Lin (3-598-11066-9)
  - Bd. 7 / Vol. 7. 1994. XLIII, 347 S. / Gb (3-598-11191-6)
- Teil III / Part III: Verzeichnis der Museen mit Ihren Bildern / Museums and Their Paintings
  - Bd. 8 / Vol. 8. 1994. XXXV, 444 S. / Gb (3-598-11192-4)
  - Bd. 9 / Vol. 9. 1994. XXXV, 395 S. / Gb (3-598-11193-2)
  - Bd. 10 / Vol. 10. 1994. XXXV, 475 S. / Gb (3-598-11194-0)

Schweers, Hans: Genrebilder in deutschen Museen. Verzeichnis der Künstler und Werke / 1986. XX, 373 S., 932 Ill / Lin (3-598-10517-7)

Schweiz-Index → *Marburger Index. Schweiz-Index.*

Schwerin, Kurt: Bibliographie rechtswissenschaftlicher Schriftenreihen / A Bibliography of German-Language Legal Monograph Series / 1978. XVI, 383 S. / Lin (3-7940-7037-2)

Schwinge, Gerhard: Bibliographische Nachschlagewerke zur Theologie und ihren Grenzgebieten. Systematisch geordnete Auswahl / 1975. 232 S. / Lin (3-7940-3224-1)

Science and Technology and the Future. Proceedings and joint report of World Future Studies Conference and DSE-Preconference, held in Berlin (West), 4.-10. May 1979. Ed.: World Futures Studies Fed.; Hans Buchholz und Wolfgang Gmelin / 2 Bde. 1979. 1439 S. (3-598-10193-7)

The **SCOLMA** Directory of Libraries and Special Collections on Africa in the United Kingdom and Western Europe. Publ. on behalf of the Standing Conference on Library Materials on Africa (SCOLMA). Ed. by Harry Hannam / 4th rev. and enl. ed. 1983. 183 p. / Hard (3-598-10502-9) [Hans Zell Publishers]

Seelbach, Dieter: Computerlinguistik und Dokumentation. Key-Phrases in Dokumentationsprozessen / 1975. 151 S., ca. 30 Tab. u. Schemata / Kst. (UTB 468) (3-7940-2643-8)

Seifert, Siegfried / Albina A. Volgina: Heine-Bibliographie 1965-1982 / 1986. XIV, 427 S. / Ln (3-598-07246-5)

Seminarberichte der Deutschen UNESCO-Kommission. Hrsg.: Deutsche UNESCO-Kommission, Köln [bis Bd. 29]; Deutsche UNESCO-Kommission, Bonn [ab Bd. 30]
- 6: Der Zugang Jugendlicher zu den Zeugnissen außereuropäischer Kulturen / 1967. 120 S. / Br (3-7940-5172-6)
- 7: Film im Museum. Bericht über ein internationales Seminar der Deutschen UNESCO-Kommission, veranstaltet in Zusammenarbeit mit dem Museum Folkwang vom 11. bis 14. Oktober 1966 in Essen / 1967. 223 S. / Br (3-7940-5171-8)
- 11: Europäische Konferenz über das Sonderschulwesen / 1970. 172 S. / Br (3-7940-5173-4)
- 12: Fernsehen und Museum. Bericht über ein internationales Seminar der Deutschen UNESCO-Kommission, veranstaltet in Zusammenarbeit mit dem Museum Folkwang und dem Deutschen Nationalkomitee des Internationalen Museumsrates (ICOM) vom 21. bis 24. Juni 1969 in Folkwang / 1970. 114 S. / Br (3-7940-5170-X)
- 14: Elektronische Datenverarbeitung in Bibliotheken. Electronic data processing in libraries Bericht über ein internationales Seminar der Deutschen UNESCO-Kommission in Zusammenarbeit mit dem Internationalen Verband der Bibliothekar-Vereine (IFLA) u. d. Universität Regensburg vom 13. bis 18. April 1970 in Regensburg / 2., unveränd. Aufl. 1975. XXIV, 283 S., 7 Übersichten / Br (3-7940-5214-6)
- 15: Ausbildungsprogramme für Unterrichtstechnologien. Trainung Programs for Educational Technologies. Berichte und Materialien eines Werkstattseminars, 18.-20. Juni 1970 in Konstanz / 1971. 53 S. / Br (3-7940-5177-7)
- 16: Forschung und Massenmedien. Bericht über ein internationales Seminar, 7. - 9. Sept. 1970 in Konstanz / 1972. 108 S. (3-7940-5216-1)

- 17: Transfer of educational materials. Report of an expert meeting on „Possibilities of international co-operation in the exchange of education materials", Hannover 1972. Red.: Hans-Wolf Rissom / 1973. 128 S. / Br (3-7940-5217-X)
- 18: Museologie. Bericht über ein internationales Symposium der Deutschen Sektion des Internationalen Museumsrates (ICOM) und der Deutschen UNESCO-Kommission, veranstaltet vom 8. bis 13. März 1971 in München. / 1973. 210 S. / Br (3-7940-5218-8)
- 19: Heutige Probleme der Volksmusik. Bericht über e. internat. Seminar d. Dt. UNESCO-Kommission . 1971 in Hindelang / Allgäu / 1973. 186 S. / Br (3-7940-5219-6)
- 20: Der Mensch und die Biosphäre. Bericht über e. internat. Symposium, veranst. von d. Dt. UNESCO-Komm. u.d. Bundesanst. für Vegetationskunde, Naturschutz u. Landschaftspflege vom 14. bis 19. Juni 1972 in Bonn-Bad Godesberg / 1974. 234 S. / Br (3-7940-5220-6)
- 21: Die Praxis der Museumsdidaktik. Bericht über e. internat. Seminar d. Dt. UNESCO-Kommission, veranst. in Zusammenarbeit mit d. Museum Folkwang vom 23. bis 26. Nov. 1971 in Essen. Red. Hans-Dieter Dyroff / 1974. 171 S. / Br (3-7940-5221-8)
- 22: Implementation of curricula in science education. Report of an International Seminar on „The Implementation of Curricula in Science Education with Special Regard to the Teaching of Physics". Organ. by the German Commission for UNESCO and the Institut for Science Education at the University of Kiel, Kiel, March 16-18, 1982. Bearb.: Karl Hecht; Klaus Blänsdorf / 1974. 207 S. / Br (3-7940-5222-6)
- 23: Ingenieurausbildung und soziale Verantwortung. Bericht über d. internat. Symposium „Die Ausbildung von Ingenieuren unter besonderer Berücksichtigung ihrer sozialen Verantwortung", veranstaltet von der Deutschen UNESCO Kommission und dem Verein Deutscher Ingenieure (VDI) vom 29. bis 31. Mai 1972 in München. Hrsg.: Alois Huning / 1974. VIII, 240 S. (3-7940-5223-4) [Koprod. m. d. VDI-Vlg., Düsseldorf]
- 24: Arbeitnehmer im Ausland. Bericht über ein Internationales Seminar der Deutschen. UNESCO-Kommission und der Friedrich Ebert-Stiftung über Probleme der Ausbildung und der kulturellen Integration ausländischer Arbeitnehmer unter besonderer Berücksichtigung der Jugendlichen, veranstaltet vom 5.-8. Dez. 1972 in Bergneustadt / 1974. 125 S. / Br (3-7940-5224-2)
- 25: Symposion Leo Frobenius. Perspectives des études africains contemporains Rapport final d'un symposium internat. organisé par les Comm. Allemande A amerounais pour l'UNESCO du 3 au 7 décembre 1973 à Yaoundé / 1974. 371 S. / Br (3-7940-5225-0)
- 26: Allgemeine und berufliche Bildung. Bericht über ein internationales Seminar der Deutschen UNESCO-Kommission und des Deutschen Volkshochschul-Verbandes über „Allgemeine und berufliche Bildung, Fragen der Weiterbildung nach der 3. Weltkonferenz über Erwachsenenbildung, Tokio 1972" veranstaltet vom 23.-26. Okt. 1973 in der Akademie Sankelmark bei Flensburg / 1975. 107 S. / Br (3-7940-5226-9)
- 27: Die Spannweite des Humanen. Span of man Bericht über e. Kolloquium, veranst. von d. Dt. UNESCO-Komm. in Zusammenarbeit mit d. Fak. für Biologie d. Univ. Freiburg am 23. u. 24. Mai 1874 in Freiburg / Breisgau / 1975. 248 S. / Br (3-7940-5227-7)
- 28: Die soziale Dimension der Museumsarbeit. Bericht über ein internationales Seminar der Deutschen UNESCO-Kommission, veranstaltet in Zusammenarbeit mit dem Museum Folkwang vom 20. bis 23. Mai 1974 in Essen / 1976. 170 S. / Br (3-7940-5228-5)
- 29: Geoscientific studies and the potential of the natural enviroment. Report of an Internat. Training Seminar, organized by the German Comm. for UNESCO in cooperation with the Carl-Duisberg-Ges. and the Federal Inst. for Geosciences and Mineral Resources, Hannover, April 28 - May 23, 1975 / 1975. 312 S. / Br (3-7940-5229-3)
- 30: Stadtökologie. Bericht über e. Kolloquium d. Dt. Uneco-Kommission, veranstaltet in Zusammenarbeit mit d. Werner-Reimers-Stiftung, vom 23.-26. Februar 1977 in Bad Homburg Red.: Folkert Precht / 1978. 204 S. / Br (3-7940-5230-7)
- 31: Die Vereinten Nationen im Unterricht. Bericht über eine Arbeitstagung für Pädagogen u. Sozialwissenschaftler, veranst. vom 3. -8. Jan. 1977 in d. Europ. Akad. Berlin. Bearb. von Klaus Hüfner / 1979. 119 S. / Br (3-598-20421-3)
- 32: Kind und Spiel im öffentlichen Raum. Bericht über e. internat. Fachtagung, veranst. von d. Dt. UNESCO-Komm. u. d. Stadt Erlangen. Hrsg. von Bernhard Meyer / 1980. 271 S. / Br (3-598-20422-1)
- 33: Symposium Leo Frobenius II. Le rôle des traditions dans le développement de l'Afrique. Die Rolle der Traditionen für die Entwicklung Afrikas. Organisé par les Comm. Allemande et Sénégalaise pour l'UNESCO du 14 au 17 mars à Dakar / 1980. 417 S. / Br (3-598-20423-X)
- 34: Die junge Generation in der Randgruppengesellschaft. Bericht über e. Fachtagung vom 17. bis 19. Sept. 1980 in Bergisch Gladbach, veranst. von d. Dt. UNESCO-Komm. u. d. Europ. Akademie. Hrsg. von Bernhard Meyer / 1983. 159, LXIII S. / Br (3-598-20424-8)
- 35: Wandlung von Verantwortung und Werten in unserer Zeit / Evolution of Responsibilities and Values Today. Bericht über ein internationales Kolloquium veranstaltet von der Deutschen UNESCO Kommission vom 2. bis 4. Juni 1982 in Freiburg / Breisgau / Report ofa Colloquium organized by the German Commission for UNESCO, Freiburg / Breisgau, June 2-4, 1982 / 1983. 328 S. / Br (3-598-20425-6)
- 36: Motivation for Adult Education. Working papers presented to the European Conference on Motivation for Adult Education organised by the German Commission for UNESCO and the UNESCO-Institute for Education, Hamburg 1983. Ed. by Joachim H. Knoll / 1985. 243 S. / Br (3-598-20426-4)
- 37: Die Multikulturellen. Über die Chancen im Zusammenleben mit Ausländern. Bericht über die Migranten Tagungen, veranstaltet von der Deutschen UNESCO Kommission in Zusammenarbeit mit dem Evangelischen Industrie- und Sozialpfarramt Gelsenkirchen und der Evangelischen Akademie / 1985. 184 S. / Br (3-598-20427-2)

Senckenbergische Bibliothek Frankfurt am Main. Mikrofiche-Edition / 2 Tle.
- Alphabetischer Katalog der Zeitschriften und Serien. Stand 01. 03. 1985 / 1985. 73 Fiches. Lesefaktor 42x / Diazofiche (3-598-30322-X)

- Schlagwortkatalog der Monographien und Periodika bis Erwerbungsjahr 1964. Hrsg. v. H. Burkhard / 1983. 161 Fiches. Lesefaktor 42x / Silberfiche (3-598-30321-1) / Diazofiche (3-598-30320-3)

Serials and Microforms: Patron-Oriented Management. Proceedings of the Second Annual Serials and Eighth Annual Microform Conference. Ed. by Nancy M. Nelson / 1983. 160 P / Hard (3-598-10949-0)

SfB-Systematik für Bibliotheken. Hrsg.: Stadtbüchereien Hannover; Stadtbibliothek Bremen; Büchereizentrale Schleswig-Holstein / 1997. IV, 787 S., incl. Diskette / Br (3-598-11358-7)

Sicherheitsadressbuch Deutschland, Österreich, Schweiz 1996/97 / 1996. 352 S. / Gb (3-598-23076-1)

Siebel, Wigand: Grundlagen der Logik. Eine Einführung in Aufbau und praktische Anwendungen der Begriffslogik / 1975. 198 S., 14 Abb., 18 Tab. / Kst. (UTB 515) (3-7940-2645-4)

Signale für die Musikalische Welt. 1843-1941. Mikrofiche-Edition / 1993. 315 Fiches, Lesefaktor 42x / Silberfiche (3-598-33193-2) / Diazofiche (3-598-33192-4)
- Bibliographie und Indizes zur Mikrofiche-Edition. Bearb. v. Michael Peschke / 1995. XII, 582 S. / Gb (3-598-33194-0)

Simon, Elisabeth: Bibliotheks- und Informationssysteme in Spanien und den lateinamerikanischen Ländern. Eine Einführung. Unter Mitarb. v. Susan Aramayo; Ofelia Amador / 1992. 178 S. / Br (3-598-11086-3)
- Bibliothekswesen in den USA. Eine Einführung. Unter Mitarb. v. Jolande Goldberg; Hannelore Rader; Sabine Wieczorek; Ursula Wolfart / 1988. 155 S. / Br (3-598-10667-X)
- Bibliothekswesen in England. Eine Einführung. Unter Mitarb. v. Blaise Cronin; Derek Handley; Jennifer Marshmann; Michael Wells / 1985 74 S. / Br (3-598-10567-3)
- Bibliothekswesen in Frankreich. Eine Einführung / 1985. 89 S. / Br (3-598-10598-3)

Simon, Konstantin Romanovič: Bibliographische Grundbegriffe und Fachtermini / 1973. 149 S. (3-7940-4200-X)

Singer, Samuel: Schweizer Märchen. Anfang e. Kommentars zu d. veröffentl. Schweizer Märchenliteratur. Mit e. Nachw. von Max Lüthi. Unveränd., berecht. Nachdr. d. Ausg. Bern 1903 - 1906 / 1971 77, VI, 166 S. (3-7940-3398-1)

Das **SKK**-Statut. Zur Geschichte der Sowjetischen Kontrollkommission in Deutschland 1949-1953. Eine Dokumentation. Im Auftr. d. Instituts f. Zeitgeschichte zsgst. u. eingel. v. Elke Scherstjanoi (Texte und Materialien zur Zeitgeschichte Bd. 11. Red.: Werner Röder u. Christoph Weisz) / 1998. X, 239 S. / Gb (3-598-11333-1)

Slavens, Thomas P.: A Great Library through Gifts / 1986. X, 355 p. / Hard (3-598-10621-1)
- Theological Libraries at Oxford / 1984. VIII, 197 p. / Hard (3-598-10563-0)

Smith, Clifford Neal / Anna Pisczan-Czaja Smith: American Genealogical Resources in German Archives (AGRIGA). A Handbook / 1977. 336 S. / Lin (3-7940-5180-7)

Smith, Clifford N. / Anna Pisczan-Czaja Smith: Encyclopedia of German-American Genealogical Research / 1976. 273 S. / Ln (3-7940-5179-3)

SMM. Systematik des Musikschrifttums und der Musikalien für Öffentliche Musikbibliotheken. Erarb. von der Kommission für Musiksystematik des Arbeitskreises Öffentliche Musikbibliotheken bei der Arbeitsstelle für das Bibliothekswesen / 2., unveränd. Aufl. 1976. 52 S. / Br (3-7940-5141-6)

So mache ich meine Werbung. Hilfen für den Werbealltag von Praktikern aus der Wirtschaft.. Hrsg. v. Horst Dohm und Hugo Müller-Vogg / 1979. 196 S. / Br (3-598-10197-X)

Solon, M. L.: Ceramic Literature. An analytical index to the works published in all languages on the history and the technology of the ceramic art; also to the catalogues of public museums, private collections and of auction sales in which the descriptions of ceramic objects occupy an improtant place; and to the most important price lists of the ancient and modern manufactories of pottery and porcelain. - Nachdr. d. Ausg. London 1910 / 1985. XVIII, 660 p. / Hard (3-598-07227-9)

Songe, Alice H.: The Land-Grant Idea in American Higher Education. A Guide to Information Sources / 1980. XII, 62 p. (3-598-40019-5)

Songe, Alice H. / Pauline B. Mangin: Educator's source book on China. A selcted List of Information Resources / 1981. VII, 80 p. (3-598-40020-9)

South-East Asian Biographical Archive / Südostasiatisches Biographisches Archiv (SEABA). Microfiche Edition. Comp. by Berend Wispelwey / 1997 ff. Approx. 500 fiches in 12 instalments. Reader factor 24x / Silverfiche (3-598-34211-X) / Diazofiche (3-598-34210-1)

The **Soviet** Union / Die Sowjetunion. Figures - Facts - Data / Zahlen - Fakten - Daten. Ed. by Borys Lewytzkyj / 1979. XXXVI, 614 p. / Hard (3-598-07040-3)

Sozialpädagogik - Institution, Partizipation, Selbstorganisation. Tagung der Kommission Sozialpädagogik der Deutschen Gesellschaft (DGfE) vom 3. und 5. Oktober 1977 in der Universität Bremen. Grusswort von Walter Franke; Alexander Wittkowsky. Hrsg. von Christian Marzahn / 1978. II, 120 S. [Minerva Publikation] (3-597-10044-9)

Sozialstrategien der Deutschen Arbeitsfront (DAF). Reprint und Mikrofiche-Edition. Hrsg.: Hamburger Stiftung für Sozialgeschichte des 20. Jahrhunderts. Bearb. u. eingel. v. Michael Hepp; Karsten Linne und Karl H. Roth / Silberfiche (3-598-31570-8)
- Tl. A. 1936-1940 / 6 Bde. in 9 Teilbdn / Lin (3-598-31572-4)
  - Bd. 1. Jahrbuch 1936 / 1986. VI, 338 S. / Lin (3-598-31573-2)
  - Bd. 2. Jahrbuch 1937 / 1986. IV, 460 S. / Lin (3-598-31574-0)

- Bd. 3. Jahrbuch 1938 / (2 Teilbde) 1986. Zus. 1.320 S. / Lin (3-598-31575-9)
- Bd. 4. Jahrbuch 1939 / (2 Teilbde) 1987. Zus. IV, 1.144 S. / Lin (3-598-31576-7)
- Bd. 5. Jahrbuch 1940/41 / (2 Teilbde) 1987. Zus. IV, 1.779 S. / Lin (3-598-31577-5)
- Bd. 6. Kommentar und Register / 1991. 226 S. / Lin (3-598-31578-3)
- Tl. B. Denkschriften, Gutachten und Veröffentlichungen des Arbeitswissenschaftlichen Instituts der Deutschen Arbeiterfront (DAF). Mikrofiche-Edition / 1987-1992. Zus. 42.758 S. auf 498 Fiches, Polarität: negativ, Lesefaktor 24x / Silberfiche (3-598-31579-1)
- Begleitband mit Kommentar und Register / 1992. 211 S. (3-598-31585-6)

Soziokulturelle Herausforderungen - Sozialpolitische Aufgaben: Aspekte moderner Sozialarbeit. Hrsg. v. Ernst von Kardorff; Hubert Oppl [Minerva Publikation]
- 1: Sozialarbeit für und mit alten Menschen / 1989. 134 S. / Br (3-597-10682-X)
- 2: Selbsthilfe und Krise der Wohlfahrtsgesellschaft / 1989 196 S. / Br (3-597-10683-8)
- 3: Soziale Beschäftigungsformen. Zur Zukunft der Arbeit. Hrsg. v. Dietmar Radke; Hubert Oppl; Bernd Stiller / 1991. 240 S. / Br (3-597-10684-6)
- 4: Oliva, Hans / Hubert Oppl; Rudolf Schmid: Rolle und Stellenwert Freier Wohlfahrtspflege. Forschungsbericht im Auftrag des Bayerischen Staatsministeriums für Arbeit, Familie und Sozialordnung / 1991. 179 S. / Br (3-597-10685-4)
- 5: Sozialstation 2000. Zur Zukunft der Pflege. Hrsg.: Caritasverband d. Erzdiözese München u. Freising; Innere Mission München / 1992. 184 S. / Br (3-597-10686-2)

Spanien- und Portugal-Index. Bilddokumentation zur Kunst in Spanien und Portugal. Mikrofiche-Edition. Hrsg.: Bildarchiv Foto Marburg; Deutsches Dokumentationszentrum f. Kunstgeschichte; Philipps-Universität Marburg / 1995. ca. 303.933 Fotos auf 298 Fiches, Lesefaktor 24x / Silberfiche (3-598-33633-0)

Sparck Jones, Karen / Martin Kay: Linguistik und Informationswissenschaft / 1976. 261 S. / Kst. (UTB 571) (3-7940-2653-5)

Spiess, Volker: Verzeichnis deutschsprachiger Hochschulschriften zur Publizistik 1885 - 1967 / 1969. 459 S. / Br (Vlg.-Nr. 04259)

Die **Staatsbürgerin**. Offenbach a. M. 1886. - Originalgetreuer Nachdruck der ersten Arbeiterinnenzeitschrift Deutschlands. - Hrsg. u. erl. v. Hartwig Gebhardt; Ulla Wischermann / 1988. 55, 114 S. / Ln (3-598-10694-7)

Staeck, Erich: Rechenbuch für das graphische Gewerbe. Unter Mitarb. von Heinz Kraus und Helmut Seichter / 10., überarb. u. erw. Aufl. 1968. 378 S. / Ln (Vlg.-Nr. 08709) [Vorher im Ullstein Vlg., Berlin]
- 11., überarb. u. erw. Aufl. 1970. 378 S. / Ln (Vlg.-Nr. 08709)
- 12., unveränd. Aufl. 1974. 387 S. (3-7940-8709-7)
- 13., neubearb. u. erw. Aufl. Rechenbuch für die Druckindustrie. Begr. von Erich Staeck. Bearb. v. Helmut Fromm; Helmut Seichter und Konrad Wächter / 1977. 378 S. / Lin (3-7940-8710-0) [Anschließend im Vlg. Beruf + Schule, Itzehoe]

Standards for public libraries. Ed. by International Federation of Library Associations, Section of Public Libraries / 1973. 53 S. / Br (3-7940-4310-3)

Standortkatalog der Sammlung Welding. Vorw. v. Hans. A. Koch. Hrsg.: Staats- u. Universitätsbibliothek Bremen. Bearb. v. Armin Hetzer / 1990. XI, 214 S. / Lin (3-598-10931-8)

Standortverzeichnis ausländischer Zeitungen und Illustrierten in Bibliotheken und Instituten der Bundesrepublik Deutschland und Berlin (West) (SAZI) / Catalogue of Foreign Newspapers and Illustrated Periodicals in German Libraries. Hrsg.: Staatsbibliothek Preussischer Kulturbesitz. Redaktion und Bearb. von Martin Winckler / 1975. 334 S. (3-7940-3055-9)

Stauder, Peter: Die Hochschulschriften der alten Kölner Universität 1583-1798. Ein Verzeichnis. Einl. v. Peter Stauder, / 1990. XIV, 351 S. / Lin (3-598-10859-1)

Steady-State, Zero Growth and the Academic Library. A Collection of Essays. Ed. by Colin Steele / 1978. 148 S. (0-85157-243-X) [Clive Bingley]

Sterbebücher von Auschwitz / Death Books from Auschwitz / Ksiegi zgonow z Auschwitz. Fragmente / Remnants / Fragmenty. Hrsg.: Staatliches Museum Auschwitz-Birkenau
- Deutsche Ausg.: Bd. 1. Berichte / 1995. 532 S. / Gb (3-598-11263-7)
- English Edition: Vol. 1. Reports / 1995. 506 p. / Hard (3-598-11272-6)
- Wydanie polskie: Tom 1. Relacje / 1995. 520 S. / Gb (3-598-11274-2)
- Bd. 2/3 / Vol. 2/3 / Tom 2/3: Namensverzeichnis A-Z. Annex / Index of Names A-Z. Annex / Indeks nazwisk A-Z. Aneks / 2 Bde. 1995. 1.661 S. / Gb (3-598-11275-0)

Steuernagel, D. Carl: Lehrbuch der Einleitung in das Alte Testament. Mit einem Anhang über die Apokryphen und Pseudoepigraphen. Autoris. Reprint / 1978. XVI, 869 S. [Minerva Publikation] (3-597-20002-8)

Stevens, Henry: The Humboldt Library. A Catalogue of the Library of Alexander von Humboldt. With a Bibliographical and a Biographical Memoir. Unveränd. Nachdruck der 1863 in London erschienenen Ausgabe. Mit einem Vorwort von Erwin Stresemann und einer Einführung von Fritz G. Lange / 1968. XIII, VIII, 791 S. (3-7940-5028-2)

Stiehl, Ulrich: Satzwörterbuch des Buch- und Verlagswesens. Dictionary of Book Publishing Dt.-Engl. With 12,000 sample sentences and phrases / 1977. XX, 538 S. / Lin (3-7940-4147-X)

Stock, Karl F.: Einführung in die elektronische Datenverarbeitung für Archivare, Bibliothekare und Dokumentare / 1983. IV, 169 S. (3-598-10498-7)
- 2. überarb. Aufl. 1985. 200 S. / Br (3-598-10602-5)

Stock, Karl F.: Einführung in die elektronische Datenverarbeitung mit BASIC für Bibliothekare und Dokumentare. 1979. X, 286 S. (3-598-10062-0)

Stock, Karl F. / Rudolf Heilinger; Marylène Stock: Personalbibliographien österreichischer Dichter und

Schriftsteller von den Anfängen bis zur Gegenwart. Mit Auswahl einschlägiger Bibliographien, Nachschlagewerke, Sammelbiographien, Literaturgeschichten und Anthologien / 1972. XXIV, 703 S. / Lin (3-7940-3185-7)

Stock, Marylène / Karl F. Stock Bibliographie der Programmiersprachen. Bibliography of programming languages. Bibliographie des langages de programmation. Bücher, Manuals u. Aufsätze vom Plankalkül bis PL / l. / 1973. LVII, 393 S.

Stoffe - Formen - Strukturen. Studien zur deutschen Literatur. Festschrift für Hans Heinrich Borcherdt zum 75. Geburtstag / Hrsg. von Albert Fuchs; Helmut Motekat / 1962. XVI, 521 S. / Ln (3-597-06-1) [Minerva Publikation. 1977 v. Max Hueber Vlg., München übernommen]

Stone, Elizabeth / Eileen Sheahan; Katherine J. Harig: Model continuing education recognition system in library ans information science / 1979. IX, 313 S. (3-598-40003-9)

Sträter, Hans H.: Thesaurus Pädagogik. Unter Mitarb. von Doris Friedrich. Hrsg. vom Dokumentationsring Pädagogik (DOPAED) / 1973. 243 S. (3-7940-3211-X)

Struktur des Thesaurus für Parlamentsmaterialien PARTHES. Formaler Aufbau und Datenverarbeitungskonzept. Dem Leiter der Abteilung Wissenschaftliche Dokumentation Heinz Matthes zum 65. Geburtstag. Hrsg.: Deutscher Bundestag, Gruppe Datenverarbeitung. Bearb. v. Jörg Hansis; Wolfgang Mausberg; Johannes Tappertzhofen / 1985. 146 S. / Lin (3-598-10584-3)

Strzolka, Rainer: Anwendersoftware für Bibliothekare und Dokumentare. 2000 Programme für PC, Micro-, Mini- und Mainframecomputer. Vorw. v. Werner Schwuchow / 1991. 313 S. / Lin (3-598-11052-9)

Studien des Forschungsinstitutes für Wirtschaftspolitik an der Universität Mainz. Hrsg.: Forschungsinstitut für Wirtschaftspolitik an der Universität Mainz [Minerva Publikation]
- 35: Morschhäuser, Berthold: Strukturwandel ohne Staatshilfe. Das Beispiel der Holzverarbeitenden Industrie in der Bundesrepublik Deutschland (1976-1985) / 1988. XIV, 506 S. / Br (3-597-10622-6)
- 36: Hamm, Walter: Deregulierung im Verkehr als politische Aufgabe / 1989. X, 106 S. / Br (3-597-10623-4)
- 37: Piffl, Stefan: Eigenbeteiligung in der ambulanten Gesundheitsversorgung. Möglichkeiten gesundheitsökonomischer Ausgestaltung / 1990. XV, 208 S. / Br (3-597-10624-2)
- 38: Kopp, Wolfgang: Der Markt für Übertragungseinrichtungen der Telekommunikation. Struktur, Entwicklung, Perspektiven / 1990. XXII, 340 S. / Br (3-597-10625-0)
- 39: Beckers, Stefan: Die Bankdienste der Deutschen Bundespost in ordnungspolitischer Sicht / 1990. XX, 258 S. / Br (3-597-10626-9)
- 40: Metzger, Michaela M.: Realisierungschancen einer Privatisierung öffentlicher Dienstleistungen / 1990. XVIII, 256 S. / Br (3-597-10627-7)
- 41: Weber, Lukas: Die deutsche Rundfunkordnung nach der Zulassung privater Anbieter. Absehbare Entwicklungen und medienpolitischer

Handlungsbedarf / 1990. XXII, 323 S. / Br (3-597-10628-5)
- 42: Cantzler, Florian: Quantitative und qualitative Beschäftigungswirkungen neuer Technologien. Eine Analyse für Rheinland-Pfalz / 1991. XXVI, 433 S. / Br (3-597-10629-3)
- 43: Beyer, Andrea: Tarifpolitik in strukturschwachen Branchen. Eine empirische Untersuchung der Druckerei- und der Textilindustrie in der Bundesrepublik Deutschland / 1991. XX, 309 S. / Br (3-597-10630-7)
- 44: Kern, Dieter: Marktwirtschaftliche Innovationspolitik durch Output-Orientierung staatlicher Förderung / 1991. XVI, 212 S. / Br (3-597-10695-1)

Studien zum Buch- und Bibliothekswesen
- Bd. 4.. Hrsg. v. Friedhilde Krause; Hans Erich Teitge. Im Auftr. d. Deutschen Staatsbibliothek Berlin / 1986. 114 S., 16 Abb. / Br (3-598-07035-7)
- Bd. 5.. Hrsg. v. Friedhilde Krause; Hans Erich Teitge. Im Auftr. d. Deutschen Staatsbibliothek Berlin / 1987. 96 S., 16 Taf / Br (3-598-07036-5)
- Bd. 6.. Hrsg. v. Friedhilde Krause; Hans Erich Teitge. Im Auftr. d. Deutschen Staatsbibliothek Berlin / 1988. 108 S., 8 Abb. / Br (3-598-07037-3)
- Bd. 7.. Hrsg. v. Friedhilde Krause; Hans Erich Teitge. Im Auftr. d. Deutschen Staatsbibliothek Berlin / 1989. 85 S., 8 Taf / Br (3-598-07038-1)
- Bd. 8.. Hrsg. v. Richard Landwehrmeyer; Hans Erich Teitge. Hrsg. im Auftr. d. Staatsbibliothek zu Berlin - Preussischer Kulturbesitz / 1993. 96 S., davon 8 S. schw.-w / Br (3-598-11161-4)
- Bd. 9. Hrsg. v. Richard Landwehrmeyer; Hans Erich Teitge. Hrsg. im Auftr. d. Staatsbibliothek zu Berlin - Preussischer Kulturbesitz / 1994. 93 S., 9 Abb. / Br (3-598-11223-8)

Studien zur Publizistik. Bremer Reihe / Deutsche Presseforschung. Hrsg. v. Elger Blühm. [Vorher im Schünemann Universitätsvlg., Bremen]
- 13: Weller, B. Uwe: Maximilian Harden und die „Zukunft" / 1970. 486 S. (3-7940-4583-1 / 3-7940-4513-0)
- 14: Plewnia, Margarete: Auf dem Wege zu Hitler. Der „völkische" Publizist Dietrich Eckart / 1970. 155 S. (3-7940-4514-9)
- 15: Fischer, Heinz D.: Parteien und Presse in Deutschland seit 1945. / 1971. 600 S., zahlr. Abb. / Br (3-7940-4515-7)
- 16: Mauersberger, Volker: Rudolf Pechel und die „Deutsche Rundschau" 1919 - 1933. Eine Studie zur konservativ-revolutionären Publizistik in der Weimarer Republik / 1971. 344 S. (3-7940-4516-5)
- 17: Die deutschen Zeitungen des 17. Jahrhunderts. Ein Bestandsverzeichnis mit historischen und bibliographischen Angaben. Zsgst. v. Else Bogel; Elger Blühm / 2 Bde. 1971. Zus. XXXII, 629 S., zahlr. Abb. / Ln (3-7940-4517-3)
- Bd. 3. Nachtrag. Zsgst. v. Elger Blühm; Brigitte Kolster; Helga Levin / 1985. 308 S., 103 Abb. / Lin (3-598-21625-4)
- 18: Siegel, Christian E.: Egon Erwin Kisch. Reportage und politischer Journalismus / 1973. 384 S. (3-7940-4518-1)
- 19: Bogel, Else: Schweizer Zeitungen des 17. Jahrhunderts. Beiträge zur frühen Pressegeschichte von Zürich, Basel, Bern, Schaffhausen, St. Gallen und Solothurn / 1973. 152 S. (3-7940-4519-X)
- 20: Griesche, Hans D.: Die Bremer Hochschulreform und die Presse. Eine Analyse der Berichterstattung 1970/71 / 1974. 144 S. / Br (3-7940-4520-3)
- 21: Brandes, Helga: Die „Gesellschaft der Maler" und ihr literarischer Beitrag zur Aufklärung. Eine Untersuchung zur Publizistik des 18. Jahrhunderts / 1974. 274 S. (3-7940-4521-1)

- 22: Hilscher, Elke: Die Bilderbogen im 19. Jahrhundert. / 1977. 378 S., 24 Taf / Br (3-7940-4522-X)
- 23: Presse und Geschichte. Beitr. zur histor. Kommunikationsforschung. Referate e. internat. Fachkonferenz d. Dt. Forschungsgemeinschaft u. d. Dt. Presseforschung, Univ. Bremen, 5. - 8. Okt. 1976 in Bremen / 1977. 240 S. (3-7940-4523-8)
- 24: Wischermann, Ulla: Frauenfrage und Presse. Frauenarbeit und Frauenbewegung in der illustrierten Presse des 19. Jahrhunderts / 1983. 222 S. / Br (3-598-21624-6)

Studiengruppe für Systemforschung e.V., Heidelberg. Berichte [1970 übernommen]
- Nr. 12: Hirsch, Hans J.: Programmstrukturen für komplexe Informationssysteme / 1973. V, 39 S. / Br (3-7940-3712-X)
- Nr. 20 Rittel, Horst: Hierarchie oder Team? Betrachtungen zu den Kooperativformen in Forschung und Entwicklung. - Sonderdr. a.: Forschungsplanung. Einen Studie über Ziele und Strukturen amerikanischer Forschungsinstitute [R. Oldenbourg Vlg., München] / 1964. 52 S. / Br (Vlg.-Nr. 03520)
- Nr. 24 Rittel, Horst: Überlegungen zur wissenschaftlichen und politischen Bedeutung der Entscheidungstheorie. - Sonderdr. a.: Forschungsplanung. Einen Studie über Ziele und Strukturen amerikanischer Forschungsinstitute [R. Oldenbourg Vlg., München] / 1963. 28 S. / Br (Vlg.-Nr. 03524)
- Nr. 38: Wessel, Andrew E.: Some Implications of Strategic Concepts for Western European Nuclear Weapons / 1966. 17 S. / Br (3-7940-3538-0)
- Nr. 40 Bahrdt, Hans Paul: Die wissenschaftspolitische Entscheidung / 1966. 20 S. / Br (3-79403540-2)
- Nr. 42: Krauch, Helmut: Kritik an der Expertokratie / 1965. 21 S. / Br (3-7940-3542-9)
- Nr. 44: Krauch, Helmut: Wissenschaftspolitik in demokratischer Gesellschaft / 1966. 16 S. / Br (3-7940-3544-5)
- Nr. 45: Krauch, Helmut: Science Policy in a Democratic Society / 1966. 16 S. / Br (3-7940-3545-3)
- Nr. 46: Wheeler, Harvey: The Political Implications of the Scientific Revolution / 1965. 25 S. / Br (3-7940-3546-1)
- Nr. 49: Die Statistik der Forschungs- und Entwicklungskapazitäten (FE-Statistik). Teil A. Von Ingrid Herrmann; Frederike Müller-Köppern; Reinhard Coenen u.a / 1964. 38 S. / Br (3-7940-3549-6)
- Nr. 58: Die Statistik der Forschungs- und Entwicklungskapazitäten (FE-Statistik). Teil B. Von Ingrid Herrmann; Frederike Müller-Köppern; Reinhard Coenen u.a / 1964. 38 S. / Br (Vlg.-Nr. 03558)
- Nr. 59: Die Statistik der Forschungs- und Entwicklungskapazitäten (FE-Statistik). Teil C. Von Ingrid Herrmann; Frederike Müller-Köppern; Reinhard Coenen u.a / 1965. 36 S. / Br (3-7940-3559-3)
- Nr. 65: Paas, Dieter: Einstellung von Lehrern an Höheren Schulen zum Programmierten Unterricht / 1967. 49 S. / Br (Vlg.-Nr. 03565)
- Nr. 84: Hornke, Gertrud: Maschinelle Dokumentation in der organischen Chemie / 1967. 164 S. / Br (3-7940-3584-4)
- Nr. 85: Hornke, Gertrud / Werner Jacob; Werner Kunz: Umfrage zum Stand der maschinellen Dokumentation der Chemie / 1968. 34 S. / Br (3-7940-3585-2)
- Nr. 92: Meyer-Uhlenried, Karl-Heinrich / Uta Krischker: Die Auswertung von Dokumenten als Voraussetzung für ein integriertes automatisiertes Informationssystem / 1970. 75 S. / Br (Vlg.-Nr. 03592)
  - 2., überarb. Aufl. 1971. 75 S. / Br (3-7940-3592-5)

- Nr. 93: Rothkirch und Trach, Karl Christoph Graf von: Prinzipien der Thesauruserstellung. Dargestellt am Beispiel: I. des Aufbaues eines internen Informationssystems, II. der Erstellung von Fachthesauri für das Projekt Bundestagsverwaltung der Studiengruppe für Systemforschung / 2. Aufl. 1970. 83 S., 8 Abb., 2 Übersichten / Br (3-7940-3593-3)
- Nr. 94: Schwuchow, Werner: Zur Messung von Wirtschaftlichkeit von Dokumentnachweissystemen („Information Retrieval Systems") / 1970. 67 S. / Br (3-7940-3594-1)
- Nr. 95: Meindl, Ulrich: Informationsbedarfsanalyse bei der Entwicklung von Informationssystemen / 1972. 61 S. / Br (3-7940-3595-X)
- Nr. 96: Czayka, Lothar: Die Bedeutung der Graphentheorie für die Forschungsplanung / 1970. 44 S. / Br
  - 2., überarb. Aufl. 1971. IV, 44 S. / Br (3-7940-3596-8)
- Nr. 97: Meyer-Uhlenried, Karl-Heinrich: Systemanalyse und Entwurf eines integrierten, automatisierbaren Informations- und Dokumentationssystems. Dargest. am Projekt „Entwicklung e. Dokumentationssystems f. d. Parlamentsmaterialien" im Auftr.d. Dt. Bundestages / 1970. VII, 67 S. / Br
  - 2., überarb. Aufl. 1971. VII, 67 S. / Br (3-7940-3597-6)
- 98: Czayka, Lothar: Grundzüge der Aussagenlogik / 1971 IV; 40 S. / Br
  2., wesentl. erw. Aufl. → *Czayka, Lothar: Grundzüge der Aussagenlogik*
- 99: Meyer-Uhlenried, Karl-Heinrich / Uta Krischker: Die Entwicklung eines Datenerfassungsschemas für komplexe Informationssysteme. Dargest. am Projekt „Entwurf e. Dokumentationssystems in d. Abt. Wissenschaftl. Dokumentation d. Dt. Bundestages" / 1971 XV; 194 S. / Br (3-7940-3599-2)
- 100: Zimmermann, Werner: Prinzipien zur Erstellung von Instruktionen für die Erfassung dokumentarischer Bezugseinheiten. Dargestellt am Projekt „Entwurf e. Dokumentationssystems in d. Abt. Wissenschaftl. Dokumentation d. Dt. Bundestages" / 1971 VIII, 116 S. / Br (3-7940-3700-6)
- 101: Erler, Brigitte: Die Entwicklung von Hilfsmitteln für die Dokumentation von Vorgängen. Dargest. am Projekt „Entwurf e. Dokumentationssystems in d. Abt: Wissenschaftl. Dokumentation d. Dt. Bundestages" / 1971 X, 34 S. / Br (3-7940-3701-4)
- 102: Wenzel, Uwe: Die Arbeitsablauf-Organisation in einem intergrierten, automatisierbaren. Dargestellt am Projekt: „ Entwurf e. Dokumentationssystems in d. Abt. Wissenschaftl. Dokumentation d. Dt. Bundestages". / 1971. VIII, 56 S. / Br (3-7940-3702-2)
- 103: Fischer, Peter: Datenstrukturierung zur Übertragung auf maschinenlesbare Datenträger zur EDV-unterstützten Verarbeitung. Dargest. am Projekt: „Entwurf e. Dokumentationssystems in d. Abt. Wissenschaftl. Dokumentation d. Dt. Bundestages / 1971. VI, 42 S. / Br (3-7940-3703-0)
- 104: Rolland, Maria Theresia: Grundriß eines Thesaurus als funktionsfähiges Hilfsmittel für Indexierung und Recherche. Erarb. f. d. Deutschen Bundestag im Rahmen d. Projektes: „Entwicklg e. Dokumentationssystems f. d. Parlamentsmaterialien" / 1971. 68 S. / Br (3-7940-3704-9)
- 105: Berger, Albrecht: Entwurf eines Systems zur Dokumentation von expliziten Verweisungen in gesetzlichen Vorschriften. Dargestellt am Projekt „Entwicklung e. Dokumentationssystems in d. Abt. Wissenschaftl. Dokumentation d. Dt. Bundestages" / 1971. 28, 204 S. / Br (3-7940-3705-7)

- 106: Dehlinger, Hans E.: Verweisungsbezogene Textdokumentation von gesetzlichen Vorschriften. Entwurf e. Systems zur maschinellen Verarb. expliziter Verweisgn. Dargest. am Projekt „Entwicklung e. Dokumentationssystems in d. Abt. Wissenschaftl. Dokumentation d. Dt. Bundestages" / 1971. 120 S. / Br (3-7940-3606-9)
- 107: Der Umweltschutz und seine chemisch-toxikolgischen Probleme. T. A. Problemanalysen über d. chemisch-toxische Gefährdung d. Umwelt. Ergebnisse einer Befragungsaktion. Von Reinhard Coenen u.a / 1972. 104 S. / Br (3-7940-3707-6)
- 108 / 109: Der Umweltschutz und seine chemisch-toxikolgischen Probleme. T. B. Forschungsvorhaben auf dem Gebiet des Umweltschutzes Dokumentation abgeschlossener, laufender u. geplanter Vorhaben auf d. Gebiet „chem.-tox. Gefährdung d. Umweltschutzes". Von Reinhard Coenen u.a / 1974. 148 S. / Br (3-7940-3708-1)
- 110: Gresser, Klaus: Das Planning-Programming-Budgeting-System Probleme d. Anwendung bei d. staatl. Aufgaben- u. Finanzplanung / 1972. X, 153 S. / Br (3-7940-3710-3)
- 111: Meyer-Uhlenried, Karl-Heinrich: Forschungsplanung als Informationsproblem / 1972. VIII, 159 S. / Br (3-7940-3711-1)
- 112: Hirsch, Hans-Joachim: Programmstrukturen komplexer Informationssysteme / 1973. V, 39 S. / Br (37940-3712-X)
- 113: Rittel, Horst W. J.: Die Entwicklung der Technik Konsequenten f. Bildg. u. Wissenschaft / 1972. VII, 31 S. / Br (3-7940-3713-8)
- 114: Herbst, Jörn: Möglichkeiten zur Wirtschaftlichkeitskontrolle von Verwaltungsabläufen und Verwaltungsverfahren im öffentlichen Dienst / 1972. VIII, 167 S. / Br (3-7940-3714-6)

Studies in Library Management [Clive Bingley]
- Vol. 5. 1979. 168 S. (0-85157-265-0)
- Vol. 6. The coming of age of library management 1960-1980 / 1980. 214 S. (0-85157-301-0)

Studies on the organizational structure and services in national and university libraries in the Federal Republic of Germany and in the United Kingdom. Papers presented at a joint meeting of British and German librarians at the University of Bristol in September 1978. Ed.: Deutsche Forschungsgemeinschaft / 1980. 227 p. / Soft (3-598-10094-9)

Studium in Europa und in den Vereinigten Staaten. Hrsg.: UNESCO, Paris / 1973. 324 S. / Kst. (UTB 70) (3-7940-2614-4)

Subject *Guide to Microforms in Print* → ***Guide*** *to Microforms in Print*

Suchier, Wolfram: Gottscheds Korrespondenten. Aus „Kleine Gottsched-Halle". - Nachruck der Ausg. Leipzig 1910-1912 / 1971. 88 S. (3-7940-5012-6)

Südasien. Alphabetischer Katalog der Erwerbungen 1961 - 1976. Hrsg.: Universitätsbibliothek Tübingen, Orientabteilung / 2 Bände 1977. 1340 S. (3-598-10137-6)

Süddeutsche Monatshefte München und Leipzig 1904-1936. Mikrofiche-Edition / 1985. ca. 404.028 S. auf 99 Fiches. Lesefaktor 42x / Diazofiche (3-598-30670-9)

Südosteuropäisches Biographisches Archiv / South-East-European Biographical Archive (SOBA).

Eine Kumulation der ca. 150 wichtigsten biographischen Nachschlagewerke zu Albanien, Bulgarien, Rumänien und ehemaligem Jugoslawien erschienen zwischen 1711 und 1996. Mikrofiche-Edition. Bearb. v. Ulrike Kramme und Zelmíra Urra Muena / 1998 ff. Ca. 450 Fiches in 12 Lfgn, Lesefaktor 24x / Silberfiche (3-598-34121-0) / Diazofiche (3-598-34120-2)

SWI Schlagwortindex. Zu SfB Systematik für Bibliotheken, ASB Allgemeine Systematik für Öffentliche Bibliotheken, SSD Systematik Stadtbibliothek Duisburg / 1986. VIII, 424 S. / Br (3-598-10632-7)
- Teil 2. Zu Systematik für Bibliotheken SFB, Allgemeine Systematik für öffentliche Bibliotheken ASB, Systematik Stadtbibliothek Duisburg SSD Bearb.: Verein der Bibliothekare an Öffentlichen Bibliotheken e.V. (VBB); Marion Beaujean. Vorw. v. Thomas Bündgen / 1990. 245, VII S. / Br (3-598-10830-3)

Sydow, Achim: Numerische Analyse des Hybridrechnens / 1970. 120 S. / Br (3-7940-4285-9)

Sykes, Paul: The Public Library in Perspective. An Examination of its Origins and Modern Role / 1979. 184 S. (0-85157-284 7) [Clive Bingley]

Systematik der Amerika-Gedenkbibliothek, Berliner Zentralbibliothek. (3-598-10084)
- Geschichte, Wirtschafts- und Sozialwissenschaften, Militär, Sport. Lfg. 1 / 2., überarb. Ausg 1979. 111 S. (3-598-10084-1)

Systematik der Stadtbibliothek Hannover (SSH). Lfg. 1: Erd- und Völkerkunde, Volkskunde, Geschichte, Pädagogik, Philosophie, Literatur. (3-7940-5140-8); Lfg. 2: Medizin, Psychologie, Sprache, Technik, Veterinärmedizin (3-7940-5144-0); Lfg. 3: Betriebswirtschaftslehre. Religion. (3-7940-5178-5); Lfg. 4: Mathematik, Sport, Theater, Wirtschaft. (3-7940-7043-7. Hrsg.: Verein der Bibliothekare an Öffentlichen Bibliotheken e.V. / 4 Teile 1975-1977. 101 S; 72 S.; 77 S.

Systematik der Stadtbücherei Duisburg. Buchaufstellung und Ordnung des systematischen Katalogs. Hrsg.: Verein der Bibliothekare an Öffentlichen Büchereien / 1971. 131 Bl.; 42 S. (3-7940-5143-2)

Systematik für Bibliotheken. Hrsg.: Verein der Bibliothekare an Öffentlichen Bibliotheken e.V.
- Lfg. 1. Erd- und Völkerkunde, Volkskunde, Geschichte, Pädagogik, Philosophie, Literatur / 2. Aufl. 1978. 157 S. / Gh (3-7940-5241-2)
- Lfg. 2. Medizin, Psychologie, Sprache, Technik, Veterinärmedizin / 2. Aufl. 1978. 78 S. / Gh (3-7940-5242-0)
- Lfg. 3. Betriebswirtschaftslehre, Politik, Staat, Verwaltung, Religion. 1976. 77 S.
  - 2. verb. Aufl. 1979. 80 S. / Gh (3-598-20570-8)
- Lfg. 4. Kunst, Musik, Sport, Theater / 1979. 99 S. / Gh (3-598-20571-6)
- Lfg. 5. Biologie, Naturwissenschaften, Recht, Wirtschaft / 1984 / Gh (3-598-20572-4)
- Lfg. 6. Hauswirtschaft, Militärwesen, Sozialpolitik und Soziologie / 1986. 86 S. / Br (3-598-20573-2)
- Lfg. 7. Allgemeines, Bibliothekswesen, Elektronische Datenverarbeitung, Mathematik, Land- und Forstwirtschaft / 1987. 83 S. / Br (3-598-20575-9)
- Sonderlfg. Musikalien / 1985. 29 S. / Gh (3-598-20574-0)

Systematik für Kinder- und Jugendbibliotheken SKJ. Erarb. vom Arbeitskreis „Kinder- und Jugendbüchereiarbeit in städt. u. ländl. Büchereisystemen" bei d. Arbeitsstelle für d. Bibliothekswesen. Hrsg. v. Verein der Bibliothekare an Öffentlichen Bibliotheken e. V. / 2., unveränd. Aufl. 1974. 20 S. / Gh (3-7940-5142-7)

Systemtechnik und Innovationsforschung. Zweites Jahreskolloquium des Instituts für Systemtechnik und Innovationsforschung (ISI) der Frauenhofer-Gesellschaft Karlsruhe, 15. Mai 1974. Hrsg. von Helmar Krupp / 1975. 153 S. (3-79404242-5)
- Drittes Jahreskolloquium des Instituts für Systemtechnik und Innovationsforschung (ISI) der Frauenhofer-Gesellschaft Karlsruhe, 7. November 1975. Hrsg. von Helmar Krupp / 1976. 95 S. (3-7940-4243-3)

Die **Tagebücher** von Joseph Goebbels. Sämtliche Fragmente. Teil I: Die handschriftlichen Tagebücher. Juli 1924 bis Juli 1941. Hrsg. v. Elke Fröhlich, im Auftr. d. Instituts f. Zeitgeschichte München u. in Verb. mit d. Bundesarchiv / 4 Bde. u. 1 Interimsreg. 1987. CXXIII, 3.193 S. / Ln (3-589-21915-6)
- Bd. 1: 27. 6. 1924 - 31. 12. 1930 / 1987. CVIII, 654 S. / Ln (3-589-21916-4)
- Bd. 2: 1. 1. 1931 - 31. 12. 1936 / Bd. 2 / 1987. V, 764 S. / Ln (3-598-21917-2)
- Bd. 3: 1. 1. 1937. - 31. 12. 1939 / Bd. 3 / 1987. V, 682 S. / Ln (3-598-21918-0)
- Bd. 4: 1. 1. 1940. - 8. 7. 1941 / Bd. 4 / 1987. V, 741 S. / Ln (3-598-21919-9)
- Interimsregister / 1987. 352 S. / Ln (3-598-21914-8)

Die **Tagebücher** von Joseph Goebbels. Teil I: Aufzeichnungen 1923 bis 1941. Hrsg. v. Elke Fröhlich, im Auftr. d. Instituts f. Zeitgeschichte mit Unterstützung d. Staatlichen Archivdienstes Russlands / 9 Bde. Vollst. überarb. Neuaufl. 1997 ff. / Ln (3-598-23730-8)
Bde. 1 - 5 [i.Vb.]
- Bd. 6: August 1938-Juni 1939 / Bearb. v. Jana Richter / 1998. 425 S. / Ln (3-598-23736-7)
- Bd. 7: Juli 1939-März 1940 / Bearb. v. Elke Fröhlich / 1998. 399 S. / Ln (3-598-23737-5)
- Bd. 8: April-November 1940 / Bearb. v. Jana Richter / 1998. 470 S. / Ln (3-598-23738-3)
- Bd. 9: Dezember 1940-Juli 1941 / 1997. 462 S. / Ln (3-598-23739-1)

Die **Tagebücher** von Joseph Goebbels. Teil II: Diktate 1941 bis 1945. Hrsg. v. Elke Fröhlich, im Auftrag d. Instituts f. Zeitgeschichte München mit Unterstützung d. Staatlichen Archivdienstes Russlands / 15 Bde. u. 1 Reg.-Bd. 1993 ff. Zus. 9.337 S. / Ln (3-598-21920-2)
- Bd. 1: Juli-September 1941. Bearb. v. Elke Fröhlich / 1996. 546 S. / Ln (3-598-21921-0)
- Bd. 2: Oktober-Dezember 1941. Bearb. v. Elke Fröhlich / 1996. 642 S. / Ln (3-598-21922-9)
- Bd. 3: Januar-März 1942. Bearb. v. Elke Fröhlich / 1994. 613 S. / Ln (3-598-21923-7)
- Bd. 4: April-Juni 1942. Bearb. v. Elke Fröhlich / 1995. 676 S. / Ln (3-598-21924-5)
- Bd. 5: Juli-September 1942. Bearb. v. Angela Stüber / 1995. 636 S. / Ln (3-598-22136-3)
- Bd. 6: Oktober-Dezember 1942. Bearb. v. Hartmut Mehringer / 1996. 556 S. / Ln (3-598-22137-1)
- Bd. 7: Januar-März 1943. Bearb. v. Elke Fröhlich / 1993. 702 S. / Ln (3-598-22138-X)
- Bd. 8: April-Juni 1943. Bearb. v. Hartmut Mehringer / 1993. 591 S. / Ln (3-598-22304-8)

- Bd. 9: Juli-September 1943. Bearb. v. Manfred Kittel / 1993. 655 S. / Ln (3-598-22305-6)
- Bd. 10: Oktober-Dezember 1943. Bearb. v. Volker Dahm / 1994. 600 S. / Ln (3-598-22306-4)
- Bd. 11: Januar-März 1944. Bearb. v. Dieter M. Schneider / 1994. 616 S. / Ln (3-598-22307-2)
- Bd. 12: April-Juni 1944. Bearb. v. Hartmut Mehringer / 1994. 614 S. / Ln (3-598-22308-0)
- Bd. 13: Juli-September 1944. Bearb. v. Jana Richter / 1995. 622 S. / Ln (3-598-22309-9)
- Bd. 14: Oktober-Dezember 1944. Bearb. v. Hermann Graml; Jana Richter / 1995. 531 S. / Ln (3-598-22310-2)
- Bd. 15: Januar-April 1945. Bearb. v. Maximilian Gschaid / 1995. 737 S. / Ln (3-598-22311-0)
- Gesamtregister / 1996. 737 S. / Ln (3-598-21925-3)

Tammelo, Ilmar / Helmut Schreiner: Grundzüge und Grundverfahren der Rechtslogik Band 2. / 1977. 208 S., zahlr. Formeln / Kst. (UTB 685) (3-7940-2659-4)

Tarnschriften 1933 bis 1945. Mikrofiche-Edition. Hrsg. in Zus.-Arb. mit der Stiftung Archiv der Parteien und Massenorganisationen der DDR im Bundesarchiv / 1997. 1.008 Tarnschriften mit ca. 40.500 S. auf 175 Fiches, Lesefaktor 24x, Begleitbroschüre / Silberfiche (3-598-32016-7)
*siehe auch: → **Gittig**, Heinz: Bibliographie der Tarnschriften*

Tarnschriften der KPD aus dem antifaschistischen Widerstandskampf Originalgetreue Reproduktion von 12 Heften aus den Jahren 1935 / 1936. Hrsg.: Institut für Marxismus-Leninismus beim ZK der SED. Bearb. v. Gerhard Nitzsche; Margot Pikarski / 1986. 996 S. (3-598-07229-5)

Taschenbuch 1954. Dokumentationen der Technik. 3. Jahrgang / 1954. 140 S. / Br
*1. u. 2. Jahrgang → **Jahrbuch** des Internationalen Literaturdienstes für alle Gebiete der Technik*

The **Teaching** of Reading. A collection undertaken by the International Reading Association on behalf of UNESCO. Hrsg. von Ralph C. Staiger / 1973. 213 S. (3-7940-5127-0)

Techne - Technik - Technologie. Philosophische Perspektiven. Hrsg. v. Hans Lenk; Simon Moser / 1973. 247 S., 4 Abb. / Kst. (UTB 289) (3-7940-2622-5)

Technik und Gesellschaft. Hrsg. v. Hans Sachsse / 3 Bde
- Bd. 1: Ein Literaturführer. Hrsg. u. eingef. v. Hans Sachsse, in Zus.-Arb. mit Herbert Fein u.a / 1974 309 S. / Kst. (UTB 413) (3-7940-2626-8)
  - 2., unveränd. Aufl. 1983. 309 S. / Lin (3-598-10484-7)
- Bd. 2: Die Darstellung der Technik in der Literatur. Organisationsformen technischer Zusammenarbeit. Hrsg. v. Hans Sachsse, unter Mitarb. v. Manfred Kunzelmann; Fritz Winterling / 1976. 260 S. / Kst. (UTB 570) (3-7940-2649-7)
  - 2., unveränd. Aufl. 1983 / Lin (3-598-10485-5)
- Bd. 3: Selbstzeugnisse der Techniker. Philosophie der Technik. Hrsg. v. Hans Sachsse, unter Mitarb. v. Alois Huning; Heinz R. Spiegel / 1976. 260 S. / Kst. (UTB 622) (3-7940-2650-0)
  - 2., unveränd. Aufl. 1983 / Lin (3-598-10486-3)

Teichler, Ulrich: Der Arbeitsmarkt für Hochschulabsolventen. Zum Wandel der Berufsperspektiven im Zuge der Hochschulexpansion / 1981. 182 S., 40 Tab. / Br (3-598-10337-9)

Teichler, Ulrich: Probleme der Hochschulzulassung in den Vereinigten Staaten. Zulassungstests u. offener Hochschulzugang in e. differenzierten Bildungssystem / 1978. 217 S. / Br (3-598-07083-7)

Teichler, Ulrich / Friedrich Voss: Bibliography on Japanese education. Postwar publications in Western languages / Bibliographie zum japanischen Erziehungswesen / 1974. 294 S. (3-598-03183-1)

Television and Socialization Processes in the Family. Special English Issue of „Fernsehen und Bildung. Internationale Zeitschrift für Medienpsychologie und Medienpraxis" Vol. 9. 1975, Heft 2 / 3. 1976. 192 S. (3-7940-3368-X)

Terminologie als angewandte Sprachwissenschaft. Gedenkschrift für Univ.-Prof. Dr. Eugen Wüster. Hrsg. v. Helmut Felber; Friedrich Lang; Gernot Wersig / 1979. 240 S. / Lin (3-598-10028-0)

Terminology of documentation. A selection of 1200 basic terms published in English, French, German, Russian and Spanish. Zsgst. von Gernot Wersig; Ulrich Neveling / 1976. 274 S. (3-7940-5150-5) [Koprod. m. d. UNESCO, Paris]

Terplan, Kornel: Leistungsfähigkeit der Datenverarbeitung. Steuerung u. Methoden d. Effektivitätsmessung für Computersysteme / 1977. 319 S. (3-7940-5190-4)

Texte - Dokumente - Berichte. Zum Bildungswesen ausgewählter Industriestaaten. Hrsg.: Marburger Forschungsstelle für vergleichende Erziehungswissenschaft [Minerva Publikation]
- 6: Blumenthal, Viktor von / Helmut Weber: Materialien zur Arbeitsplatz- und Qualifikationsstruktur: Italien / 1974. VIII, 97 S. (3-597-00013-4)
- 7: Beratung in der Schule. Zur Situation in Frankreich, Italien, Schweden, USA. Von Roland Hein; Viktor von Blumenthal; Bodo Willmann; Herbert F. Bode / 1975. VI, 69 S. (3-597-00014-2)
- 8: Bode, Herbert F.: Materialien zur Arbeitsplatz- und Organisationsstruktur in den USA / 1975. VI, 95 S. (3-597-00015-0)
- 9/10: Beiträge zum mathematisch-naturwissenschaftlichen und polytechischen Unterricht in der DDR. Von Günter Dannwolf; Ursula Reinermann; Wolfgang Palm; Rainer Brämer / 1976. VI, 148 S. (3-597-00016-9)
- 11: Stübig, Heinz / Herbert F. Bode; Uwe Zänker: Arbeitswelt und Schule. Veränderung in der Sekundarschule: England, USA, UdSSR. Hrsg.: Marburger Forschungsstelle für Vergleichende Erziehungswissenschaft / 1977. VIII, 103 S. (3-597-00017-7)
- 12: Blumenthal, Viktor von: Mitbestimmung in der Schule: Italien / 1976. VI, 72 S. (3-597-00018-5)
- 14: Froese, Leonhard / Herbert F. Bode; Uwe Zänker: Materialien zur Vergleichenden Bildungsforschung / 1978. VIII, 92 S. (3-597-00019-3)
- 22: Berufliche Weiterbildung - Zweiter Bildungsweg: Zur Situation in England, Frankreich, Italien, Schweden, USA / 1978. XII, 98 S. (3-597-00020-7)
- 23: Die Arbeitswelt als Lernbereich in der Sekundarschule Zur Situation in England, Frankreich, Italien, Schweden, USA. Hrsg. v. Viktor von Blumenthal; Bruno Nieser; Heinz Stübig; Bodo Willmann / 1979. X, 108 S. / Br (3-597-10174-7)
- 24: Formen und Funktionen von Schulabschlüssen in England, Italien und Schweden. Hrsg. v. Viktor von Blumenthal; Bruno Nieser; Heinz

Stübig; Bodo Willmann / 1980. VIII, 100 S. / Br
(3-597-10225-5)
- 25: Grundfragen der Vergleichenden Erziehungswissenschaft. Beitr. v. Viktor von Blumenthal;
Rainer Brämer; Leonhard Froese; Erika Gartmann; Götz Hillig; Bruno Nieser / 1981. VIII, 136
S. / Br (3-597-10226-3)
- 26: Matzke, Peter: Funktionaler Analphabetismus
in den USA. Zur Bildungsbenachteiligung in In-
dustriegesellschaften / 1982. VIII, 125 S. / Br
(3-597-10227-1)
- 27: Schulkrise - international? Von Leonhard
Froese; Torsten Husén; Wolfgang Mitter; Achim
Leschinsky; Wolfgang Klafki / 1983. VI, 89 S. / Br
(3-597-10228-X)
- 28: Ergebnisse und Perspektiven Vergleichender
Bildungsforschung. Zur Funktion des internationalen Bildungstransfers. Von Heinz Stübig; Viktor
von Blumenthal; Horst Messmer; Rainer Brämer;
Siegfried Weitz; Götz Hillig; Gert Meyer / 1984.
VIII, 343 S. / Br (3-597-10229-8)
- 29: Blumenthal, Viktor von: Grundlegende Bildung für alle. Die Reform der Sekundarstufe I in
Italien / 1983. VI, 263 S. / Br (3-597-10230-1)
- 30: Buttlar, Annemarie: Die Reform der amerikanischen Gesamtschule. Neue Ansätze für die Entwicklung von Curricula / 1985. VI, 134 S. / Br
(3-597-10231-X)
- 31: Nieser, Bruno: Gesamtschulreform in Frankreich. Zum Verhältnis von organisatorischer und
curricularer Erneuerung der Sekundarstufe 1 /
1984. VI, 96 S. / Br (3-597-10476-2)
- 32: Stübig, Heinz: Aspekte der englischen Sekundarschulreform Leistungsdifferenzierung, Fächer-
angebot und Curriculumplanung / 1983 VI, 102 S.
/ Br (3-597-10477-0)
- 33: Gartmann, Erika: Pädagogik zwischen Men-
schen- und Güterproduktion. Zur Sittlichkeits-
erziehung V. A. Suchomlinskijs / 1984. X, 201 S. /
Br (3-597-10478-9)
- 34: Die Gesamtschule in England, Frankreich,
Italien, Schweden und den USA. Eine Zwischenbilanz. Von Viktor von Blumenthal; Annemarie
Buttlar; Heinz Stübig; Bodo Willmann / 1987. VIII,
103 S. / Br (3-597-10479-7)

Texte und Daten zur Hochschulplanung. Hrsg. vom
Sonderforschungsbereich 63 - Hochschulbau, Universität Stuttgart [1975 übernommen. Vorher im
Eigenvlg. erschienen]
- 4: Informationseinrichtungen an Hochschulen.
Projektstudie. Von Projektgruppe E. 4. Hrsg.: Sonderforschungsbereich 63 - Hochschulbau Universität Stuttgart / 1972. 158 S. (3-7940-5401-6)
- 6, 1. 2: Böhm, Wolfgang / Peter Dietze; Uwe
Schüler: Denkschriften zur Universitätsgründung.
Analyse und Vergleich der Denkschriften der
Städte Ulm, Konstanz, Trier, Bamberg, Passau. T.
1: Analyse; T. 2: Dokumentation / Hrsg. Sonderforschungsbereich 63 - Hochschulbau Universität
Stuttgart / 2 Bde. / 1973. Getr. Pag.; 1973 S.
(3-7940-5403-2 / 3-7940-5404-0B)
- 7: Kleinefenn, Andreas / Horst Küsgen; Christof
Riepl: Kostenplanung für Hochschulinstitute.
Berechnungsgleichung mit Bauelementen / 1973.
14 S. (3-7940-5405-9)
- 9: Informationseinrichtungen an Hochschulen Bestandsaufnahme in Baden-Württemberg. Mitarb.
Jürgen Herr / 1974. 92 S.
- 10 / 1 u. 2: Simulationsmodell einer Hochschul-
entwicklung. Ein Prognosemodell zur Hochschulplanung und zu den Auswirkungen auf die Stadtentwicklungsplanung. Von Eckart Müller u.a / 2
Teile / 1973. 266 S. (3-7940-5407-5 / 3-7940-5408-3)

- 12: Standortbestimmung von zentralen Einrichtungen der Hochschule über ein Zuordnungsmo-
dell, dargestellt am Beispiel der Speiseversorgung.
Allgemeine Problem- und Modellentwicklung. Von
Ulrich Fleming u.a / 1974. 146 S. (3-7940-5413-X)
- 13: Standortbestimmung von zentralen Einrichtungen der Hochschule über ein Zuordnungmo-
dell dargestellt am Beispiel der Speiseversorgung.
Anwendung e. interaktiven Programms. Unter
Mitarb. von Eberhard Bappert / 1977. 158 S.
(3-7940-5413-X)
- 14: Heuck, Kathrin: Planerausbildung. Literaturanalyse, Bibliographie / 1974. 84 S. (3-7940-5414-8)
- 15: Stoffl, Heinrich: Die Lösung von Problemen
der Personal- und Raumzuordnung zu Lehrver-
anstaltungen an Hochschulen auf der Basis von
Kapazitätsmodellen. Ein Lehrorganisations- und
Studentenplanprogramm / 1975. 81 S.
(3-7940-5415-6)
- 16: Gutachten zur Entwicklung einer Gesamthochschule Heilbronn Anwendung eines Simul-
tanmodells zur Hochschul- und Stadtentwicklungsplanung u. e. Verfahrens zur Fächerausbau-
planung von Hochschulen Mitarb. Eckart Müller /
1975. 264 S. (3-7940-5416-4)
- 17: König, Hartmut / Robert Schmittmann: Zur
Ökologie der Schule. E. öko-psychologische Untersuchung zum Einfluß von Schulbauten auf Lehr-
und Lernprozesse / 1976. 285 S. (3-7940-5417-2)
- 18: Forschungsplanung im Sonderforschungsbereich. Beispiele zur Planung universitärer Forschung mit Anwendungsorientierung. Zsgst. von
Sibylle Heeg / 1975. 115 S. (3-7940-5418-0)
- 19: Bautechnische Flächenarten Ergebnisbericht
Ingenieurwissenschaften: Maschinenbau, Elektrotechnik, Bauingenieurwesen. Mitarb. von Tamás
Görhely / 1976. 230 S. (3-7940-5419-9)
- 20: Möglichkeiten der Planung anwendungsorien-
tierter Forschungsprozesse. Vortr. d. Workshopver-
anstaltung „Forschungsproduktion auf Halde?" /
1976. 107 S. (3-7940-542o-2)
- 21: Becker, Ruth / Marianne Heinemann-Knoch;
Rotraut Weeber: Zur Rolle einer Universität in
Stadt und Region Univ. als regionaler Wirtschafts-
faktor, Mobilität von Hochschulabsolventen, Verhalten von Hochschulangehörigen im Reproduk-
tionsbereich / 1976. 186 S. (3-7940-5421-0)
- 22: Arbeiten und Arbeitswelt in der Universität.
Ergebnisse von Befragungen an 2 dt. Univ. Mitarb.
Volkmar Trepte / 1976. 366 S. (3-7940-5422-9)
- 23: Computersimulation in der regionalen und in
der Stadtentwicklung Anwendung und neuere
Ansätze / 2 Bde. 1978. 206, 108 S. (3-7940-5423-7)
- 24: Interimsbauten für Hochschulen. Überblick
über kurzlebige u. nichtortsfeste baul. Lösungen,
Kostenuntersuchungen interimsartiger Bauten.
Zsgst. von Jaim Aveco; Helmut Paschmann / 1976.
199 S. (3-7940-5424-5)
- 25: Gritzka, Christoph: Anwendung der heuristischen Systemanalyse bei Standortentscheidungen
/ 1976. 281 S. (3-7940-5425-3)

Texte und Materialien zur Zeitgeschichte →
Ämter, Abkürzungen ... [Bd. 5]
Deutsche und Polen ... [Bd. 9]
Deutschland im ersten Nachkriegsjahr [Bd. 10]
Inventar archivalischer Quellen des NS-Staates
[Bd. 3]
Inventar der Befehle [Bd. 8]
Das **SKK-Statut** [Bd. 11]
Widerstand als „Hochverrat" ... [Bd. 7]

Thelert, Gustav: Supplement zu Heinsius', Hinrich's
und Kaysers Bücher-Lexikon. Verzeichnis einer

Anzahl von Schriften, welche seit der Mitte des 19.
Jahrhunderts in Deutschland erschienen, in den
genannten Katalogen aber gar nicht oder fehlerhaft
aufgeführt sind. - Nachdruck der Ausgabe Grossen-
hain und Leipzig 1893 / 1973 / Gb (3-7940- 5033-9)

Theoretische Literatur zum Kinder- und Jugendbuch. Bibliograph. Nachweis von den Anfängen im
18. Jahrh. bis zur Gegenwart. Bearb. nach den
Beständen der Deutschen Staatsbibliothek, Berlin,
von Heinz Wegehaupt. Vorw. von Horst Kunze /
1972. XVIII, 448 S. / Ln (3-7940-3189-X)

Thesaurus. Hrsg.: Kommission der Europäischen
Gemeinschaften 1980 (3-598-10096-5)
- Agrarökonomie und ländliche Soziologie Mehr-
sprachiger Thesaurus / Agricultural Economics
and Rural Sociology Multilingual Thesaurus /
Economie Agricole et Sociologie Rurale Thesaurus Multilingue / Economia Agraria e Sociologia
Rurale Thesaurus Multilingue 5 Bde.
(3-598-10097-3)
  - Deutsche Ausgabe / 1979. XX, 80 S. / Br
(3-598-10098-1)
  - Edizione italiana / 1979. XX, 78 p. / Br
(3-598-10101-5)
  - English Edition / 1979. XX, 84 p. / Soft
(3-598-10099-X)
  - Version française / 1979. XX, 84 p. / Br
(3-598-10100-7)
  - Viersprachiges Register und Mikrofiches /
Quadrilingual Index and Microfiches / Index
quadrilingue et microfiches / Indice quadrilingue e microschede / 1979. VIII, 120 S., 16
Fiches / Br (3-598-10102-3)
- Lebensmittel. Mehrsprachiger Thesaurus / 5 Bde.
(3-598-10103-1)
  - Deutsche Ausgabe / 1979. XX, 129 S. / Br
(3-598-10104-X)
  - Edizione italiana / 1979. XX, 132 p. / Br
(3-598-10107-4)
  - English Edition / 1979. XX, 145 p. / Soft
(3-598-10105-8)
  - Version française / 1979. XX, 144 p. / Br
(3-598-10106-6)
  - Viersprachiges Register und Mikrofiche / Quadrilingual Index and Microfiches / Index quadrilingue et microfiches / Indice quadrilingue e
microschede / 1979. VIII, 168 S., 24 Fiches / Br
(3-598-10108-2)
- Veterinärwissenschaft. Mehrsprachiger Thesaurus
/ 5 Bde. (3-598-10109-0)
  - Deutsche Ausgabe / 1979. XX, 214 S. / Br
(3-598-10110-4)
  - Edizione italiana / 1979. XX, 204 p. / Br
(3-598-10113-9)
  - English Edition / 1979. XX, 244 p. / Soft
(3-598-10111-2)
  - Version française / 1979. XX, 218 p. / Br
(3-598-10112-0)
  - Viersprachiges Register und Mikrofiches / Quadrilingual Index and Microfiches / Index quadrilingue et microfiches / Indice quadrilingue e
microschede / 1979. VIII, 240 S., 36 Fiches / Br
(3-598-10114-7)

Thesaurus Bildungsforschung. Verzeichnis d. De-
skriptoren u.Nichtdeskriptoren in d. Literaturdokumentation des Max-Planck-Instituts für Bildungsforschung. Von Edgar Guhde / 1972. 471 S.
(3-7940-3204-7)

Thesaurus für Parlamentsmaterialien. PARTHES.
Microfiche-Edition. Hrsg. vom Deutschen Bundestag
und Bundesrat

- 1986. 35 COM-Fiches (3-598-32286-0)
- 1989. Jährlich neu: ca. 30 COM-Fiches, 3 Suppl., Begleitbr. (3-598-32287-9)

Thesaurus Pädagogik. Hrsg.: Dokumentationsring Pädagogik (DOPAED). Bearb.: Dokumentation d. Pädagogischen Zentrum, Berlin, unter Mitarb. der Thesaurus Arbeitsgr. DOPAED / 2. völlig neu bearb. Aufl. 1982. 385 S. / Lin (3-598-10380-8)

Thieme-Becker / Vollmer. Gesamtregister. Register zum „Allgemeinen Lexikon der Bildenden Künstler von der Antike bis zur Gegenwart" und zum „Allgemeinen Lexikon der bildenden Künstler des XX. Jahrhunderts" / 2 Tle. in 6 Bdn 1996 / 1997 / Hld (3-598-23640-9)
- Tl. 1. Gesamtregister Länder / 3 Bde. 1996. Zus. XX, 2.390 S. / Hld (3-598-23641-7)
- Tl. 2. Gesamtregister Künstlerische Berufe / 3 Bde. 1997. Zus. X, 2.284 S. / Hld (3-598-23642-5)

Thompson, Godfrey: Planning and Design of Library Buildings / 2. Aufl. 1977. 189 S. (3-7940-7051-8)

Thompson, James: An Introduction to University Library Administration / 3., überarb. Aufl. 1979. (0-85157-288-X) [Clive Bingley]

TIB - Katalog der Technischen Informationsbibliothek Hannover / TIB - Catalogue of the Technical Information Library Hannover. CD-ROM-Edition. Hrsg.: Universitätsbibliothek und Technische Informationsbibliothek Hannover / 2 CD-ROM
- 1. Ausg.: Katalog der Universitätsbibliothek Hannover und Technischen Informationsbibliothek. 1996 (3-598-40348-8)
- 2. Ausg.: TIB - Katalog der Technischen Informationsbibliothek Hannover / TIB - Catalogue of the Technical Information Library Hannover. 1997 (3-598-40368-2)
- 3. Ausg. 1998 (3-598-40389-5)

Kataloge der Technischen Informationsbibliothek Hannover (TIB). Microfiche-Edition ➔ *Kataloge der Technischen ...*

Tolzmann, Don H.: Catalog of the German-Americana Collection, University of Cincinnati / 2 vols. 1990. Cplt. XXXVI, 749 p. / Hard (3-598-41241-X)

Topitsch, Ernst: Gottwerdung und Revolution. Beiträge zur Weltanschauungsanalyse und Ideologiekritik / 1973. 260 S. / Kst. (UTB 288)

Topographie der Osteuropa- Südosteuropa- und DDR-Sammlungen. Von der Zentralbibliothek der Wirtschaftswissenschaften in der Bundesrepublik Deutschland / Arbeitsgemeinschaft der Bibliotheken und Dokumentationsstellen der Osteuropa-, Südosteuropa- und DDR-Forschung in der Bundesrepublik Deutschland und West-Berlin. Hrsg. von Gerhard Teich / 1978. 388 S. (3-7940-7039-9)

Toussaint, Ingo: Die Universitätsbibliothek Freiburg im Dritten Reich / 1982. XII, 235 S. (3-598-10465-0)

Toussaint, Ingo: Die Universitätsbibliothek Freiburg im Dritten Reich / 2. verb. u. erw. Aufl. 1984. XIV, 272 S. / Lin (3-598-10547-9)

Toward International Descriptive Standards for Archives / Projet de normes internationales de description en archivistique. Papers presented at the ICA Invitational Meeting of Experts on Descriptive Standards. National Archives of Canada, Ottawa 4-7 October 1988 / Communications présentées à la réunion restreinte d'experts en normes de description. Archives nationales du Canada, Ottawa, du 4 au 7 octobre 1988. Comp. and ed. with the financial assist. of the Toronto Area Archivists Group Education Foundation / 1993. XIV, 177 p. / Soft (3-598-11163-0)

Transkriptionen der chinesischen und japanischen Sprache / Transliteration Tables of Chinese and Japanese. Zsgst. v. Arthur Matthies / 1989. 77 S. / Ln (3-598-10789-7)

Trotsky Bibliography. List of separately published titles, periodical articles and titles in collections treating L. D. Trotsky and Trotzkyism / Bibliographie selbständiger und unselbständiger Schriften über L. D. Trockij und den Trotzkismus. Ed. by Wolfgang Lubitz / 1982. 458 p. / Hard (3-598-10469-3)
- 2nd totally rev. and expanded ed. A classified list of published items about Leon Trotshy and Trotskyism. Ed. by Wolfgang Lubitz / 1988. XXXI, 581 p. / Hard (3-598-10754-4)
- 3rd completely rev. and enl. ed. An International Classified List of Publications about Leon Trotsky and Trotskyism 1905-1998 / Comp. and ed. by Wolfgang Lubitz and Petra Lubitz / 2 vols. 1998. Cplt. XXVIII, 840 p. / Hard (3-598-11391-9)

Trotskyist Serials Bibliography 1927-1991. With Locations and Indices. Preface by Paolo Casciola. Ed. by Wolfgang Lubitz and Petra Lubitz / 1992. XXVII, 475 p. / Hard (3-598-11157-6)

Truhart, Peter: Historical Dictionary of States / Lexikon der historischen Staatennamen. States and State-like Communities from their Origins to the Present / Staaten und staatenähnliche Gemeinwesen von den Ursprüngen bis zur Gegenwart / 1995. XXXIV, 872 p. / Hard (3-598-11292-0)

Truhart, Peter: International Directory of Foreign Ministers 1589-1989 / Internationales Verzeichnis der Aussenminister 1589-1989 / Répertoire international des Ministres des Affaires Etrangères 1589-1989 / Repertorio internacional de los Ministros de Relaciones Exteriores 1589-1989 / Repertorio internazionale dei Ministri degli Esteri 1589-1989. Engl. / Dt. / Franz. / Span. / Ital. / 1989. LIII, 475 p. / Hard (3-598-10823-0)
- Supplement 1945-1995 / Ergänzungsband 1945-1995 / Supplément 1945-1995 / Supplemento 1945-1995 / 1995. IXV, 292 p. / Hard (3-598-11276-9)

Truhart, Peter: Regents of Nations / Regenten der Nationen. Systematic Chronology of States and their Political Representatives in Past and Present. A Biographical Reference Book / Systematische Chronologie der Staaten und ihrer politischen Repräsentanten in Vergangenheit und Gegenwart. Ein biographisches Nachschlagewerk / 3 parts in 4 vols. 1984-1986. Cplt. 4258 p. / Hard (3-598-10491-X)
- Part I: Africa / America / 1984 / Hard
- Part II: Asia / Australia-Oceania / 1985 / Hard
- Part III/1: Middle-, Eastern-, Northern-, Southern-, Southeastern Europe / 1986 / Hard
- Part III/2: Western Europe, Bilbiography, Addenda to Part I-III / 1986 / Hard

UBCIM Publication Series. Ed.: International Federation of Library Associations and Institutions (IFLA), Den Haag.
- Form and Structure of Corporate Headings. Approved by the Standing Committees of the IFLA Section on Cataloguing; IFLA Section on Official Publications / 1980. X, 15 p. / Soft (3-598-10962-8)
- Guidelines for the Application of the ISBDs to the Description of Component Parts. Approved by the Standing Committees of the IFLA Section on Cataloguing and the IFLA Section of Serial Publications / 1988. VIII, 22 p. / Soft (3-598-10956-3)
- International Target Audience Code (ITAC): a proposal and report on its development and testing. Prepared by Russell Sweeney / 1977. III, 12 p. / Soft (3-598-10963-6)
- International Transfers of National MARC Records Guidelines for Agreements Relating to the Transfer of National MARC Records Between National Bibliographic Agencies. Advisory Committee reporting jointly to the IFLA and the Conference of Directors of National Libraries / 1987. 48 p. / Soft (3-598-10959-8)
- ISBD (CF): International Standard Bibliographic Description for Computer Files / 1990. V, 98 p. / Soft (3-598-10983-0)
- ISBD (CM): International Standard Bibliographic Description for Cartographic Materials. Approved by the Section on Cataloguing and the IFLA Section of Geography and Map Libraries / Rev. ed. 1987. VII, 55 p. / Soft (3-598-10952-0)
- ISBD (CM): International Standard Bibliographic Description for Monographic Publications. Approved by the Standing Committees of the IFLA Section on Cataloguing / Rev. ed. 1987. VIII, 62 p. / Soft (3-598-10953-9)
- ISBD (NBM): International Standard Bibliographic Description for Non-Book Materials. Approved by the Standing Committees of the IFLA Section on Cataloguing / Rev. ed. 1987. VII, 74 p. / Soft (3-598-10954-7)
- ISBD (S): International Standard Bibliographic Description for Serials. Approved by the Standing Committees of the IFLA Section on Cataloguing and the IFLA Section on Serial Publications / Rev. ed. 1988. VIII, 76 p. / Soft (3-598-10955-5)
- Komorous, Hanna / Robert B. Harriman: International Guidelines for the Cataloguing of Newspapers. Ed.: For the IFLA Section on Serial Publications. Working Group on Newspapers / 1989. VI, 28 p. / Soft (3-598-10982-2)
- List of Uniform Titles for Liturgical Works of the Latin Rites of the Catholic Church. Recommended by the Working Group on Uniform Headings for Liturgical Works / 2nd rev. ed. 1981. X, 17 p. / Soft (3-598-10957-1)
- Manuel Unimarc Version française. Traduit par Mireille Chauveinc; Thierry Cloarec; Suzanne Jouguelet. Ed: Federation Internationale des Associations des Bibliothecaires et d'Institutions / 1991. IX, 419 S. / Br (3-598-10984-9)
  - 2ème ed. Traduit par Marc Chauveinc / 1996. VIII, 396 S. / Br (3-598-11308-0)
  - 3ème ed. Traduit par Marc Chauveinc / 1999. IX, 427 S. / Br (3-598-11417-6)
- Names of Persons - Supplement. Ed.: IFLA International Office for UBC, London / 3rd ed. 1980. 49 p. / Soft (3-598-11046-4)
- Names of Persons National Usages for Entry in Catalogues. Ed.: IFLA International Office for UBC, London / 3rd. ed. 1977. 55 p. / Soft (3-598-11045-6)

- Names of States: an Authority List of Language Forms for Catalogue Entries. Ed.: IFLA International Office for UBC / 1981. VIII, 49 p. / Soft (3-598-10958-X)
- Proceedings of the National Bibliographies Seminar Brighton, 18 August 1987 held under the Auspices of the IFLA Division of Bibliographic Control. Ed.: Winston D. Roberts / 1988. IV, 76 p. / Soft (3-598-10966-0)
- Recommended Standards for Cataloguing-in-Publication: the CIP Data Sheetand the CIP Record in the Book / 1986. III, 30 p. / Soft (3-598-10964-4)
- Serial Holdings Statements at the Summary Level: Recommendations. Ed.: The IFLA International Programme for UBC. Prepared by Marjorie E. Bloss / 1985. I, 56 p. / Soft (3-598-10965-2)
- Standard Practices in the Preparation of Bibliographic Records. Ed.: IFLA Universal Bibliographic Control and International MARC Program. Comp. by Dorothy Anderson / Rev. ed. 1989. VIII, 62 p. / Soft (3-598-10967-9)
- UNIMARC in Theory and Practice. Papers from the Unimarc Workshop Sydney, Australia August 1988. Ed. by Sally H. McCallum; Winston D. Roberts / 1989. VIII, 90 p. / Soft (3-598-10981-4)
- UNIMARC Manual. Ed.: International Federation of Library Associations and Institutions; Brian P. Holt, with the assistance of Sally H. McCallum; A. B. Long, / 1987. V, 498 p. / Soft (3-598-10960-1)
- Woods, Elaine W.: The MINISIS / UNIMARC Project. Final Report A Study conducted for the IFLA Programme Management Committee / 1988. VI, 18 p. / Soft (3-598-10961-X)

UBCIM Publications - New Series. Ed. by Marie F. Plassard; International Federation of Library Associations and Institutions (IFLA), Den Haag.
- 1: ISBD (PM) International Standard Bibliographic Description for Printed Music. Recommended by the Project Group on ISBD (PM) of the International Association of Music Libraries, Archives and Documentation Centres (IAML). Approved by the Standing Committee of the IFLA Section on Cataloguing / 2nd rev. ed. 1991. 73 p. / Hard (3-598-10985-7)
- 2: UNIMARC / Authorities Universal Format for Authorities. Recommended by the IFLA Steering Group ona UNIMARC Format for Authorities. Ed.: Approved by the Standing Committee of the IFLA Sections on Cataloguing and Information Technology / 1991. 80 p. / Hard (3-598-10986-5)
- 3: ISBD (A) International Standard Bibliographic Description for Older Monographic Publications (Antiquarian). Recommended by the Project Group on the International Standard Bibliographic Description for Older Monographic Publication (Antiquarian). Ed.: International Federation of Library Associations and Institutions. Approved by the Standing Comittes of the IFLA Section on Cataloguing and the IFLA Section on Rare Books and Manuscripts / 2nd rev. ed. 1991. XIV, 100 p. / Hard (3-598-10988-1)
- 4: Dewey: An International Perspective. Papers from a Workshop on the Dewey Decimal Classification and DDC 20. Presented at the General Conference of the International Federation of LibraryAssociations and Institutions (IFLA) August 24, 1989, Paris, France. Ed. by Robert P. Holley / 1991. 173 p. / Hard (3-598-11060-X)
- 5: Management and Use of Name Authority Files Personal Names, Corporate Bodies and Uniform Titles. Evaluation and Prospects. Ed. by Marcelle

Beaudiquez; Françoise Bourdon / 1991. 90 p. / Hard (3-598-10989-X)
- 6: ISBD (G): General International Standard Bibliographic Description. Annotated Text. Prepared by the ISBD Review Committee Working Group. Set up by the IFLA Committee on Cataloguing / Rev. ed. 1992. VIII, 36 p. / Hard (3-598-11084-7)
- 7: Seminar on Bibliographic Records. Proceedings of the Seminar held in Stockholm, 15-16 August 1990 and sponsored by the IFLA UBCIM Programme and the IFLA Division of Bibliographic Control. Ed. by Ross Bourne / 1992. VIII, 147 p. / Soft (3-598-11085-5)
- 8: Lambrecht, Jay H.: Minimal Level Cataloguing by National Bibliographic Agencies / 1992. VIII, 73 p. / Hard (3-598-11102-9)
- 9: International Guide to MARC Databases and Services National Magnetic Tape, Online and CD-ROM, Services. Ed. by the IFLA UBCIM Programme with the assistance of Kathleen McBride; Paula Jones-Fuller / 3rd rev. enl. ed. 1993. 307 p. / Hard (3-598-10987-3)
- 10: UNIMARC / CCF Proceedings of the Workshop held in Florence, 5-7 June 1991. Ed. by Marie F. Plassard; Diana McLean Brooking. Sponsored by IFLA and UNESCO / 1993. X, 157 p. / Hard (3-598-11140-1)
- 11: Bourdon, Françoise: International Cooperation in the Field of Authority Data. An Analytical Study with Recommendations. Trans. from the French by Ruth Webb / 1993. VI, 135 p. / Hard (3-598-11169-X)
- 12: Guidelines for Subject Authority and Reference Entries / 1993. 62 p. / Hard (3-598-11180-0)
- 13: UNIMARC and CDS / ISIS Proceedings of the Workshops held in Budapest, 21.-22. June 1993 and Barcelona, 26. August 1993. Ed. by Marie F. Plassard; Marvin Holdt / 1994. VIII, 84 p. / Hard (3-598-11210-6)
- 14: UNIMARC Manual Bibliographic Format. Ed. by Marie F. Plassard / 2nd ed. 1994. IV, 496 p. / Loose leaf (3-598-11211-4)
  - Update 1. 1996 120 p. / Loose leaf (3-598-11212-2)
  - Update 2. 1998 160 p. / Loose leaf (3-598-11213-0)
- 15: Principles and Practices in the 90's. Proceedings of the IFLA Satellite Meeting Held in Lisbon, Portugal, 17-18 August 1993, and Sponsored by the IFLA Section on Classification and Indexing and the Instituto da Biblioteca Nacional e do Livro, Lisbon, Portugal / 1995. X, 302 p. / Hard (3-598-11251-3)
- 16: Names of Persons. National Usages for Entry in Catalogues. Ed.: Federation Internationale des Associations des Bibliothecaires et d'Institutions / 4th rev. and enl. ed. 1996. XII, 263 p. / Hard (3-598-11342-0)
- 17: ISBD (ER): International Standard Bibliographic Description for Electronic Resources. Revised from the ISBD (CF) International Standard Bibliographic Description for Computer Files. Recommended by the ISBD (CF) Review Group / 1997. VI, 109 p. / Hard (3-598-11369-2)
- 18: Bell, Barbara L.: An Annotated Guide to Current National Bibliographies / 2nd comp. rev. ed. 1997. XXXIII, 487 p. / Hard (3-598-11376-5)
- 19: Functional Requirements for Bibliographic Records Final Report. Ed.: IFLA Study Group on the Functional Requirements for Bibliographic Records. Approved by the Standing Commitee of the IFLA Section on Cataloguing / 1998. VIII, 136 p. / Hard (3-598-11382-X)

UFITA Archiv für Urheber-, Film-, Funk- und Theaterrecht. Hrsg. v. Georg Roeber / Bd. 51/1969 - Bd. 55/1970 (3 - 4 Bde. jährl.)

Umstätter, Walther / Margarete Rehm: Einführung in die Literaturdokumentation und Informationsvermittlung Medizin - Biologie - Chemie - Physik / 1981. 208 S. / Br (3-598-10390-5)

UNESCO-Konferenzberichte. Hrsg.: Deutsche UNESCO-Kommission, Köln (Bde. 1 - 3) u. Bonn (Bde. 4 - 6) und die UNESCO Kommisionen Österreichs und der Schweiz.
- 1: Dritte Weltkonferenz über Erwachsenenbildung. Schlußbericht der internationalen UNESCO-Konferenz, 25. Juli bis 7. August 1972 in Tokyo / 1973. 124 S. / Br (3-7940-5301-X)
- 2: Zwischenstaatliche Konferenz über Kulturpolitik in Europa. Schlußbericht d.v.d.UNESCO 1972 in Helsinki veranst. internat. Konferenz / 1973. 103 S. (3-7940-5302-8)
- 3: Hochschulbildung in Europa. Schlußbericht u. Arbeitsdokumente der von d. UNESCO vom 26. Nov. bis 3. Dez. 1973 in Bukarest veranstalteten 2. Konferenz d. europ. Erziehungsminister / 1974. 303 S. (3-7940-5303-6)
- 4: Zwischenstaatliche Konferenz über Umwelterziehung. Schlußbericht u. Arbeitsdokumente d. von d. UNESCO in Zusammenarbeit mit d. Umweltprogramm d. Vereinten Nationen (UNEP) vom 14. bis 26. Oktober 1877 in Tiflis (UdSSR) veranstalteten Konferenz / 1979. 251 S. (3-598-21324-7)
- 5: Weltkonferenz über Kulturpolitik / World Conference on Cultural Politics. Schlußbericht von der internationalen UNESCO Konferenz. Mexico-Stadt / 1982. Übers. v. Brigitte Scheer / 1983. 226 S. / Br (3-598-21325-5)
- 6: Weltkonferenz über Erwachsenenbildung (4.) Ergebnisse der von der UNESCO vom 19. bis 29. März 1985 in Paris veranstalteten Konferenz / 1989. 105 S. / Br (3-598-21326-3)

Ungarische Drucke und Hungarica 1480-1720 / Magyar Es Magyar Vonatkoz sú Nyomtatv nyok 1480-1720 Katalog der Herzog August Bibliothek Wolfenbüttel / A wolfenbütteli Herzog August Könyvtár katalógusa. Bearb. v. S. Katalin Németh / 3 Bde. 1993. Zus. LXV, 919 S. / Ln (3-598-32813-3)

Ungarisches Biographisches Archiv / Magyar Eletrazi Archivum / Hungarian Biographical Archive (UBA). Eine alphabetische Kumulation von biographischen Einträgen aus ca. 127 Nachschlagewerken von 1559 bis 1990. Mikrofiche-Edition. Bearb. v. Ulrike Kramme und Zelmíra Urra Muena / 1994-1998. 141 Fiches in 12 Lfgn, Lesefaktor 24x / Silberfiche (3-598-33780-9) / Diazofiche (3-598-33766-3)

The **Unification** of private Law and Law and Legal Literature in Italy. International Association of Law Libraries. Courses in Law Librarianship. 4th Course in Rome, September 4-8, 1972. Edited by IALL, Marburg / 1974. X, 120 S. (3-7940-5145-9)

Union List of African Censuses Development Plans, and Statistical Abstracts. Comp. by Victoria K. Evalds / 1985. XIV, 232 S. / Hard (3-598-10576-2) [Hans Zell Publishers]

United Nations Resolutions on Palestine 1947-1972. Compiled and classifiedby Sami Musallam. Ed.: Institute for Palestine Studies, Beirut, and Center for

Documentation and Studies, Abu Dhabi / 1974. 242 S. (3-7940-5159-9)

Universalbibliographie Technik und Wirtschaft. Jahresbände in zwei Halbjahresausgaben Juli und Januar mit den neuesten Nachweisen der einschlägigen Fachliteratur über den Stand von Technik und Wirtschaft in über 150 Ländern / Von 37 Fachausgaben 1959 (monatl.) bis 50 Fachausgaben / 1964 (halbjährl.)
• Fachausgaben:
  - Technik allgemein - Technik- und Wirtschaftsgeschichte
  - Atom - Mathematik - Physik - Thermotechnik - Medizin. Technik - Pharmazie - Unfall- Katastrophen- Gesundheitsschutz - Akusitk, Optik, Feinmechanik - Meteorologie, Astronomie - Chemie
  - Meßtechnik, Technologie - Holz - Kunststoffe - Textil und Bekleidung - Gummi - Farben, Lacke - Glas, Keramik
  - Maschinenbau - Regelungstechnik - Motoren, Pumpen - Fertigung, Werkzeugmaschinen - Wehrtechnik
  - Innerbetriebl. Transport - Kraftfahrzeuge - Eisenbahntechnik - Schiffe, Seefahrt - Luftfahrt - Verkehr
  - Energie - Feuerungstechnik, Brennstoffe - Starkstromtechnik, Lichttechnik - Nachrichtentechnik, Elektronik
  - Bergbau, Geologie - Eisenhüttenwesen - Metall
  - Bau allgemein - Baustoffe, -maschinen - Hochbau - Tiefbau - Wasserbau
  - Nahrungsmitteltechnik - Land- u. Forstwirtschaft
  - Gewerbliche Technik - Druck, graphische Technik - Betriebswirtschaft - Werbung u. Marktforschung - Industrie und Volkswirtschaft - Patentwesen

Die **Universität** Göttingen unter dem Nationalsozialismus. Das verdrängte Kapitel ihrer 250jährigen Geschichte. Hrsg. v. Heinrich Becker; Hans J. Dahms; Cornelia Wegeler / 1987. 523 S., 16 S. Abb. / Lin (3-598-10676-9)
• 2. erw. Ausg. 1998. 759 S. / Lin (3-598-10853-2)

Universität zu Köln Institut für Theater-, Film- und Fernsehwissenschaft. Theatersammlung Findbuch der szenischen Graphik / University of Cologne Institute of Theatre-, Film- and Television Studies Theatre-Collection Index of Scenographic Design. Hrsg.: Universität zu Köln, Institut f. Theater-, Film u. Fernsehwissenschaft, Theatersammlung / 4 Bde. 1993. XXXVIII, 1.069 S. / Gb (3-598-11148-7)

Universitätsbibliothek der Freien Universität Berlin. Alphabetischer Katalog / Monographien. Mikrofiche-Edition. Stand: Juli 1981 / 1982. ca. 1.1504.289 Karten auf 846 Fiches. Lesefaktor 42x. Diazofiche (3-598-30340-8)

Der **vergeudete** Reichtum. Über die Partizipation von Frauen im öffentlichen Leben. Hrsg. v. Claudia Bernardoni u. Vera Werner; Deutsche UNESCO-Kommission, Bonn. Red.: Hans-Dieter Dyroff / 2., berb. Aufl. 1987. 245 S. / Br (3-598-07523-5)

Vergleichende Erziehungswissenschaft. Texte zur Methodologie-Diskussion. Hrsg., übers. u. eingel. v. Adelheid Busch; Friedrich W. Busch; Bernd Krüger; Marianne Krüger-Potratz / 1974. 309 S. / Kst. (UTB 410) (3-7940-2628-4)

Verhandlungen des Deutschen Bundestages und des Bundesrates 1949-1998 ff. Protokolle, Drucksachen, Register. Mikrofiche-Edition / 1980 ff. Lesefaktor 48x (Koprod. mit C. H. Beck, München) (3-598-30200:2)
• 1. Wahlperiode
  - Verhandlungen des Deutschen Bundestages 1949-1953 / 1981. 38.220 S. auf 91 Fiches / Silberfiche (3-598-30850-7) / Diazofiche (3-598-30201-0)
  - Verhandlungen des Bundesrates 1949-1953 / 1982. 53.000 S. auf 128 Fiches / Silberfiche (3-598-30860-4) / Diazofiche (3-598-30202-9)
  Sachregister → 2. Wahlperiode
• 2. Wahlperiode
  - Verhandlungen des Deutschen Bundestages 1953-1957 / 1981. 54.000 S. auf 113 Fiches / Silberfiche (3-598-30851-5) / Diazofiche (3-598-30203-7)
  - Verhandlungen des Bundesrates 1954-1957 / 1983. 46.000 S. auf 111 Fiches / Silberfiche (3-598-30861-2) / Diazofiche (3-598-30204-5)
  - Sachregister zu den Verhandlungen des Deutschen Bundestages 1. und 2. Wahlperiode (1949-1957) und den Verhandlungen des Bundesrates (1949-1957) / 1984. XVI, 289 S. / Ln (3-598-30236-3)
• 3. Wahlperiode
  - Verhandlungen des Deutschen Bundestages 1957-1961 / 1982. 410.00 S. auf 96 Fiches / Silberfiche (3-598-3085-3) / Diazofiche (3-598-30205-3)
  - Verhandlungen des Bundesrates 1958-1961 / 1984. 38.220 S. auf 91 Fiches / Silberfiche (3-598-30862-0) / Diazofiche (3-598-30206-1)
  Sachregister → 4. Wahlperiode
• 4. Wahlperiode
  - Verhandlungen des Deutschen Bundestages 1961-1965 / 1982. 504.387 S. auf 112 Fiches / Silberfiche (3-598-30853-1) / Diazofiche (3-598-30207-X)
  - Verhandlungen des Bundesrates 1962-1965 / 1985. 42.000 S. auf 1.00 Fiches / Silberfiche (3-598-30863-9) / Diazofiche (3-598-30208-8)
  - Sachregister zu den Verhandlungen des Deutschen Bundestages 3. und 4. Wahlperiode (1957-1965) und den Verhandlungen des Bundesrates (1958-1965) / 1986. XV, 244 S. / Ln (3-598-30237-1)
• 5. Wahlperiode
  - Verhandlungen des Deutschen Bundestages 1965-1969 / 1982. 47.000 S. auf 153 Fiches / Silberfiche (3-598-30854-X) / Diazofiche (3-598-30209-6)
  - Verhandlungen des Bundesrates 1966-1969 / 1985. 36.000 S. auf 120 Fiches / Silberfiche (3-598-30864-7) / Diazofiche (3-598-30210-X)
  Sachregister → 6. Wahlperiode
• 6. Wahlperiode
  - Verhandlungen des Deutschen Bundestages 1969-1972 / 1982. 62.000 S. auf 148 Fiches / Silberfiche (3-598-30855-8) / Diazofiche (3-598-30211-8)
  - Verhandlungen des Bundesrates 1970-1972 / 1986. 24.000 S. auf 122 Fiches / Silberfiche (3-598-30865-5) / Diazofiche (3-598-30212-6)
  - Sachregister zu den Verhandlungen des Deutschen Bundestages 5. und 6. Wahlperiode (1965-1972) und den Verhandlungen des Bundesrates (1966-1972) / 1987. XV, 258 S. / Ln (3-598-30238-X)
• 7. Wahlperiode
  - Verhandlungen des Deutschen Bundestages 1972-1976 / 1980. 104.000 S. auf 248 Fiches / Silberfiche (3-598-30856-6) / Diazofiche (3-598-30213-4)

  - Verhandlungen des Bundesrates 1973-1976 / 1981. 76.000 S. auf 187 Fiches / Silberfiche (3-598-30866-3) / Diazofiche (3-598-30214-2)
  - Sachregister und Konkordanzliste zu den Verhandlungen des Deutschen Bundestages 7. Wahlperiode (1972-1976) und zu den Verhandlungen des Bundesrates (1973-1976) Bearb. v. Beate Heiss; Barbara Kantenwein-Pabst; Detlef Scharff. Gesamtredakt.: Stefan Tischler / 1982. IX, 219 S. / Ln (3-598-30233-9)
• 8. Wahlperiode
  - Verhandlungen des Deutschen Bundestages 1976-1980 / 1981. 78.000 S. auf 196 Fiches / Silberfiche (3-598-30857-4) / Diazofiche (3-598-30215-0)
  - Verhandlungen des Bundesrates 1977-1980 / 1981. 71.000 S. auf 177 Fiches / Silberfiche (3-598-30867-1) / Diazofiche (3-598-30216-9)
  - Sachregister und Konkordanzliste zu den Verhandlungen des Deutschen Bundestages 8. Wahlperiode (1976-1980) und den Verhandlungen des Bundesrats (1977-1980) Bearb. v. Günther Hagen; Barbara Kantenwein-Pabst; Paula Scharff. Gesamtredakt.: Stefan Tischler / 1982. XVI, 173 S. / Ln (3-598-30234-7)
• 9. Wahlperiode
  - Verhandlungen des Deutschen Bundestages 1980-1983 / 1981-1983. 48.300 S. auf 115 Fiches / Silberfiche (3-598-30858-2) / Diazofiche (3-598-30217-7)
  - Verhandlungen des Bundesrates 1981-1982 / 1983. 40.720 S. auf 97 Fiches / Silberfiche (3-598-30868-X) / Diazofiche (3-598-30225-8)
  - Sachregister und Konkordanzliste zu den Verhandlungen des Deutschen Bundestages 9. Wahlperiode (1980-1983) und zu den Verhandlungen des Bundesrates (1981-1982) Bearb. v. Günther Hagen; Barbara Kantenwein-Pabst; Paula Scharff. Gesamtredakt.: Stefan Tischler / 1983. XIII, 106 S. / Ln (3-598-30235-5)
• 10. Wahlperiode
  - Verhandlungen des Deutschen Bundestages 1983-1987 / 1983-1987. 240 Fiches u. Konkordanzliste / Silberfiche (3-598-30859-0) / Diazofiche (3-598-30901-5)
  - Verhandlungen des Bundesrates 1983-1987 / 1984-1987. 185 Fiches u. Konkordanzliste / Silberfiche (3-598-30869-8) / Diazofiche (3-598-30916-3)
  - Sachregister zu den Verhandlungen des Deutschen Bundestages 10. Wahlperiode (1983-1987) und zu den Verhandlungen des Bundesrates (1983-1986) Bearb. v. Susanne Bauder; Günther Hagen; Rosina Heck. Gesamtredakt.: Stefan Tischler / 1987 / Ln (3-598-30239-8)
• 11. Wahlperiode
  - Verhandlungen des Deutschen Bundestages 1987-1991 / 1992. 316 Fiches u. Konkordanzliste / Silberfiche (3-598-30870-1) / Diazofiche (3-598-30924-4)
  - Verhandlungen des Bundesrates 1987-1991 / 1992. 250 Fiches / Silberfiche (3-598-30871-X) / Diazofiche (3-598-30938-4)
  - Sachregister zu den Verhandlungen des Deutschen Bundestages 11. Wahlperiode (1987-1991) und zu den Verhandlungen des Bundesrates (1987-1990) Bearb. v. Susanne Bauder; Dorothea Wüst. Gesamtredakt.: Günther Hagen / 1991 / Ln (3-598-30979-1)
• 12. Wahlperiode
  - Verhandlungen des Deutschen Bundestages 1991-1994 / 1991-1995. 385 Fiches / Silberfiche

(3-598-30872-8) / Diazofiche (3-598-31929-0)
- Verhandlungen des Bundesrates 1991-1994 /
1992-1995. 334 Fiches / Silberfiche
(3-598-3083-6) / Diazofiche (3-598-31949-5)
- Sachregister zu den Verhandlungen des
Deutschen Bundestages 12. Wahlperiode
(1991-1994) und zu den Verhandlungen des
Bundesrates (1991-1994) Bearb. v. Günther
Hagen / 1995. 191 S. / Ln (3-598-31944-4)
• 13. Wahlperiode
- Verhandlungen des Deutschen Bundestages
1995-1998 / 1995-1999 / Silberfiche
(3-598-30874-4) / Diazofiche (3-598-31957-6)
- Verhandlungen des Bundesrates 1995-1998 /
1996-98 / Silberfiche (3-598-30875-2) /
Diazofiche (3-598-31974-6)
Sachregister zu den Verhandlungen des Deut-
schen Bundestages 13. Wahlperiode (1995-1998)
und zu den Verhandlungen des Bundesrates
(1995-1998) [i.Vb.]
siehe auch: → ***Protokolle*** *des*
Vermittlungsausschusses des Deutschen
Bundestages und des Bundesrates

Verteidigungs-Dokumentation. Beitr. zur Aufgabe,
Organisation u. Methodik d. Dokumentation im
Geschäftsbereich d. Bundesministers d. Verteidigung
d. Bundesrepublik Deutschland mit Hinweisen auf
d. Verteidugungs-Dokumentation anderer Staaten.
Hrsg. von Rudolf Harbeck / 1976. IX, 537 S.
(3-7940-3222-5)

Verzeichnis der im Rahmen der Europäischen
Gemeinschaften zusammengeschlossenen land- und
ernährungswirtschaftlichen Verbände. Directory of
nongovernmental agricultural organizations set up
at Community level. / Répertoire des organizations
agricoles non gouvernementales groupées dans le
cadre des Communautés européennes. Fortegnelse
over ikke-statslige landbrugsorganisationer inden
for De europaeiske Faelleskaber; Repertorio delle
organizzazioni agricole non governative reggruppate
nel quadro delle Comunitá europee. Lijst van niet-
gouvernementele organisaties op landbouwgebied
binnen het kader van de Europese Gemeenschappen
/ . Hrsg.: Kommission der Europäischen Gemeinschaften / 5. Aufl. 1977. 466 S. (3-7940-3021-4)

Verzeichnis der Land- und ernährungswirtschaftlichen Verbände, zusammengeschlossen im Rahmen
der EG / Directory of nongovernmental agricultural
organizations, set up at European Community level /
Repertoire des organisations agrioles et alimentaire.
Hrsg.: Kommission der Europäischen Gemeinschaften / 1980. 402 S. (3-598-10147-3)
• 7. Ausg. 1986. 319 S. / Br (3-598-10504-5)
• 8. Ausg. Hrsg.: Generaldirektion Wissenschaftliche
und Technische Information und Informationsmanagement der EG. Zsgst. v. d. Generaldirektion
Landwirtschaft der EG / 1989. 347 S. / Br
(3-598-10801-X)

Verzeichnis der Nobelpreisträger 1901-1987. Mit
Preisbegründungen, Kurzkommentaren, literarischen Werkbibliographien und einer Biographie
Alfred Nobels. Hrsg. v. Werner Martin / 1985. X, 362
S. (3-598-10578-9)
• 2. überarb. u. erw. Ausg. 1988. XI, 382 S. / Lin
(3-598-10721-8)

Verzeichnis deutscher Datenbanken, Datenbank-
Betreiber und Informationsvermittlungsstellen.
Bundesrepublik Deutschland und Berlin (West)
(ISSN 0178-1685)

• Ausgabe 1988. Hrsg.: Gesellschaft f. Mathematik
u. Datenverarbeitung mbH (GMD), St. Augustin /
1988. IX, 424 S. / Br (3-598-10765-X)
Ausg. 1985 → ***Informationsdienste****: Verzeichnis*
deutscher Datenbanken

Verzeichnis deutscher Informations- und Doku-
mentationsstellen. Bundesrepublik Deutschland und
Berlin (West) (ISSN 0344-3647)
• Ausgabe 3 - 1979. Hrsg.: Gesellschaft f. Informa-
tion u. Dokumentation mbH (GID), Frankfurt a.
M. / 1979. 463 S. / Br (3-10013-2))
Ausg. 4 → ***Informationsdienste****: Verzeichnis*
deutscher Informations- und
Dokumentationsstellen
• Ausgabe 5 - 1989. Hrsg.: Gesellschaft f. Mathema-
tik u. Datenverarbeitung mbH (GMD), St. Augu-
stin / 1990. VIII, 518 S. / Lin (3-598-10622-X)

Verzeichnis in Deutschland gelaufener Filme.
Entscheidungen der Filmzensur 1911-1920 Berlin,
Hamburg, München, Stuttgart. Hrsg. v. Herbert Birett
/ 1980. IV, 918 S. / Lin (3-598-10067-1)

Verzeichnis lieferbarer Bücher / German Books in
Print (VLB) [Vertrieb u. Auslieferung für die Buchhändler-Vereinigung, Frankfurt]
• Verzeichnis lieferbarer Bücher / 2 Bde. 1971. Zus.
1.986 S. / Ln (3-7657-0378-8)
• 1972/73
- Verzeichnis lieferbarer Bücher 1972/73. Titelverzeichnis nach dem Autorenalphabet, alpha-
numerisches Verlagsregister, numerisches
verzeichnis der Reihenschlüssel, numerisches
Register der ISBN, Titel- und Stichwortver-
zeichnis / 2 Bde. 2. Ausg. 1972. Zus. 4.080 S. /
Lin (3-7657-0433-4)
- Ergänzungsband Frühjahr 1973 / 1973. VIII,
411 S. / Lin (3-7657-0450-4)
• 1973/74
- 3. Ausg. 3 Bde. 1973. Zus. 4.385 S. / Skinoid
(3-7657-0503-9)
- Ergänzungsband Frühjahr 1974 / 2. Ausg. 1974.
339 S. / Skinoid
• 1974/75
- 4. Ausg. 3 Bde. 1974. Zus. 4.385 S. / Skinoid
(3-7657-0503-9)
- Ergänzungsband Frühjahr 1975 / 3. Ausg. 1975.
339 S. / Skinoid
• 1975/76
- 5. Ausg. 3 Bde. 1975. Zus. 4.564 S. / Skinoid
(3-7657-0552-7)
- Ergänzungsband Frühjahr 1976 / 4. Ausg. 1976.
447 S. / Skinoid
• 1976/77
- 6. Ausg. 3 Bde. 1976. Zus. 4.849 S. / Skinoid
(3-7657-0603-5)
- Ergänzungsband Frühjahr 1977 / 5. Ausg. 1977.
416 S. / Skinoid (3-7657-0656-6)
• 1977/78
- Verzeichnis lieferbarer Bücher / German Books
in Print (VLB). Bücherverzeichnis im Autorenalphabet, kumuliert mit Titel- und Stichwortregister, mit Verweisung auf den Autor /
Authors - Titles - Catchwords / 3 Bde. 7. Ausg.
1977. Zus. 5.213 S. / Kst. (3-7657-0658-2)
- ISBN Register / 7. Ausg. 1977. 494 S. / Kst.
(3-7657-0695-7)
- Ergänzungsband Frühjahr 1978 / Supplement
Spring 1978 / 6. Ausg. 1978. 450 S. / Kst.
(3-7657-0758-9)
- Ergänzungsband Frühjahr 1978. ISBN Register
/ Supplement Spring 1978. ISBN-Index / 6.
Ausg. 1978. 38 Bl / Kst. (3-7657-0773-2)

• 1978/79
- Bücherverzeichnis im Autorenalphabet, kumu-
liert mit Titel- und Stichwortregister, mit
Verweisung auf den Autor / Authors - Titles -
Catchwords / 3 Bde. 8. Ausg. 1978. Zus. 5.536 S.
/ Kst. (3-7657-0779-1)
- ISBN Register / 8. Ausg. 1978. 554 S. / Kst.
(3-7657-0880-1)
- Schlagwortverzeichnis / Subject Guide / 2 Bde.
1978. Zus. 2.289 S. / Kst. (3-7657-0881-X)
- Ergänzungsband Frühjahr 1979 / Supplement
Spring 1979 / 7. Ausg. 1979. 474 S. / Kst.
(3-7657-0825-39
- Ergänzungsband Frühjahr 1979. ISBN Register
/ Supplement Spring 1979. ISBN-Index / 7.
Ausg. 1979. 78 S. / Kst. (3-7657-0827-5)
• 1979/80
- Bücherverzeichnis im Autorenalphabet, kumu-
liert mit Titel- und Stichwortregister, mit Ver-
weisung auf den Autor / Authors - Titles -
Catchwords / 3 Bde. 9. Ausg. 1979. Zus. 5.915 S.
/ Kst
- ISBN Register / 9 Ausg. 1979 / Kst
- Schlagwortverzeichnis / Subject Guide / 3 Bde.
2. Ausg. 1979. Zus. 4.858 S. / Kst
- Ergänzungsband Frühjahr 1980 / Supplement
Spring 1980 / 8. Ausg. 1980 / Kst
- Ergänzungsband Frühjahr 1980. ISBN Register
/ Supplement Spring 1980. ISBN-Index / 8.
Ausg. 1980 / Kst
• 1980/81
- Bücherverzeichnis im Autorenalphabet, kumu-
liert mit Titel- und Stichwortregister, mit Ver-
weisung auf den Autor / Authors - Titles -
Catchwords / 4 Bde. 10. Ausg. 1980 / Kst
- ISBN Register / 10. Ausg. 1980 / Kst
- Schlagwortverzeichnis / Subject Guide / 3 Bde.
3. Ausg. 1980. Zus. 4.858 S. / Kst.
(3-7657-0987-5)
- Ergänzungsband Frühjahr 1981 / Supplement
Spring 1981 / 9. Ausg. 1981 / Kst.
(3-7657-0988-3)
- Ergänzungsband Frühjahr 1981. ISBN Register
/ Supplement Spring 1981. ISBN-Index / 9.
Ausg. 1981 / Kst. (3-7657-0989-1)
• 1981/82
- Bücherverzeichnis im Autorenalphabet, kumu-
liert mit Titel- und Stichwortregister, mit Ver-
weisung auf den Autor / Authors - Titles -
Catchwords / 4 Bde. 11. Ausg. 1981. Zus. 7.000
S. / Kst. (3-7657-1046-6)
- ISBN Register / 11. Ausg. 1981. 650 S. / Kst.
(3-7657-1047-4)
- Schlagwortverzeichnis / Subject Guide / 3 Bde.
4. Ausg. 1981 / Kst
- Ergänzungsband Frühjahr 1982 / Supplement
Spring 1982 / 10. Ausg. 1982 / Kst
- Ergänzungsband Frühjahr 1982. ISBN Register
/ Supplement Spring 1982. ISBN-Index / 10.
Ausg. 1982 / Kst
• 1982/83
- Bücherverzeichnis im Autorenalphabet, kumu-
liert mit Titel- und Stichwortregister, mit Ver-
weisung auf den Autor / Authors - Titles -
Catchwords / 4 Bde. 12. Ausg. 1982. Zus. 7.000
S. / Kst. (3-7657-1132-2)
- ISBN Register / 12. Ausg. 1982. 650 S. / Kst.
(3-7657-1133-0)
- Schlagwortverzeichnis / Subject Guide / 3 Bde.
5. Ausg. 1982. Zus. 6.000 S. / Kst.
(3-7657-1134-9)
- Ergänzungsband Frühjahr 1983 / Supplement
Spring 1983 / 11. Ausg. 1983. 600 S. / Kst.
(3-7657-1135-7)

- Ergänzungsband Frühjahr 1983. ISBN Register / Supplement Spring 1983. ISBN-Index / 11. Ausg. 1983. 40 S. / Kst. (3-7657-1136-5)
• 1983/84
- Bücherverzeichnis im Autorenalphabet, kumuliert mit Titel- und Stichwortregister, mit Verweisung auf den Autor / Authors - Titles - Catchwords / 4 Bde. 13. Ausg. 1983. Zus. 7.500 S. / Kst. (3-7657-1217-5)
- ISBN Register / 13. Ausg. 1983. 680 S. / Kst. (3-7657-1218-3)
- Schlagwortverzeichnis / Subject Guide / 3 Bde. 6. Ausg. 1983. Zus. 6.200 S. / Kst. (3-7657-1219-1)
- Ergänzungsband Frühjahr 1984 / Supplement Spring 1984 / 12. Ausg. 1984. 600 S. / Kst. (3-7657-1220-5)
- Ergänzungsband Frühjahr 1984. ISBN Register / Supplement Spring 1984. ISBN-Index / 12. Ausg. 1984. 80 S. / Kst. (3-7657-1221-3)
• 1984/85
- Bücherverzeichnis im Autorenalphabet, kumuliert mit Titel- und Stichwortregister, mit Verweisung auf den Autor / Authors - Titles - Catchwords / 4 Bde. 14. Ausg. 1984. Zus. 7.500 S. / Kst. (3-7657-1256-6)
- ISBN Register / 14. Ausg. 1984. 770 S. / Kst. (3-7657-1257-4)
- Schlagwortverzeichnis / Subject Guide / 3 Bde. 7. Ausg. 1984. Zus. 6.200 S. / Kst. (3-7657-1258-2)
- Ergänzungsband Frühjahr 1985 / Supplement Spring 1985 / 13. Ausg. 1985. 727 S. / Kst. (3-7657-1259-0)
- Ergänzungsband Frühjahr 1985. ISBN Register / Supplement Spring 1985. ISBN-Index / 13. Ausg. 1985. 140 S. / Kst. (3-7657-1260-4)
• 1985/86
- Bücherverzeichnis im Autorenalphabet, kumuliert mit Titel- und Stichwortregister, mit Verweisung auf den Autor / Authors - Titles - Catchwords / 4 Bde. 15. Ausg. 1985. Zus. 8.200 S. / Kst. (3-7657-1294-9)
- ISBN Register / 15. Ausg. 1985. 800 S. / Kst. (3-7657-1295-7)
- Schlagwortverzeichnis / Subject Guide / 4 Bde. 8. Ausg. 1985. Zus. 6.500 S. / Kst. (3-7657-1296-5)
- Ergänzungsband Frühjahr 1986 / Supplement Spring 1986 / 14. Ausg. 1986. 785 S. / Kst. (3-7657-1297-3)
- Ergänzungsband Frühjahr 1986. ISBN Register / Supplement Spring 1986. ISBN-Index / 14. Ausg. 1986. 140 S. / Kst. (3-7657-1298-1)
• 1986/87
- Bücherverzeichnis im Autorenalphabet, kumuliert mit Titel- und Stichwortregister, mit Verweisung auf den Autor / Authors - Titles - Catchwords / 5 Bde. 16. Ausg. 1986. Zus. 9.300 S. / Kst. (3-7657-1353-8)
- ISBN Register / 16. Ausg. 1986. 890 S. / Kst. (3-7657-1354-6)
- Schlagwortverzeichnis / Subject Guide / 4 Bde. 9. Ausg. 1986. Zus. 7000 S. / Kst. (3-7657-1355-4)
- Ergänzungsband Frühjahr 1987 / Supplement Spring 1987 / 15. Ausg. 1987. 700 S. / Kst. (3-7657-1356-2)
- Ergänzungsband Frühjahr 1987. ISBN Register / Supplement Spring 1987. ISBN-Index / 15. Ausg. 1987. 140 S. / Kst. (3-7657-1357-0)
• 1987/88
- Bücherverzeichnis im Autorenalphabet, kumuliert mit Titel- und Stichwortregister, mit Verweisung auf den Autor / Authors - Titles -

Catchwords / 5 Bde. 17. Ausg. 1987. Zus. 9.300 S. / Kst. (3-7657-1392-9)
- ISBN Register / 17. Ausg. 1987. 890 S. / Kst. (3-7657-1393-7)
- Schlagwortverzeichnis / Subject Guide / 4 Bde. 10. Ausg. 1987. Zus. 7.000 S. / Kst. (3-7657-1394-5)
- Ergänzungsband Frühjahr 1988 / Supplement Spring 1987 / 16. Ausg. 1988. 700 S. / Kst. (3-7657-1395-3)
- Ergänzungsband Frühjahr 1988. ISBN Register / Supplement Spring 1988. ISBN-Index / 16. Ausg. 1988. 140 S. / Kst. (3-7657-1396-1)
• 1988/89
- Bücherverzeichnis im Autorenalphabet, kumuliert mit Titel- und Stichwortregister, mit Verweisung auf den Autor / Authors - Titles - Catchwords / 5 Bde. 18. Ausg. 1988. Zus. 10.100 S. / Kst. (3-7657-1476-3)
- ISBN Register / 18 Ausg. 1988. 970 S. / Kst. (3-7657-1477-1)
- Schlagwortverzeichnis / Subject Guide / 4 Bde. 11. Ausg. 1988. Zus. 8.800 S. / Kst. (3-7657-1478-X)
- Ergänzungsband Frühjahr 1989 / Supplement Spring 1989 / 17. Ausg. 1989. 900 S. / Kst. (3-7657-1479-8)
- Ergänzungsband Frühjahr 1989. ISBN Register / Supplement Spring 1989. ISBN-Index / 17. Ausg. 1989. 190 S. / Kst. (3-7657-1480-1)
• 1989/90
- Bücherverzeichnis im Autorenalphabet, kumuliert mit Titel- und Stichwortregister, mit Verweisung auf den Autor / Authors - Titles - Catchwords / 6 Bde. 19. Ausg. 1989. 10.100 S. / Kst. (3-7657-1523-9)
- ISBN Register / 19. Ausg. 1989. 970 S. / Kst. (3-7657-1524-7)
- Schlagwortverzeichnis / Subject Guide / 4 Bde. 12. Ausg. 1989. 8.800 S. / Kst. (3-7657-1525-5)
- Ergänzungsband Frühjahr 1990 / Supplement Spring 1990 / 18. Ausg. 1990. 900 S. / Kst. (3-7657-1526-3)
- Ergänzungsband Frühjahr 1990. ISBN Register / Supplement Spring 1990. ISBN-Index / 18. Ausg. 1990. 190 S. / Kst. (3-7657-1527-1)
• 1990/91
- Bücherverzeichnis im Autorenalphabet, kumuliert mit Titel- und Stichwortregister, mit Verweisung auf den Autor / Authors - Titles - Catchwords / 6 Bde. 20. Ausg. 1990. Zus. 11.600 S. / Kst. (3-7657-1567-0)
- ISBN Register / 20. Ausg. 1990. 1.100 S. / Kst. (3-7657-1568-9)
- Schlagwortverzeichnis / Subject Guide / 4 Bde. / 13. Ausg. 1990. Zus. 9.350 S. / Kst. (3-7657-1569-7)
- Ergänzungsband Frühjahr 1991 / Supplement Spring 1991 / 19. Ausg. 1991. 1.100 S. / Kst. (3-7657-1570-0)
- Ergänzungsband Frühjahr 1991. ISBN Register / Supplement Spring 1991. ISBN- 1991 / 19. Ausg. 1991. 160 S. / Kst. (3-7657-1571-9)
• 1991/92
- Bücherverzeichnis im Autorenalphabet, kumuliert mit Titel- und Stichwortregister, mit Verweisung auf den Autor / Authors - Titles - Catchwords / 6 Bde. 21. Ausg. 1991. Zus. 11.600 S. / Kst. (3-7657-1616-2)
- ISBN Register / 21. Ausg. 1991. 1.180 S. / Kst. (3-7657-1617-0)
- Schlagwortverzeichnis. Anh.: Schlagwortübersicht / Subject Guide / 5 Bde. 14. Ausg. 1991. 9.950 S. / Kst. (3-7657-1618-9)
- Ergänzungsband Frühjahr 1992. Anh.: Verlagsverzeichnis / Supplement Spring 1992 /

20. Ausg. 1992. 1.050 S. / Kst. (3-7657-1619-7)
- Ergänzungsband Frühjahr 1992. ISBN Register / Supplement Spring 1992. ISBN-Index / 20. Ausg. 1992. 165 S. / Kst. (3-7657-1620-0)
• 1992/93
- Bücherverzeichnis im Autorenalphabet, kumuliert mit Titel- und Stichwortregister, mit Verweisung auf den Autor / Authors -Titles - Catchwords / 6 Bde. 22. Ausg. 1992. Zus. 11.700 S. / Kst. (3-7657-1670-7)
- ISBN Register. Anhang: Verlagsverzeichnis / 22. Ausg. 1992. 1.180 S. / Kst. (3-7657-1671-5)
- Schlagwortverzeichnis / Subject Guide / 5 Bde. 15. Ausg. 1992. Zus. 10.105 S. / Kst. (3-7657-1672-3)
- Ergänzungsband Frühjahr 1993 / Supplement Spring 1993 / 21. Ausg. 1993. 1.100 S. / Kst. (3-7657-1673-1)
- Ergänzungsband Frühjahr 1993. ISBN Register. Anhang: Verlagsverzeichnis / Supplement Spring 1993 / 21. Ausg. 1993. 170 S. / Kst. (3-7657-1674-X)
• 1993/94
- Bücherverzeichnis im Autorenalphabet, kumuliert mit Titel- und Stichwortregister, mit Verweisung auf den Autor / Authors -Titles - Catchwords / 7 Bde. 23. Ausg. 1993. Zus. 15.000 S. / Kst. (3-7657-1748-7)
- ISBN Register. ISBN-Index / 23. Ausg. 1993. 1.200 S. / Kst. (3-7657-1749-5)
- Schlagwortverzeichnis / Subject Guide / 6 Bde. 16. Ausg. 1993. Zus. 11.000 S. / Kst. (3-7657-1750-9)
- Ergänzungsband Frühjahr 1994 / Supplement Spring 1994 / 22. Ausg. 1994. 1.200 S. / Kst. (3-7657-1751-7)
- Ergänzungsband Frühjahr 1994. ISBN Register / Supplement Spring 1994. ISBN-Index / 22. Ausg. 1994. 180 S. / Kst. (3-7657-1752-5)
• 1994/95
- Bücherverzeichnis mit Autorenalphabet, kumuliert mit Titel- und Stichwortregister, mit Verweisung auf den Autor. Authors -. Titles - Catchwords / 7 Bde. 24. Ausg. 1994. Zus. 15.000 S. / Kst. (3-7657-1817-3)
- ISBN Register. Anh.: Verlagsverzeichnis / 24. Ausg. 1994. 1.200 S. / Kst. (3.7657-1818-1)
- Schlagwortverzeichnis. Anh.: Schlagwortübersicht, Verlagsverzeichnis / Subject Guide / 6 Bde. 17. Ausg. 1994. Zus. 12.000 S. / Kst. (3-7657-1819-X)
- Ergänzungsband Frühjahr 1995. Anhang: Verlagsverzeichnis / Supplement Spring 1995 / 23. Ausg. 1995. 1.300 S. / Kst. (3-7657-1820-3)
- Ergänzungsband Frühjahr 1995. ISBN Register. Anh.: Verlagsverzeichnis / Supplement Spring 1995. ISBN-Index / 23. Ausg. 1995. 200 S. / Kst. (3-7657-1821-1)
• 1995/96
- Bücherverzeichnis mit Autorenalphabet, kumuliert mit Titel- und Stichwortregister, mit Verweisung auf den Autor. Authors -. Titles - Catchwords / 8 Bde. 25. Ausg. 1995. Zus. 14.500 S. / Kst. (3-7657-1886-6)
- ISBN Register. Anh.: Verlagsverzeichnis / 25. Ausg. 1995. 1.350 S. / Kst. (3.7657-1887-4)
- Schlagwortverzeichnis. Anh.: Schlagwortübersicht, Verlagsverzeichnis / Subject Guide / 7 Bde. 18. Ausg. 1995. Zus. 12.500 S. / Kst. (3-7657-1888-2)
- Ergänzungsband Frühjahr 1995. Anhang: Verlagsverzeichnis / Supplement Spring 1995 / 24. Ausg. 1996. 1.350 S. / Kst. (3-7657-1889-0)
- Ergänzungsband Frühjahr 1995. ISBN Register. Anh.: Verlagsverzeichnis / Supplement Spring

1995. ISBN-Index / 24. Ausg. 1996. 264 S. / Kst. (3-7657-1890-4)
• 1996/97
  - Bücherverzeichnis im Autorenalphabet, komuliert mit Titel- und Stichwortregister, mit Verweisung auf den Autor / Authors - Titles - Catchwords / 8 Bde. 26. Ausg. 1996. Zus. 15.600 S. / Kst. (3-7657-1937-4)
  - ISBN Register. Anh.: Verlagsverzeichnis / 26. Ausg. 1996. 1.300 S. / Kst. (3-7657-1938-2)
  - Schlagwortverzeichnis. Anh.: Schlagwortübersicht. Verlagsverzeichnis / Subject Guide / 7 Bde. 19. Ausg. 1996. 13.600 S. / Kst. (3-7657-1939-0)
  - Ergänzungsband Frühjahr 1997 / Supplement Spring 1997 / 25. Ausg. 1997. 1.400 S. / Kst. (3-7657-1940-4)
  - Egänzungsband Frühjahr 1997. ISBN Register. Anh.: Verlagsverzeichnis / Supplement Spring 1997. ISBN-Index / 25. Ausg. 1997. 260 S. / Kst. (3-7657-1941-2)
• 1997/98
  - Bücherverzeichnis im Autorenalphabet, kumuliert mit Titel- und Stichwortregister, mit Verweisung auf den Autor / Author - Titles - Catchwords / 8 Bde. 27. Ausg. 1997. Zus. 15.600 S. / Kst. (3-7657-2000-3)
  - ISBN Register. Anh.: Verlagsverzeichnis / 27. Ausg. 1997. 1.300 S. / Kst. (3-7657-2001-1)
  - Schlagwortverzeichnis. Anh.: Schlagwortübersicht, Verlagsverzeichnis / Subject Guide / 7 Bde. 20. Ausg. 1997. 13.600 S. / Kst. (3-7657-2002-X)
  - Ergänzungsband Frühjahr 1998. Anh.: Verlagsverzeichnis / Supplement Spring 1998 26. Ausg. 1998. 1.400 S. / Kst. (3-7657-2003-8)
  - Ergänzungsband Frühjahr 1998. ISBN Register. Anh.: Verlagsverzeichnis / Supplement Spring 1998. ISBN-Index / 26. Ausg. 1998. 260 S. / Kst. (3-7657-2004-6)
• 1998/99
  - Bücherverzeichnis im Autorenalphabet, kumuliert mit Titel- und Stichwortregister, mit Verweisung auf den Autor / Authors - Titles - Catchwords / 8 Bde. 28. Ausg. 1998. Zus. 15.600 S. / Kst. (3-7657-2072-0)
  - ISBN Register / 28. Ausg. 1998. 1.300 S. / Kst. (3-7657-2073-9)
  - Schlagwortverzeichnis / Subject Guide / 7 Bde. 21. Ausg. 1998. Zus. 13.600 S. / Kst. (3-7657-2074-7)
  - Ergänzungsband Frühjahr 1999 / Supplement Spring 1999 / 27. Ausg. 1999. 1.500 S. / Kst. (3-7657-2075-5)
  - Ergänzungsband Frühjahr 1999. ISBN Register / Supplement Spring 1999. ISBN-Index / 27. Ausg. 1999. 280 S. / Kst. (3-7657-2076-3)

Verzeichnis lieferbarer Bücher (VLB Aktuell) CD-ROM-Ausgabe / German Books in Print (VLB Aktuell) CD-ROM-Edition. Autoren, Titel, Stichworte, Schlagworte. Verlage, ISBN / Authors, Titles, Catchwords, Subjects, Publishers, ISBN. Dt. / Engl. / Franz. / 1988 - 1993 [Vertrieb u. Auslieferung für die Buchhändler-Vereinigung, Frankfurt] [Danach: Aufteilung in Deutsche und Schweizer Version (s.d)]
• Schinzel, Wilfried H.: Verzeichnis Lieferbarer Bücher (VLB Aktuell) CD-ROM-Ausgabe / German Books in Print (VLB Aktuell) CD-ROM-Edition. Benutzer-Dokumentation / 1990. 131 S. / Br

Verzeichnis lieferbarer Bücher (VLB Aktuell) CD-ROM-Ausgabe (Deutsche Version) / German Books in Print (VLB Aktuell) CD-ROM-Edition (German Version). Autoren, Titel, Stichworte, Schlagworte, Verlage, ISBN / Authors, Titles, Catchwords, Subjects, Publishers, ISBN. Dt / Engl. / Franz. / seit 1993 [Vertrieb u. Auslieferung für die Buchhändler-Vereinigung, Frankfurt]

Verzeichnis lieferbarer Bücher (VLB Aktuell) CD-ROM-Ausgabe (Schweizer Version) / German Books in Print (VLB Aktuell) CD-ROM-Edition (Swiss Version). Autoren, Titel, Stichworte, Schlagworte. Verlage, ISBN / Authors, Titles, Catchwords, Subjects, Publishers, ISBN. Dt. / Engl. / Franz. / seit 1993 [Vertrieb u. Auslieferung für die Buchhändler-Vereinigung, Frankfurt]

Verzeichnis medizinischer und naturwissenschaftlicher Drucke 1472 - 1830. Reihe A - D. Hrsg.: Herzog August Bibliothek Wolfenbüttel / 14 Bde. / Ln (3-598-31680-1)
• Reihe A: Alphabetischer Index 4 Bde. Bearb. v. Ursula Zachert, unter Mitarb. v. Ursel Zeidler / 1987/1988. Zus. XII, 1.857 S. / Ln (3-598-31681-X)
• Reihe B: Chronologischer Index 3 Bde. Bearb. v. Ursula Zachert / 1987/1988. Zus. VI, 1.300 S. / Ln (3-598-31686-0)
• Reihe C: Ortsindex 3 Bde. Bearb. v. Ursula Zachert / 1987/1988. Zus. XXII, 1.345 S. / Ln (3-598-31690-9)
• Reihe D: Systematischer Index 4 Bde. Bearb. v. Werner Arnold, unter Mitarb. v. Gisela Kabisch / 1987. Zus. 1.767 S. / Ln (3-598-31694-1)

Verzeichnis rechtswissenschaftlicher Zeitschriften und Serien (VRZS) / Union List of Legal Serials. In ausgewählten Bibliotheken der Bundesrepublik Deutschland einschliesslich Berlin (West) / In Selected Libraries including Berlin (West). Zugl. Nachtr. zum „Zeitschriftenverz. der jurist. Max-Planck-Institute (ZVJM)". Union list of legal serials in selected libraries of the Federal Republic of Germany including Berlin (West). Hrsg. von d. Staatsbibl. Preuß. Kulturbesitz, Berlin. Bearb. im Fachreferat Rechtswiss. / 1978. IX, 350 S. / Lin (3-598-07079-9)
• 2. Ausg. Führt das „Zeitschriftenverzeichnis der juristischen Max Planck-Institute (ZVJM)" fort / Continues the „Zeitschriftenverzeichnis der juristischen Max Planck-Institute (ZVJM)". Hrsg.: Staatsbibliothek Preussischer Kulturbesitz. Fachreferat Rechtswiss. / 2 Bde. 1984. Zus. X, 1.896 S. / Lin (3-598-10525-8)
• 3., wes. erw. Aufl. / 2 Bde. 1990. Zus. 2.628 S. / Lin (3-598-10819-2)

Verzeichnis vergriffener Bücher (VVB). CD-ROM-Ausgabe / German Books Out-of-Print (VVB). CD-ROM-Edition [Vertrieb u. Auslfg / Orig.-Vlg.: Buchhändler-Vereinigung] (3-7657-1742-8)

Verzeichniss eines Theils der Kupferstich- und Büchersammlung, Johann Wilhelm Ludewig Gleims (Halberstadt 1804). Mit einem Kommentar v. Reinhard Selz / 1989. 284 S. / Gb (3-598-07262-7)

Vickery, Brian Campell: Dokumentationssysteme. Einführung in d. Theorie d. Dokumenationssysteme / 1971. VIII, 247 S., 36 Tab., Übersichten u. Flußdiagr / Kst. (UTB 25) (3-7940-2602-0)

Vickery, Brian Campell: Facettenklassifikation / 1969. 72 S. / Lin (3-7940-3329-9)

Vickery, Brian Campell: Zur Theorie von Dokumentationssystemen. Hrsg. v. d. Deutschen Gesellschaft für Dokumentation, Frankfurt a. M. Übersetzung d. 2. Aufl. aus dem Englischen / 1970. XV, 247 S. / Lin (3-7940-3329-9)

Der **Vielseher** Herausforderung für Fernsehforschung und Gesellschaft. (Fernsehen und Bildung Jg. 15/1981, 1-3). Hrsg. v. Helmut Oeller; Hertha Sturm. Redakt.: Marianne Grewe-Partsch; Käthe Nowacek / 1982. 173 S. / Br (3-598-00155-X)

Violence and Peace. Building in the Middle East. Proceedings of a Symposium held by the Israeli Institute for the Study of International Affairs I. I. S. I. A. Ed. by Mari'on Mushkat / 1981. VIII, 194 p. / Soft (3-598-10355-7)

Vollnhals, Otto: Multilingual Dictionary of Electronic Publishing. English - German - French - Spanish - Italian / 1996. 384 p. / Hard (3-598-11295-5)

VomEnde, Rudolf: Criminology and Forensic Sciences / Kriminologie und Kriminalistik An international bibliography / Eine internationale Bibliographie 1950-1980 / 3 vols. 1981 / 1982. XV, 2,389 p. / Hard (3-598-10374-3)

Von der Arbeiterbewegung zum modernen Sozialstaat. Festschrift für Gerhard A. Ritter zum 65. Geburtstag. Hrsg. v. Jürgen Kocka; Hans J. Puhle; Klaus Tenfelde. Gedruckt mit Unterstützung d. Friedrich-Ebert-Stiftung / 1994. XI, 866 S. / Ln (3-598-11201-7)

Von der Wirkung des Buches. Festgabe für Horst Kunze zum 80. Geburtstag. Gewidmet von Schülern und Freunden. Hrsg.: Deutsche Staatsbibliothek Berlin. Bes. v. Friedhilde Krause / 1990. 252 S. / Kt (3-598-07048-9) [Lizenzausg. d. Deutschen Staatsbibliothek, Berlin, DDR]

Vorschulkinder und Fernsehen. Empirische Untersuchungen in drei Ländern. Bearb. v. Erentraud Hömberg. Hrsg.: Stiftung Prix Jeunesse München / 1978. 78 S. / Br (3-7940-7049-6)

Voss, Friedrich: Die studentische Linke in Japan. Geschichte, Organisation u. hochschulpolitischer Kampf / 1976 (3-7940-3171-7)

Wahlstatistik in Deutschland. Bibliographie der deutschen Wahlstatistik von 1848 bis 1975. Hrsg. u. bearb. v. Nils Diederich u.a. / 1976. X, 206 S. / Lin (3-7940-3220-9)

Waibl, Elmar / Philip Herdina: Dictionary of Philosophical Terms / Wörterbuch philosophischer Fachbegriffe German-English / Deutsch-Englisch - English-German / Englisch-Deutsch / 2 vols. 1997. Cplt. XLVII, 885 p. / Gb (3-598-11329-3)

Waldow, Alexander: Illustrierte Encyclopädie der graphischen Künste und der verwandten Zweige. (Buch-, Stein- und Kupferdruck, Lithographie, Photolithographie, Chemietypie, Zinkographie.) Leipzig 1884 / 1993. XII, 918 S., 581 Abb., 2 Taf / Gb (3-598-07250-3)

Walk, Joseph: Kurzbiographien zur Geschichte der Juden in Deutschland 1918-1945. Hrsg.:

Leo-Baeck-Institut, Jerusalem / 1988. XVIII, 452 S. / Ln (3-598-10477-4)

Wallas, Armin A.: Zeitschriften und Anthologien des Expressionismus in Österreich. Analytische Bibliographie und Register / 2 Bde. 1995. Zus. 1.200 S. / Ln (3-598-11222-X)

Walter, Erhard / Karl Wiegel: Kleinod der Buchkunst / 1979. 382 S. (3-598-10008-6)

Walther, Karl Klaus: Die deutschsprachige Verlagsproduktion von Pierre Marteau/Peter Hammer, Köln. Zur Geschichte eines fingierten Impressums. (Beiheft 93 zum Zentralblatt für Bibliothekswesen) / 1983. 151 S., 4 Tafeln / Br (3-598-07041-1) [Lizenzausg. m. Genehm. v. VEB Bibliographisches Institut, Leipzig]

Walther, Wilhelm: Luthers deutsche Bibel. Festschrift zur Jahrhundertfeier der Reformation. - Reprint / 1978. VI, 218 S. (3-597-20003-6) [Minerva Publikation]

Wandlungen im Film Junge deutsche Produktion. Texte: Theo Fürstenau. Hrsg. vom Deutschen. Institut für Filmkunde / 1969. 97 S.

Ward, Charles A.: Moscow and Leningrad. A Topographical Guide to Russian Cultural History / 2 vols.
• Vol. 1: Buildings and Builders / 1989. X, 214 p. / Hard (3-598-10833-8)
• Vol. 2: Writers, Painters, Musicians and Their Gathering Places / 1992. IX, 309 p. / Hard (3-598-10834-6)

Was wissen Fernsehproduzenten von ihren jungen Zuschauern? Empirische Untersuchungen in vier Ländern. Hrsg.: Stiftung Prix Jeunesse, München / 1979. 170 S. / Br (3-598-10091-4)

Wasilewitsch, Gennadi: Wörterbuch des Bibliothekswesens. Russisch-Deutsch / Deutsch-Russisch. Hrsg.: VEB Vlg. Enzyklopädie Leipzig / 1988. 253 S. / Gb (3-598-07268-6)

Wasted Wealth. The Participation of Women in Public Life. Ed. by Claudia Bernardoni and Vera Werner; German Commission for UNESCO, Bonn. Trans. by Ruth Stanley, assist. by Mary Hess. Ed. Hans-Dieter Dyroff / 1985. 238 p. / Soft (3-598-10603-3)

Weber, Angelika: Bibliographie Architektur. Zeitschriftenaufsätze zu den Berichten Planen und Bauen 1974 - 1979 / 1979. 399 S. (3-598-07066-7)

Weber, Angelika: Fachinformation Bauingenieurwesen. Hrsg. von Reinhard Horn; Wolfram Neubauer / 1991. 200 S. (3-598-10974-1)

Weeks, David J.: Eccentrics. The Scientific Investigation. In collab. with Kate Ward / 1988. X, 259 p. / Hard (3-598-10677-7)

Weimann, Karl H.: Bibliotheksgeschichte. Lehrbuch zur Entwicklung und Topographie des Bibliothekswesens / 1975. 254 S. / Ln (3-7940-3179-2)

Weischenberg, Siegfried: Journalismus in der Computergesellschaft. Informatisierung, Medientechnik und die Rolle der Berufskommunikatoren / 1982. 256 S., 23 Abb., 14 Tab. / Br (3-598-10416-2)

Weiss, Johannes: Max Webers Grundlegung der Soziologie. Eine Einführung / 1975. 240 S. / Kst. (UTB 517) (3-7940-2644-6)
• 2. überarb. u. erw. Aufl. 1992. 256 S. / Br (3-598-11092-8)

Weiss, Wisso: Historische Wasserzeichen / 1987. 136 S. / Lin (3-598-07228-7)

Welke, Klaus M.: Einführung in die Valenz- und Kasustheorie / 1988. 233 S. / Ln (3-598-07046-2)

Wendland, Ulrike: Biographisches Handbuch deutschsprachiger Kunsthistoriker im Exil. Leben und Werk der unter dem Nationalsozialismus verfolgten und vertriebenen Wissenschaftler / 2 Tlbde. / 1998. Zus. XLVI, 813 S. / Gb (3-598-11339-0)

Wennrich, Peter: Anglo-American and German Abbreviations in Data Processing / Anglo-Amerikanische und deutsche Abkürzungen für den Bereich der Datenverarbeitung / 1984 VI, 736 p. / Hard (3-598-20524-4)

Wennrich, Peter: Anglo-American and German Abbreviations in Environmental Protection Anglo-amerikanische und deutsche Abkürzungen für den Bereich Umweltschutz / 1979. 550 p. / Hard (3-598-10060-4)

Wennrich, Peter: Dictionary of Electronics and Information Processing / Wörterbuch der Elektronik und Informationsverarbeitung English-German / Englisch-Deutsch, German-English / Deutsch-Englisch / 2 vols. 1990. Cplt. X, 770 p. / Hard (3-598-10885-0)

Wennrich, Peter: International Dictionary of Abbreviations and Acronyms of Electronics, Electrical Engineering, Computer Technology and Information Processing / Internationales Verzeichnis der Abkürzungen und Akronyme der Elektronik, Elektrotechnik, Computer / 2 vols. 1992. Cplt. 960 p. / Hard (3-598-10977-6)

Wennrich, Peter / Paul Spillner: International Encyclopedia of Abbreviations and Acronyms of Organizations / Internationale Enzyklopädie der Abkürzungen und Akronyme von Organisationen Part I: Vol. 1-6 (Abbreviations and Acronyms); Part II: Vol. 7-10 (Organizations and Institutions) / Teil I: Bd. 1-6 (Abkürzungen und Akronyme); Teil II: Bd. 7-10 (Organisationen und Institutionen) / 10 Vols. 3rd ed. 1990-1993. Cplt. CLX, 3,768 p. / Hard (3-598-22160-6)

Werkstatt des Buches. Hrsg. v. Horst Kliemann; Fritz Hodeige u. Werner Adrian [1969 v. Carl Ernst Poeschel Vlg., Stuttgart übernommen]
• Hirsch, Martin: Arbeitsrecht / 1968. 68 S. / Br (3-7940-8817-4)
• Bibliographie des Buchhandels / 1970 112 S. / Br (3-7940-8823-9)
• Bibliographie der Buchherstellung / 1970. 112 S. / Br (3-7940-8822-0)
• Delp, Ludwig: Kleines Praktikum für Urheber- und Verlagsrecht / 2. erw. Aufl. 1966 150 S. / Br (3-7940-8815-8)
• Ensslin, Werner: Werbung im Sortiment / 1958. 76 S., 8 Muster / Br (3-7940-8806-9)
• Glock, Karl Borromäus: Der Buchhändler als Kaufmann / 1953. 110 S. / Br (3-7940-8802-6)

• Grieshaber, Ernst L.: Wider die Druckfehler. Betrachtungen über das Korrekturlesen / 1961. VII, 82 S. (3-7940-8808-5)
• Heyd, Werner Paul: Der Korrektor. Versuch eines Berufsbildes / 1971. 83 S. / Br (3-7940-8824-7)
• Kärcher, Gustav F.: Warenkunde des Buches. Ein Leitfaden für Buchhändler und Bücherfreunde / 2., unveränd. Aufl 1965 120 S. m. 4 Abb. / Br (3-7940-8814-X)
• Kliemann, Horst: Arbeitshilfen, Tabellen und Fauszahlen für die buchhändlerische Praxis. Überarb. Neuaufl v. Horst Kliemanns „Arbeitshilfen für Buchhändler".Durchges. u. neu zsgst. v. Ernst Metelmann / 1970. VI, 63 S., 4 Einlagen / Br (3-7940-8821-2)
• Lerch, Gerda: Kleines Praktikum des Buchexports / 1956. 125 S. / Br (3-7940-8804-2)
• Olzog, Günter: Deutsche Staatsbürgerkunde in 150 Fragen und Antworten / 1964. 96 S. / Br (3-7940-8811-5)
• Schauer, Kurt G.: Wege der Buchgestaltung. Erfahrungen und Ratschläge / 1953. XXXII, 111 S. / Br (3-7940-8803-4)
• Schulz, Gerd: Zeugnisse und Programme zur Geschichte des deutschen Buchhandels (von 1794 - 1951) / 1964. 112 S. / Br (3-7940-8813-1)
• Uhlig, Friedrich: Fachrechnen des Buchhändlers / 2. überarb. u. erw. Aufl. 1970. 109 S. / Br (3-7940-8809-3)
• Unwin, Philipp: Das Berufsbild des Verlagsbuchhändlers. Arbeitsaufgaben und Entwicklungsmöglichkeiten / 1968. VII, 136 S. / Br (3-7940-8816-6)
• Vatter, Arnold: Schreibwaren als Verkaufsartikel. Kleine Warenkunde. / 3. Aufl. 1964. VIII, 76 S. / Br (3-7940-8812-3)

Werner, Andreas / Margot Wiesner; Peter Heydt: Mikroformen. Ein Leitfaden für Einkauf und Bearbeitung in Bibliotheken / 1980. 224 S. / Br (3-598-10061-2)

Das **Westfälische** Wirtschaftsarchiv und seine Bestände. Hrsg. v. Ottfried Dascher. Bearb. v. Wilfried Reininghaus; Gabriele Unverferth; Horst Pradler; Ottfried Dascher / 1990. XXXVIII, 706 S., 78 Abb. / Gb (3-598-10904-0)

Westphal, Margarete: Die besten deutschen Memoiren. Lebenserinnerungen und Selbstbiographien aus sieben Jahrhunderten. M. ein. Abh. ü. d. Entwicklung der deutschen Selbstbiographie v. Hermann Ulrich - Nachdr. d. Ausg. Leipzig 1923 / 1971. 423 S. / Ln (3-7940-3410-4) [Lizenzausg. m. Genehm. v. Vlg. Koehler und Volckmar, Leipzig]

Wettstein, Hermann: Dietrich Buxtehude (1637-1707). Bibliographie zu seinem Leben und Werk / 1988. 109 S., 7 Abb. / Lin (3-598-10786-2)

What do TV producers know about their young viewers? Four studies carried out in four countries. Ed.: Stiftung Prix Jeunesse, München / 1979. 153 p. / Br (3-598-10092-2)

Where to Order Foreign Books. A Directory for Booksellers, Librarians and Readers. Produced for the International Booksellers Federation by Archie Rugh; Bernhard Hefele, with assistance of Karl H. Strasser / 1989. 117 p. / Soft (3-598-10891-5)

Who's Who at the Frankfurt Book Fair. An International Publisher's Guide. Ed. by the Frankfurt

Book Fair, Ausstellungs- und Messe-GmbH des Börsenvereins des Deutschen Buchhandels - Frankfurter
Buchmesse / 1970. 399 p. / Soft (3-7940-3396-5)
- 2nd ed. 1971. 397 p. / Soft (3-7940-3381-7)
- 3rd ed. 1972. 383 p. / Soft (3-7940-3382-5)
- 4th ed. 1973. 370 p. / Soft (3-7940-3383-3)
- 5th ed. 1974. 386 p. / Soft (3-7940-3384-1)
- 6th ed. 1975. 324 p. / Soft (3-7940-3385-X)
- 7th ed. 1976. 344 p. / Soft (3-7940-3386-8)
- 8th ed. 1977. 353 p. / Soft (3-7940-7022-4)
- 9th ed. 1978. 399 p. / Soft (3-7940-07067-5)
- 10th ed. 1979. 425 p. / Soft (3-598-10029-9)
- 11th ed. 1980. 422 p. / Soft (3-598-21880-X)
- 12th ed. 1981. 447 p. / Soft (3-598-21881-8)
- 13th ed. 1982. 449 p. / Soft (3-598-21882-6)
- 14th ed. 1983. 489 p. / Soft (3-598-21883-4)
- 15th ed. 1984. 492 p. / Soft (3-598-21884-2)
- 16th ed. 1985. 436 p. / Soft (3-598-21885-0)
- 17th ed. 1986. 434 p. / Soft (3-598-21886-9)
- 18th ed. 1987. 455 p. / Soft (3-598-21887-7)
- 19th ed. 1988. 498 p. / Soft (3-598-21888-5)
- 20th ed. 1989. 487 p. / Soft (3-598-21889-3)
- 21st ed. 1990. 517 p. / Soft (3-598-21890-7)
- 22nd ed. 1991. 514 p. / Soft (3-598-21891-5)
- 23rd ed. 1992. 500 p. / Soft (3-598-21892-3)
- 24th ed. 1993 521 p. / Soft (3-598-21893-1)
- 25th ed. 1994. 529 p. / Soft (3-598-21894-X)
- 26th ed. 1995. 583 p. / Soft (3-598-21895-8)
- 27th ed. 1996. 618 p. / Soft (3-598-21896-6)
- 28th ed. 1997. 635 p. / Soft (3-598-21897-4)
- 29th ed. 1998. 650 p. / Soft (3-598-21898-2)

Who's who der Herz- und Kreislaufmedizin.
Redakt.: Gertrud Conrad-Bergweiler; Marianne
Stumpfoll / 1982. XXII, 275 S. / Ln (3-598-10427-8)

Who's who in continuing education. Human
resources in continuing library-information-media
education. Comp. by CLENE, inc., the Continuing
Library Education Network and Exchange / 1979. V,
304 S. (3-598-08016-6)

Who's Who in der Politik? Ein biograph. Verz. von
6.000 Politikern in d. Bundesrepublik Deutschland.
Bearb. v. Karl Otto Saur / 1971. X, 342 S. / Lin
(3-7940-3219-5)

Who's Who in International Organizations. A Biographical Encyclopedia of more than 1.200 leading
Personalities. Ed. by Jon C. Jenkins; Union of International Associations / 3 vols. 1992. Cplt. XXXV,
1150 p. / Hard (3-598-10908-3)
- 2nd ed. A Biographical Encyclopedia of more than
13,000 leading Personalities in International
Organizations. Ed.: Union of International
Associations. Comp. by Nancy Carfrae, in assist. by
Cécile Vanden Bloock / 3 vols. 1995 Cplt. XX,
1280 p. / Hard (3-598-11239-4)

Who's Who in Mass Communication. Foreword by
Jörg Becker; Walery Pisarek. Ed.: The Press Research
Center, Cracow on behalf of the International
Association for Mass Communication Research,
Bibliographic Section; Sylwester Dziki; Jarina
Maczuga; Walery Pisarek. Database indexe / 2nd
rev. enl. ed. 1990 XI, 191 p. / Hard (3-598-10884-2)

Who's Who in the Socialist Countries. A bio-
graphical encyclopedia of 10,4621 leading per-
sonalities in 16 communist countries. Ed. by Borys
Lewytzkyj; Juliusz Stroynowski / 1977. XVI, 736 p. /
Hard (3-7940-3193-8)
- 2nd ed. Who's Who in the Socialist Countries of
Europe. Albania, Bulgaria, Czechoslovakia,
German Democratic Republic, Hungary, Poland,
Romania, Yugoslavia. A biographical encyclopedia
of more than 12,600 leading personalities. Ed. by

Juliusz Stroynowski / 3 vols. 1989. LVIII, 1369 p. /
Hard (3-598-10636-X)

Who's Who in the Soviet Union. A biographical
encyclopedia of 5,000 leading personalities in the
Soviet Union. Comp. by Elwine Sprogis. Ed. by
Borys Lewytzkyi / 1984. XI, 408 p. / Hard
(3-598-10467-7)

Widerstand als „Hochverrat" 1933-1945. Die Verfahren gegen deutsche Reichsangehörige vor dem
Reichsgericht, dem Volksgerichtshof und dem
Reichskriegsgericht. Mikrofiche-Edition. Hrsg.: Institut für Zeitgeschichte, München. Bearb. v. Jürgen
Zarusky und Hartmut Mehringer / 1994-1997. ca.
74.000 S. auf 754 Fiches incl. Registerbd. Lesefaktor
24x / Silberfiche (3-598-33673-X) / Diazofiche
(3-598-33670-5)
- Erschliessungsband zur Mikrofiche-Edition (Texte
und Materialien zur Zeitgeschichte Bd. 7. Red.:
Werner Röder u. Christoph Weisz) / 1998. 790 S. /
Gb (3-598-33676-4)

Wieder, Joachim: Frankreich und Goethe. Das Goe-
thebild d. Franzosen / 1976. 47 S. / Br (3-7940-3033-8)

Wilbert, Gerd: Topographie audiovisueller Materialien (AVM) an wissenschaftlichen Einrichtungen der
Bundesrepublik Deutschland / 1987. 234 S. / Lin
(3-598-10722-6)

Wildbihler, Hubert / Sonja Voelklein: The Musical.
An International Annotated Bibliography / Eine internationale annotierte Bibliographie. Frwd. by Thomas Siedhoff / 1986. 320 p. / Hard (3-598-10635-1)

Wilson, Alexander: Library Policy for Preservation
and Conservation in the European Community.
Principles, Practices and the Contribution of New
Information Technologies. Frwd by Ariane Iljon. Ed.:
Commission of the European Communities / 1988.
141 p. / Hard (3-598-10766-8)

Winter, Ursula: Die europäischen Handschriften der
Bibliothek Diez in der Deutschen Staatsbibliothek
Berlin. Erster Teil: Die Manuscripta Dieziana B
Santeniana. Zweiter Teil: Die Libri impressi cum
notis manuscriptis der Bibliotheca Dieziana / 1986.
267 S., 18 Taf / Gb (3-598-07217-1)

Wirtschaftsinformatik und Quantitative Betriebs-
wirtschaftslehre. Hrsg. v. Dieter B. Pressmar; August
W. Scheer; Ch. Schneeweiss; H. Wagner [Minerva
Publikation]
- 1: Jahnke, Bernd: Gestaltung leistungsfähiger
Nummernsysteme für die DV-Organisation.
Geleitw. v. Dieter B. Pressmar / 1979. XIV, 321 S.
/ Br (3-597-10073-2)
- 2: Schünemann, Thomas M.: Wirtschaftliche
Organisation von Speicherhierarchien. Geleitw. v.
Dieter B. Pressmar / 1979. XIV, 263 S. 20, 9 x 14,
6 cm / Br (3-597-10074-0)
- 3: Martens, Jan-Mathias.: Anwendungsbezogene
Datenorganisation formatierter Datenbanken.
Geleitw. v. Dieter B. Pressmar / 1979. XIV, 360 S.
20, 9 x 14, 6 cm / Br (3-597-10075-9)
- 4: Helber, Claus: Entscheidungen bei der
Gestaltung optimaler EDV-Systeme Geleitw. v. A.
W. Scheer / 1981. XIV, 256 S. / Br (3-597-10327-8)
- 5: Günther, Hans O.: Mittelfristige Produktionsplanung. Konstruktion und Vergleich quantitativer Modelle / 1982. XVIII, 262 S. / Br
(3-597-10328-6)

- 6: Gabriel, Roland: Optimierungsmodelle bei
logischen Verknüpfungen Modellaufbau und
Modellösung von Mixed-Integer-Problemen bei
qualitativen Anforderungen / 1982. 249 S. / Br
(3-597-10329-4)
- 7: Wiedey, Gustav: Anleihekonversion. Eine
empirische und entscheidungsorientierte Analyse
/ 1982. XVIII, 250 S. / Br (3-597-10330-8)
- 8: Desiere, Dirk A.: Entwicklung und Betrieb von
Computerverbundsystemen / 1982. 174 S. / Br
(3-597-10331-6)
- 9: Zacharias, Claus O.: EDV-Einsatz in hierarchi-
schen Lagerverbundsystemen / 1982. VIII, 220 S. /
Br (3-597-10399-5)
- 10: Brandenburg, Volker: Simulation von Com-
puter-am-Arbeitsplatz-Systemen / 1983 234 S. / Br
(3-597-10400-2)
- 11: Plüschke, B. Roderich: Anwendung von
Stichprobenverfahren bei der Stichtagsinventur
und der permanenten Inventur / 1982. XII, 224 S.
/ Br (3-597-10401-0)
- 12: Stadtler, Hartmut: Interaktive Lösung schlecht-
strukturierter Entscheidungsprobleme. Methoden
und Ergebnisse bei der Stauung von Chemika-
lientankern / 1983. XII, 252 S. / Br
(3-597-10402-9)
- 13: Oehlmann, Rainer: Produktionsplanung auf
der Grundlage von Loszyklen / 1983. 177 S. / Br
(3-597-10403-7)
- 14: Weigert, Peter M.: Benutzerorientierte Daten-
allokation in verteilten Informationssystemen /
1983. X, 308 S. / Br (3-597-10429-0)
- 15: Krcmar, Helmut: Gestaltung von Computer-
am-Arbeitsplatz-Systemen. Entwicklung von
Alternativen und deren Bewertung durch Simulation / 1984. XIV, 281 S. / Br (3-597-10430-4)
- 16: Werners, Brigitte: Interaktive Entscheidungs-
unterstützung durch ein flexibles mathematisches
Programmierungssystem / 1984. VI, 248 S. / Br
(3-597-10431-2)
- 17: Ausbildung in Betriebsinformatik. Arbeitstagung der Wissenschaftlichen Kommission Betriebsinformatik des Verbandes der Hochschullehrer für Betriebswirtschaft e.V. / 1985. VIII, 147
S. / Br (3-597-10432-0)
- 18: Papke, Thomas: Datenallokationsplanung in
verteilten Informationssystemen Konzeption und
Realisation eines Planungsmodelles auf der Basis
eines heuristischen Lösungsansatzes / 1985. IV,
228 S. / Br (3-597-10433-9)
- 19: Bergholter, Volker: Beschreibung betrieblicher
Informationssysteme. Entwurf und Implementation eines operationalen Modells / 1988. ca. 400 S.
/ Br (3-597-10556-4)
- 20: Reepmeyer, Jan A.: Rasterentwurfsmethode.
Ein Ansatz zur methodischen Softwareerstellung /
1985. 215 S. / Br (3-597-10557-2)
- 21: Hall, B.: Kapazitätsfestlegung in Distributions-
systemen / 1986. XIV, 266 S. / Br (3-597-10558-0)
- 22: Hansmann, Rolf: Ansätze zur systemtheore-
tischen Modellierung lernfähiger und adaptiver
betriebswirtschaftlicher Entscheidungsprozesse /
1986. VIII, 337 S. / Br (3-597-10559-9)
- 23: Pocsay, Alexander: Datenermittlungsverfahren
zur Unterstützung von Wirtschaftlichkeitsbe-
rechnungen beim Ersatz von Software / 1987. XII,
315 S. / Br (3-597-10560-2)
- 24: Eberhard, Ulrich: Mehr-Depot-Tourenplanung
/ 1987. XII, 207 S. / Br (3-597-10561-0)
- 25: Schröder, Helmut: Integration von Planung
und Steuerung in Reparatursystemen / 1988. XIV,
183 S. / Br (3-597-10562-9)

Wissenschaftliche und kulturelle Institutionen der
Bundesrepublik Deutschland. Bibliotheken, Universitä-

ten, Museen, wiss. Ges., Verlage, Kunsthandel, Kunstvereine, Kunstpreise / 1976. 758 S. (3-7940-2247-5)

Wissenschaftsgeschichte [Minerva Publikation]
• Arithmos - Arrythmos. Skizzen aus der Wissenschaftsgeschichte. Festschrift für Joachim Otto Fleckenstein zum 65. Geburtstag. Hrsg. von Karin Figala; Ernst H. Berninger / 1979. II, 331 S. (3-597-10181-X)
• Carl Friedrich Gauß (1777-1855). Sammelband von Beiträgen zum 200. Geburtstag von C. F. Gauß. Hrsg. von Ivo Schneider / 1980. 245 S. (3-597-10280-8)
• Vogel, Kurt: Beiträge zur Geschichte der Arithmetik. Zum 90. Geburtstag des Verfassers am 30. 9. 78 mit Lebensbeschreibung und Schriftenverzeichnis. Hrsg.: Forschungsinstitut des Deutschen Museums für die Geschichte der Naturwissenschaften und Technik und von Ernst H. Berninger / 1978. 96 S. (3-597-10036-8)

Wissensveränderung durch Medien. Theoretische Grundlagen und empirische Analysen. Hrsg. v. Karin Böhme-Dürr; Jürgen Emig; Norbert M. Seel / 1990. 319 S. / Br (3-598-10896-6)

Witkowski, Georg: Geschichte des literarischen Lebens in Leipzig (Reprint). Nachw. v. Christl Förster / 1994. 520 S. / Ln (3-598-11195-9)

Die **Wörishöfer** Küche. Kochbuch im Sinne Kneipps erprobt und verfaßt auf Grund beinahe 10jähriger, diesbezüglich praktischer Erfahrungen, nach den Vorträgen und Vorschriften Sr. Hochwürden Herrn Prälaten Sebastian Kneipp und Frau Agathe Haggenmiller. - Originalgetreuer Nachdruck der Ausgabe von 1897 / 1963. XVI, 351 S. / Gb (3-7940-3023-0)

Wörterbuch der Reprographie. Dictionary of Reprography / Dictionnaire de la reprographie. Begriffe und Definitionen. Terms and Definitions / Termes et definitions / . Hrsg.: Deutsches Komitee für Reprographie / 2. Nachdr. der 3., völlig neu bearb. Aufl. 1976. 273 S. (3-7940-3259-4)
• 6. Aufl. Wörterbuch der Reprographie deutsch-englisch-chinesisch / Dictionary of Reprography english-german-chinese. Deutsch-englisch-chinesisch / chinesisch-englisch-deutsch / englisch-chinesisch-deutsch. Vorw. v. Georg Thiele. Hrsg.: Gesellschaft für Information und Dokumentation mbH (GID) / 1987. 247 S. / Br (3-598-10642-4)
5. Aufl. → Informationsdienste 4: Wörterbuch der Reprografie.

Wörterbuch des Bibliothekswesens. Chinesisch-Deutsch / Deutsch-Chinesisch Von L. Zhijian; Z. Langfang; Y. Pingfen; W. Zie / 1996. XXII, 386 S. / Ln (3-598-11332-3)

Wörterbuch des Verlagswesens in zwanzig Sprachen → The Publisher's Practical Dictionary in 20 Languages

Wörterbücher und Lexika. Internationale bibliographische Verzeichnisse
• Berg- und Hüttenwesen, Geologie und Geographie / 1969. 28 S. (3-7940-2379-X)
• Buch- und Bibliothekswesen, Publizistik und Abkürzungen / 1969. 32 S. (3-7940-2389-7)
• Geisteswissenschaften, Archäologie, Geographie, Pädagogik, Geschichte, Philosophie, Theologie,

Sprachwissenschaften / 1969. 52 S. (3-7940-2329-3)
• Kunst, Film, Musik, Architektur und Bauwesen / 1969. 28 S. (3-7940-2439-7)
• Landwirtschaft und Nahrungsmittel, Veterinärmedizin / 1969. 24 S. (3-7940-2459-1)
• Maschinenbau, Elektrotechnik und allgemeine Technik / 1969. 92 S. (3-7940-2349-8)
• Medizin, Biologie, Gesundheitswesen, Pharmazie und Veterinärmedizin / 1969. 36 S. (3-7940-2339-0)
• Physik und Chemie, Astronomie und Geodäsie / 1969. 76 S. (3-7940-2369-2)
• Rechts- und Staatswissenschaften, Politik, Recht, Soziologie und Abkürzungen / 1969. 32 S. (3-7940-2359-5)
• Verkehrswesen, Luftfahrt, Transport und Fremdenverkehr / 1969. 24 S. (3-7940-2449-4)
• Wehrwesen, Luftfahrt und Geodäsie / 1969. 24 S. (3-7940-2429-X)
• Wirtschaftswissenschaften, Betriebs- und Volkswirtschaft, Werbung, Handel und Handwerk / 1969. 52 S. (3-7940-2319-6)

Wohnen in der Stadt. Neue Wohnformen. Arbeitspapiere d. 1. Architektenkongresses, den die Akademie der Architektenkammer Nordrhein-Westfalen u. die Deutsche UNESCO-Kommission vom 31. Mai - 3. Juni 1978 in Westerland, Sylt veranstaltet hat. Hrsg.: Akademie der Architekenkammer Nordrhein-Westfalen; Deutsche UNESCO-Kommission, Köln / 1979. 64 S. / Br (3-598-21070-1)

Wolf, Steffen: Kinderfilm in Europa. Darstellung der Geschichte, Struktur und Funktion des Spielfilmschaffens für Kinder in der Bundesrepublik Deutschland, CSSR, Deutschen Demokratischen Republik und Großbritannien 1945-1965 / 1969. 475 S. (3-7940-3279-9)

Wolfenbütteler Bibliographie zur Geschichte des Buchwesens im deutschen Sprachgebiet 1840-1980 (WBB). Hrsg.: Herzog August Bibliothek, Wolfenbüttel. Vorw. v. Paul Raabe. Bearb. v. Erdmann Weyrauch, unter Mitarb. v. Cornelia Fricke / 8 Bde, 4 Reg.-Bde. 1990-1998 / Ln (3-598-30323-8)
• Bd. 01: Der Autor / 1990. XVIII, 552 S. / Gb (3-598-30324-6)
• Bd. 02: Buchhertstellung / 1991. XXIV, 397 S. / Gb (3-598-30325-4)
• Bd. 03: Buchgattungen / 1992. XXIV, 397 S. / Gb (3-598-30326-2)
• Bd. 04: Verlagswesen und Buchhandel / 1994. XXIII, 549 S. / Gb (3-598-30327-0)
• Bd. 05: Bibliothekswesen Tl. 1 / 1996. XXII, 478 S. / Gb (3-598-30328-9)
• Bd. 06: Bibliothekswesen Tl. 2 / 1996. XVI, 473 S. / Gb (3-598-30329-7)
• Bd. 07: Der Leser / 1998. XVII, 353 S. / Gb (3-598-30393-9)
• Bd. 08: Supplement / 1998. XIX, 489 S. / Gb (3-598-30394-7)
• Bd. 09: Register d. Verf. u. anonymen Titel A-K / 1998. VI, 438 S. / Gb (3-598-30395-5)
• Bd. 10: Register d. Verf. u. anonymen Titel L-Z / 1998. VI, 404 S. / Gb (3-598-30396-3)
• Bd. 11: Register der Länder u. Orte / 1998. VI, 431 S. / Gb (3-598-30407-2)
• Bd. 12: Register d. Personen. Reg. d. Firmen, Gesellschaften, Vereine u. Institutionen. Ges.-Inh / 1998. VII,547 S. / Gb (3-598-30408-0)

World Biographical Dictionary of Artists (IKD) → Allgemeines Künstlerlexikon - Internationale Künstlerdatenbank

World Biographical Index → Internationaler Biographischer Index / World Biographical Index. CD-ROM-Edition

World Guide to Foundations → Handbuch der Internationalen Dokumentation und Information Bd. 19

World Guide to Libraries → Handbuch der Internationalen Dokumentation und Information Bd 8

World Guide to Libraries PLUS. CD-ROM-Edition (3-598-40289-9)
• 2nd ed. 1998 (3-598-40362-3)
• 3rd ed. 1998 (3-598-40398-4)

World Guide to Logotypes, Emblems and Trademarks of International Organizations. Ed.: Union of International Associations. Comp. by Nancy Carfrae / 3 vols. 1997. Cplt. XXXVI, 1097 p. / Hard (3-598-11345-5)

World Guide to Religious and Spiritual Organizations. Ed.: Union of International Associations / 1996. X, 471 p. / Hard (3-598-11296-3)

World Guide to Scientific Associations and Learned Societies → Handbuch der Internationalen Dokumentation und Information Bd. 13

World Guide to Special Libraries → Handbuch der Internationalen Dokumentation und Information Bd. 17

World Guide to Trade Associations → Handbuch der Internationalen Dokumentation und Information Bd. 12

World Patent Information. International journal for patent information and documentation. A joint periodical of the Commission of the European Communities and the World Intellectual Property Organization / Vol. 1/1979 - Vol. 2/1980 (4 issues annually) (ISSN 0172-2190) [Ab 1980 bei Pergamon Press, Tarrytown, NY]

Wüst, Ernst: Lexicon Aristophaneum. Ein handschriftliches Spezialwörterbuch zu den Kommödien des Aristophanes. Mikrofiche-Edition. Vorw. v. Günther Pflug. Hrsg. v. Karl Wüst / 1984. VIII, 1457 S. auf 37 Fiches. Lesefaktor 24x / Silberfiche (3-598-30460-9)

Wüster, Eugen: The road to Infoterm / 2 reports 1974. IX, 141 S. (3-7940-5501-2)

Wulf, Josef: Das Dritte Reich und seine Vollstrecker. Die Liquidation von 500.000 Juden im Ghetto Warschau. - Nachdruck der Ausg. 1961 / 1978. 283 S. (3-7940-4603-X)

Years Work in Serials 1985. An annual publication in conjunction with the United Kingdom Serials Group. Ed. by David Woodworth / 1986. VIII, 143 p. / Lin (3-598-10662-9)

Yearbook of International Organizations. Ed.: Union of International Associations
• 20th ed. 1983/84 / 3 vols. 1983 / Hard (3-598-21855-9)
  - Vol. 1: Organization Descriptions and Index / 1300 p. (3-598-21856-7)

- Vol. 2: Geographic Volume / 1000 p.
  (3-598-21857-5)
- Vol. 3: Subject Volume with bibliography of the
  publications of the International Association /
  500 p. (3-598-21858-3)
• 21st ed. 1984/85 / 3 vols. 1984 / Hard
  (3-598-21861-3)
  - Vol. 1: Organization Descriptions and Index /
    1400 p. (3-598-21860-5)
  - Vol. 2: Geographic Volume. International
    Organisation Participation, Country Directory of
    Secretariats and Membership / 1250 p.
    (3-598-21864-8)
  - Vol. 3: Subject Volume. Global Action Net-
    works. Classified Directory by Subject and
    Region / 900 p.
• 22nd ed. 1985/86 / 3 vols. 1985 / Hard
  (3-598-21868-0)
  - Vol. 1 / 1651 p. (3-598-21865-6)
  - Vol. 2 / 1453 p. (3-598-21866-4)
  - Vol. 3 / 1070 p. (3-598-21867-2)
• 23rd ed. 1986/87 / 3 vols. 1986 / Hard
  (3-598-21869-9)
  - Vol. 1 / 1656 p. (3-598-21870-2)
  - Vol. 2 / 1500 p. (3-598-21871-0)
  - Vol. 3 / 1100 p. (3-598-21872-9)
• 24th ed. 1987/88 / 3 vols. 1987 / Hard
  (3-598-21873-7)
  - Vol. 1 / 1609 p. (3-598-21874-5)
  - Vol. 2 / 1660 p. (3-598-21875-3)
  - Vol. 3 / 1637 p. (3-598-21876-1)
• 25th ed. 1988/89 / 3 vols. 1988 / Hard
  (3-598-22140-1)
  - Vol. 1 / XX, 1643 p. (3-598-22141-X)
  - Vol. 2 / 1496 p. (3-598-22142-8)
  - Vol. 3 / 1637 p. (3-598-22143-6)
• 26th ed. 1989/90 / 3 vols. 1989 / Hard
  (3-598-22200-9)
  - Vol. 1 / XVI, 1620 p. (3-598-22201-7)
  - Vol. 2 / 1632 p. (3-598-22202-5)
  - Vol. 3 / 1460 p. (3-598-22203-3)
• 27th ed. 1990/91 / 3 vols. 1990 / Hard
  (3-598-22204-1)
  - Vol. 1 / 1792 p. (3-598-22205-X)
  - Vol. 2 / 1759 p. (3-598-22206-8)
  - Vol. 3 / 1684 p. (3-598-22207-6)
• 28th ed. 1991/92 / 3 vols. 1991 / Hard
  (3-598-22208-4)
  - Vol. 1 / 1798 p. (3-598-22209-2)
  - Vol. 2 / 1703 p. (3-598-22210-6)
  - Vol. 3 / XII, 1685 p. (3-598-22211-4)
• 29th ed. 1992/93 / 3 vols. 1992 / Hard
  (3-598-22212-2)
  - Vol. 1 / XII, 1780 p. (3-598-22213-0)
  - Vol. 2 / VIII, 1655 p. (3-598-22214-9)
  - Vol. 3 / VIII, 1576 p. (3-598-22215-7)
• 30th ed. 1993/94 / 3 vols. 1993 / Hard
  (3-598-22216-5)
  - Vol. 1 / XII, 1818 p. (3-598-22217-3)
  - Vol. 2 / VI, 1750 p. (3-598-22218-1)
  - Vol. 3 / VIII, 1781 p. (3-598-22219-X)
• 31st ed. 1994/95 / 3 vols. 1994 / Hard
  (3-598-22220-3)
  - Vol. 1 / XII, 1747 p. (3-598-22221-1)
  - Vol. 2 / VI, 1,727 p. (3-598-22222-X)
  - Vol. 3 / VIII, 1781 p. (3-598-22223-8)

• 32nd ed. 1995/96 / 3 vols. 1995 / Hard
  (3-598-22224-6)
  - Vol. 1 / XII, 1,796 p. (3-598-22225-4)
  - Vol. 2 / VI, 1731 p. (3-598-22226-2)
  - Vol. 3 / VIII, 1775 p. (3-598-22227-0)
• 33rd ed. 1996/97 / 4 vols. 1996 / Hard
  (3-598-23351-5)
  - Vol. 1: Organization Descriptions and Index /
    XII, 1,705 p. (3-598-23352-3)
  - Vol. 2: Geographic Volume. International
    Organisation Participation, Country Directory of
    Secretariats and Membership / VI, 1,682 p.
    (3-598-23353-1)
  - Vol. 3: Subject Volume. Global Action Net-
    works. Classified Directory and Index / VIII,
    1,702 p. (3-598-23354-X)
  - Vol. 4: International Organisation Bibliography
    and Resources / VI, 487 p. (3-598-23355-8)
• 34th ed. 1997/98 / 4 vols. 1997 / Hard
  (3-598-23356-6)
  - Vol. 1: Organization Descriptions and
    Cross-References / XII, 1,788 p. (3-598-23357-4)
  - Vol. 2: Geographic Volume. International
    Organisation Participation, Country Directory of
    Secretariats and Membership / VI, 1,785 p.
    (3-598-22358-2)
  - Vol. 3: Subject Volume. Global Action Net-
    works. Classified Directory and Index / VIII,
    1,787 p. (3-598-23359-0)
  - Vol. 4: International Organisation Bibliography
    and Resources / 574 p. (3-598-23360-4)
• 35th ed. 1998/99 / 4 vols. 1998 / Hard
  (3-598-23361-2)
  - Vol. 1 / X, 1,782 p. (3-598-23362-0)
  - Vol. 2 / XII, 1,796 p. (3-598-23363-9)
  - Vol. 3 / VIII, 1,787 p. (3-598-22364-7)
  - Vol. 4 / X, 566 p. (3-598-22365-5)

Yearbook PLUS - International Organizations and
Biographies. CD-ROM-Edition. Ed: Union of International Associations / 1995 (3-598-40224-4)
• 2nd ed. 1996 (3-598-40337-2)
• 3rd ed. 1997 (3-598-40363-1)

Yiddish Books from the Harvard College Library.
Microfiche edition / 1994-1999. Ca. 5,500 fiches.
Reader factor 24x / Silverfiche (3-598-33184-3) /
Diazofiche (3-598-33176-2)

Young, George: Accomodation Services in Britain
1970 - 1980 / 1980. 771 S. (0-85157-508-0) [Clive
Bingley]

Zapf, Georg W.: Augsburgs Buchdruckergeschichte
nebst den Jahrbüchern derselben. 2 Bde. - Nachdruck der Ausg. Augsburg 1786 und 1791 / 1968.
XLVIII, 220 S.; XVI, 276 S. (3-7940-5098-3)

Zeitung und Bibliothek. Ein Wegweiser zu Sammlungen und Literatur. Hrsg. von Gert Hagelweide /
1974. 302 S. (3-7940-3054-0)

Zeitungen und zeitschriftenähnliche Periodika -
ihre Beschreibung und Erfassung in der Zeitschrif-
tendatenbank. Hrsg. v. Hartmut Walravens. Bearb. v.

Marieluise Schillig / 1998. 313 S. / Gb
(3-598-11385-4)

Zeitungs-Index. Verzeichnis wichtiger Buchbespre-
chungen aus deutschsprachigen Zeitungen. Hrsg.
Willi Gorzny / Jg. 1/1974 - Jg. 17/1990 (4 Hefte u.
Reg. jährl.) (ISSN 0340-0107) [Ab 1991 im Vlg. Willi
Gorzny, Pullach]

Zeitungs-Index. Buchrezensionen 1974. Verzeichn.
wichtiger Buchbesprechungen aus deutschsprachigen Zeitungen. Hrsg. von Willi Gorzny (Beiheft 1
zum Jg. 1/1974) / 1975. 102 S. (3-7940-3230-6)
• Buchrezensionen 1975. (Beiheft 2 zum Jg. 2/1975)
  / 1978. 189 S. (3-7940-3242-X)

Zeitungs-Index 1982-1989. Verzeichnis wichtiger
Aufsätze aus 19 deutschsprachigen Zeitungen.
CD-ROM-Edition. Hrsg. v. Willi Gorzny / 1992
(3-598-40242-2)

Zeltner, Gustav G.: Kurtz-gefasste Historie des
Lebens und Fatorum Hanns Luffts, berühmten
Buchdruckers zu Wittenberg 1727. - Reprint d.
Orig.-Ausg. 1727. - Nachw. v. Elke Stiegler / 1989.
114, XVI S. / Gb (3-598-07244-9)

Zentralblatt für Bibliothekswesen (Hauptwerk:
Jahrgang 61-102, Leipzig 1947-1988, Fiche Nr.
1-301). Generalregister: für Jahrgang 51-70 und
Jahrgang 71-80, Leipzig 1962 und 1968, Fiche Nr.
1-4. (Beihefte: Jahrgang 74-95, Leipzig 1948-1987,
Fiche Nr. 1-56). Mikrofiche-Edition / 1990. CMLVIII,
29.162, (1.067) S. auf 361 Fiches. Lesefaktor 24x /
Silberfiche (3-598-33001-4) / Diazofiche
(3-598-33000-6)

Die **Zukunft** automatisierter Bibliotheksnetze in der
Bundesrepublik Deutschland. Möglichkeiten u.
Grenzen aus techn. u. bibliothekar. Sicht. Bericht e.
Symposiums, veranst. von d. Arbeitsstelle für Biblio-
thekstechnik am 1. / 2. Dez. 1976. Hrsg.:
Arbeitsstelle für Bibliothekstechnik bei d.
Staatsbibliothek Preussischer Kulturbesitz / 1977.
221 S. (3-7940-7021-6)

Zur Benutzerforschung in Bibliotheken. Die
Situation in d. Bundesrepublik Deutschland. Ein
Reader. Hrsg. von Gunter Bock / 1972. 236 S.
(3-7940-3202-0)

Zur Theorie und Praxis des modernen Bibliothekswesens. Band 1: Gesellschaftliche Aspekte. Band 2:
Technologische Aspekte. Band 3: Betriebswirtschaftliche Aspekte. Register. Hrsg. v. Wolfgang
Kehr; Karl W. Neubauer und Joachim Stoltzenburg /
3 Bde. 1976. Zus. 1184 S. / Br (3-7940-4205-0) / Ln
(3-7940-4201-0)

Chronologisches Register / Chronological Index

1949

Literaturzusammenstellung Wasserbau

1951

Internationaler Literaturdienst. Fachausgaben

1952

Internationaler Literaturdienst. Fachausgaben
Jahrbuch 1952 des Internationalen Literaturdienstes
  für alle Gebiete der Technik

1953

Internationaler Literaturdienst. Fachausgaben
Jahrbuch 1953 des Internationalen Literaturdienstes
  für alle Gebiete der Technik

1954

Indexbücher der Technik. Human Relations
- Regelungstechnik
- Schul- und Kirchenbau
- Textilprüfung
- Ultraschall
Internationaler Literaturdienst. Fachausgaben
Taschenbuch 1954. Dokumentationen der Technik

1955

Dokumentation - ein Rationalisierungsfaktor?
Indexbücher der Technik. Human Relations. 1. Erg
- Innerbetrieblicher Transport
- Leistungslohn
- Regelungstechnik. 1. Erg
- Schul- und Kirchenbau. 1. Erg
- Textilprüfung. 1. Erg
- Ultraschall. 1. Erg
Internationaler Literaturdienst. Fachausgaben

1956

Indexbücher der Technik. Human Relations. 2. Erg
- Innerbetrieblicher Transport. 1. Erg
- Leistungslohn. 1. Erg
- Regelungstechnik. 2. Erg
- Schul- und Kirchenbau. 2. Erg
- Textilprüfung. 2. Erg
- Ultraschall. 2. Erg
Internationaler Literaturdienst. Fachausgaben

1957

Handbuch der technischen Dokumentation und
  Bibliographie. Bd. 1
- Bd. 2
Indexbücher der Technik. Human Relations. 3. Erg
- Innerbetrieblicher Transport. 2. Erg
- Leistungslohn. 2. Erg
- Regelungstechnik. 3. Erg
- Regelungstechnik. 4. Erg
- Schul- und Kirchenbau. 3. Erg
- Textilprüfung. 3. Erg
- Ultraschall. 3. Erg
Internationaler Literaturdienst. Fachausgaben

1958

Handbuch der technischen Dokumentation und
  Bibliographie. Bd. 2 / 2. erw. Aufl.
Indexbücher der Technik. Human Relations. 4. Erg
- Innerbetrieblicher Transport. 3. Erg
- Leistungslohn. 3. Erg
- Regelungstechnik. 5. Erg
- Regelungstechnik. 6. Erg
- Schul- und Kirchenbau. 4. Erg
- Textilprüfung. 4. Erg
- Ultraschall. 4. Erg
Internationaler Literaturdienst. Fachausgaben

1959

Atom - Kartei
Handbuch der technischen Dokumentation und
  Bibliographie. Bd. 2 / 3. erw. Aufl.
- Bd. 3
Indexbücher der Technik. Innerbetrieblicher
  Transport. 4. Erg
- Leistungslohn. 4. Erg
Universalbibliographie Technik und Wirtschaft

1960

Handbuch der technischen Dokumentation und
  Bibliographie. Bd. 4
Internationaler Generalkatalog der Zeitschriften für
  ... Ausg. A
- Ausg. B
- Ausg. C
Universalbibliographie Technik und Wirtschaft

1961

Handbuch der technischen Dokumentation und
  Bibliographie. Bd. 1 / 2. Aufl.
- Bd. 2 / 4. Aufl.
- Bd. 4. 1. Ergänzungsbd
Universalbibliographie Technik und Wirtschaft

1962

Handbuch der technischen Dokumentation und
  Bibliographie. Bd. 3 / 4. Ausg.
- Bd. 4. 2. Ergänzungsbd
- Bd. 5
- Bd. 6
- Bd. 6. Zeitschriftenverzeichnis
Universalbibliographie Technik und Wirtschaft

1963

Handbuch der technischen Dokumentation und
  Bibliographie. Bd. 2 / 5. Aufl.
- Bd. 3. Ergänzungsbd
- Bd. 4 / 2. Aufl.
- Bd. 5. Ergänzungsbd
- Bd. 6 / 2. Ausg.
Kneipp, Sebastian: Meine Wasser-Kur, durch ...
Universalbibliographie Technik und Wirtschaft
Die Wörishöfer Küche

1964

Handbuch der technischen Dokumentation und
  Bibliographie. Bd. 1 / 3. Aufl.
- Bd. 3 / 5. Ausg.
- Bd. 4 / 2., erw. Aufl. Ergänzungsbd
- Bd. 5 / 2. Ausg.
- Bd. 6 / 3. Ausg.
- Bd. 7
Universalbibliographie Technik und Wirtschaft

1965

Buchhandel. Internationale Bibliographie
Handbuch der technischen Dokumentation und
  Bibliographie. Bd. 2 / 6. Ausg.
  - Bd. 3 B
  - Bd. 3 / 7. Ausg.
  - Bd. 6 / 4. Ausg.
  - Bd. 7 / 2. Ausg.
Pharmazie. Jahresfachbibliographie 1964
Politische Dokumentation poldok. Jg. 1

1966

Handbuch der technischen Dokumentation und
  Bibliographie. Bd. 2 / 7. Ausg.
  - Bd. 3 / 8. Ausg.
  - Bd. 4 / 3. Ausg.
  - Bd. 5 / 3. Ausg.
  - Bd. 8
Kunze, Horst: Über das Registermachen
Politische Dokumentation poldok. Jg. 2
Ruppert, Fritz: Initials

1967

Bibliographie der Bibliotheksadressbücher
Buchwissenschaftliche Beiträge. 1
  - Bd. 2
Deutsches Universitäts-Handbuch
Fugger, Wolfgang: Ein nutzlich und wolgegrundt
  Formular ...
Handbuch der technischen Dokumentation und
  Bibliographie. Bd. 2 / 8. Ausg.
  - 4 / 3. Ausg. Ergänzungsbd
  - Bd. 6 / 5. Ausg.
  - Bd. 7 / 3. Ausg.
Information Theory, Statistical Decision Functions, ...
  Transactions of the Fourth ...
Internationaler Restauratorentag
Internationales Verlags- und Bezugsquellen-
  Verzeichnis ...
Ladenpreistabelle für die Mehrwertsteuer / 1. - 3.
  Aufl.
Lexikon der graphischen Technik
Politische Dokumentation poldok. Jg. 3
Rusch, Gerhard: Einführung in die Titelaufnahme
Samurin, Evgenij I.: Geschichte der bibliothekarisch
  - bibliographischen ...
Scheele, Martin: Wissenschaftliche Dokumentation
Scholl, Hanns Karl: Wegmarken der Entwicklung ...
Seminarberichte der Deutschen UNESCO-
  Kommission Nr. 6
  - Nr. 7

1968

Abhandlungen und Sitzungsberichte der Deutschen ...
Abrechnungstabelle für das Gastgewerbe. 10 %
  Mehrwertsteuer
Actes de la conférence internationale ... (10e)
  Copenhague 1967
Apianus, Peter (Petrus): Abbreviationes Vetustorum ...
Appell, Johann Wilhelm: Die Ritter-, Räuber- und ...
Aus alten Börsenblättern

Beiträge zur Bedarfsmessung an wissenschaftlichen
  Hochschulen. 15
  - Bd. 16
Bibliographie der Buchherstellung
  - 2. Ausg.
Bibliographie des Bibliothekswesens
  - 2. Ausg.
Bibliographie des Buchhandels
  - 2. Ausgabe
Debes, Dietmar: Das Figurenalphabet
Dictionarium Bibliothecarii Practicum / 2. Aufl.
Dußler, Sepp / Fritz Kolling: Moderne Setzerei
Egenolff, Christian: Sprichwörterlexikon
Europäisches Bibliotheks- Adressbuch
Fachbegriffe und Sinnbilder der Datenverarbeitung
Feuereisen, Fritz / Ernst Schmacke: Die Presse in
  Afrika
~: Die Presse in Asien und Ozeanien
~: Die Presse in Lateinamerika
Genzmer, Fritz: Das Buch des Setzers
Gesamttabelle für die Mehrwertsteuer. 5,5%
  - 11%
Handbuch der technischen Dokumentation und
  Bibliographie. Bd. 8 / 2. Ausg.
  - Bd. 9
Kalkulationstabelle für Mehrwertsteuer. 5,5%
  - 11%
Krämer, Louise: Die Fachbücherei in Wirtschaft, ...
Krüger, Karl-Heinz: Wörterbuch der
  Datenverarbeitung
Krumholz, Walter: Die politische Dokumentation in ...
Kybernetische Analysen geistiger Prozesse
Ladenpreistabelle für die Mehrwertsteuer / 4. Aufl.
  - 5. Aufl.
Liebich, Ferdinand Karl: Die Kennedy-Runde
Lingenberg, Jörg: Das Fernsehspiel in der DDR
Mostar, Gerhart Herrmann: In diesem Sinn ...
Muziol, Roman: Pressedokumentation
Oppermann, Alfred: Wörterbuch der modernen
  Technik
Politische Dokumentation poldok. Jg. 4
Rabattabelle für die Mehrwertsteuer. 5,5%
  - 11%
Roloff, Heinrich: Lehrbuch der Sachkatalogisierung
  - Unveränd. Nachdr
Samurin, Evgenij I.: Geschichte der bibliothekarisch
  - bibliographischen ...
Staeck, Erich: Rechenbuch für das graphische
  Gewerbe
Stevens, Henry: The Humboldt Library
Zapf, Georg W.: Augsburgs Buchdruckergeschichte ...

1969

Afrika 69/70
Anekdoten-Lexikon
Beiträge zur Bedarfsmessung an wissenschaftlichen
  Hochschulen. 13
  - Bd. 18
Beiträge zur Universitätsplanung. 5
Berlin-ABC
Deutsches Universitäts-Handbuch / 2. Ausg.
Dictionarium Bibliothecarii Practicum / 3. Aufl.
Empirische Sozialforschung. Bd. 1968
Funke, Fritz: Buchkunde / 3. Aufl.
Handbuch der technischen Dokumentation und
  Bibliographie. Bd. 1 / 4. erw. u. neubearb. Ausg.
  - Bd. 3 B / 2. Ausg.
  - Bd. 3 / 9. Ausg.
  - Bd. 4 / 4. Ausg.

  - Bd. 7 / 4. Ausg.
Heimeran, Ernst: Von Büchern und von ...
Das Hochschulwesen in der UdSSR 1917-1967
Internationale Bibliographie zur Geschichte der ... Tl. 1
Kataloge der Internationalen Jugendbibliothek. 2
Kerner, Immo Ottomar / Gerhard Zielke:
  Einführung in die ...
Krumholz, Walter: Die politische Dokumentation in
  ... / 2. Aufl.
Kunze, Horst: Grundzüge der Bibliothekslehre
Lülfing, Hans: Johannes Gutenberg und das ...
Oppermann, Alfred: Wörterbuch Kybernetik
Politische Dokumentation poldok. Jg. 5
Proverbien-Kodex
Rusch, Gerhard: Einführung in die Titelaufnahme /
  2., unveränd. Aufl.
Schriftenreihe der UFITA H. 33
  - H. 34
  - H. 35
  - H. 36
  - H. 37
Spiess, Volker: Verzeichnis deutschsprachiger
  Hochschulschriften ...
UFITA Archiv für Urheber-, Film-, Funk- und ...
Vickery, Brian Campell: Facettenklassifikation
Wandlungen im Film Junge deutsche Produktion
Werkstatt des Buches
  - Delp, Ludwig: Kleines Praktikum ...
  - Ensslin, Walter: Werbung im Sortiment
  - Glock, Karl B.: Der Buchhändler als Kaufmann
  - Grieshaber, Ernst L.: Wider die Druckfehler
  - Hirsch, Martin: Arbeitsrecht
  - Kärcher, Gustav F.: Warenkunde des Buches
  - Lerch, Gerda: Kleines Praktikum des
    Buchexports
  - Olzog, Günter: Deutsche Staatsbürgerkunde ...
  - Schauer, Kurt G.: Wege der Buchgestaltung
  - Schulz, Gerd: Zeugnisse ...
  - Unwin, Philipp: Das Berufsbild des
    Verlagsbuchhändlers
  - Vatter, Arnold: Schreibwaren als
    Verkaufsartikel
Wörterbücher und Lexika. Berg- ...
  - Buch- ...
  - Geisteswissenschaften ...
  - Kunst ...
  - Landwirtschaft ...
  - Maschinenbau ...
  - Medizin ...
  - Physik ...
  - Rechts- ...
  - Verkehrswesen ...
  - Wehrwesen ...
  - Wirtschaftswissenschaften ...
Wolf, Steffen: Kinderfilm in Europa

1970

Actes de la conférence internationale ... (11e et 12e)
  Bucarest 1969 / Jérusalem 1970
Alisch, Alfred: Richtlinien für den ...
Beiträge zur Bedarfsmessung an wissenschaftlichen
  Hochschulen. 14
Beiträge zur Informations- und Dokumentations-
  wissenschaft. Folge 1
  - Folge 2
  - Folge 3
Beiträge zur Universitätsplanung. 6
  - Bd. 7
Bericht über Projekte ...

1971

1972

Fischer, Konrad: Geschichte des Deutschen Volks-
schullehrerstandes
Forschungsarbeiten in den Sozialwissenschaften.
1971
Funke, Klaus-Detlef: Innere Pressefreiheit
Handbuch der technischen Dokumentation und
Bibliographie. Bd. 4 / 5., neu bearb. Ausg.
- Bd. 7 / 5., überarb. Ausg. Teil 1
- Bd. 7 / 5., überarb. Ausg. Teil 2
- 9 / 2. Aufl.
- Bd. 9. Spillner, Paul: Internationales ... Tl. 3
- Bd. 10. Tl. 4
Heinsius, Wilhelm: Alphabetische Verzeichnis der
von ...
Kleine philosophische Bibliographien Bd. 1. Hen-
richs, Norbert: Bibliographie der Hermeneutik ...
Kleine philosophische Bibliographien Bd. 2.
Hogrebe; Kamp; König: Periodica Philosophica
Internationale Bibliographie zur Geschichte der ... Tl.
2/2
Kayser, Christian Gottlob: Vollständiges Verzeichnis
Köster, Hermann Leopold: Geschichte der deutschen
... Unveränd. Nachdruck
Kommunikation und Politik. 5
- Bd. 6
Krüger, Henning Anders: Betriebsabrechnung für die
Druckindustrie. Bd. 2
Lengrand, Paul: Permanente Erziehung
Liebich, Ferdinand Karl: Kultur ohne Handels-
schranken
Linschmann, Theodor: Ludwig Bechsteins Schriften
Moderne Unterrichtstechnologie Situationsanalyse u.
Basisinformationen ...
Neubauer, K. W. / G. Schwarz; W. Schwuchow:
Kommunikation in ...
Oppermann, Alfred: Wörterbuch der modernen
Technik. Bd. 1 / 3. Aufl.
Pocztar, Jerry: The Theory and Practice of ...
Politikwissenschaftliche Forschung. Bd. 1970
Politische Dokumentation poldok. Jg. 8
Pulizistik-historische Beiträge. Bd. 2
Roloff, Heinrich: Sachkatalogisierung auf neuen
Wegen
Schenk, Elisabeth M.: Zwanzig Jahre demokratische ...
Seminarberichte der Deutschen UNESCO-
Kommission Nr. 16
Stock, Karl F. / Rudolf Heilinger; Marylène Stock:
Personalbibliographien ...
Studiengruppe für Systemforschung ... Bericht Nr. 95
- Bericht Nr. 107
- Bericht Nr. 110
- Bericht Nr. 111
- Bericht Nr. 113
- Bericht Nr. 114
Theoretische Literatur zum Kinder- und Jugendbuch
Thesaurus Bildungsforschung
Verzeichnis lieferbarer Bücher 1972/73
Who's Who at the Frankfurt Book Fair / 3rd ed.
Zur Benutzerforschung in Bibliotheken

1973

Allgemeine Grundlagen der marxistischen
Pädagogik
Arndt, Alfred: Kleines Formellexikon
Arndt, Karl J. / May E. Olson: The German Language
Press ... Vol. 2
Beiträge zur Informations- und Dokumentations-
wissenschaft. Folge 6
Bergk, Johann Adam: Die Kunst zu denken

Bibliographie der Sozialforschung in der
Sowjetunion 1960 - 1970
Bibliographie Pädagogik. Jg. 8
Bibliographie Programmierter Unterricht. Jg. 9
Bibliography of Publications issued by ...
Bibliotheksforum Bayern (BFB) Jg. 1
Bibliothekspraxis. 9
- Bd. 10
- Bd. 11
The book hunger
Cigánik, Marek: Informationsfonds in Wissenschaft,
Technik ...
Deutschsprachige Lehrprogramme
Die deutschsprachige UNESCO-Literatur
DGD-Schriftenreihe Bd. 2
Dortmunder Beiträge zur Zeitungsforschung Bd. 17
Dußler, Sepp / Fritz Kolling: Moderne Setzerei / 3.
Aufl.
Empirische Sozialforschung. Bd. 1972
Feuereisen, Fritz / Ernst Schmacke: Die Presse in
Afrika / 2. Ausg.
~: Die Presse in Asien und Ozeanien / 2. Aufl.
~: Die Presse in Lateinamerika / 2. Aufl.
Forschungsarbeiten in den Sozialwissenschaften.
1972
Gesamtplan für das wissenschaftliche
Bibliothekswesen. Bd. 1
Geschichts- und Romanen-Litteratur der Deutschen ...
Glossary on Educational Technologie
Götz von Olenhusen: Handbuch der Raubdrucke
Hacker, Rupert: Bibliothekarisches Grundwissen / 2.,
unveränderte Aufl.
Handbuch der Internationalen Dokumentation und
Information Bd. 2 / 10. Ausg.
- Bd. 5 / 5. Ausg.
- Bd. 11
- Bd. 12 Tle. 1 u. 2
Hermann, Peter: Informationsrecherchesysteme
HIS-Brief Nr. 34
- Nr. 35
- Nr. 36
- Nr. 37
- Nr. 38
- Nr. 39
- Nr. 40
Hochschulplanung. 13
- Nr. 14
- Nr. 15
- Nr. 16
- Nr. 17
Holl, Oskar: Wissenschaftskunde
IFLA Annual 1972
Information Theory, Statistical Decision Functions, ...
Transactions of the Sixth ...
Informationssysteme - Grundlagen und Praxis der ... 6
- Bd. 7
International Bibliography of Historical Sciences.
Vol. 39/40
The International Exchange of publications
International Journal of Law Libraries. Jg. 1
Jahreskolloquium des Instituts für Systemtechnik ...
Körperschaftsnamendatei Index of corporate bodies
Krüger, Henning Anders: Betriebsabrechnung für die
Druckindustrie. Bd. 1
- Kalkulationsleitfaden Buchdruck
Laubinger, Hans-Dieter: Materialien über die
Bibliotheksplanung ...
Lernsysteme
MacKenzie, Norman / Michael Eraut; Hywel C.
Jones: Lehren und Lernen
Museums of the World / 1st ed.
Neue Verfahren für die Dateneingabe und ...

Oppermann, Alfred: Wörterbuch der Daten-
verarbeitung
Politikwissenschaftliche Forschung. Bd. 1971
Politische Dokumentation poldok. Jg. 9
Pulizistik-historische Beiträge. Bd. 3
Roeber, Georg / Gerhard Jacoby: Handbuch der
filmwirtschaftlichen .
Rusch, Gerhard: Einführung in die Titelaufnahme /
4., unveränd. Aufl.
Samson, Benvenuto: Urheberrecht
Seminarberichte der Deutschen UNESCO-
Kommission Nr. 17
- Nr. 18
- Nr. 19
Simon, Konstantin Romanovič: Bibliographische
Grundbegriffe ...
Standards for public libraries
Stock, Marylène / Karl F. Stock Bibliographie der
Programmiersprachen
Sträter, Hans H.: Thesaurus Pädagogik
Studien zur Publizistik 18
- Bd. 19
Studiengruppe für Systemforschung ... Bericht Nr. 12
- Bericht Nr. 112
Studium in Europa und in den Vereinigten Staaten
The Teaching of Reading
Techne - Technik - Technologie
Thelert, Gustav: Supplement zu Heinsius', Hinrich's
und ...
Topitsch, Ernst: Gottwerdung und Revolution
UNESCO-Konferenzberichte Bd. 1
- Bd. 2
Verzeichnis lieferbarer Bücher 1972/73
Ergänzungsband
Verzeichnis lieferbarer Bücher 1973/74
Who's Who at the Frankfurt Book Fair / 4th ed.

1974

Anderson, Dorothy: Universal Bibliographic Control
Archäographie. Jg. 3
Bautechnische Flächenarten
Beiträge zur Informations- und Dokumentations-
wissenschaft. Folge 5 / 2. Aufl.
Bibliographie Pädagogik. Jg. 9
Bibliotheksforum Bayern (BFB) Jg. 2
Bibliothekspraxis. 8
- Bd. 12
- Bd. 13
Bibliotheksstudien. 1B
- Bd. 2
Burghardt, Anton: Allgemeine Wirtschaftssoziologie
Catalogue of Reproductions of Paintings. Catalogue
of Reproductions of Paintings 1860-1973
Czayka, Lothar: Systemwissenschaft
Deutsche Nationalbibliographie
- Ergänzung 2
Deutscher Dokumentartag 1973
Deutsches Bibliotheksadressbuch / 1. Ausgabe
DGD-Schriftenreihe Bd. 3
Dictionarium Bibliothecarii Practicum / 6., verb. und
erw. Auflage
Dortmunder Beiträge zur Zeitungsforschung Bd. 18
- Bd. 19
Dußler, Sepp / Fritz Kolling: Moderne Setzerei / 4.
Aufl.
Empirische Sozialforschung. Bd. 1973
Europäischer Kongreß über Dokumentations-
systeme und -netze. (1.) Erste ...
- First ...

1975

Seelbach, Dieter: Computerlinguistik und
   Dokumentation
Seminarberichte der Deutschen UNESCO-
   Kommission Nr. 14
- Nr. 26
- Nr. 27
- Nr. 29
Siebel, Wigand: Grundlagen der Logik
Standortverzeichnis ausländischer Zeitungen und ...
Systematik der Stadtbibliothek Hannover (SSH)
Systemtechnik und Innovationsforschung
Television and Socialization Processes in the Family
Texte und Daten zur Hochschulplanung Bd. 4
- Bd. 6
- Bd. 7
- Bd. 9
- Bd. 10
- Bd. 12
- Bd. 14
- Bd. 15
- Bd. 16
- Bd. 18
Verzeichnis lieferbarer Bücher 1974/75
   Ergänzungsband
Verzeichnis lieferbarer Bücher 1975/76
Weimann, Karl H.: Bibliotheksgeschichte
Weiss, Johannes: Max Webers Grundlegung
Who's Who at the Frankfurt Book Fair / 6th ed.
Zeitungs-Index. Buchrezensionen 1974
Zeitungs-Index. Verzeichnis wichtiger ... Jg. 2

1976

Aktuelle Probleme des EDV-Einsatzes in ...
Arbeitsbuch Geschichte. Neuzeit. 1. 16.-18.
   Jahrhundert. Repetitorium
- 3., überarb. Aufl.
Archäographie. Jg. 5
Archivum. Special Vol. 1
- Vol. XXIV
Arndt, Karl J. / May E. Olson: The German Language
   Press ... Vol. 1
Beiträge zur Informations- und
   Dokumentationswissenschaft. Folge 8
Berichte und Arbeiten aus der Universitätsbibliothek
   Gießen. 26
Berichte und Materialien des Zentralinstituts ... Bd. 4
Bibliographia Cartographica. Nr. 2
Bibliographie der Zeitschriftenliteratur zum Stand ...
   Jg. 1
Bibliographie Pädagogik. Jg. 11
Bibliotheksforum Bayern (BFB) Jg. 4
Bibliothekspraxis. 19
Bibliotheksstudien. 1 C
Boßmeyer, Christine: Fortschritte des EDV-
   Verbundes ...
Brauer, Werner: Graphik + (und) Design
Brendel, Detlev / Bernd Grobe: Journalistisches
   Grundwissen
Buch und Bibliothek Jg. 28
Bulletin of Reprints. Vol. XI
Busshoff, Heinrich: Kritische Rationalität und Politik
Computer-based aids to parliamentary work
Crass, Hanns M.: Bibliotheksbauten des 19.
   Jahrhunderts in ...
Dauenhauer, Erich: Curriculumforschung
Deutscher Dokumentartag 1975
Deutsches Bibliotheksadressbuch / 2. Ausgabe
Dokumentation Westeuropa. 1
Dortmunder Beiträge zur Zeitungsforschung Bd. 22

- Bd. 23
Duic, Walter Z.: Europa-Administration
Einführung in die Kommunikationswissenschaft
Empirische Sozialforschung. Bd. 1975
Englert, Ludwig / Otto Mair; Siegfried Mursch:
   Georg Kerschensteiner
Europäischer Kongreß über Dokumentations-
   systeme und -netze. (2.) Zweiter ...
- Deuxième ...
- Second ...
Fachbegriffe und Sinnbilder der Datenverarbeitung /
   2. Aufl.
Faust-Bibliographie. Tl. III
Fenlon, Iain: Catalogue of the printed music ...
Fernsehen und Bildung Jg. 10
Gesamtkatalog der Malik-Buchhandlung ...
Gesamtverzeichnis der Kongress-Schriften ...
   Hauptband mit ...
- Registerband
Gesamtverzeichnis der Zeitschriften und Serien ...
   Stand: Nov. 1975
- Stand: Juni 1976
Gesamtverzeichnis des deutschsprachigen
   Schrifttums 1911-1965
Grundzüge des Marketing
Güdter, Bernd: Verstehen üben Bilden in ...
Hacker, Rupert: Bibliothekarisches Grundwissen / 3.,
   neubearb. Aufl.
Haller, Klaus: Titelaufnahme nach RAK
Handbuch der baubezogenen Bedarfsplanung. 1.
   Ergänzung
Handbuch der Internationalen Dokumentation und
   Information Bd. 2 / 11. Ausg.
- Bd. 10 / 2. Ausg. Tl. I
- Bd. 14 Tl. 1: A - E
HIS-Brief Nr. 58
- Nr. 59
Hitzeroth, Christiane / Dagmar Marek; Isa Müller:
   Leitfaden für die formale ...
Hochschulplanung Bd. 22
- Bd. 23
- Bd. 24
- Bd. 25
- Bd. 26
Hörning, Karl: Gesellschaftliche Entwicklung und ...
Hüfner, Klaus / Jens Naumann: The United Nations
   System. Vol. 1
IFLA Annual 1975
IFLA Journal Vol. 2
IFLA Publications Vol. 4
- Vol. 5
- Vol. 6
- Vol. 7
- Vol. 8
Information et documentation en matière de brevets ...
Infoterm Series. 3
International Bibliography of Historical Sciences.
   Vol. 42
Internationale Bibliographie der Reprints. Bd. 1
International classification. Vol. 3
International Journal of Law Libraries. Jg. 4
Internationaler Kongreß für Reprographie und
   Information ...
Jahrbuch des Instituts für Deutsche Geschichte. Bd. V
Kataloge der Internationalen Jugendbibliothek. 3 /
   2., erw. Aufl.
Kommunikation und Politik. 8
- Bd. 9
Lehrer und Lernprozess. Bd. 1
- Bd. 2
Leuchtmann, Horst: Wörterbuch Musik / 2. ergänzte
   Aufl.

Literaturinformationen zur Berufsbildungs-
   forschung. Jg. 3
Lübeck-Schrifttum
Nachrichten für Dokumentation (NfD) Jg. 27
A New International Economic Order
Patentinformation and Documentation in Western
   Europe
Patentinformation und Patentdokumentation in
   Westeuropa
Politikwissenschaftliche Forschung. Bd. 1973
- Bd. 1974
Politische Dokumentation poldok. Jg. 12
Publizistikwissenschaftlicher Referatedienst (prd).
   Bd. 9
Roloff, Heinrich: Lehrbuch der Sachkatalogisierung /
   4., überarb. Aufl.
Schneider, Klaus: Computer aided subject index ...
Schriftenreihe Internationales Zentralinstitut für das
   Jugend- ... 9
Seminarberichte der Deutschen UNESCO-
   Kommission Nr. 28
Smith, Clifford N. / Anna Pisczan-Czaja Smith:
   Encyclopedia of German-American ...
SMM. Systematik des Musikschrifttums ...
Sparck Jones, Karen / Martin Kay: Linguistik und
   Informationswissenschaft
Systematik der Stadtbibliothek Hannover (SSH)
Systematik für Bibliotheken. Lfg. 3
Systemtechnik und Innovationsforschung. Drittes ...
Technik und Gesellschaft. Bd. 2
- Bd. 3
Television and Socialization Processes in the Family
Terminology of documentation
Texte und Daten zur Hochschulplanung Bd. 17
- Bd. 19
- Bd. 20
- Bd. 21
- Bd. 22
- Bd. 24
- Bd. 25
Verteidigungs-Dokumentation
Verzeichnis lieferbarer Bücher 1975/76
   Ergänzungsband
Verzeichnis lieferbarer Bücher 1976/77
Voss, Friedrich: Die studentische Linke in Japan
Wahlstatistik in Deutschland
Who's Who at the Frankfurt Book Fair / 7th ed.
Wieder, Joachim: Frankreich und Goethe
Wissenschaftliche und kulturelle Institutionen der
   Bundesrepublik ...
Wörterbuch der Reprographie
Zeitungs-Index. Verzeichnis wichtiger ... Jg. 3
Zur Theorie und Praxis des modernen
   Bibliothekswesens

1977

Actes de la conférence internationale ... (15e) Ottawa
   1974
The African Book World & Press
Allgemeine Systematik der öffentlichen Büchereien
Arbeitsbuch Geschichte. Mittelalter / 5., verb. Aufl.
- Neuzeit. 1. 16.-18. Jahrhundert. Quellen
Archäographie. Jg. 6
Beiträge zur Informations- und
   Dokumentationswissenschaft. Folge 9
Bibliographia Cartographica. Nr. 3
Bibliographischer Alt-Japan-Katalog 1542-1853
Bibliothek Forschung und Praxis Jg. 1
Bibliotheksforum Bayern (BFB) Jg. 5

1978

1979

1980

1981

1982

1983

1984

1985

1988

1989

Loh, Gerhard: Geschichte der Universitätsbibliothek ...
Lorck, Carl B.: Handbuch der Geschichte der ...
Mácha, Karel: Glaube und Vernunft. Tl. III
Main Catalog of the Library of Congress ...
Marburger Beiträge zur Vergleichenden ... 21
Marburger Index
Marburger Index. Heusinger, Lutz: Marburger
    Informations-, ...
Marburger Index. Text-Inventar ...
Minerva-Fachserie Wirtschafts- und Sozialwissen-
    schaften. Erfolgschancen ausserschulischer ...
    - Geserick, Rolf: 40 Jahre ...
Museologie
The Nazi Holocaust Historical Articles on the
    Destruction ... Part 1
    - Part 2
    - Part 3
    - Part 4
    - Part 5
    - Part 6
    - Part 7
    - Part 8
    - Part 9
Nestler, Friedrich: Bibliographie
Ostdeutsches Kulturgut in der Bundesrepublik
    Deutschland
Paenson, Isaac: Manual of the Terminology of the
    Law of Armed ...
Panskus, Hartmut: 40 Jahre K. - G.- Saur - Verlag
Plaul, Hainer: Illustrierte Karl-May-Bibliographie
Politique de Préservation du Patrimoine
    Archivistique
Protokolle des Vermittlungsausschusses des
    Deutschen ...
Presse-, Rundfunk- und Filmarchive ... 10
Prinz-Albert-Studien Bd. 7
Publikation der Vereinigung linksgerichteter
    Verleger ...
Publishers' International ISBN-Directory (PIID) /
    16th ed.
The Pulitzer Prize Archive Part A. Vol. 3
Rassler, Gerda: Pariser Tageblatt - Pariser
    Tageszeitung ...
Roller, Franz A.: Systematisches Lehrbuch der
    bildenden ...
Scandinavian Biographical Archive
Schilfert, Sabine: Die Simon-Protokolle ...
Schlieder, Wolfgang: Riesaufdrucke
Schmidt, Ralph: Informationsagenturen im ... Bd. 12
Schriftenreihe der Deutschen Gesellschaft für
    Photographie e.V. 3
Schriftenreihe der Georg-von-Vollmar-Akademie. 2
Schriftenreihe Internationales Zentralinstitut für das
    Jugend- ... 23
Soziokulturelle Herausforderungen ... 1
    - Bd. 2
Studien des Forschungsinstitutes für Wirtschafts-
    politik ... 36
Studien zum Buch- und Bibliothekswesen. Bd. 7
Thesaurus für Parlamentsmaterialien. 1989
Transkriptionen der chinesischen und japanischen
    Sprache
Truhart, Peter: International Directory of Foreign
    Ministers 1589-1989
UBCIM Publication Series. Komorous, Hanna /
    Robert B. Harriman: International ...
    - Standard Practices ...
    - UNIMARC in Theory ...
UNESCO-Konferenzberichte. 6
Verzeichnis der Land- und
    ernährungswirtschaftlichen .../ 8. Ausg.

Verzeichnis lieferbarer Bücher 1988/89 Ergänzungs-
    band
    - Ergänzungsband. ISBN Register
Verzeichnis lieferbarer Bücher 1989/90 Bücher-
    verzeichnis
    - ISBN Register
    - Schlagwortverzeichnis
Verzeichnis lieferbarer Bücher (VLB Aktuell) CD-
    ROM-Ausgabe ...
Verzeichniss eines Theils der Kupferstich- und ...
Von der Wirkung des Buches
Ward, Charles A.: Moscow and Leningrad. Vol. 1
Where to Order Foreign Books
Who's Who at the Frankfurt Book Fair / 20th ed.
Who's Who in the Socialist Countries / 2nd ed.
World Guide to Libraries 9th ed.
Yearbook of International Organizations / 26th ed.
    Vol. 1
    - 26th ed. Vol. 2
    - 26th ed. Vol. 3
Zeitungs-Index. Verzeichnis wichtiger ... Jg. 16
Zeltner, Gustav G.: Kurtz-gefasste Historie des ...

1990

The African Book Publishing Record Vol. 16
Akten der Prinzipalkommission des Immer-
    währenden ... Mikrofiche-Ed.
Akten der Prinzipalkommission des Immer-
    währenden ...
American Biographical Archive
Arbeiten zur sozialwissenschaftlich orientierten
    Freiraumplanung. 10
Arbeitsbuch Geschichte. Mittelalter / 9. durchges.
    Aufl.
Archives Biographiques Françaises
Archivio Biografico Italiano
Archivo Biográfico de España, Portugal e
    Iberoamérica. Indice Biográfico de España,
    Portugal e Iberoamérica
Archivum. Vol. XXXVI
Arntz, Helmut: Die Kognakbrenner
Australasian Biographical Archive
Bartke, Wolfgang: Biographical Dictionary and ...
    - Who's who in the ... / 3rd ed.
Bayerische Staatsbibliothek. Alphabetischer Katalog
    - Katalog der Musikzeitschriften
Beiträge zur Bibliothekstheorie und Bibliotheks-
    geschichte 4
    - Bd. 5
Beiträge zur Kommunalwissenschaft. 33
    - Bd. 34
Berichte z. Erziehungstherapie u. Eingliederungs-
    hilfe. 40: Bd. I
    - Bd. 41: Bd. II
    - Bd. 54
    - Bd. 55
Berliner China-Studien. 18
Bibliographia Cartographica. Nr 16
Bibliographie Pädagogik. Jg. 21. Reihe/Series A 1986
    - Reihe/Series B 1986
Bibliographie und Berichte
Bibliographie zum Antisemitismus
Bibliotheca Palatina
Bibliothek der Deutschen Literatur
Bibliothek Forschung und Praxis Jg. 14
Bibliotheksforum Bayern (BFB) Jg. 18
Bibliothekspraxis. 28
    - Bd. 29
Bibliotheksstudien. 5

Briefe Deutscher Philosophen (1750-1850)
The British Library General Catalogue of Printed
    Books 1988-1989
Buch und Bibliothekswissenschaft im
    Informationszeitalter
Chitnis, A. C.: Scotland's Age of Equilibrium
Communication Research and Broadcasting. 9
Conservation in Archives ...
Corpus Librorum Emblematum. Series B. No. 2
Cumulative Microform Reviews, 1972-76
DeCoster, Jean de: Dictionary for automotive
    engineering / 3rd rev. and enl. ed.
Der Deutsch-israelische Dialog Tl. III / Bd. 7
    - Tl. III / Bd. 8
Deutsche Drucke des Barock 1600-1720. Abt. A. Bd. 9
    - Abt. B. Bd. 10
    - Abt. B. Bd. 11
    - Abt. B. Bd. 12
    - Abt. B. Bd. 13
    - Abt. B. Bd. 14
    - Abt. B. Bd. 9
Deutsches Biographisches Archiv. Neue Folge
DGD-Schriftenreihe Bd. 9
Directory of Administrative and Judical ...
Döring, Detlef: Die Bestandsentwicklung der ...
Dortmunder Beiträge zur Zeitungsforschung Bd. 44
    - Bd. 47
Encyclopedic Dictionary of Electronic, Electronical ...
    Vol. 01
EUDISED R & D Bulletin
François Villon - Bibliographie ... Bibliographie
    - Materialien zu ...
Die Frauenfrage in Deutschland. Neue Folge. Bd. 4
German Yearbook on Business History. 1987
    - 1988
Gesamtverzeichnis des deutschsprachigen
    Schrifttums ausserhalb des Buchhandels
Gesamtverzeichnis Deutschsprachiger
    Hochschulschriften 1966-1980
Glossarium Artis. Bd. 7 / 2., vollst. neu bearb. u. erw.
    Aufl.
Grundwissen Buchhandel - Verlage Bd. 8
Guide to Microforms in Print 1990. Author-Title
    - Subject Guide
Hadamitzky, Wolfgang / Marianne Rudat-Kocks:
    Japan-Bibliografie. Reihe A. Bd. 1
Handbuch der Tonstudiotechnik / 5. völlig neubearb.
    Aufl. Bd. 2
Handbuch der Universitäten und Fachhochschulen /
    5. Ausg.
Health Information for All ...
Hebrew Books from the Harvard College ...
Hessische Bibliographie. Bd. XII
IFLA Annual 1989
IFLA Journal Vol. 16
IFLA Publications Vol. 51
    - Vol. 52/53
... in Szene gesetzt
Index der Antiken Kunst und Architektur
Index of Conference Proceedings Received ...
Innenpolitik in Theorie u. Praxis. 21
International Bibliography of Plant Protection 1965-
    1987. Vol. 20
    - Vol. 21
    - Vol. 22
    - Vol. 23
    - Vol. 24
    - Vol. 25
    - Vol. 26
    - Vol. 27
    - Vol. 28
    - Vol. 29

- Vol. 30
- Vol. 31

International Books in Print 1990. Part I
- Part II

International Books in Print. CD-ROM-Ed. 1990
International Congress Calendar. Vol. 30
Journal of Commonwealth literature
Katalog der Graphischen Porträts ... Reihe A. Bd. 12
- Reihe A. Bd. 13
- Reihe A. Bd. 14
- Reihe A. Bd. 15

Kimminich, Otto: Einführung in das Völkerrecht / 4. erg. u. verb. Aufl.
Klassifikation für Allgemeinbibliotheken. Bd. 1
Die Kolonne Zeitung der jungen Gruppe Dresden
Kommunale Sozialpolitik. 7
Kommunikation und Politik. 22
Konrad Mellerowicz
Laminski, Adolf: Die Kirchenbibliotheken zu St. Nicolai ...
Ludes, Peter: Bibliographie zur Entwicklung des Fernsehens
Management of Recorded Information
Marburger Beiträge zur Vergleichenden ... 22
- Bd. 23
- Bd. 24

Marburger Index
Marburger Index. Text-Inventar ...
Die Matrikel der Universität Jena. Bd. III. Lfg. 9
Microform Review Vol. 19
Microform Review. Cumulative Index
Microform Review Series in Library Micrographics Management Vol. 1 - 13
Minerva-Fachserie Medizin. Prävention im Betrieb
Minerva-Fachserie Wirtschafts- und Sozialwissenschaften. Greca, Rainer: Die Grenzen ...
The New ICC World Directory of Chambers ...
Patentinformation und Patentdokumentation in Westeuropa / 4. überarb. u. erw. Aufl.
Presse-, Rundfunk- und Filmarchive ... 11
Prinz-Albert-Studien Bd. 8
Protokolle des Vermittlungsausschusses des Deutschen ...
Publishers' International ISBN-Directory (PIID) / 17th ed.
The Pulitzer Prize Archive Part B. Vol. 4
Rundfunkstudien Bd. 4
Scandinavian Biographical Archive
Schauspieltexte im Theatermuseum der Universität ...
Schmidt, Ralph: Informationsagenturen im ... Bd. 3
Schriftenreihe der Georg-von-Vollmar-Akademie. 3
Schriftenreihe Internationales Zentralinstitut für das Jugend- ... 24
Standortkatalog der Sammlung Welding
Stauder, Peter: Die Hochschulschriften der alten ...
Studien des Forschungsinstitutes für Wirtschaftspolitik ... 37
- Bd. 38
- Bd. 39
- Bd. 40
- Bd. 41

SWI Schlagwortindex. Teil 2
Tolzmann, Don H.: Catalog of the German-Americana ...
UBCIM Publication Series. ISBD (CF) ...
Verzeichnis deutscher Informations- und Dokumentationsstellen ... / Ausg. 5
Verzeichnis lieferbarer Bücher 1989/90 Ergänzungsband
- Ergänzungsband. ISBN Register

Verzeichnis lieferbarer Bücher 1990/91 Bücherverzeichnis

- ISBN Register
- Schlagwortverzeichnis

Verzeichnis lieferbarer Bücher (VLB Aktuell) CD-ROM-Ausgabe ...
- Schinzel, Wilfried H.: Verzeichnis ...

Verzeichnis rechtswissenschaftlicher Zeitschriften ... / 3., wes. erw. Aufl.
Wennrich, Peter: Dictionary of Electronics and Information ...
Wennrich, Peter / Paul Spillner: International Encyclopedia of Abbreviations ...
Das Westfälische Wirtschaftsarchiv und seine Bestände
Who's Who at the Frankfurt Book Fair / 21st ed.
Who's Who in Mass Communication
Wissensveränderung durch Medien
Wolfenbütteler Bibliographie zur Geschichte des ... Bd. 01
World Guide to Scientific Associations and Learned Societies / 5th ed.
World Guide to Special Libraries / 2nd ed.
Yearbook of International Organizations / 27th ed. Vol. 1
- 27th ed. Vol. 2
- 27th ed. Vol. 3

Zeitungs-Index. Verzeichnis wichtiger ... Jg. 17
Zentralblatt für Bibliothekswesen ...

1991

The African Book Publishing Record Vol. 17
Akten der Prinzipalkommission des Immerwährenden ... Mikrofiche-Ed.
Alphabetischer Musikalienkatalog der Pfälzischen Landesbibliothek ...
American Biographical Archive
Antisemitism. Vol. 2
Arbeitsbuch Geschichte. Neuzeit. 1. 16.-18. Jahrhundert. Repetitorium / 6. durchges. Aufl.
Archives Biographiques Françaises
Archivio Biografico Italiano. Nuova Serie
Archivo Biográfico de España, Portugal e Iberoamérica. Nueva Serie
Australasian Biographical Archive
Bayerische Staatsbibliothek. Katalog der Geschichtszeitschriften
Beiträge zur Bibliothekstheorie und Bibliotheksgeschichte 3
Beiträge zur empirischen Kriminologie 12
Beiträge zur Kommunalwissenschaft. 35
Bergmann, Joachim: Die Schaubühne - ...
Bericht der 16. Tonmeistertagung
Berliner China-Studien. 19
Bibliographia Cartographica. Nr 17
Bibliographie der deutschsprachigen Lyrikanthologien ...
Bibliographie Sport und Freizeit
Bibliographie zum Antisemitismus
Bibliographien zur deutschen Barockliteratur Bd. 1 - 3
Bibliographie zur deutschen Literaturgeschichte des Barockzeitalters Tl. 1
- Tl. 2

Bibliographie zur lateinischen Wortforschung Bd. 1
- Bd. 2
- Bd. 3

Bibliographie zur Zeitgeschichte 1953-1995. Bd. IV
Bibliographien zur Regionalen Geographie und ... 8
Bibliotheca Palatina
Bibliothek der Deutschen Literatur
Bibliothek Forschung und Praxis Jg. 15

Bibliotheksforum Bayern (BFB) Jg. 19
Bibliothekspraxis. 30
Bild- und Tonträger-Verzeichnisse. 19
Bild- und Tonträger-Verzeichnisse. 20
Bild- und Tonträger-Verzeichnisse. 21
The Black Women Oral History Project. Vol. 1
- Vol. 2
- Vol. 3
- Vol. 4
- Vol. 5
- Vol. 6
- Vol. 7
- Vol. 8
- Vol. 9
- Vol. 10
- Vol. 11

Biographisches Wörterbuch zur deutschen Geschichte
British Biographical Archive. British Biographical Index
British Biographical Archive. Series II
The British Library General Catalogue of Printed Books 1988-1989
The British Library General Subject Catalogue 1986-1990
Computers in the Humanities and ...
Corpus Librorum Emblematum. Series B. No. 1
Crime & Justice in American History. Vol. 1
- Vol. 2
- Vol. 3
- Vol. 4

Deutsche Drucke des Barock 1600-1720. Abt. B. Bd. 15
- Abt. B. Bd. 16
- Abt. B. Bd. 17

Deutsche Presseforschung 27
Deutsches Biographisches Archiv. Neue Folge
Deutsches Literatur-Lexikon Bde. 1 - 12
- Bd. 13

Deutsches Theater-Lexikon Bde. I u. II
Deutschsprachige Exilliteratur seit 1933 Bd. 1
- Bd. 2

Deutschsprachige Literatur Prags und der ...
Documents that move and speak ...
Dortmunder Beiträge zur Zeitungsforschung Bd. 48
- Bd. 49

Eichhoff, Jürgen: Wortatlas der deutschen Umgangssprachen Bd. 1
- Bd. 2

Die Emigration der Wissenschaften nach 1933
Encyclopedia of World Problems and ... / 3rd ed.
Estermann, Alfred: Die deutschen Literatur-Zeitschriften 1815-1850
EUDISED R & D Bulletin
Film, Television, Sound Archives Series. 1
- Bd. 2

Filmkultur zur Zeit der Weimarer Republik
Fokus. 2
Franz Kafka. Eine kommentierte Bibliographie der Sekundärliteratur
Franz Kafkas Werke. Eine Bibliographie der Primärliteratur (1908-1980)
Die Frauenfrage in Deutschland. Neue Folge. Bd. 5
Genealogische Bestände der Universitätsbibliothek Düsseldorf
Gesamtverzeichnis der Übersetzungen deutschsprachiger Werke Bd. 1
Grund, Uwe: Indices zur sprachlichen und ... Bd. 1
Grundwissen Buchhandel - Verlage Bd. 5 / 2. verb. Ausg.
Guide to Microforms in Print 1990. Supplement
Guide to Microforms in Print 1991. Author-Title
- 1991. Subject Guide

1992

Crime & Justice in American History. Vol. 5
- Vol. 6
- Vol. 7
- Vol. 8
- Vol. 9
- Vol. 10
- Vol. 11
Deutsche Drucke des Barock 1600-1720. Abt. B. Bd. 18
- Abt. B. Bd. 19
- Abt. B. Bd. 20
Deutsche literarische Zeitschriften 1945-1970
Deutsche Presseforschung 28
Deutsches Biographisches Archiv. Neue Folge
Deutsches Literatur-Lexikon. Bd. 14
Deutsches Theater-Lexikon Bd. III
Dictionarium Museologicum. by ICOM ...
Dictionary of Dictionaries
Dittmar, Peter: Die Darstellung der Juden in ...
Documents for the History of Collecting. Italian
    Inventories. Vol. 1
Dortmunder Beiträge zur Zeitungsforschung Bd.
    50/1. Tle. A, B u. C
EUDISED R & D Bulletin
Fachkatalog Afrika. Bd. 12/2
Film, Television, Sound Archives Series. 3
- Bd. 4
Fokus. 3
Forsyth, Michael: Bauwerke für Musik
Funke, Fritz: Buchkunde / 5., neubearb. Aufl.
German Yearbook on Business History. 1988-92
Gesamtverzeichnis der Übersetzungen
    deutschsprachiger Werke Bd. 2
- Bd. 3
- Bd. 4
- Bd. 5
- Bd. 6
Glossarium Artis Bd. 2 / 3. vollst. neubearb. u. erw.
    Aufl.
Grosse deutsche Lexika. Allgemeine deutsche ...
- Hederich, Benjamin: Gründliches Antiquitäten-
    Lexicon
- Hederich, Benjamin: Gründliches Lexicon ...
- Hederich, Benjamin: Reales Schul-Lexicon
- Hübner, Johann: Reales Staats- Zeitungs- ...
- Hübner, Johann: Reales Staats-, Zeitungs- und ...
- Universal-Lexikon ...
- Walch, Johann Georg: Philosophisches Lexicon
Grundwissen Buchhandel - Verlage Bd. 7
Guide to Microforms in Print 1991. Supplement
Guide to Microforms in Print 1992. Author-Title
- 1992. Subject Guide
- 1992. Supplement
Guide to the Archival Materials of ... Vol. 2
Hacker, Rupert: Bibliothekarisches Grundwissen / 6.,
    völlig neubearb. Aufl.
Hebrew Books from the Harvard College ...
Heiber, Helmut: Universität unterm Hakenkreuz. Tl. II
Hessische Bibliographie. Bd. XIII
History of Women in the United States. Vol. 1
- Vol. 2
- Vol. 3
- Vol. 4
- Vol. 5
IFLA Annual 1991
IFLA Journal Vol. 18
IFLA Publications Vol. 57
- Vol. 59
- Vol. 60
- Vol. 61
International Bibliography of Historical Sciences.
    Vol. 57
International Books in Print 1992. Part I

- Part II
International Books in Print. CD-ROM-Ed. / 2nd. ed.
    1992
International Congress Calendar. Vol. 32
International Directory of Cinematographers ... Vol. 12
Inventar zu den Nachlässen der ...
Italien-Index
Jablonska-Skinder, Hanna / Ulrich Teichler:
    Handbook of Higher ...
Journal of Commonwealth literature
Katalog der Graphischen Porträts ... Reihe A. Bd. 20
- Reihe A. Bd. 21
Knaus, Hermann: Studien zur Handschriftenkunde
Kommunale Sozialpolitik. 8
Kopper, Gerd G.: Medien- und
    Kommunikationspolitik der ...
Koschnick, Wolfgang J.: Standard Dictionary of the ...
    Vol. 2
Kunze, Horst: Über das Registermachen / 4. erw. u.
    verb. Aufl.
Leesch, Wolfgang: Die deutschen Archivare ... Bd. 2
Leuchtmann, Horst: Dictionary of Terms in Music /
    4th rev. and enl. ed.
Lexikon deutsch-jüdischer Autoren. Bd. 1
Libretti in deutschen Bibliotheken
Literatur und Archiv. 4 / 2. überarb. u. erw. Aufl.
Literatur und Archiv. 6
Longerich, Peter: Hitlers Stellvertreter
Marburger Index
Marburger Index. Heusinger, Lutz: DV-Anleitung
Market Economy and Planned Economy
Microform Review. Vol. 21
- parallel Microfiche ed.
Modern American Protestantism and its ... Vol. 1
- Vol. 2
- Vol. 3
- Vol. 4
- Vol. 6
- Vol. 7
Münnich, Monika: PC-Katalogisierung mit RAK
Museum und Denkmalpflege
Museums of the World / 4th ed.
Die neue Weltbühne 1933-1939
Personal Work and Training for Library Managers
Peters, Klaus: Fachinformation Rechtswissenschaften
PIK - Praxis der Informationsverarbeitung und ... Jg. 15
Plakatsammlung des Instituts für Zeitungsforschung
    ... Bd. 1
- Bd. 2
The Political Risk Yearbook 1992
Polnische Drucke und Polonica 1501-1700. Band 1
Polskie Archiwum Biograficzne
Prinz-Albert-Studien Bd. 9
Publishers' International ISBN-Directory (PIID) /
    19th ed.
The Pulitzer Prize Archive Part B. Vol. 6
Richter, Brigitte: Precis de Bibliothéconomie / 5ème
    ed. revue et ...
Ritter, Paul: Frans Masereel
Samurin, Evgenij I.: Geschichte der bibliothekarisch
    - bibliographischen ... 2. unveränderter Nachdr
Sawoniak, Henryk / Maria Witt: New International
    Dictionary of ... / 2nd rev. and enl. ed.
Schriftenreihe der Georg-von-Vollmar-Akademie. 4
- Bd. 5
Schulz-Torge, Ulrich J.: Who was Who in the Soviet
    Union
Simon, Elisabeth: Bibliotheks- und
    Informationssysteme in ...
Sozialstrategien der Deutschen Arbeitsfront. Tl. B
Sozialstrategien der Deutschen Arbeitsfront. Tl. B.
    Begleitband ...

Soziokulturelle Herausforderungen ... 5
Trotskyist Serials Bibliographie 1927-1991
UBCIM Publications - New Series Vol. 6
- Vol. 7
- Vol. 8
Verhandlungen des Deutschen Bundestages ... 11.
    Wahlperiode. Verhandlungen des Bundesrates ...
- 11. Wahlperiode. Verhandlungen des Deutschen
    Bundestages ...
- 12. Wahlperiode. Verhandlungen des
    Bundesrates ...
- 12. Wahlperiode. Verhandlungen des Deutschen
    Bundestages ...
Verzeichnis lieferbarer Bücher 1991/92 Ergänzungs-
    band
- Ergänzungsband. ISBN Register
Verzeichnis lieferbarer Bücher 1992/93 Bücher-
    verzeichnis
- ISBN Register
- Schlagwortverzeichnis
Verzeichnis lieferbarer Bücher (VLB Aktuell) CD-
    ROM-Ausgabe ...
Ward, Charles A.: Moscow and Leningrad. Vol. 2
Weiss, Johannes: Max Webers Grundlegung ... / 2.
    überarb. u. erw. Aufl.
Wennrich, Peter: International Dictionary of
    Abbreviations and ...
Wennrich, Peter / Paul Spillner: International
    Encyclopedia of Abbreviations ...
Who's Who at the Frankfurt Book Fair / 23rd ed.
Who's Who in International Organizations
Wolfenbütteler Bibliographie zur Geschichte des ...
    Bd. 03
Yearbook of International Organizations / 29th ed.
    Vol. 1
- 29th ed. Vol. 2
- 29th ed. Vol. 3
Zeitungs-Index 1982-1989. CD-ROM-Ed.

1993

Adam, Paul: Der Bucheinband
The African Book Publishing Record Vol. 19
Akten der Britischen Militärregierung in ...
Akten der Prinzipalkommission des Immer-
    währenden ... Begleitbd.
Allensbacher Jahrbuch der Demoskopie Bd. IX
Allgemeines Künstlerlexikon - Internationale
    Künstlerdatenbank
Allgemeines Künstlerlexikon Bd. 7
- Bd. 8
American Biographical Archive. American
    Biographical Index
American Biographical Archive. Series II
Analytische Bibliographien deutschsprachiger
    literarischer Zeitschriften. 14
L'Anarchisme. Vol. II
Archives Biographiques Françaises. Deuxième Série
Archives Biographiques Françaises. Index
    Biographique Français
Archivio Biografico Italiano. Indice Biografico
    Italiano
Archivio Biografico Italiano. Nuova Serie
Archivo Biográfico de España, Portugal e
    Iberoamérica. Nueva Serie
Archivum. Vol. XXXIX
Australasian Biographical Archive
Bayerische Staatsbibliothek. Katalog der
    Osteuropazeitschriften
Der befragte Leser

Beiträge zur Bibliothekstheorie und
  Bibliotheksgeschichte 8
Beiträge zur Kommunalwissenschaft. 38
- Bd. 39
Bericht der 17. Tonmeistertagung
Berichte z. Erziehungstherapie u.
  Eingliederungshilfe. 42
- Bd. 56
Berliner China-Studien. 20
- Bd. 22
- Bd. 24
Berthold, Werner / Brita Eckert; Frank Wende:
  Deutsche Intellektuelle ...
Bibliographia Cartographica. Nr 19
Bibliographie Bauwesen - Architektur - Städtebau
Bibliographie Medizin / Bibliography of Medicine
Bibliographie Militärwesen
Bibliographie Politik und Zeitgeschichte
Bibliographie Religion und Philosophie
Bibliographie zum Antisemitismus
Bibliographien zur deutsch-jüdischen Geschichte. 5
Bibliography of American Imprints to ... Vol. 43-56
- Vol. 57-71
- Vol. 72-82
- Vol. 83-92
Bibliography of American Imprints to 1901
Bibliography on Peace Research and ...
Bibliotheca Palatina
Bibliothek der Deutschen Literatur
Bibliothek Forschung und Praxis Jg. 17
Bibliotheksforum Bayern (BFB) Jg. 21
Bibliothekspraxis. 33
- Bd. 34
Bibliotheksstudien. 6
Bild- und Tonträger-Verzeichnisse. 24
Biografisch Archief van de Benelux
Brenneke, Adolf / Wolfgang Leesch: Archivkunde
- Bd. 2
British Biographical Archive. Series II
The British Library General Catalogue of Printed
  Books 1990-1992
Catalogue général des ouvrages en langue française
  1930-1933. Auteurs
Český biografický archiv a Slovenský biografický
  archiv
Communication Research and Broadcasting. 11
Communications. Vol. 18
Current contents Africa
Deutsche Drucke des Barock 1600-1720. Abt. D. Bd. 1
Deutsches Biographisches Archiv. Neue Folge
Deutsches Literatur-Lexikon. Bd. 15
Deutschsprachige Literatur Prags und der ... / 2.
  überarb. u. erw. Aufl.
Dietze, Joachim: Texterschliessung: Lexikalische
  Semantik ...
Dokumente zur Geschichte der kommunistischen ...
  Bd. 1
Dortmunder Beiträge zur Zeitungsforschung Bd. 35/4
- Bd. 51
Eichhoff, Jürgen: Wortatlas der deutschen
  Umgangssprachen Bd. 3
Einzelveröffentlichungen der Historischen
  Kommission zu Berlin. 76
Encyclopedia of Film Directors in the ... Vol. 1
Encyclopedic Dictionary of Electronic, Electronical ...
  Vol. 02
EUDISED R & D Bulletin
Film, Television, Sound Archives Series. 5
Flemig, Kurt: Karikaturisten-Lexikon
Fokus. 4
- Bd. 5
- Bd. 7

- Bd. 8
- Bd. 9
France - Allemagne. II
German Yearbook on Business History. 1993
Gesamtverzeichnis der Übersetzungen
  deutschsprachiger Werke Bd. 7
- Bd. 8
Global Books in Print PLUS
Grundwissen Buchhandel - Verlage Bd. 2
- . Bd. 5 / 3. überarb. Aufl.
Guide to Microforms in Print 1993. Author-Title
- 1993. Subject Guide
- 1993. Supplement
Hadamitzky, Wolfgang / Marianne Rudat-Kocks:
  Japan-Bibliografie. Reihe A. Bd. 2
Handbuch der Bibliotheken Bundesrepublik
  Deutschland, ... / 3. Aufl.
Handbuch der Universitäten und Fachhochschulen /
  6. Aufl.
Henning, Hans: Faust-Variationen
Hermann Hesse. Supplement ...
Hessische Bibliographie. Bd. XIV
History of Women in the United States. Vol. 6
- Vol. 7
- Vol. 8
- Vol. 9
- Vol. 10
- Vol. 11
Hitler. Reden, Schriften, Anordnungen ... Bd. III / Tl. 1
- Bd. IV / Tl. 1
IFLA Annual 1992
IFLA Journal Vol. 19
IFLA Publications Vol. 62
- Vol. 63
- Vol. 64
- Vol. 65
- Vol. 66/67
Information Handling in Offices and Archives
International Bibliography of Historical Sciences.
  Vol. 58
International Books in Print 1993. Part I
- Part II
International Books in Print. CD-ROM-Ed. / 3rd. ed.
  1993
International Congress Calendar. Vol. 33
International Directory of Arts 1993/94
International Directory of Cinematographers ... Vol. 11
International Yearbook of Library Service ... Vol. 1
Internationale Zeitungsbestände in Deutschen
  Bibliotheken
Jahrbuch für die Geschichte Mittel- ... 41
Japanese American World War II. Part V
Jewish Immigrants of the Nazi Period ... Vol. 4
Journal of Commonwealth literature
Katalog der Graphischen Porträts ... Reihe A. Bd. 22
- Reihe A. Bd. 23
- Reihe A. Bd. 24
- Reihe A. Bd. 25
Knigge, Adolph von: Sämtliche Werke
Knowledge for Europe: Librarians and ...
Koschnick, Wolfgang J.: Standard Dictionary of the ...
  Vol. 2
Kutsch, Karl J. / Leo Riemens: Grosses
  Sängerlexikon / 2. Aufl.
Leleu-Rouvray, Geneviève / Gladys Langevin:
  Bibliographie Internationale de la Marionette
Lexikon deutsch-jüdischer Autoren. Bd. 2
Marburger Index
Die Matrikel der Universität Jena. Bd. III. Lfg. 10
Memorial Book
Microform Review. Vol. 22
- parallel Microfiche ed.

Microform Market Place 1992-1993
Milano, Ernesto / Renzo Margonari; Mauro Bini:
  Xilografia dal Quattrocento ...
Minerva-Fachserie Politik. Riescher, Gisela /
  Raimund Gabriel: Die Politikwissenschaft ...
- Verstehen und Erklären ...
Modern American Protestantism and its ... Vol. 8
- Vol. 10
- Vol. 11
- Vol. 12
- Vol. 13
- Vol. 14
Multilingual Lexicon of Higher Education. Vol. 1
NS-Presseanweisungen der Vorkriegszeit Bd. 4
Paenson, Isaac: Handbuch der Terminologie des
  Völkerrechts
Paul, Hans-Holger: Inventar zu den Nachlässen der ...
PIK - Praxis der Informationsverarbeitung und ... Jg. 16
Polskie Archiwum Biograficzne
Prinz-Albert-Studien Bd. 10
Publishers' International ISBN-Directory (PIID) /
  20th ed.
The Pulitzer Prize Archive Part C. Vol. 7
- Part C. Vol. 8
Roth, Karl H.: Intelligenz und Sozialpolitik im ...
Rundfunkstudien Bd. 5
- Bd. 6
- Bd. 7
Schriften der Herbert und Elsbeth Weichmann
  Stiftung. Schicksale ...
Die Schriften der Mainzer Jakobiner
Die Schriften der Mainzer Jakobiner. Bibliographie
Schriftenreihe Internationales Zentralinstitut für das
  Jugend- ... 26
Schulz-Torge, Ulrich J.: Who's Who in Russia Today
Signale für die Musikalische Welt
Studien zum Buch- und Bibliothekswesen. Bd. 8
Die Tagebücher von Joseph Goebbels. Teil II:
  Diktate 1941 bis 1945. Bd. 7
- Bd. 8
- Bd. 9
Toward International Descriptive Standards for
  Archives
UBCIM Publications - New Series Vol. 9
- Vol. 10
- Vol. 11
- Vol. 12
Ungarische Drucke und Hungarica 1480-1720
Universität zu Köln Institut für Theater-, Film- ...
Verhandlungen des Deutschen Bundestages ... 12.
  Wahlperiode. Verhandlungen des Bundesrates ...
- 12. Wahlperiode. Verhandlungen des Deutschen
  Bundestages ...
Verzeichnis lieferbarer Bücher 1992/93 Ergänzungs-
  band
- Ergänzungsband. ISBN Register
Verzeichnis lieferbarer Bücher 1993/94 Bücher-
  verzeichnis
- ISBN Register
- Schlagwortverzeichnis
Verzeichnis lieferbarer Bücher (VLB Aktuell) CD-
  ROM-Ausgabe ...
Verzeichnis lieferbarer Bücher (VLB Aktuell) CD-
  ROM-Ausgabe (Deutsche Version)
Verzeichnis lieferbarer Bücher (VLB Aktuell) CD-
  ROM-Ausgabe (Schweizer Version)
Waldow, Alexander: Illustrierte Encyclopädie der
  graphischen ...
Wennrich, Peter / Paul Spillner: International
  Encyclopedia of Abbreviations ...
Who's Who at the Frankfurt Book Fair / 24th ed.
World Guide to Libraries / 11th ed.

- 12. Wahlperiode. Verhandlungen des Deutschen
  Bundestages ...
- 13. Wahlperiode. Verhandlungen des Deutschen
  Bundestages ...
Verzeichnis lieferbarer Bücher 1994/95 Ergänzungs-
band
- Ergänzungsband. ISBN Register
Verzeichnis lieferbarer Bücher 1995/96 Bücher-
verzeichnis
- ISBN Register
- Schlagwortverzeichnis
Verzeichnis lieferbarer Bücher (VLB Aktuell) CD-
ROM-Ausgabe (Deutsche Version)
Verzeichnis lieferbarer Bücher (VLB Aktuell) CD-
ROM-Ausgabe (Schweizer Version)
Wallas, Armin A.: Zeitschriften und Anthologien des
Expressionismus ...
Who's Who at the Frankfurt Book Fair / 26th ed.
Who's Who in International Organizations / 2nd ed.
World Guide to Libraries / 12th ed.
World Guide to Special Libraries / 3rd ed.
World Guide to Trade Associations / 4th ed.
Yearbook of International Organizations / 32nd ed.
Vol. 1
- 32nd ed. Vol. 2
- 32nd ed. Vol. 3
Yearbook PLUS

1996

African Biographical Archive
Allgemeines Künstlerlexikon - Internationale
Künstlerdatenbank / 3. Ausg.
Allgemeines Künstlerlexikon Bd. 13
- Bd. 14
- Bd. 15
American Biographical Archive. Series II
Arab-Islamic Biographical Archive
Archives Biographiques Françaises. Deuxième Série
Archivio Biografico Italiano. Nuova Serie
Archivo Biográfico de España, Portugal e
Iberoamérica 1960-1995
Archivum. Vol. XLII
- Vol. XLIII
L'Art et ses adresses en France 1996/97
August Bebel. Ausgewählte Reden und Schriften Bd. 10
Australasian Biographical Archive. Australasian
Biographical Index
Baltisches Biographisches Archiv
Bayerische Staatsbibliothek. Katalog
Bayerisches Jahrbuch. 76. Jg
Berlin-Bibliographie.1991
Der Bestand Preussische Akademie der Künste. Tl. 2.
Findbuch
Bibliographia Cartographica. Nr 22
Bibliographie zur Zeitgeschichte 1953-1995. Bd. V
Bibliographien zur Regionalen Geographie und ... 9
Bibliothek Forschung und Praxis Jg. 20
Bibliotheksforum Bayern (BFB) Jg. 24
Biografisch Archief van de Benelux. Biografische
Index van de Benelux
Biographisches Archiv der Antike
Biographisches Handbuch der SBZ
Die Bundesrepublik Deutschland und Frankreich:
Dokumente 1949-1963. Bd. 1
Catalogue of Dated and Datable Manuscripts
Český biografický archiv a Slovenský biografický
archiv
Chinese Biographical Archive
Deutsche Biographische Enzyklopädie (DBE). Bd. 3

- Bd. 4
- Bd. 5
Deutsche Drucke des Barock 1600-1720. Abt. A. Bd. 15
- Namenregister
- Register der Drucker, ...
- Titelregister
Deutsche und Polen zwischen den Kriegen
Deutsches Museum - Bildarchiv
Digitales Informationssystem für Kunst und
Sozialgeschichte ... Tl. 006: Die Gemälde ...
- Tl. 007: Wallraf-Richartz-Museum ...
- Tl. 008: Plakate ...
- Tl. 009: Politische Allegorien ...
- Tl. 010: Politische Abzeichen ...
Documents for the History of Collecting. Italian
Inventories. Vol. 2
Dokumentation zur jüdischen Kultur in
Deutschland 1840-1940. Abt. I. Teil 1
- Abt. II
Dokumente zur Geschichte der kommunistischen ...
Bd. 5
Dortmunder Beiträge zur Zeitungsforschung Bd. 35/7
Encyclopedic Dictionary of Electronic, Electronical ...
Vol. 03
- Vol. 04
Ersch, Johann S. / Johann G. Gruber: Allgemeine
Encyclopädie der Wissenschaften ...
Estermann, Alfred: Inhaltsanalytische Bibliographien
deutscher Kulturzeitschriften ... Bd. 9
- Bd. 10
EUDISED. European Educational Research Yearbook
1994/95
- Yearbook 1995/96
Fokus. 14
Frank Wedekind. A Bibliographic Handbook
Gittig, Heinz: Bibliographie der Tarnschriften ...
Global Books in Print PLUS
Grundlagen der praktischen Information und ...
Guggenheimer, Eva / Heinrich Guggenheimer:
Etymologisches Wörterbuch ...
Guide to Microforms in Print 1996. Author-Title
- 1996. Subject Guide
- 1996. Supplement
Guide to the Archival Materials of ... Vol. 3
Guides to the Sources for the History ... 3rd Series.
Vol. 5. Part 1
Haller, Klaus / Hans Popst: Katalogisierung nach
den RAK-WB / 5. überarb. Aufl.
Handbuch der Bibliotheken Bundesrepublik
Deutschland, ... / 4. Aufl.
Handbuch der Universitäten und Fachhochschulen /
7. Aufl.
Handbuch zur „Völkischen Bewegung" 1871-1918
Hebrew Books from the Harvard College ... Index to ...
Hessische Bibliographie. Bd. XVII
- Bd. XVIII
Hitler. Reden, Schriften, Anordnungen ... Bd. IV / Tl. 2
- Bd. V / Tl. 1
IFLA Annual 1995
IFLA Journal Vol. 22
IFLA Publications Vol. 74
- Vol. 75
- Vol. 76
- Vol. 77
International Bibliography of Historical Sciences.
Vol. 61
International Books in Print 1996. Part I
- Part II
International Books in Print PLUS / 6th ed. 1996
International Directory of Arts 1997/98
International Encyclopedia of Abbreviations and ...
Vol. 2

- Vol. 3
- Vol. 4
- Vol. 5
Internationale Bibliographie zur Deutschen Klassik
1750-1850. Flg. 41
Internationale Bibliographie zur deutschsprachigen
Presse ...
Internationaler Biographischer Index / 2. Ausg.
Internationaler Biographischer Index der Bildung ...
Internationaler Biographischer Index der
Geisteswissenschaften
Internationaler Biographischer Index der Politik ...
Internationaler Biographischer Index des
Militärwesens
Internationaler Biographischer Index des Rechts ...
Internationaler Nekrolog. Jg. 1994
Inventar zur Geschichte der deutschen ... Reihe B.
Bd. 1
Jahrbuch des Vereins „Gegen Vergessen ..." 1
Jahrbuch für die Geschichte Mittel- ... 44
Jüdisches Biographisches Archiv
Katalog der Bibliothek des schiitischen .../ 2. Aufl.
Katalog der Graphischen Porträts ... Reihe A. Bd. 29
Koschnick, Wolfgang J.: Standard-Lexikon für
Marketing ... / 2. Aufl.
~: Standardlexikon Werbung-Verkaufsförderung-
Öffentlichkeitsarbeit
Kunstadressbuch Deutschland, Österreich, Schweiz /
10. Ausg.
Lexikon deutsch-jüdischer Autoren. Bd. 5
Literatur und Archiv. 7
- Bd. 8
Manuscrits musicaux après 1600 / 4th cumulative ed.
Marburger Index
Marburger Index. Österreich-Index
Marburger Index. Wegweiser zur Kunst ... / 2. Ausg.
Microform Review. Vol. 25
- parallel Microfiche ed.
Microform Market Place 1996-1997
Multilingual Lexicon of Higher Education. Vol. 2
Music Publishers' International ISMN Directory /
1995/96 ed.
Neue Informations- und
Kommunikationstechnologien für
wissenschaftliche ...
ÖVK-WB Österreichischer Verbund-Katalog ... / 4.
Ausg.
- 5. Ausg.
PIK - Praxis der Informationsverarbeitung und ... Jg. 19
Pressearchiv zur Geschichte Deutschlands sowie ...
Findbuch ...
Prinz-Albert-Studien Bd. 13
Publishers' International ISBN-Directory (PIID) /
23rd ed.
Publishers' International ISBN-Directory PLUS 1996
The Pulitzer Prize Archive Part D. Vol. 10
Russisches Staatsarchiv für Literatur und Kunst ...
Sauppe, Eberhard: Dictionary of Librarianship ... /
2nd rev. and enl. ed.
Sicherheitsadressbuch Deutschland, Österreich,
Schweiz ...
Die Tagebücher von Joseph Goebbels. Teil II:
Diktate 1941 bis 1945. Bd. 1
- Bd. 2
- Bd. 6
- Gesamtregister
Thieme-Becker / Vollmer. Gesamtregister. Tl. 1
TIB - Katalog der Technischen
Informationsbibliothek Hannover. / 1. Ausg.
UBCIM Publication Series. Manuel Unimarc Version
française / 2ème ed.
UBCIM Publications - New Series Vol. 14 / Update 1

1997

Verzeichnis lieferbarer Bücher (VLB Aktuell) CD-
ROM-Ausgabe (Deutsche Version)
Verzeichnis lieferbarer Bücher (VLB Aktuell) CD-
ROM-Ausgabe (Schweizer Version)
Waibl, Elmar / Philip Herdina: Dictionary of
Philosophical Terms
Who's Who at the Frankfurt Book Fair / 28th ed.
Widerstand als „Hochverrat" 1933-1945
World Guide to Foundations
World Guide to Logotypes, Emblems ...
World Guide to Scientific Associations and Learned
Societies / 7th ed.
Yearbook of International Organizations / 34th ed.
Vol. 1
- 34th ed. Vol. 2
- 34th ed. Vol. 3
- 34th ed. Vol. 4
Yearbook PLUS / 3rd ed.

1998

African Biographical Archive. African Biographical
Index
Allgemeines Künstlerlexikon - Internationale
Künstlerdatenbank / 6. Ausg.
- 7. Ausg.
Allgemeines Künstlerlexikon Bd. 19
- Bd. 20
- Bd. 21
- Reg. zu d. Bdn 11-20 Teil 1
- Reg. zu d. Bdn 11-20 Teil 2
American Biographical Archive. Series II. American
Biographical Index / 2nd cumulative and enl. ed.
Arab-Islamic Biographical Archive
Archivio Biografico Italiano sino al 1996
Archivio Biografico Italiano. Nuova Serie
Archivo Biográfico de España, Portugal e
Iberoamérica 1960-1995
Baltisches Biographisches Archiv
Die Bayerische Staatsbibliothek in historischen ... /
2. durchges. Aufl.
Bayerisches Jahrbuch. 78. Jg
Der Bestand Preussische Akademie der Künste. Tl. 3.
Findbuch
Bibliografía General Española Siglo XV ...
Bibliografia Generale Italiana dal XV secolo ...
Bibliographia Cartographica. Nr 24
Bibliographie française du XVe siècle ...
Bibliographie zu den Biographischen Archiven. / 2.
korr. u. erw. Aufl.
Bibliotheca Palatina. Bd. 1
- Bd. 2
- Bd. 3
- Bd. 4
- Katalog und Register ...
Bibliothek Forschung und Praxis Jg. 22
Bibliotheksforum Bayern (BFB) Jg. 26
Biographisches Archiv der Antike
British Biographical Archive. Series II. British
Biographical Index
Český biografický archiv a Slovenský biografický
archiv
Chinese Biographical Archive
DeCoster, Jean de: Dictionary for automotive
engineering / 4th rev. ed.
Deutsche Biographische Enzyklopädie (DBE). Bd. 8
- Bd. 9
Deutsches Biographisches Archiv. Neue Folge.
Deutscher Biographischer Index. CD-ROM-Edition
Deutsches Literatur-Lexikon. Bd. 18

- Ergänzungsband V
- Ergänzungsband VI
Deutsches Theater-Lexikon. Bd. IV
Deutschland im ersten Nachkriegsjahr
Deutschland, Armenien und die Türkei 1895-1925.
Tl. 1
Dietze, Joachim: B. Travens Wortschatz
Digitales Informationssystem für Kunst und Sozial-
geschichte ... Tl. 011: Filmplakate ...
- Tl. 012: 1848 - Politk, Propaganda ...
- Tl. 013: Gemäldegalerie ...
Dortmunder Beiträge zur Zeitungsforschung Bd. 35/9
Edhofer, Ingrid: Die Biogenese Chl-bindender
Proteine
Einzelveröffentlichungen der Historischen
Kommission zu Berlin. 82
English Bibliography 15th Century ...
English Bibliography 1901 to 1945
Estermann, Alfred: Kontextverarbeitung.
Buchwissenschaftliche Studien
EUDISED. European Educational Research Yearbook
1997/98
European Music Directory
Fokus. 16
Funke, Fritz: Buchkunde / 6., überarb. u. erg. Aufl.
German-Americans in the World Wars Vol. V
Global Books in Print PLUS
Glossarium Artis Bd. 3 / 3., neu bearb. u. erw. Aufl.
Griechisches Biographisches Archiv
Grundwissen Buchhandel - Verlage Bd. 5 / 4.
überarb. Aufl.
Guide to Microforms in Print 1998. Author-Title
- 1998. Subject Guide
- 1998. Supplement
Hadamitzky, Wolfgang: Japanese, Chinese, and ...
Hadamitzky, Wolfgang / Marianne Rudat-Kocks:
Japan-Bibliografie. Reihe B. Bd. 1/Tle. 1 u. 2
Haller, Klaus: Katalogkunde / 3. erw. Aufl.
Handbuch der Bibliotheken Bundesrepublik
Deutschland, ... / 5. Ausg.
Handbuch der Universitäten und Fachhochschulen /
8. Ausg.
Handbuch des deutschsprachigen Exiltheaters ... Bd. I
- Bd. II
Hegel Bibliography. Part II
Hessische Bibliographie. Bd. XX
- CD-ROM-Edition
Hitler. Reden, Schriften, Anordnungen ... Bd. V / Tl. 2
- Der Hitler-Prozess 1924. Tl. 3
IFLA Journal Vol. 24
IFLA Publications Vol. 84
- Vol. 85
- Vol. 86
- Vol. 87
Indian Biographical Archive
International Bibliography of Maps and ...
International Books in Print 1998. Part I
- Part II
International Directory of Arts 1998/99
International Encyclopedia of Abbreviations and ...
Vol. 9
- Vol. 10
- Vol. 11
Internationale Bibliographie der Bibliographien
1959-1988. Bd. 01
- Bd. 02
Internationale Bibliographie zur Deutschen Klassik
1750-1850. Flg. 44
Internationale Bibliographie zur Deutschen Klassik
1750-1850. Gesamtregister ...
Internationaler Biographischer Index / 4. Ausg.
Internationaler Biographischer Index / 5. Ausg.

Internationaler Biographischer Index der
Darstellenden ...
Internationaler Biographischer Index der
Naturwissenschaften
Internationaler Biographischer Index der Publizistik ...
Inventar zur Geschichte der deutschen ... Reihe B.
Bd. 2
- Reihe B. Bd. 3
Jahrbuch des Vereins „Gegen Vergessen ..." 2
Jüdisches Biographisches Archiv. Jüdischer
Biographischer Index
Jüdisches Biographisches Archiv. Supplement
Katalog der Graphischen Porträts ... Reihe A. Bd. 31
Koenig, Michael E.: Information Driven
Management ...
Kürschners Deutscher Literatur-Kalender 1922-1988
Kürschners Deutscher Literatur-Kalender 61. Jg
Kunstadressbuch Deutschland, Österreich, Schweiz /
11. Ausg.
Leistungsmessung in wissenschaftlichen
Bibliotheken
Lexikon deutsch-jüdischer Autoren. Bd. 6
- Bd. 7
LIBER QUARTERLY. Vol. 8
Libri. Vol. 48
Lion Feuchtwanger Vol. I
Manuscrits musicaux après 1600 / 6th cumulative ed.
Marburger Index
Marburger Index. Wegweiser zur Kunst ... / 4. Ausg.
Microform & Imaging Review Vol. 27
Multi-script, Multi-lingual, Multi-character
Music Publishers' International ISMN Directory
Musikhandschriften der Staatsbibliothek zu Berlin
1: Die Bach-Sammlung
Nachschlagewerke u. Quellen z. Kunst.
Internationale Bibliographien zur ...
Nachschlagewerke zur Musik. Internationale Musik-
Sachlexika
The New Library Legacy
NS-Presseanweisungen der Vorkriegszeit Bd. 5
- Registerband
ÖVK-WB Österreichischer Verbund-Katalog ... / 8.
Ausg.
- 9. Ausg.
PIK - Praxis der Informationsverarbeitung und ... Jg. 21
Polskie Archiwum Biograficzne. Polski Indeks
Biograficzny
Prinz-Albert-Studien Bd. 15
Publishers' International ISBN-Directory (PIID) /
25th ed.
Publishers' International ISBN-Directory PLUS / 3rd ed.
The Pulitzer Prize Archive Part D. Vol. 12
Restaurator. Vol. 19
Russian National Bibliography PLUS / 4th ed.
Russisches Biographisches Archiv
Sawoniak, Henryk / Maria Witt: International
Bibliography of ... Vol. 2
Schritte zur Neuen Bibliothek
Das SKK-Statut
South-East Asian Biographical Archive
Südosteuropäisches Biographisches Archiv
Die Tagebücher von Joseph Goebbels. Teil I:
Aufzeichnungen 1923 bis 1941. Bd. 6
- Bd. 7
- Bd. 8
TIB - Katalog der Technischen Informations-
bibliothek Hannover / 3. Ausg.
Trotsky Bibliography / 3rd ed.
UBCIM Publications - New Series Vol. 14 / Update 2
- Vol. 19
Ungarisches Biographisches Archiv

1999

Register der Personen / Index of Persons

[Aufgeführt werden Personen, die in den bibliographischen Einträgen genannt werden:
Verfasser, Herausgeber, Bearbeiter, Übersetzer, Verfasser von Vorworten und Personen,
die Gegenstand von Festschriften, Bibliographien und Abhandlungen sind /
Listing persons named in the bibliographic entries:
authors, editors, compilers, translators, authors of forewords and
persons featuring as the subject of festschrifts, bibliographies and treatises]

Becker, Ruth
Texte und Daten zur Hochschulplanung. 21
Beckers, Stefan
Studien des Forschungsinstitutes für
Wirtschaftspolitik ... 39
Beckmann, Jochen
Hochschulplanung. 15
Beckmann, Michael
Minerva-Fachserie Wirtschafts- und Sozialwissen-
schaften. ~: Theorie ...
Bedi, Dina N.
IFLA Publications Vol. 30
Beer, Johann
Bibliographien zur deutschen Barockliteratur Bd. 2
Behr, Barbara
Minerva-Fachserie Medizin. ~: Zum
Stoffwechselverhalten ...
Behrbalk, Erhard
Dortmunder Beiträge zur Zeitungsforschung Bd. 1
Behrend, Siegfried
Schwarz, Werner: Guitar Bibliography
Behrendt, Ethel Leonore
~: Recht auf Gehör
Behrens, Christoph
Fürsten-Postkarten ...
Behrmann, Jörn
Directory of International Cooperation in Science ...
Belda, Luis S.
Guides to the Sources for the History ... / 3rd
Series. Vol. 7
Beling, Gerd
Beiträge zur Informations- und Dokumentations-
wissenschaft. Folge 6
Bell, Barbara L.
UBCIM Publications - New Series Vol. 18
Below, Elke
Minerva-Fachserie Psychologie. ~:
Metatheoretische ...
Bemann, Rudolf
~ / Jakob Jatzwauk: Bibliographie der
sächsischen Geschichte
Ben-Ari, Jitzhak
Der Deutsch-israelische Dialog
Bender, Wolfgang F.
~ / Siegfried Bushuven / Michael Huesmann:
Theaterperiodika des ...
Benjamin, Walter
Rundfunkstudien Bd. 1
Bennack, Jürgen
Minerva-Fachserie Pädagogik. ~: Schule ...
Bennwitz, Hanspeter
Deutsches Theater-Lexikon
Benser, Günter
Dokumente zur Geschichte der kommunistischen ...
Berberich, Volker
Minerva-Fachserie Rechts- und Staatswissen-
schaften. ~: Rechtsschutz ...
Berg, Gunnar
HIS-Brief. 44
Berger, Albrecht
Informationssysteme - Grundlagen und Praxis der ... 3
Studiengruppe für Systemforschung ... Nr. 105
Berger, Bruno
Deutsches Literatur-Lexikon. Bd. 1
- Bd. 2
- Bd. 3
Bergholter, Volker
Wirtschaftsinformatik und Quantitative ... 19
Bergk, Johann Adam
~: Die Kunst, Bücher ...
~: Die Kunst zu denken

Bergmann, Hans J.
Informationsmanagement. 5
Bergmann, Joachim
~: Die Schaubühne - ...
Bergner, Walter
Adam, Paul: Der Bucheinband
Berkemann, I.
Normenlogik
Berkemeier, Beate
Berichte z. Erziehungstherapie u. Eingliederungs-
hilfe. 4
Berlin, Charles
Catalog of the Hebrew Collection of the Harvard ...
Hebrew Books from the Harvard College ...
- Index to ...
Bermann, Richard A.
Richard A. Bermann alias Arnold Höllriegel
Bernardoni, Claudia
Ohne Seil und Haken
Der vergeudete Reichtum
Wasted Wealth
Berndt, Helmut
HIS-Brief. 48
Bernhard, Paulette
IFLA Publications Vol. 79
Bernin-Israel, Ingrid
Die Sammlung Hobrecker der
Universitätsbibliothek Braunschweig
Berninger, Ernst H.
Wissenschaftsgeschichte. Arithmos ...
Wissenschaftsgeschichte. Vogel, Kurt: Beiträge zur ...
Berns, Jörg J.
Rundfunkstudien Bd. 5
Berthold, Werner
~ / Brita Eckert; Frank Wende: Deutsche
Intellektuelle ...
Berthoud, Maurice
Glossarium Artis Bd. 10
Bertram, Hans
Moderne Unterrichtstechnologie Situationsanalyse
u. Basisinformationen ...
Besch, J. Dietrich
Planen und Bauen. 15
Beske, Anneliese
August Bebel. Ausgewählte Reden und Schriften
Bd. 2
- Bde. 3-5
- Bd. 7
- Bd. 8
- Bd. 9
- Bd. 10
Bessai, Burghardt
Hochschulplanung. 14
Best, Heinrich
Computers in the Humanities and ...
Bethge, Ulrike
Schuster-Schmah, Sigrid: Der Leserattenfänger
Betten, Anne
Eichstätter Hochschulreden 64
Beyer, Andrea
Studien des Forschungsinstitutes für
Wirtschaftspolitik ... 43
Beyer, Marga
August Bebel. Ausgewählte Reden und Schriften
Bd. 2
Bickel, Horst
Minerva-Fachserie Pädagogik. ~: Diagnose ...
- Trost, Günter / ~: Studierfähigkeit ...
Bieber, Horst
Dortmunder Beiträge zur Zeitungsforschung Bd. 16

Bienhold, Holger
Minerva-Fachserie Naturwissenschaften. ~:
Beteiligung ...
Bierkandt, Rudolf
Minerva-Fachserie Wirtschafts- und Sozialwissen-
schaften. ~: Der soziale ...
- Lawrence, Peter A.: Technische ...
Biernacki, Stanislaw
Einzelveröffentlichungen der Historischen
Kommission zu Berlin. 80
Bigler-Marschall, Ingrid
Deutsches Theater-Lexikon
Billerbeck, Rudolf
Innenpolitik in Theorie u. Praxis. 4
Bilsky, H.-D.
Hochschulplanung. 21
Binder, Annette
Minerva-Fachserie Pädagogik. Keller, Gustav /
Rolf D. Thiel; ~: Studienerfolg ...
Binding, Günther
Glossarium Artis Bd. 3 / 3., neu bearb. u. erw.
Aufl.
- Bd. 10
Bingel, Marie A.
Künstler der jungen Generation
Bingert, Alfred
Beiträge zur Bedarfsmessung an wissenschaft-
lichen Hochschulen. 18
Bini, Mauro
Milano, Ernesto / Renzo Margonari; ~: Xilografia
dal Quattrocento ...
Birau, Nikolaus
Minerva-Fachserie Medizin. ~: Melatonin ...
Bircher, Martin
Deutsche Drucke des Barock 1600-1720
Birett, Herbert
Verzeichnis in Deutschland gelaufener Filme
Birk, Lothar
Hochschulplanung. 28
Birke, Adolf M.
Akten der Britischen Militärregierung in ...
Prinz-Albert-Studien Bd. 1
- Bd. 2
- Bd. 3
- Bd. 4
- Bd. 5
- Bd. 6
- Bd. 7
- Bd. 8
- Bd. 9
- Bd. 10
- Bd. 11
- Bd. 12
- Bd. 13
Birr, Ewald
Analytische Bibliographien deutschsprachiger
literarischer Zeitschriften. 14
Birreck, Manfred
Beiträge des Instituts für Zukunftsforschung. 3
Bisbrouck, Marie-Françoise
IFLA Publications Vol. 88
Bischof, Werner
Göttinger Atomrechtskatalog. Internationale
Bibliographie ...
Bischof-Peters, Alma
HIS-Brief. 50
Bischoff, Dietmar
Minerva-Fachserie Psychologie. ~: Grundlagen ...
Biser, Eugen
Integrale Anthropologie. 3
Bittlinger, Ludwig
Minerva-Fachserie Pädagogik. ~: Elemente ...

Blaar Ute
Deutschland, Armenien und die Türkei 1895-
1925. Tl. 2

Blänsdorf, Klaus
Seminarberichte der Deutschen UNESCO-
Kommission Nr. 22

Blaes, Ruth
Minerva-Fachserie Wirtschafts- und Sozialwissenschaften. ~: Qualifikationsstruktur ...

Blahusch, Friedrich
HIS-Brief. 39

Blaicher, Günther
Eichstätter Hochschulreden 18

Blana, Hubert
~ / Hermann Kusterer; Peter Fliegel: Partner im ...
Grundwissen Buchhandel - Verlage Bd. 5

Blandow-Wechsung, Steffi
Minerva-Fachserie Pädagogik. ~: Heimerziehung ...

Blaum, Verena
Minerva-Fachserie Wirtschafts- und Sozialwissenschaften. ~: Marxismus-Leninismus ...

Bleek, Wilhelm
Bibliographie der geheimen DDR-Dissertationen

Bleuler, Eugen
Dementia Praecox oder ...

Blome, Gerhard
Minerva-Fachserie Wirtschafts- und Sozialwissenschaften. ~: Gesellschaftliche ...

Blondel, Jean
Directory of European Political Scientists

Bloock, Cécile Vanden
International Biographical Dictionary of Religion
Who's Who in International Organizations / 2nd ed.

Bloss, Marjorie E.
UBCIM Publication Series. Serial Holdings ...

Blühm, Elger
Studien zur Publizistik 17
- 17. Bd. 3
Studien zur Publizistik

Blum, Askan
Beiträge zur Bibliothekstheorie und Bibliotheksgeschichte 8

Blum, Franz
Minerva-Fachserie Pädagogik. ~: Studieneignung ...

Blumenbach, Dedo
International Bibliography of Plant Protection
1965-1987

Blumenthal, Bert
Minerva-Fachserie Geisteswissenschaften. ~: Der
Weg ...

Blumenthal, Viktor von
Marburger Beiträge zur Vergleichenden ... 9
- Bd. 10
- Bd. 11
- Bd. 12
- Bd. 16
- Bd. 17
- Bd. 23
- Bd. 25
Qualifizierung und wissenschaftlich-technischer
Fortschritt ... Bd. 1
Texte - Dokumente - Berichte Bd. 6
- Bd. 7
- Bd. 12
- Bd. 23
- Bd. 24
- Bd. 25
- Bd. 28
- Bd. 29
- Bd. 34

Bluthardt, Manfred
Krüger, Henning Anders: Kalkulationsleitfaden
Buchdruck

Boberach, Heinz
Ämter, Abkürzungen, Aktionen des NS-Staates
In der Gemeinschaft der Völker
Inventar archivalischer Quellen des NS-Staates. Tl. 1
- Tl. 2
Inventar zur Geschichte der deutschen ...
Inventar zur Geschichte der deutschen ... Reihe B.
Bd. 1
- Reihe B. Bd. 2
- Reihe B. Bd. 3
Schriften der Herbert und Elsbeth Weichmann
Stiftung. Quellen zur ...

Boberg, Gisela
HIS-Brief. 46
Hochschulplanung. 16

Bobeth, Johannes
~: Die Zeitschriften der ...

Bock, Gunter
Benutzerverhalten an deutschen Hochschul-
bibliotheken
Zur Benutzerforschung in Bibliotheken

Bock-Werthmann, Werner
Directory of International Cooperation in Science ...

Bode, Herbert F.
Marburger Beiträge zur Vergleichenden ... 11
Minerva-Fachserie Wirtschafts- und Sozialwissen-
schaften. ~: Arbeit ...
Qualifizierung und wissenschaftlich-technischer
Fortschritt ... Bd. 3
Texte - Dokumente - Berichte Bd. 7
- Bd. 8
- Bd. 11
- Bd. 14

Böcker, Franz
Grundzüge des Marketing

Bödey, Jósef
Dictionarium Bibliothecarii Practicum / 6., verb.
und erw. Auflage

Böhm, Wolfgang
~: Biographisches Handbuch zur ...

Böhm, Wolfgang
Texte und Daten zur Hochschulplanung. 6

Böhme-Dürr, Karin
Wissensveränderung durch Medien

Boekhorst, Peter
Leistungsmessung in wissenschaftlichen
Bibliotheken

Bölhoff, Reiner
Bibliographie zur deutschen Literaturgeschichte
des Barockzeitalters Tl. 1
- Gesamtregister

Böning, Holger
Deutsche Presseforschung 28

Bönsch, Manfred
~: Beiträge zu einer ...

Bötte, Gerd J.
The First Century of German Language ...

Böttger, Joachim
Gekeler, Otto / Klaus-Dieter Herdt; Walter
Oberender: Warenkatalogisierung und
Kommunikation ...

Boetticher, Heiner von
Minerva-Fachserie Naturwissenschaften. ~:
Gamma- ...

Böve, Bernd
Minerva-Fachserie Rechts- und Staatswissen-
schaften. ~: Unfallforschung ...

Bogel, Else
Studien zur Publizistik 17

- Bd. 19

Bohmüller, Lothar
Beiträge zur Bibliothekstheorie und Bibliotheksgeschichte 2

Bohne, Werner
Minerva-Fachserie Geisteswissenschaften. ~ /
Felix Semmelroth: Die künstlerische ...

Bohrmann, Hans
Dortmunder Beiträge zur Zeitungsforschung
- Bd. 43
- Bd. 48
- Bd. 49
Handbuch der Pressearchive
Kommunikation und Politik
- Bd. 4
NS-Presseanweisungen der Vorkriegszeit

Bolle, Michael
Beiträge zur Sozialökonomik der Arbeit
- Bd. 3
- Bd. 4
- 14

Bongartz, Klaus
Berichte z. Erziehungstherapie u. Eingliederungshilfe. 38: Tl. I

Bongartz, Walter
Minerva-Fachserie Psychologie. ~: Selektive ...

Bonik, Klaus
Eichstätter Hochschulreden 20

Bonin, Hinrich
HIS-Brief. 54
Hochschulplanung. 20

Bonnemann, Arwed
Minerva-Fachserie Pädagogik. ~: Studenten ...

Bonness, Elke
Bibliotheksstudien. 1 A
- Bd. 1B
- Bd. 3
- Bd. 6

Bonse, Mechthilde
Haller, Klaus: Titelaufnahme nach RAK / 2. Aufl.

Booms, Hans
Akten der Britischen Militärregierung in ...
Akten der Prinzipalkommission des
Immerwährenden ...

Boorstin, Daniel S.
International Librarianship Today and Tomorrow

Borchardt, D. H.
IFLA Publications Vol. 17

Borcherdt, Hans Heinrich
Stoffe - Formen - Strukturen

Borck, Karin
Einzelveröffentlichungen der Historischen
Kommission zu Berlin. 80

Bormann, Martin
Akten der Partei-Kanzlei der NSDAP
Longerich, Peter: Hitlers Stellvertreter

Born, Jan
Minerva-Fachserie Psychologie. ~: Effekte ...

Born, Jürgen
Deutschsprachige Literatur Prags und der ...

Bornstedt, Adalbert von
Dortmunder Beiträge zur Zeitungsforschung Bd. 51

Bosbach, Franz
Prinz-Albert-Studien Bd. 14
- Bd. 15

Bosl, Karl
Biographisches Wörterbuch zur deutschen
Geschichte

Boss, Richard
Microform Review Series in Library
Micrographics Management Vol. 7

Boßmeyer, Christine
~: Fortschritte des EDV-Verbundes ...
IFLA Publications Vol. 38
Bossow, Manfred
Minerva-Fachserie Wirtschafts- und Sozialwissen-
schaften. ~: Öffentlicher ...
Bossuat, Marie L.
IFLA Publications Vol. 3
Bott-Bodenhausen, Karin
Minerva-Fachserie Geisteswissenschaften. Sinti in
der Grafschaft ...
Bouché, Reinhard
~: Datenerhebungskatalog für die
Dokumentation ...
Kind, Friedbert: Bausysteme im Hochschulbau
Bourdon, Françoise
UBCIM Publications - New Series Vol. 5
- Vol. 11
Bourke, Thomas A.
Microform & Imaging Review
Bourne, Ross
UBCIM Publications - New Series Vol. 7
Boutros Ghali, Boutros
Hüfner, Klaus / Jens Naumann: The United
Nations System. Vol. 5. Part A
Bowden, Russell
IFLA Publications Vol. 20
- Vol. 68
- Vol. 71
Bower, Alan
Journal of Commonwealth literature
Boyle, Leonard
Bibliotheca Palatina
Brachmann, Botho
Archivwesen in der DDR
Bradley, Susan
Archives Biographiques Françaises
Brämer, Rainer
Marburger Beiträge zur Vergleichenden ... 18
Texte - Dokumente - Berichte. 9/10
- Bd. 25
- Bd. 28
Bradburn, Norman
Demoskopie und Aufklärung
Braesel, Michaela
Prinz-Albert-Studien Bd. 15
Bramann, Klaus W.
Grundwissen Buchhandel - Verlage
- Bd. 2
Bramley, Gerald
~: Outreach
Brammerts, Hermann
Minerva-Fachserie Pädagogik. ~: Die
Bildungsarbeit ...
Brandenburg, Uwe
Minerva-Fachserie Medizin. Prävention im Betrieb
Brandenburg, Volker
Wirtschaftsinformatik und Quantitative ... 10
Brandes, Helga
Studien zur Publizistik 21
Brandt, Ulrich
Ludes, Peter: Bibliographie zur Entwicklung ...
Brandtstädter, Susanne
Berliner China-Studien Bd. 25
Brauer, Werner
~: Graphik + (und) Design
Braun, Bernd
Quellen zur Geschichte der Juden in den ... Bd. 1
- Bd. 2
Braun, Traute
Bibliothekspraxis. 34

Braun-Falco, O.
Minerva-Fachserie Medizin. Hirsch, Rolf D.: Das
Basaliom
Braune-Steiniger, Franz
Minerva-Fachserie Rechts- und Staatswissen-
schaften. ~: Die Bedeutung ...
Brechtken, Magnus
Prinz-Albert-Studien Bd. 12
- Bd. 13
Breckwoldt, Michael
Arbeiten zur sozialwissenschaftlich orientierten
Freiraumplanung. 14
Brede, Werner
Funke, Klaus-Detlef: Innere Pressefreiheit
Breitbach, Irmgard
Minerva-Fachserie Pädagogik. ~:
Identitätsentwicklung ...
Breitkopf, Johann G.
~: Versuch den Ursprung der ...
Bremer, Joachim
Minerva-Fachserie Technik. ~: Binäre ...
Brendel, Detlev
~ / Bernd Grobe: Journalistisches Grundwissen
Brenneke, Adolf
~: Archivkunde
~ / Wolfgang Leesch: Archivkunde
Bresemann, Hans J.
Schriftenreihe der Deutschen Gesellschaft für
Photographie e.V. 1
Bressem, Lydia
Bibliographie zum Antisemitismus
Breton, Jacques
Le livre, les bibliothèques et la documentation
La pratique du catalogage
Brettschneider, Frank
Innenpolitik in Theorie u. Praxis. 22
Breuer, Günter
~: Literatur des gesamten Eisenbahnwesens
Brewer, Gordon J.
The Literature of Geography
Brian, Rob
IFLA Publications Vol. 83
Brick, Lennart
Politische Information
Briesenick, Christa
Bibliotheksstudien. 1 C
Brimer, M. A.
Studies and surveys in comparative education
Brito, Rui
International Directory of Cinematographers ...
Vol. 8
Brocke, Michael
Bibliographien zur deutsch-jüdischen Geschichte
Brockhoff, Evamaria
Geschichte und Kultur der Juden in Bayern.
Aufsätze
- Lebensläufe
Brockstedt, Jürgen
HIS-Brief. 51
- Bd. 55
Brodner, Gerhard
Minerva-Fachserie Psychologie. ~: Beiträge ...
Bröcher, Achim
Berichte z. Erziehungstherapie u. Eingliederungs-
hilfe Bd. 48
- Bd. 49
- Bd. 51
- Bd. 53
Brogiato, Heinz P.
Bibliographien zur Regionalen Geographie und ... 8
Broxis, Peter F.
~: Organising the Arts

Bruce, Anthony
~: A Bibliography of the British Army ...
~: A Bibliography of the British Military ...
Bruderer, Herbert E.
~: Handbuch der maschinellen und ...
~: Sprache -Technik - Kybernetik
Brücker-Steinkuhl, Kurt
Minerva-Fachserie Technik. ~: Spatial ...
Brünn, Dieter
Bibliotheca Trinitariorum. Bd. I
Brugger, Gustav
Schriftenreihe der UFITA H. 40
Brummund, Peter
Dortmunder Beiträge zur Zeitungsforschung Bd.
40/41 / Tl. 1
Brunner, Karl
Arbeitsbuch Geschichte. Mittelalter / 4., überarb.
Aufl.
Brunner, Klaus
~: Katalog der Ritter-Waldauf-Bibliothek
Bubb, Heiner
Eichstätter Hochschulreden 60
Buber, Martin
Martin Buber
Buber, Rafael
Martin Buber
Buchanan, Brian
~: Bibliothekarische Klassifikationstheorie
Bucher, Alexius J.
Eichstätter Hochschulreden 43
Buchholz, Hans
Beiträge des Instituts für Zukunftsforschung. 3
- Bd. 5
Science and Technology and the Future
Buchholz, Rüdiger
Arbeiten zur sozialwissenschaftlich orientierten
Freiraumplanung. 4
Buck, Bernhard
Baeyer, Alexander von / Bernhard Buck:
Wörterbuch Kommunikation ...
Buder, Marianne
Beiträge zur Informations- und Dokumentations-
wissenschaft. Folge 10
DGD-Schriftenreihe Bd. 9
Grundlagen der praktischen Information und ...
Bückmann, Walter
Beiträge des Instituts für Zukunftsforschung. 19
- Bd. 20
Bückner, Jens A.
Handbuch Vereinte Nationen
Büllesbach, Alfred
Minerva-Fachserie Wirtschafts- und Sozialwissen-
schaften. Politikwissenschaft und ...
Bündgen, Thomas
SWI Schlagwortindex. Teil 2
Bürger, Thomas
Deutsche Drucke des Barock 1600-1720
Büsch, Otto
Jahrbuch für die Geschichte Mittel- ...
Büssem, Eberhard
Arbeitsbuch Geschichte
- Neuzeit. 1. 16.-18. Jahrhundert. Repetitorium
- Neuzeit. 3/ 2. 1871-1914 / 2., völlig überarb.
Aufl.
Büter, Frauke
Beiträge zur Bibliothekstheorie und Bibliotheks-
geschichte 10
Büttner, Frank
Prinz-Albert-Studien Bd. 15
Büttner, Jana
Deutschland, Armenien und die Türkei 1895-
1925. Tl. 2

Büttner, Otto
Planen und Bauen. 4
Bulling, Karl
Keudell, Elisev / ~: Goethe als Benutzer ...
Bundy, Carol
African Book World & Press / 3. ed.
The Book Trade of the World Vol. IV
Bura, Josef
Minerva-Fachserie Wirtschafts- und Sozialwissen-
schaften. ~: Obdachlosigkeit ...
Burbridge, John
International Exposition of Rural Development. 3
Burghardt, Anton
~: Allgemeine Wirtschaftssoziologie
Burgmer, Brigitte
Minerva-Fachserie Kunst. ~: Ausdrucksformen ...
Burkhard, H.
Senckenbergische Bibliothek Frankfurt ...
Schlagwortkatalog ...
Busch, Adelheid
Vergleichende Erziehungswissenschaft
Busch, Friedrich W.
Vergleichende Erziehungswissenschaft
Busch, Günter
Beiträge zur Bedarfsmessung an wissenschaft-
lichen Hochschulen. 16
Busch, Wilhelm
Minerva-Fachserie Psychologie. Pietzcker, Frank:
Wilhelm Busch - Schuld und Strafe ...
Bushuven, Siegfried
Bender, Wolfgang F. / Siegfried Bushuven /
Michael Huesmann: Theaterperiodika des ...
Busse-Steffens, Meggy
Politik - Recht - Gesellschaft. 3
Busshoff, Heinrich
~: Kritische Rationalität und Politik
~: Systemtheorie als Theorie der Politik
Butchard, Ian
Fothergill, Richard / Ian Butchard: Non-Book
Materials ...
Butt, Irene
Bibliographie Bildende Kunst
Bibliographie Bildung, Erziehung, Unterricht
Bibliographie Buch- und Bibliothekswesen,
Medienkunde, ...
Bibliographie Geographie, Kartographie, Reisen
Bibliographie Geschichte, Volkskunde,
Völkerkunde
Bibliographie Land-, Forst- und
Ernährungswirtschaft
Bibliographie Medizin / Bibliography of Medicine
Bibliographie Militärwesen
Bibliographie Politik und Zeitgeschichte
Bibliographie Religion und Philosophie
Bibliographie Sport und Freizeit
Bibliographie Sprache und Literatur
Bibliographie Veterinärmedizin
Buttlar, Annemarie
Marburger Beiträge zur Vergleichenden ... 15
- Bd. 16
- Bd. 17
Texte - Dokumente - Berichte. 30
- Bd. 34
Buxtehude, Dietrich
Wettstein, Hermann: Dietrich Buxtehude (1637-
1707)
Byrum, John D. jr.
IFLA Publications Vol. 85
Multi-script, Multi-lingual, Multi-character

C

Cadmus, Manuel
Minerva-Fachserie Rechts- und Staatswissen-
schaften. ~: Zivilrechtliche ...
Calic, Edouard
Der Reichstagsbrand. Eine wissenschaftliche
Dokumentation. Bd. 1
- Bd. 2
Calvo, Gabriel
~ / Eberhard Sauppe: Diccionario de
Biblioteconomía
Cambio, Edward P.
IFLA Publications Vol. 11
Canibol, Hans P.
Minerva-Fachserie Wirtschafts- und Sozialwissen-
schaften. ~: Zur Erklärung ...
Cantzler, Florian
Studien des Forschungsinstitutes für
Wirtschaftspolitik ... 42
Capellan, Frank
Rundfunkstudien Bd. 7
Caplan, Hannah
International Biographical Dictionary of Central
European Émigrés ... Vol. II
Capurro, Rafael
~: Information
Caputo-Mayr, Marie L.
Franz Kafka. Eine kommentierte Bibliographie
der Sekundärliteratur
Franz Kafkas Werke. Eine Bibliographie der
Primärliteratur (1908-1980)
Carfrae, Nancy
Who's Who in International Organizations / 2nd ed.
World Guide to Logotypes, Emblems ...
Carls, Hans G.
Minerva-Fachserie Geisteswissenschaften. ~: Alt-
Hormoz ...
Carrington, David K.
IFLA Publications Vol. 31
Cartarius, Ulrich
Bibliographie 'Widerstand'
Casada, James A.
~: An Annoted Bibliography of Exploration ...
Casasus, Gilbert
Beiträge zur Kommunalwissenschaft. 16
Chaplin, A. H.
IFLA Publications Vol. 6
Chapman, Michael
British Catalogue of Music 1957-1985
Charlton, Michael
Communication Research and Broadcasting. 9
Schriftenreihe Internationales Zentralinstitut für
das Jugend- ... 24
Chauveinc, Marc
IFLA Publications Vol. 88
UBCIM Publication Series. Manuel Unimarc
Version française / 2ème ed.
- 3ème ed.
Chauveinc, Mireille
UBCIM Publication Series. Manuel Unimarc
Version française
Cheng Ying
Berliner China-Studien Bd. 20
Chew, Shirley
Journal of Commonwealth literature
Chirgwin, F. John
~ / Phyllis Oldfield: The Library Assistant's ...
Chitnis, A. C.
~: Scotland's Age of Equilibrium

Chmielewski, Horst von
Bestandverzeichnis der deutschen
Heimatvertriebenenpresse
Chon, Tuk Chu
Minerva-Fachserie Rechts- und Staatswissen-
schaften. ~: Die Beziehungen ...
Christes, Hans
Beiträge zur Bedarfsmessung an wissenschaft-
lichen Hochschulen. 15
Christian, Hatto
Minerva-Fachserie Pädagogik. ~: Studierfähigkeit ...
Christophersen, Kurt P.
Dortmunder Beiträge zur Zeitungsforschung Bd. 28
Chuguev, Victor
Biographical Dictionary of the Soviet Union ...
Cigánik, Marek
~: Informationsfonds in Wissenschaft, Technik ...
Ciolek-Kümper, Jutta
Kommunikation und Politik. 8
Clain-Stefanelli, Elvira E.
~: Numismatic Bibliography
Clark, Leslie
IFLA Publications Vol. 30
Clark-Mazo, Georgina
Scandinavian Biographical Archive. Scandinavian
Biographical Index
Claus, Helmut
Rahmen deutscher Buchtitel im 16. Jahrhundert
Claus, Paul
Persönlichkeiten der Weinkultur deutscher
Sprache ...
Claus, Sybille
Biographisches Handbuch der deutschsprachigen
Emigration ... Bd. III
Clauss, Manfred
Eichstätter Hochschulreden 48
Cloarec, Thierry
UBCIM Publication Series. Manuel Unimarc
Version française
Clodius, Heinrich J.
~: Primae liniae Bibliothecae Lusoriae ...
Cloes, Henning
HIS-Brief. 43
Cloonan, Michele V.
IFLA Publications Vol. 69
Coburger, Marlies
Protokoll der „Brüsseler Konferenz" der KPD 1935
Coenen, Reinhard
Studiengruppe für Systemforschung ... Nr. 49
- Nr. 58
- Nr. 59
- Nr. 107
- Nr. 108/109
Coenenberg, Gerhard
Internationaler betriebswirtschaftlicher
Zeitschriftenreport (IbZ)
Cohen, Susan Sarah
Antisemitism
Cohn, Margot
Martin Buber
Comenius, Johann A.
~: Informatorium Maternum, Mutter Schul
Commynes, Philippe de
Minerva-Fachserie Geisteswissenschaften.
Baumann, Heidrun: Der Geschichtsschreiber ...
Conen, Marie L.
Minerva-Fachserie Wirtschafts- und Sozialwissen-
schaften. ~: Mädchen ...
Conrad-Bergweiler, Gertrud
Libraries, Information Centers and Databases for
Science ...
Who's who der Herz- und Kreislaufmedizin

Geldsetzer, Lutz
Kleine philosophische Bibliographien Bd. 4. ~: In
Honorem
- Bd. 5: Bibliography of the International
Congresses

Gellert, Christian Fürchtegott
Minerva-Fachserie Geisteswissenschaften.
Freudenreich, Carla: Zwischen ...

Gemkow, Heinrich
August Bebel. Ausgewählte Reden und Schriften
Bd. 1
- Bd. 2

Gemtos, R.
Glossarium Artis Bd. 3

Genzel, Peter
IFLA Publications Vol. 13
- Vol. 18

Genzmer, Fritz
~: Das Buch des Setzers

Georg-Shirley, Leonore
Ludes, Peter: Bibliographie zur Entwicklung ...

Gerasch, Sabine
Dortmunder Beiträge zur Zeitungsforschung Bd. 52

Gerd, Reinhold
Minerva-Fachserie Geisteswissenschaften.
Kawashima, Takeyoshi: Die japanische ...

Gerken, Horst
HIS-Brief. 38
- Bd. 47
- Bd. 64

Geserick, Rolf
Kommunikation und Politik. 20
Minerva-Fachserie Wirtschafts- und Sozialwissen-
schaften. ~: 40 Jahre ...

Gessenharter, Wolfgang
Innenpolitik in Theorie u. Praxis. 11

Gessner, Dietrich
Inventar archivalischer Quellen des NS-Staates. Tl. 1

Gessulat, Siegfried
Minerva-Fachserie Psychologie. ~:
Selbstmordverhalten ...
- Gessulat, Siegfried: Therapie ...

Geyer, F.
Beiträge zur Universitätsplanung. 6

Ghai, O. P.
International Publishing Today.

Gibb, Jan P.
IFLA Publications Vol. 45/46

Gibbs, James
Handbook for African Writers, A

Gibbs, Nanette
American Biographical Archive

Giesen Frank
Minerva-Fachserie Wirtschafts- und Sozialwissenschaften. ~: Hochschulunterricht ...

Giessner, Klaus
Eichstätter Hochschulreden 50

Gikas, Michael
Minerva-Fachserie Wirtschafts- und Sozialwissen-
schaften. ~ / Wolfgang Vierke: Methodologische ...
- Gikas, Michael: Funktion ...

Gilder, Lesley
Research Libraries and Collections in the ...

Gill, John M.
IFLA Publications Vol. 30

Gillespie, Paul D.
~ / Paul Katzenberger; John Page: Problems of
document delivery ...

Gilson, Julius P.
British Museum: Catalogue of Western ...

Giseler, Hinrich
HIS-Brief. 59

Gittig, Heinz
~: Bibliographie der Tarnschriften ...
Der Malik Verlag 1916-1947

Glaßbrenner, Adolf
Dortmunder Beiträge zur Zeitungsforschung Bd. 31
Kruszynski, Gisela: Die komischen Volkskalender ...

Gleichenstein, Hans von
Minerva-Fachserie Rechts- und Staatswissen-
schaften. ~: Die Allgemeinheit ...

Glénisson, Jean
International Bibliography of Historical Sciences.
Vol. 49
- Vol. 50
- Vol. 51
- Vol. 52
- Vol. 53
- Vol. 54
- Vol. 55
- Vol. 56
- Vol. 57
- Vol. 58
- Vol. 59
- Vol. 60
- Vol. 61

Glienke, Carola
Bibliotheksstudien. 1B

Glock, Karl Borromäus
Werkstatt des Buches. ~: Der Buchhändler als
Kaufmann

Gluschkow, Viktor Michaijlovič
~: Einführung in die technische .

Gmelin, Wolfgang
Science and Technology and the Future

Gmurman, W. J.
Allgemeine Grundlagen der marxistischen
Pädagogik

Gnilka, Joachim
Eichstätter Hochschulreden 38

Gnirss, Christa
Götz von Olenhusen: Handbuch der Raubdrucke
Internationale Bibliographie der Reprints. Bd. 1

Goebbels, Joseph
Die Tagebücher von Joseph Goebbels. Sämtliche
Fragmente
Die Tagebücher von Joseph Goebbels. Teil I:
Aufzeichnungen 1923 bis 1941
Die Tagebücher von Joseph Goebbels. Teil II:
Diktate 1941 bis 1945

Göhler, Marion
Die Frauenfrage in Deutschland. Neue Folge. Bd. 5

Göhler, Wolfgang
Grundwissen Buchhandel - Verlage Bd. 1
- Bd. 1 / 3. überarb. u. erw. Ausg.
- Bd. 3
- Bd. 6

Göldner, Hans D.
Minerva-Fachserie Pädagogik. ~: Elternmeinung ...

Göpfert, Herbert G.
Heimeran, Ernst: Von Büchern und von ...

Göppinger, Hans
Beiträge zur empirischen Kriminologie

Görhely, Tamás
Texte und Daten zur Hochschulplanung. 19

Görner, Franz
~: Osteuropäische bibliographische Abkürzungen

Gössmann, Elisabeth
Eichstätter Hochschulreden 57

Goethe, Johann Wolfgang von
Faust-Bibliographie. Tl. II
Fischer, Paul: Goethe-Wortschatz
Goethes Bibliothek Katalog
Jost, Dominik: Deutsche Klassik

Keudell, Elisev / Karl Bulling: Goethe als
Benutzer ...
Musculus, Carl Theodor: Inhalts- und Namen-
Verzeichnisse über ...
Wieder, Joachim: Frankreich und Goethe

Götz von Olenhusen, Albrecht
~: Handbuch der Raubdrucke

Golb, Joel
Bibliographien zur deutsch-jüdischen Geschichte. 2

Goldberg, Jolande
Simon, Elisabeth: Bibliothekswesen in den USA

Goldberg, Marian
Herbote, Burkhard: World Tourism Directory. Part 2

Goldfriedrich, Johann
Kapp, Friedrich / Johann Goldfriedrich:
Geschichte des Deutschen ...

Goldschmidt, Dietrich
Over, Albert: Die deutschsprachige Forschung
über ...

Goldstein, Moritz
Dortmunder Beiträge zur Zeitungsforschung Bd. 25

Gollmitz, Renate
Kunze, Horst: Vom Bild im Buch

Goltz, Hermann
Deutschland, Armenien und die Türkei 1895-1925

Goluszka, Malgorzata
Deutsche Drucke des Barock 1600-1720
Polnische Drucke und Polonica 1501-1700. Band 1

Gomez de Ortega, Bruno
Berlin-Bibliographie. 1985 bis 1989

Gonschorek, Gernot
Minerva-Fachserie Pädagogik. ~: Erziehung ...

Goppel, Alfons
Föderalismus

Goppel, Thomas
Föderalismus

Gorman, Gary
~ / Maureen Mahoney: Guide to Current National ...

Gorzny, Willi
Biografisch Archief van de Benelux
Deutsches Biographisches Archiv
Die deutschsprachige UNESCO-Literatur
Gesamtverzeichnis der Übersetzungen
deutschsprachiger Werke
Gesamtverzeichnis des deutschsprachigen
Schrifttums 1700-1910
Gesamtverzeichnis des deutschsprachigen
Schrifttums 1911-1965
Gesamtverzeichnis Deutschsprachiger
Hochschulschriften 1966-1980
Internationaler Nekrolog
Zeitungs-Index 1982-1989
Zeitungs-Index. Buchrezensionen 1974
Zeitungs-Index. Verzeichnis wichtiger ...

Goschler, Constantin
Hitler. Reden, Schriften, Anordnungen ... Bd. IV /
Tl. 1

Goslich, Lorenz
~: Zeitungs-Innovationen

Gossler, Marcus
Reinitzer, Sigrid / Marcus Gossler:
Nachschlagetechniken in der ...

Gothe, Ulrich
Aminde, Hans J. / Heinrich W. Wichmann: Der
städtebauliche Entwurf ...

Gottschalk, Günther
~: Hesse Lyrik Konkordanz

Gottsched, Johann Christoph
Suchier, Wolfram: Gottscheds Korrespondenten

Gottstein, Klaus
Directory of International Cooperation in Science ...

Grab, Walter
Jahrbuch des Instituts für Deutsche Geschichte
Gracq, Julien
Minerva-Fachserie Geisteswissenschaften. ~: Les
débuts ...
Grätz, Fred
Beiträge des Instituts für Zukunftsforschung. 4
Grätz, Manfred
Knigge, Adolph von: Sämtliche Werke
Graf, Christoph
Der Reichstagsbrand. Eine wissenschaftliche
Dokumentation. Bd. 2
Graml, Hermann
Die Tagebücher von Joseph Goebbels. Teil II:
Diktate 1941 bis 1945. Bd. 14
Gramm, Gerhard
Beiträge zur Bedarfsmessung an wissenschaft-
lichen Hochschulen. 16
Gransow, Bettina
Berliner China-Studien Bd. 2
- Bd. 20
- Bd. 26
Grassmann, Antjekathrin
Lübeck-Schrifttum
Greca, Rainer
Eichstätter Hochschulreden 69
Minerva-Fachserie Wirtschafts- und Sozialwissen-
schaften. ~: Die Grenzen ...
- Greca, Rainer: Handlungsmuster ...
Grefkes-Heinz, Monika
Politische Dokumentation poldok
Politische Dokumentation (pol-dok). Poldok
Referatedienst ...
Greiner-Mai, Herbert
Lexikon fremdsprachiger Schriftsteller
Greisenegger, Wolfgang
Keller, Ulrike: Otto Zoffs dramatische Werke
Gresser, Klaus
Informationssysteme - Grundlagen und Praxis der ... 1
Studiengruppe für Systemforschung ... Nr. 110
Grewe-Partsch, Marianne
Communication Research and Broadcasting. 3
Fernsehen und Bildung
Mediendramaturgie und Zuschauerverhalten
Mensch und Medien
Der Vielseher Herausforderung für
Fernsehforschung ...
Griebach, Heinz
Hochschulplanung. 27
Griepenkerl, Heiko
Minerva-Fachserie Wirtschafts- und Sozialwissen-
schaften. ~: Führungsschulung ...
Gries, Jürgen
Minerva-Fachserie Pädagogik. ~: Umdenken ...
Griesche, Hans D.
Studien zur Publizistik 20
Grieshaber, Ernst L.
Werkstatt des Buches. ~: Wider die Druckfehler
Grieswelle, Detlef
Minerva-Fachserie Wirtschafts- und Sozialwissen-
schaften. Beiträge des Instituts für empirische ... 1
- Beiträge des Instituts für empirische ... 2
- Beiträge des Instituts für empirische ... 3
- Beiträge des Instituts für empirische ... 4
Griffel, Nicola
Berichte z. Erziehungstherapie u. Eingliederungs-
hilfe. 48
- Bd. 49
Grim, Ronald E.
IFLA Publications Vol. 31

Grimm, Claus
Henker, Michael / Karlheinz Scherr; Elmar Stolpe:
De Senefelder á ...
- Von Senefelder zu Daumier
Grimm, Dieter
Prismata
Gringmuth, Grete
Handbuch der Internationalen Dokumentation
und Information Bd. 1
- Bd. 6
Gritschneder, Otto
Hitler. Reden, Schriften, Anordnungen. Der Hitler-
Prozess 1924
Gritzka, Christoph
Texte und Daten zur Hochschulplanung. 25
Grobe, Bernd
Brendel, Detlev / ~: Journalistisches Grundwissen
Grobel, Anna
Berichte z. Erziehungstherapie u. Eingliederungs-
hilfe. 54
Grodecki, L.
Glossarium Artis Bd. 3
Groebel, Jo
Mensch und Medien
Groeling-Che, Hui-wen von
Berliner China-Studien Bd. 5
Gröning, Gert
Arbeiten zur sozialwissenschaftlich orientierten
Freiraumplanung
Minerva-Fachserie Wirtschafts- und Sozialwissen-
schaften. ~: Dauercamping ...
Grötzbach, Erwin
Eichstätter Hochschulreden 33
Groh, Hans
Planen und Bauen. 16
Groh, Helmut
HIS-Brief. 52
Gromoll, Bernhard
Minerva-Fachserie Rechts- und Staatswissen-
schaften. ~: Rechtliche ...
Gronemeyer, Reimer
~: Zigeuner in Osteuropa
Gross, Engelbert
Eichstätter Hochschulreden 66
Grosskopff, Rudolf
Dortmunder Beiträge zur Zeitungsforschung Bd. 7
Grossmann, Anton J.
Minerva-Fachserie Geisteswissenschaften. ~:
Irische ...
Grote, Bernd
Dortmunder Beiträge zur Zeitungsforschung Bd. 11
Groteloh, Elke
Minerva-Fachserie Pädagogik. ~: Kommunikation ...
Groth, Michael
Dortmunder Beiträge zur Zeitungsforschung Bd. 46
Minerva-Fachserie Geisteswissenschaften. ~: The
Road ...
~: The Road to New York ...
Grube, Lutz
Minerva-Fachserie Wirtschafts- und Sozialwissen-
schaften. ~: Hafenentwicklung ...
Grube, W. Oswald
Planen und Bauen. 2
Grube-Hell, Ingrid
Minerva-Fachserie Geisteswissenschaften. ~:
Landeskunde ...
Gruber, Johann G.
Ersch, Johann S. / ~: Allgemeine Encyclopädie
der Wissenschaften ...
Grubert, Meinhard
Minerva-Fachserie Naturwissenschaften. ~:
Mucilage ...

Gruchmann, Lothar
Hitler. Reden, Schriften, Anordnungen. Der Hitler-
Prozess 1924
Grund, Uwe
~: Indices zur sprachlichen und ...
Grune, Siegfried
~: Bildschirmarbeitsplätze
Grunsky-Peper, Konrad
Minerva-Fachserie Wirtschafts- und Sozialwissen-
schaften. ~: Deutsche ...
Grupe, O.
Beiträge zur Bedarfsmessung an wissenschaft-
lichen Hochschulen. 13
Gschaid, Maximilian
Die Tagebücher von Joseph Goebbels. Teil II:
Diktate 1941 bis 1945. Bd. 15
Gsell, Otto
Eichstätter Hochschulreden 19
Güdter, Bernd
~: Verstehen üben Bilden in ...
Güldenpfennig, Sven
Minerva-Fachserie Geisteswissenschaften. ~:
Sport ...
Günter, Wolfram
Esbeck, Friedrich J. von: Reden
Günther, Hans O.
Wirtschaftsinformatik und Quantitative ... 5
Günther, Uwe
Minerva-Fachserie Rechts- und Staatswissen-
schaften. ~: Zum Verhältnis ...
Güntzer, Ulrich
Jüttner, Gerald / ~: Methoden der Künstlichen ...
Guggenheimer, Eva
~ / Heinrich Guggenheimer: Etymologisches
Wörterbuch ...
Guggenheimer, Heinrich
Guggenheimer, Eva / ~: Etymologisches
Wörterbuch ...
Guhde, Edgar
Thesaurus Bildungsforschung
Guillaumont, Agnes
Répertoire automatisé des livres du ...
Gundert, Sibylle
Münchner Ethnologische Abhandlungen. 4
Gundert-Hock, Sibylle
Münchner Ethnologische Abhandlungen. 7
Gundlach, Rolf
Archäographie
Gurnsey, John
~: The Information Professions in ...
Gurr, Andrew
Journal of Commonwealth literature
Gutenberg, Johannes
Lülfing, Hans Johannes: Gutenberg und das ...
Guthrie, Paul
The British Library General Catalogue of Printed
Books to 1975. Supplement
Scandinavian Biographical Archive
Gutmann, Wolfgang F.
Eichstätter Hochschulreden 20
Gwozdz, Andrzej
Bibliographie der Filmbibliographien

H

Haan, Oswald
Fokus. 9
Haas, Peter
Beiträge des Instituts für Zukunftsforschung. 12
Haase, Amine
Dortmunder Beiträge zur Zeitungsforschung Bd. 20

Hecht, Karl
Seminarberichte der Deutschen UNESCO-Kommission Nr. 22
Heck, Ludwig
Planen und Bauen Bd. 11
- Bd. 18
Heck, Michael
Innenpolitik in Theorie u. Praxis. 7
Kommunale Sozialpolitik
Heck, Rita
Kommunale Sozialpolitik. 4
Heck, Rosina
Verhandlungen des Deutschen Bundestages ... 10. Wahlperiode. Sachregister ...
Hecklau, Hans
Bibliographien zur Regionalen Geographie und ... 9
Heckmann, Harald
Nachschlagewerke zur Musik
Hederich, Benjamin
Grosse deutsche Lexika. Hederich, Benjamin: Gründliches Antiquitäten-Lexicon
- Hederich, Benjamin: Gründliches Lexicon ...
- Hederich, Benjamin: Reales Schul-Lexicon
Heeg, Sibylle
Texte und Daten zur Hochschulplanung. 18
Hefele, Bernhard
~: Drogenbibliographie
~: Jazz Bibliography
~: Jazz, Rock, Pop
~: Where to Order Foreign Books
Hegel, Georg Wilhelm Friedrich
Hegel Bibliography
Minerva-Fachserie Geisteswissenschaften. Meiners, Reinhard: Methodenprobleme ...
Hegger, Manfred
Jokusch, Peter / Manfred Hegger: Betriebsanalyse und Nutzungsmessung ...
Hehl, Hans
~: Die elektronische Bibliothek
Heiber, Helmut
Akten der Partei-Kanzlei der NSDAP
- Ergänzungsbände. Tl. I
- Regestenbd. 2
~: Universität unterm Hakenkreuz
Heidegger, Martin
Minerva-Fachserie Theologie. Weclawski, Tomasz: Zwischen ...
Heidelmeyer, Wolfgang
Die Menschenrechte Erklärungen, Verfassungsartikel, internat. Abkommen
Heidtmann, Frank
Bibliothekspraxis. 3
~: Die bibliothekarische Berufswahl
Schriftenreihe der Deutschen Gesellschaft für Photographie e.V. 1
- Bd. 3
~: Zur Theorie und Praxis der ...
Heilinger, Rudolf
Stock, Karl F. / ~; Marylène Stock: Personalbibliographien ...
Heilmann, Peter
Pressekonzentration
Heim, Harro
Bibliotheksstudien
- Bd. 1 A
- Bd. 1 B
- Bd. 1 C
- Bd. 3
- Bd. 4
- Bd. 6
Die neue Bibliothek

Heimeran, Ernst
~: Von Büchern und von ...
Heimeran, Silvia
Minerva-Fachserie Wirtschafts- und Sozialwissenschaften. ~: Der Einsatz ...
Heimler, Gerhard
Grundwissen Buchhandel - Verlage Bd. 8
Hein, R.
Qualifizierung und wissenschaftlich-technischer Fortschritt ... Bd. 2
Hein, Roland
Texte - Dokumente - Berichte. 7
Heindlmeyer, Peter
HIS-Brief. 35
Hochschulplanung. 13
Heine, Günther
Glossarium Artis Bd. 10
Heinemann-Knoch, Marianne
Texte und Daten zur Hochschulplanung. 21
Heinold, Ehrhardt
Aus alten Börsenblättern
Heinrich, Alois
Almanach für Freunde der Schauspielkunst
Heinrich-Jost, Ingrid
Dortmunder Beiträge zur Zeitungsforschung Bd. 31
Heinsius, Wilhelm
~: Alphabetische Verzeichnis der von ...
Heinzer, Felix
Bücher, Menschen und Kulturen
Heiss, Beate
Verhandlungen des Deutschen Bundestages ... 7. Wahlperiode. Sachregister ...
Heitmann, Margret
Bibliographien zur deutsch-jüdischen Geschichte. 6
Helber, Claus
Wirtschaftsinformatik und Quantitative ... 4
Held, Gerhard
Empirische Sozialforschung. Bd. 1979
- Bd. 1980
- Bd. 1981
Held, Hans J.
Minerva-Fachserie Naturwissenschaften. Pfeifer, Tilo / Ulrich Schwerhoff; ~: Struktur ...
Helm, J. G.
Beiträge zur Bedarfsmessung an wissenschaftlichen Hochschulen. 15
Helmreich, Reinhard
Schriftenreihe Internationales Zentralinstitut für das Jugend- ... 10
Helmrich, Herbert
~ / Bernd Ahrndt: Schüler in der DDR
Hemje-Oltmanns, Dirk
Minerva-Fachserie Geisteswissenschaften. ~: Materielle ...
Hemmert-Halswick, Sibylle
Berichte z. Erziehungstherapie u. Eingliederungshilfe. 18-20
Hempel, Joachim
HIS-Brief Bd. 42
- Bd. 59
Hempel, Ulrich
Beiträge zur Bedarfsmessung an wissenschaftlichen Hochschulen. 15
Planen und Bauen. 9
Henkel, Martin
Deutsche Presseforschung 25
Henker, Michael
~ / Karlheinz Scherr; Elmar Stolpe: De Senefelder á ...
- Von Senefelder zu Daumier

Hennicke, Peter
Minerva-Fachserie Wirtschafts- und Sozialwissenschaften. ~: Die entwicklungstheoretischen ...
Hennig, Diethard
Schriftenreihe der Georg-von-Vollmar-Akademie. 3
Hennigs, Gustav
Beiträge zur Bibliothekstheorie und Bibliotheksgeschichte 9
Henning, Hans
Egenolff, Christian: Sprichwörterlexikon
Faust-Bibliographie
~: Faust-Variationen
Henning, Wolfram
Beiträge zur Bibliothekstheorie und Bibliotheksgeschichte 8
Henrichs, Norbert
Kleine philosophische Bibliographien Bd. 1. ~: Bibliographie der Hermeneutik ...
Briefe Deutscher Philosophen (1750-1850)
Briefwechsel deutschsprachiger Philosophen 1750-1850
Henry, Carol
IFLA Annual 1985
- 1986
- 1987
- 1988
- 1989
- 1990
- 1991
- 1992
- 1993
- 1994
- 1995
- Annual Reports
IFLA Publications
- Vol. 51
Large Libraries and New Technological ...
Hentschke, Th.
~: Allgemeine Tanzkunst Theorie und ...
Henze, Wilfried
August Bebel. Ausgewählte Reden und Schriften Bd. 2
- Bd. 6
Henzler, Rolf
Data documentation
Hepp, Michael
Die Ausbürgerung deutscher Staatsangehöriger 1933-45 ...
Sozialstrategien der Deutschen Arbeitsfront
Herbote, Burkhard
Handbuch der Internationalen Dokumentation und Information Bd. 18
- Bd. 18 / 2. überarb. Aufl.
~: World Tourism Directory
Herbst, Jörn
Studiengruppe für Systemforschung ... Nr. 114
Herdina, Philip
Waibl, Elmar / ~: Dictionary of Philosophical Terms
Herdt, Klaus-Dieter
Gekeler, Otto / ~; Walter Oberender: Warenkatalogisierung und Kommunikation ...
Herfurth, Matthias
Forschungsarbeiten in den Sozialwissenschaften. 1970
- 1971
- 1972
Herkommer, Hubert
Deutsches Literatur-Lexikon. Bd. 16
- Bd. 17
- Bd. 18
- Ergänzungsband III

J

Jablonska-Skinder, Hanna
~ / Ulrich Teichler: Handbook of Higher ...

Jachmich, Gabriele
German Yearbook on Business History. 1988-92
Prinz-Albert-Studien Bd. 14

Jacob, Werner
Studiengruppe für Systemforschung ... Nr. 85

Jacobi-Dittrich, Juliane
Marburger Beiträge zur Vergleichenden ... 20

Jacobs, Arthur D.
German-Americans in the World Wars. Vol. IV

Jacoby, Gerhard
Roeber, Georg / ~: Handbuch der
filmwirtschaftlichen ...

Jacoby, Klaus
Minerva-Fachserie Psychologie. ~: Schulangst ...

Jäger, Georg
Handbuch Lesen
Literatur und Archiv

Jahnke, Bernd
Wirtschaftsinformatik und Quantitative ... 1

Jaich, Jochen
Aminde, Hans J. / Heinrich W. Wichmann: Der
städtebauliche Entwurf ...

Jamann, Wolfgang
Berliner China-Studien Bd. 16

Janneau, G.
Glossarium Artis Bd. 3

Jannet, Pierre
Bibliothetheca Scatologica ...

Jansen, Ruth
Deutschsprachige Bilderbücher

Janus, Eligiusz
Einzelveröffentlichungen der Historischen
Kommission zu Berlin. 80

Januszewski, Bodo
Berichte z. Erziehungstherapie u. Eingliederungs-
hilfe. 24
- Bd. 51

Jaquet, Frits G.
Guides to the Sources for the History ... 3rd Series.
Vol. 4. Part 2

Jarren, Otfried
Beiträge zur Kommunalwissenschaft. 14
Dortmunder Beiträge zur Zeitungsforschung Bd. 32
- Bd. 47
Kommunikation und Politik

Jaskolla, Dieter
International Bibliography of Plant Protection
1965-1987

Jatzwauk, Jakob
Bemann, Rudolf / ~: Bibliographie der
sächsischen Geschichte

Jaworski, Rudolf
Deutsche und Polen zwischen den Kriegen

Jay, Mary
The African Book Publishing Record

Jedele, Helmut
Schriftenreihe der UFITA H. 40

Jefferson, George
The College Library

Jehle, Jörg M.
Beiträge zur empirischen Kriminologie 10

Jehle, Manfred
Einzelveröffentlichungen der Historischen
Kommission zu Berlin. 82
Quellen zur Geschichte der Juden in den ... 2

Jelinik, Mariann
Schriftenreihe Internationales Zentralinstitut für
das Jugend- ... 14

Jendrowiak, Hans W
Eichstätter Hochschulreden 45

Jenkins, Jon C.
International Biographical Dictionary of Religion
Who's Who in International Organizations

Jentsch, Marianne
Protokoll der „Brüsseler Konferenz" der KPD 1935

Jersch-Wenzel, Stefi
Quellen zur Geschichte der Juden in den ...

Jesch, Nora M.
Japanese American World War II. Part V

Jessen, Jens
~: Bibliographie der Autobiographien
- Bd. 4

Jörg, Sabine
Communication Research and Broadcasting. 4
Schriftenreihe Internationales Zentralinstitut für
das Jugend- ... 12
- Bd. 18

Jörgensen, D.
Guides to the Sources for the History ... 3rd Series.
Vol. 3. Part 2

Jöstlein, Ilse
Boßmeyer, Christine: Fortschritte des EDV-
Verbundes ...

Johansson, Eve
IFLA Publications Vol. 58

John, Nancy R.
IFLA Publications Vol. 89
Libri

John, Waltraud
Deutschsprachige Literatur Prags und der ...

Johnson, Ian M.
IFLA Publications Vol. 49/50

Johnson, Richard D.
New Horizons for Academic Libraries

Jokusch, Peter
~ / Manfred Hegger: Betriebsanalyse und
Nutzungsmessung ...

Jones, Hywel C.
MacKenzie, Norman / Michael Eraut; ~: Lehren
und Lernen

Jones, Robert A.
Frank Wedekind. A Bibliographic Handbook

Jones-Fuller, Paula
UBCIM Publications - New Series Vol. 9

Jopp, Robert K.
Kind, Friedbert: Bausysteme im Hochschulbau
Planen und Bauen. 14

Joraschky, Peter
Minerva-Fachserie Medizin. ~: Das Körperschema ...

Joseph II.
Minerva-Fachserie Geisteswissenschaften.
Moerchel, Joachim: Die Wirtschaftspolitik ...

Jost, Dominik
~: Deutsche Klassik

Jouguelet, Suzanne
UBCIM Publication Series. Manuel Unimarc
Version française

Jüres, Ernst August
Forschungsarbeiten in den Sozialwissenschaften.
1969
- 1971

Jüttner, Gerald
~ / Ulrich Güntzer: Methoden der Künstlichen ...

Jung, Rudolf
Bibliothekspraxis. 23
Die „Regeln für die alphabetische Katalogisierung
(RAK)" ...

Jung, Uli
Filmkultur zur Zeit der Weimarer Republik

Junginger-Dittel, Klaus O.
Minerva-Fachserie Wirtschafts- und Sozialwissen-
schaften. Corsten, Hans / ~: Zur Bedeutung ...
- Corsten, Hans / ~: Zur Berufs- ...

Junkers, Elke
Berliner China-Studien Bd. 4

Juntke, Fritz
Das Buch als Quelle historischer Forschung

Jurczek, Peter
Beiträge zur Kommunalwissenschaft. 21

K

Kabdebo, Thomas
Dictionary of Dictionaries

Kabisch, Gisela
Verzeichnis medizinischer und
naturwissenschaftlicher Drucke ... Reihe D

Kaegbein, Paul
Baltisches Biographisches Archiv
Beiträge zur Bibliothekstheorie und Bibliotheks-
geschichte
Bibliothek Forschung und Praxis
Bibliothekspraxis. 23
Buch und Bibliothekswissenschaft im
Informationszeitalter
Buch und Zeitschrift in Geistesgeschichte und ...

Kärcher, Gustav F.
Werkstatt des Buches. ~: Warenkunde des Buches

Kafka, Franz
Franz Kafka. Eine kommentierte Bibliographie
der Sekundärliteratur
Franz Kafkas Werke. Eine Bibliographie der
Primärliteratur (1908-1980)

Kagan, Alfred
IFLA Publications Vol. 56

Kaiser, Gert
Bücher für die Wissenschaft

Kaiser, Philipp
Eichstätter Hochschulreden 4

Kaiser, R.
Documentation et Recherches. ~ / M. T. Kaiser-
Guyot: Documentation numismatique ...

Kaiser, Rolf
Minerva-Fachserie Rechts- und Staatswissen-
schaften. ~: Die künstliche ...

Kaiser-Guyot, M. T.
Documentation et Recherches. Kaiser, R. / ~:
Documentation numismatique ...

Kaisser, Ulrich
Berichte z. Erziehungstherapie u. Eingliederungshilfe. 38: Tl. I

Kaltwasser, Franz Georg
Die Bayerische Staatsbibliothek in historischen ...
Bibliothekspraxis. 31
Deutsche Biographische Enzyklopädie (DBE)

Kaminsky, Reiner
DGD-Schriftenreihe Bd. 6

Kamiya, Kunihiro
Minerva-Fachserie Geisteswissenschaften.
Kawashima, Takeyoshi: Die japanische ...

Kamp, Rudolf
Kleine philosophische Bibliographien Bd. 2.
Hogrebe; ~; König: Periodica Philosophica

Kampe, Norbert
Der Bestand Preussische Akademie der Künste
- Tl. 1. Findbuch
- Tl. 2. Findbuch
- Tl. 3. Findbuch
Jewish Immigrants of the Nazi Period ... Vol. 4

Moske, Klaus
Forschungsarbeiten in den Sozialwissenschaften. 1972

Moss, Alfred G.
A New International Economic Order

Mostar, Gerhart Herrmann
~: In diesem Sinn ...

Motekat, Helmut
Stoffe - Formen - Strukturen

Mücke, Michael
Bibliotheksforum Bayern (BFB)

Mühlhahn, Klaus
Berliner China-Studien Bd. 23

Mühlner, Manfred
~: Julius Petzholdt

Mühlpfordt, Günther
Bärwinkel, Roland / Natalija I. Lopatina / ~: Schiller-Bibliographie ...

Müller, Annelies
The First Century of German Language ...

Müller, C. Wolfgang
Pressekonzentration

Müller, Claudius C.
Münchner Ethnologische Abhandlungen. 1

Müller, Eckart
Texte und Daten zur Hochschulplanung. 10
- Bd. 16

Müller, Eckhard
August Bebel. Ausgewählte Reden und Schriften Bde. 3-5
- Bd. 7
- Bd. 8
- Bd. 9
- Bd. 10

Müller, Egon
Minerva-Fachserie Wirtschafts- und Sozialwissen-schaften. ~: Erziehung ...

Müller, Hartmut
Informationsmanagement. 2

Müller, Hildegard
Beiträge zur Bibliothekstheorie und Bibliotheks-geschichte 2

Müller, Hubert
Eichstätter Hochschulreden 13

Müller, Isa
Hitzeroth, Christiane / Dagmar Marek; ~: Leitfaden für die formale ...

Müller, Johannes
Apianus, Peter (Petrus): Abbreviationes Vetustorum ...

Müller, Michael G.
Einzelveröffentlichungen der Historischen Kommission zu Berlin. 80

Müller, Peter
Bibliotheca Trinitariorum. Bd. I
- Bd. II

Müller, Werner
Bibliographischer Dienst. 5
- Bd. 6
Communication Research and Broadcasting. 7

Müller, Wolfgang
Minerva-Fachserie Psychologie. Aktivierungs-forschung im ...
Minerva-Fachserie Psychologie. Psychophysio-logische Aktivierungsforschung ...

Müller-Boysen, Ulrike
Minerva-Fachserie Rechts- und Staatswissen-schaften. ~: Die Rechtsstellung ...

Müller-Brettel, Marianne
Bibliography on Peace Research and ...

Müller-Köppern, Frederike
Studiengruppe für Systemforschung ... Nr. 49

- Nr. 58
- Nr. 59

Müller-Vogg, Hugo
Minerva-Fachserie Wirtschafts- und Sozialwissen-schaften. ~: Public ...
So mache ich meine Werbung

Müller-Wolf, Hans M.
Kommunikations- und Verhaltenstraining für Erziehung ...
~: Lehrverhalten an der Hochschule ...

Münch, Roger
Grundwissen Buchhandel - Verlage Bd. 2

Münnich, Monika
~: PC-Katalogisierung mit RAK

Müsse, Wolfgang
Dortmunder Beiträge zur Zeitungsforschung Bd. 53

Müsseler, Jochen
Minerva-Fachserie Psychologie. ~: Aufmerksamkeitsverlagerungen ...

Mundt, Hermann
Bio-Bibliographisches Verzeichnis von Universitäts- und ...

Mursch, Siegfried
Englert, Ludwig / Otto Mair; ~: Georg Kerschensteiner

Musallam, Sami
United Nations Resolutions on Palestine 1947-1972

Musculus, Carl Theodor
~: Inhalts- und Namen-Verzeichnisse über ...

Mushkat, Mari'on
Violence and Peace

Muszynski, Bernhard
Innenpolitik in Theorie u. Praxis. 1

Muth, Ludwig
Der befragte Leser

Muthesius, S.
Glossarium Artis Bd. 3 / 3., neu bearb. u. erw. Aufl.
- 6 / 3. vollst. neubearb. u. erw. Aufl.

Muziol, Roman
~: Pressedokumentation

Mylius, Karl Heinrich
HIS-Brief. 49

Myrtek, Michael
Minerva-Fachserie Psychologie. Aktivierungsforschung im ...
- ~: Typ-A-Verhalten ...
- Psychophysiologische Aktivierungsforschung ...

N

Nachmann, Werner
Lamm, Hans: Hans Lamm

Näslund, L.
Guides to the Sources for the History ... 3rd Series. Vol. 3. Part 2

Näther, Günter
~: Bibliothekswesen in Italien

Näther, Leonore
Näther, Günter: Bibliothekswesen in Italien

Nagel, Gerhard
Minerva-Fachserie Wirtschafts- und Sozialwissen-schaften. ~: Grundbedürfniskonzepte ...

Nagelsmeier-Linke, Marlene
Bibliothekspraxis. 25

Nahodil, Otakar
Integrale Anthropologie. 3
- Bd. 4

Nalbach, Sylvia
International Books in Print PLUS. International Books in Print Handbuch
- Manual

Napier, Paul
~: Index to Micrographics Equipment Evaluations

Nappo, Tommaso
Archives Biographiques Françaises. Deuxième Série
Archivio Biografico Italiano
Archivio Biografico Italiano. Nuova Serie
Archivio Biografico Italiano sino al 1996

Narkiss, Bezalel
Index of Jewish Art. Vol. II
- Vol. IV

Naroska, Hans J.
Minerva-Fachserie Wirtschafts- und Sozialwissen-schaften. ~: Neue ...

Nassmacher, Hiltrud
Innenpolitik in Theorie u. Praxis. 3

Natalis, Gerhardt
Biologie-Dokumentation

Naumann, Jens
Hüfner, Klaus / ~: The United Nations System

Nauta, Paul
IFLA Publications Vol. 32

Neelameghan, A.
International classification

Neher, Michael
Arbeitsbuch Geschichte
- Neuzeit. 1. 16.-18. Jahrhundert. Repetitorium

Neighbor, Tim
The Catalogue of Printed Music in the ...

Neill, Richard J.
IFLA Publications Vol. 49/50

Nelde, Peter H.
~: Wortatlas der deutschen ...

Nell, Verena
Minerva-Fachserie Psychologie. ~: Interaktives ...

Nelson, Nancy M.
Serials and Microforms ...

Németh, S. Katalin
Deutsche Drucke des Barock 1600-1720
Ungarische Drucke und Hungarica 1480-1720

Nemtschinow, W.
Informationsströme in der Wirtschaft

Nestler, Friedrich
~: Bibliographie

Neubauer, Karl Wilhelm
Benutzerverhalten an deutschen Hochschulbibliotheken
~ / G. Schwarz; W. Schwuchow: Kommunikation in ...
Zur Theorie und Praxis des modernen Bibliothekswesens

Neubauer, Wolfram
Horn, Reinhard / ~: Fachinformation Politikwissenschaft
Mette, Günther / Eva Schöppl: Fachinformation Wirtschaftswissenschaften
Peters, Klaus: Fachinformation Rechtswissenschaften
Weber, Angelika: Fachinformation Bauingenieurwesen

Neubert, Heinz
Communications

Neuheuser, Hanns P.
~: Internationale Bibliographie Liturgische Bücher

Neumaier, Herbert
Minerva-Fachserie Wirtschafts- und Sozialwissen-schaften. ~: Die Steuerung ...

Neumann, Erich P.
Allensbacher Jahrbuch der Demoskopie. Bd. I
- Bd. II
- Bd. III
- Bd. IV
- Bd. V

Neumann, Gerda
~: Das Portrait der Frau ...

Neumann, J.
Enzyklopädisches Wörterbuch Kartographie in ...

Neumann, Sieglinde
Dortmunder Beiträge zur Zeitungsforschung Bd. 55

Neumann-Bechstein, Wolfgang
Minerva-Fachserie Wirtschafts- und Sozialwissenschaften. ~: Altensendungen ...

Neumeister, Werner
Baudenkmäler in Bayern. Bd. 12

Neuss, Beate
Minerva-Fachserie Wirtschafts- und Sozialwissenschaften. ~: Europa ...

Neuwirth, Wolfgang
HIS-Brief. 34

Neveling, Ulrich
DGD-Schriftenreihe Bd. 4
Pressekonzentration
Terminology of documentation

Neverla, Irene
Minerva-Fachserie Wirtschafts- und Sozialwissenschaften. ~: Arbeitszufriedenheit ...

New, Peter G.
~: Education for Librarianship

Nicholas, David
~ / Maureen Ritchie: Literature and Bibliometrics

Nicolai, Manfred
Planen und Bauen. 12
Planen und Bauen. 17

Niederland, Daniel
Biographisches Handbuch der deutschsprachigen
Emigration ... Bd. III
Jewish Immigrants of the Nazi Period ... Vol. 3/2

Niendorf, Mathias
Deutsche und Polen zwischen den Kriegen

Nieser, Bruno
Marburger Beiträge zur Vergleichenden ... 8
- Bd. 9
- Bd. 10
- Bd. 11
- Bd. 16
- Bd. 17
- Bd. 24
Texte - Dokumente - Berichte. 23
- Bd. 24
- Bd. 25
- Bd. 31

Nietzsche, Friedrich Wilhelm
Minerva-Fachserie Geisteswissenschaften. Eberan,
Barbro: Luther ...
Nietzscheana

Niewalda, Paul
Bibliothekspraxis. 1
- Bd. 9

Niggemann, Elisabeth
Bücher für die Wissenschaft
Katalog der Thomas-Mann-Sammlung der ...

Niles, Ann
Microform Review Series in Library
Micrographics Management Vol. 11
- Vol. 13

Nissen, Ursula
Bibliographischer Dienst. 3
Communication Research and Broadcasting. 2

Nitzsche, Gerhard
Tarnschriften der KPD aus dem antifaschistischen ...

Niven, Alastair
Journal of Commonwealth literature

Nkhoma-Wamunza, Alice
Africa Index to Continental Periodical .. / Vol. 7

Noel, Wanda
IFLA Publications Vol. 21

Noelle-Neumann, Elisabeth
Allensbacher Jahrbuch der Demoskopie
Der befragte Leser
Demoskopie und Aufklärung
Eine Generation später

Nogueira, Carmen C.
ICA Handbooks Series Vol. 4

Nolde, Emil
Minerva-Fachserie Psychologie. Reuter, Helmut:
Betrachtungen ...

Norden, Arlis
IFLA Publications Vol. 65

Nossol, Alfons
Eichstätter Hochschulreden 31

Noto, Paolo
Archivio Biografico Italiano. Indice Biografico
Italiano

Nouwens, Peter
Handbuch der Universitäten und Fachhochschulen / 9. Ausg.

Nowacek, Käthe
Mediendramaturgie und Zuschauerverhalten.
Der Vielseher Herausforderung für
Fernsehforschung ...

Nowak, Günther
Computergestützte Bibliotheksarbeit

Nuthmann, Reinhard
Bildung und Beschäftigung

Nutz, Walter
Communications

Nyéki-Körösy, Maria
~: Les documents sonores ...

O

O'Casey, Sean
Minerva-Fachserie Geisteswissenschaften. Karlson,
Ilse: Die Funktion ...

O'Connor, Catherine
International Yearbook of Library Service ...

O'Leary, Michael
The Political Risk Yearbook 1992

O'Neill, James
International Journal of Archives

Oberdieck, Ernst Michael
Beiträge zur Informations- und Dokumentationswissenschaft. Folge 13

Oberender, Walter
Gekeler, Otto / Klaus-Dieter Herdt; ~:
Warenkatalogisierung und Kommunikation ...

Oberhuemer, Pamela
Kinderbücher aus Italien - Spanien ...
Kinderbücher aus Türkei - Yugoslawien -
Griechenland

Obermeier, Karl M.
Dortmunder Beiträge zur Zeitungsforschung Bd. 48

Oberschelp, Reinhard
Gesamtverzeichnis des deutschsprachigen
Schrifttums. 1911-1965
Hamberger, G. C. / J. G. Meusel: Das gelehrte
Teutschland ...

Ochi, Hisashi
Minerva-Fachserie Rechts- und Staatswissen-
schaften. ~: Der aussenpolitische ...

Ockenfeld, Marlies
Beiträge zur Informations- und Dokumentationswissenschaft. Folge 7

Oda, Wilbur H.
The First Century of German Language ...

Oehler, Christoph
HIS-Brief. 57
Hochschulplanung. 24
- Bd. 25
- Bd. 29
Over, Albert: Die deutschsprachige Forschung
über ...

Oehlmann, Rainer
Wirtschaftsinformatik und Quantitative ... 13

Öhrig, Bruno
Münchner Ethnologische Abhandlungen. 8

Oel, Hans U.
Beiträge des Instituts für Zukunftsforschung. 18
- Bd. 20

Oeller, Helmut
Fernsehen und Bildung
Mediendramaturgie und Zuschauerverhalten
Der Vielseher Herausforderung für
Fernsehforschung

Oelrichs, Johann C.
~: Entwurf einer Geschichte der ...

Oesterdickhoff, Peter
Minerva-Fachserie Wirtschafts- und Sozialwissenschaften. ~: Hemmnisse ...

Ohmland, Jörg
Berichte z. Erziehungstherapie u. Eingliederungshilfe. 13

Oiserman, T. I.
Minerva-Fachserie Geisteswissenschaften. Pitschl,
Florian: Das Verhältnis ...

Oldfield, Phyllis
Chirgwin, F. John / ~: The Library Assistant's ...

Oliva, Hans
Soziokulturelle Herausforderungen ... 4

Ollé, James G.
~: Library History

Olle, Werner
Beiträge zur Sozialökonomik der Arbeit. 6

Olson, May E.
Arndt, Karl J. / ~: The German Language Press ...

Olzog, Günter
Werkstatt des Buches. ~: Deutsche
Staatsbürgerkunde ...

Opitz, Helmut
Handbuch der Bibliotheken Bundesrepublik
Deutschland, ... / 2. Aufl. - 5. Aufl.
Handbuch der Internationalen Dokumentation
und Information Bd. 8 / 9th ed. - 13th ed.
- Bd. 13 / 4th ed.
- Bd. 16 / 4th ed.
- Bd. 17 / 2nd ed. - 4th ed.
Handbuch der Universitäten und Fachhoch-
schulen / 5. - 9. Ausg.
Libraries, Information Centers, and Databases in
Biomedical ...

Oppermann, Alfred
~: Wörterbuch der Datenverarbeitung
~: Wörterbuch der Elektronik
~: Wörterbuch der Kybernetik
~: Wörterbuch der modernen Technik

Oppl, Hubert
Soziokulturelle Herausforderungen ...
- Bd. 4

Orbanz, Eva
Film, Television, Sound Archives Series
Orgel-Köhne, Armin
Orgel-Köhne, Liselotte / ~: Staatsbibliothek Berlin
Orgel-Köhne, Liselotte
~ / Armin Orgel-Köhne: Staatsbibliothek Berlin
Ostermann, Rainer
Schriftenreihe der Georg-von-Vollmar-Akademie. 5
Ott, Franziska
German-Americans in the World Wars. Vol. I
Ott, Ulrich
Bibliotheken im Netz
Otto, Jürgen
Minerva-Fachserie Psychologie. ~:
Regulationsmuster ...
Otto, Karl F. jr.
Bibliographien zur deutschen Barockliteratur Bd. 1
Overesch, Manfred
Dortmunder Beiträge zur Zeitungsforschung Bd. 19
Oversberg, Michael P.
Berichte z. Erziehungstherapie u. Eingliederungs-
hilfe. 50
Overton, David
IFLA Publications Vol. 26

P

Paas, Dieter
Studiengruppe für Systemforschung ... Nr. 65
Pacey, Phillip
IFLA Publications Vol. 34
Paenson, Isaac
~: Handbuch der Terminologie des Völkerrechts
~: Manual of the Terminology of Public ...
~: Manual of the Terminology of the Law of
Armed ...
Page, John
Gillespie, Paul D. / Paul Katzenberger; John Page:
Problems of document delivery ...
Palm, Wolfgang
Texte - Dokumente - Berichte. 9/10
Pannenberg, Wolfhart
Eichstätter Hochschulreden 10
Panskus, Hartmut
~: 40 Jahre K. G. Saur Verlag
Panzer, Georg W.
~: Literarische Nachricht von den ...
Papke, Thomas
Wirtschaftsinformatik und Quantitative ... 18
Papritz, Johannes
~: Archivwissenschaft
Paravicini, Werner
Dokumentation Westeuropa. 4
 - Bd. 5
Parcer, Jan
Memorial Book
Pardon, Inge
Lager, Front oder Heimat
Paschen, Herbert
Informationssysteme - Grundlagen und Praxis der ... 1
Paschmann, Helmut
Texte und Daten zur Hochschulplanung. 24
Patt, Raimund
Berichte z. Erziehungstherapie u. Eingliederungs-
hilfe. 9
Patte, Geneviève
IFLA Publications Vol. 28
Paul, Hans-Holger
~: Inventar zu den Nachlässen der ...
Paulerberg, Herbert
Grundwissen Buchhandel - Verlage Bd. 4

Pauli, L.
Studies and surveys in comparative education
Paulig, Peter
Eichstätter Hochschulreden 26
Pauls, Werner
Minerva-Fachserie Technik. ~: Verfahren ...
Paus-Haase, Ingrid
Minerva-Fachserie Geisteswissenschaften. ~:
Soziales ...
Pautler, Bettina
Handbuch der Internationalen Dokumentation
und Information Bd. 8 / 8th ed.
Handbuch der Universitäten und Fachhoch-
schulen / 4. Aufl.
Libraries, Information Centers and Databases for
Science ... / 2nd ed.
Pawlowsky, Peter
Beiträge zur Sozialökonomik der Arbeit. 8
Payen, Jean François
Bibliothetheca Scatologica ...
Peeckel, Aribert
Konrad Mellerowicz
Pehle, Heinrich
Beiträge zur Kommunalwissenschaft. 17
Pelletier, Monique
IFLA Publications Vol. 3
Peltz-Dreckmann, Ute
Minerva-Fachserie Geisteswissenschaften. ~:
Nationalsozialistischer ...
Pelzer, Norbert
Göttinger Atomrechtskatalog. Internationale
Bibliographie ...
Peng Pai
Berliner Chinastudien Bd. 14
Pérez de Cuéllar, Javier
Hüfner, Klaus / Jens Naumann: The United
Nations System. Vol. 4. Part A
 - Vol. 4. Part B
Perreault, Jean M.
International classification
Perutz, Peter
Beiträge des Instituts für Zukunftsforschung. 11
Peschel, Manfred
~: Anwendung algebraischer Methoden
Gluschkow, Viktor Michaijlovič: Einführung in
die technische .
Peschke, Michael
Encyclopedic Dictionary of Electronic, Electronical ...
International Encyclopedia of Abbreviations and ...
Internationale Bibliographie der Bibliographien
1959-1988
Pressearchiv zur Geschichte Deutschlands sowie ...
Findbuch ...
Signale für die Musikalische Welt. Bibliographie ...
Peter, Gerd
Aminde, Hans J. / Heinrich W. Wichmann: Der
städtebauliche Entwurf ...
Petermann, Kurt
~: Tanzbibliographie
Peters, Dieter S.
Eichstätter Hochschulreden 14
Peters, Jan
Politische Information
Peters, Klaus
~: Fachinformation Rechtswissenschaften
Petit, Helmut
Minerva-Fachserie Wirtschafts- und Sozialwissen-
schaften. ~: Die Spezifikation ...
Petrat, Gerhardt
Deutsche Presseforschung 27
Petrick, Birgit
Kommunikation und Politik. 12

Petrie, Elaine E.
James Hogg: Scottish Pastorals, Poems ...
Petzet, Michael
Baudenkmäler in Bayern
Petzholdt, Julius
Mühlner, Manfred: Julius Petzholdt
Petzke, Ingo
Dortmunder Beiträge zur Zeitungsforschung Bd. 18
Petzold, Richard
Allgemeine deutsche Musik-Zeitung
Pfäfflin, Friedrich
Fugger, Wolfgang: Ein nutzlich und wolgegrundt
Formular ...
Pfeffer, Gottfried
Die rechtliche Regelung der internationalen
Energiebeziehungen ...
Pfeifer, Tilo
Minerva-Fachserie Naturwissenschaften. ~ /
Ulrich Schwerhoff; Hans J. Held: Struktur ...
Pfeiffer-Rupp, Rüdiger
Minerva-Fachserie Pädagogik. Hörmann, Karl / ~:
Musiklehrerausbildung ...
~: Studien zur Kapazitätsermittlung
Pflüger, Helmut
Glossarium Artis Bd. 3 / 3., neu bearb. u. erw.
Aufl.
 - Bd. 7 / 2., vollst. neu bearb. u. erw. Aufl.
Pflug, Günther
Bibliothekspraxis. 7
Die neue Bibliothek
Schriftenreihe des Landesinstituts für Arabische, ...
Deutsche Fernostbibliographie
Wüst, Ernst: Lexicon Aristophaneum
Pflugk-Hartung, J. von
Rahmen deutscher Buchtitel im 16. Jahrhundert
Philipp, Franz H.
Bibliothekspraxis. 11
Philipp, Hans J.
Bibliographien zur Regionalen Geographie und ... 4
 - Bd. 7
Philipp, Michael
Handbuch des deutschsprachigen Exiltheaters ...
Bd. I
Philippoff, Eva
Minerva-Fachserie Geisteswissenschaften. ~:
Alfred ...
 - Philippoff, Eva: Kurt ...
Pichert, Dietrich
~: Kostenprobleme der Filmproduktion
Pichlmayer, Helgomar
Beiträge des Instituts für Zukunftsforschung. 3
 - Bd. 4
Piel, Edgar
Eine Generation später
Pieritz, Wulf J.
International Bibliography of Plant Protection
1965-1987
Pietsch, Erich
Die Dokumentation und ihre Probleme
Pietsch, Roland
Integrale Anthropologie. 3
Pietzcker, Frank
Minerva-Fachserie Psychologie. ~: Wilhelm ...
Piffl, Stefan
Studien des Forschungsinstitutes für
Wirtschaftspolitik ... 37
Pikarski, Margot
Tarnschriften der KPD aus dem antifaschistischen ...
Pinfold, John R.
African Population Census Reports
Pingfen, Y.
Wörterbuch des Bibliothekswesens

Riemens, Leo
Kutsch, Karl J. / ~: Grosses Sängerlexikon
- Unvergängliche Stimmen
Riepl, Christof
Texte und Daten zur Hochschulplanung. 7
Riescher, Gisela
Minerva-Fachserie Politik. ~ / Raimund Gabriel:
Die Politikwissenschaft ...
Minerva-Fachserie Wirtschafts- und Sozialwissen-
schaften. Regionalismus '90. ...
Rieth, Renate
Glossarium Artis
Riethmüller, Christian E.
Beiträge des Instituts für Zukunftsforschung. 9
Rill, Ingo
Briefe Deutscher Philosophen (1750-1850)
Rimmer, Brenda
Information et documentation en matière de
brevets ... / 3ième éd
Patentinformation and Documentation in Western
Europe / 3rd rev. and enl. ed.
- 3. überarb. u. erw. Aufl.
Ripeanu, Bujor T.
International Directory of Cinematographers ...
Vol. 3
- Vol. 5
- Vol. 8
Rischbieter, Henning
Handbuch des deutschsprachigen Exiltheaters ...
Rischkowsky, Franziska
Hochschulplanung. 16
Riss-Fang, Josephine
IFLA Publications Vol. 32
- Vol. 52/53
- Vol. 54
- Vol. 72/73
Rissom, Hans-Wolf
Seminarberichte der Deutschen UNESCO-
Kommission Nr. 17
Rist, Fred
Minerva-Fachserie Psychologie. ~: Reaktionen ...
Ristow, Walter W.
IFLA Publications Vol. 8
The Map Librarian in the Modern World
Ritchie, Maureen
Nicholas, David / Maureen Ritchie: Literature and
Bibliometrics
Rittel, Horst W. J.
Informationssysteme - Grundlagen und Praxis der ...
- Bd. 5
Kunz, Werner / ~: Werner Schwuchow: Methods
of analysis ...
Kunz, Werner / Wolf Reuter; ~: UMPLIS,
Entwicklung eines ...
Studiengruppe für Systemforschung ... Nr. 20
- Nr. 24
- Nr. 113
Ritter, Ernst
Guides to the Sources for the History ... 3rd Series.
Vol. 6
Lageberichte (1920-1929) und Meldungen ...
Ritter, Gerhard A.
Von der Arbeiterbewegung zum modernen
Sozialstaat
Ritter, Paul
~: Frans Masereel
Ritterhoff, Claus
Knigge, Adolph von: Sämtliche Werke
Roberts, Stephan
Research Libraries and Collections in the ...
Roberts, Winston D.
UBCIM Publication Series. Proceedings of the ...

UBCIM Publication Series. UNIMARC in Theory ...
Robinson, Elizabeth
British Catalogue of Music 1957-1985
Robinson, Gertrud J.
Communication Research and Broadcasting. 3
Schriftenreihe Internationales Zentralinstitut für
das Jugend- ... 14
Robinson, William H.
IFLA Publications Vol. 87
Rockenbach, Helga G.
Minerva-Fachserie Wirtschafts- und Sozialwissen-
schaften. ~: Komponenten ...
Rodger, Eleanor J.
IFLA Publications Vol. 27
Rodriguey, Adolfo
IFLA Publications Vol. 13
Roeben, Peter
Berichte z. Erziehungstherapie u. Eingliederungs-
hilfe. 13
Roeber, Georg
~ / Gerhard Jacoby: Handbuch der film-
wirtschaftlichen ...
Schriftenreihe der UFITA
UFITA Archiv für Urheber-, Film-, Funk- und ...
Röder, Werner
Biographisches Handbuch der deutschsprachigen
Emigration ...
International Biographical Dictionary of Central
European Émigrés ...
Schriften der Herbert und Elsbeth Weichmann
Stiftung. Quellen zur ...
Texte und Materialien zur Zeitgeschichte
Röller, Wolfgang
Eichstätter Hochschulreden 58
Rönsch, Hermann
~: Itala und Vulgata
Rösch-Sondermann, Hermann
~: Bibliographie der lokalen ...
Röske, Volker
Minerva-Fachserie Wirtschafts- und Sozialwissen-
schaften. ~: Der lautlose ...
Rösler, Wolfgang
Diels, Herrmann A.: Colloquium über antikes
Schriftwesen ...
Roessingh, Marius P.
Guides to the Sources for the History ... 2nd
Series. Vol. 9
- 3rd Series. Vol. 4. Part 1
Rössler, Hellmuth
Biographisches Wörterbuch zur deutschen
Geschichte
Roether, Martin
Briefe Deutscher Philosophen (1750-1850)
Rogalla von Bieberstein, Johannes
Bibliothekspraxis. 16
Minerva-Fachserie Geisteswissenschaften. ~:
Preussen ...
Roggausch, Werner
Minerva-Fachserie Geisteswissenschaften. ~: Das
Exilwerk ...
Rohlfs, Gerhard
Sanna, Salvatore A.: Sardinien - Bibliographie
Rohr, Gabriele
Minerva-Fachserie Psychologie. ~:
Objektbedeutung ...
Rohrbach, Paul
Dortmunder Beiträge zur Zeitungsforschung Bd. 16
Rohrbough, Dennis
Jewish Immigrants of the Nazi Period ... Vol. 5
Roland, M.
Information et documentation en matière de
brevets ... 2nd éd

Patentinformation and Documentation in Western
Europe / 2nd rev. and enl. ed.
Patentinformation und Patentdokumentation in
Westeuropa / 2. neu bearb. Aufl.
Rolland, Maria Theresia
Informationssysteme - Grundlagen und Praxis der ... 6
Studiengruppe für Systemforschung ... Nr. 104
Roller, Franz A.
~: Systematisches Lehrbuch der bildenden ...
Roloff, Heinrich
~: Lehrbuch der Sachkatalogisierung
~: Sachkatalogisierung auf neuen Wegen
Roman, Viorel S.
Bibliographien zur Regionalen Geographie und ... 3
Ros, Guido
Dortmunder Beiträge zur Zeitungsforschung Bd. 51
Roschmann, Rolf
Beiträge des Instituts für Zukunftsforschung. 11
Rose, Romani
Memorial Book
Rosen, Klaus
Eichstätter Hochschulreden 34
Rost, Maritta
~: Bibliographie Arnold Zweig
Rost, Reinhold
Catalogue des manuscrits et ...
Rota, Franco P.
Minerva-Fachserie Wirtschafts- und Sozialwissen-
schaften. ~: Menschen ...
Roth, Karl H.
~: Intelligenz und Sozialpolitik im ...
Sozialstrategien der Deutschen Arbeitsfront
Rothermel, Kurt
Fokus. 5
Rothkirch und Trach, Karl Christoph Graf von
Studiengruppe für Systemforschung ... Nr. 93
Rottstock, Felicitas
Minerva-Fachserie Geisteswissenschaften. ~:
Studien ...
Rouit, Huguette
IFLA Publications Vol. 47
- Vol. 62
Rousseau, Jean Jacques
Eichstätter Hochschulreden 40
Rovelstad, Mathilde
~: Bibliotheken in den Vereinigten Staaten
Rowley, Jennifer E.
~: Mechanised in-house information ...
Rudat-Kocks, Marianne
Hadamitzky, Wolfgang / ~: Japan-Bibliografie
Rudhardt, Jean
Papyri Bodmer XXX-XXXVII
Rudolf, Günther
Kommunikation und Politik. 6
Rudolph, Bernd
German Yearbook on Business History. 1985
- 1986
- 1987
- 1988
- 1988-92
- 1993
- 1994
- 1995
Rudolph, Hermann
Plasger, Uwe: Wörterbuch zur Musik
Rudolph, Jörg M.
Berliner China-Studien Bd. 10
Rudolph, Susanne
Die Frauenfrage in Deutschland. Neue Folge. Bd. 4
Ruecker, Norbert
Fachkatalog Film. Bd. 1

- Bd. 24
- Bd. 34

Willmann, Heinz
Analytische Bibliographien deutschsprachiger
literarischer Zeitschriften. 8

Willner, Christa
Jahresbibliographie der Universität München. Bd. 11

Wilson, Alexander
~: Library Policy for Preservation ...

Wimmer, Ruprecht
Eichstätter Hochschulreden 44

Wimmer, Walter
Minerva-Fachserie Theologie. ~: Eschatologie ...

Wimmer-Webhofer, Erika
Literatur und Archiv. 3

Wimmersberg, Heidrun
Minerva-Fachserie Wirtschafts- und Sozialwissen-
schaften. Möller, Wolfgang / Heidrun
Wimmersberg: Public ...

Winckler, Martin
Standortverzeichnis ausländischer Zeitungen und ...

Windel, Gunther
Beiträge zur Informations- und Dokumentationswissenschaft. Folge 10

Winkler, Gerhard
August Bebel. Ausgewählte Reden und Schriften
Bd. 2

Winkler, Helmut
Minerva-Fachserie Wirtschafts- und Sozialwissenschaften. Winkler, Helmut: Zur Theorie ...

Winkler-Haupt, Uwe
Innenpolitik in Theorie u. Praxis. 12

Winter, Ursula
~: Die europäischen Handschriften der ...

Winterhoff, Jörg
HIS-Brief. 59

Winterling, Fritz
Technik und Gesellschaft Bd. 2

Winton, Harry N.
A New International Economic Order

Wischermann, Ulla
Die Staatsbürgerin
Studien zur Publizistik 24

Wischnewsky, Barbara
Minerva-Fachserie Wirtschafts- und Sozialwissenschaften. ~: Kommunnale ...

Wispelwey, Berend
Biografisch Archief van de Benelux. Deel II
South-East Asian Biographical Archive

Witkowski, Georg
~: Geschichte des literarischen ...

Witt, Maria
Sawoniak, Henryk / ~: International Bibliography
of ...
Sawoniak, Henryk / ~: New International
Dictionary of ...

Witte, Sylvia
Empirische Sozialforschung. Bd. 1969
- Bd. 1970
- Bd. 1971
- Bd. 1972
- Bd. 1973
- Bd. 1974

Witter, Hermann
Beiträge zur empirischen Kriminologie 8

Wittig, Horst E.
Minerva-Fachserie Pädagogik. Studien zur
Pädagogik ...

Wittke, Carl
German-Americans in the World Wars. Vol. I

Wittkowsky, Alexander
Sozialpädagogik - Institution, Partizipation,
Selbstorganisation

Wittwer, Walter
August Bebel. Ausgewählte Reden und Schriften
Bde. 3-5

Wölfel, Wolf
Minerva-Fachserie Naturwissenschaften. ~:
Kernrelaxationsuntersuchungen ...

Wojciechowski, Jercy A.
Conceptual Basis of the Classification of ...

Wojciechowski, Marian
Deutsche und Polen zwischen den Kriegen

Wolcke-Renk, Irmtraud D.
Current contents Africa
Fachkatalog Afrika

Wolf, Eva
Minerva-Fachserie Geisteswissenschaften. ~: Der
Schriftsteller ...

Wolf, Margarethe
Bibliographie Darstellende Kunst und Musik
Bibliographie Psychologie
Bibliographie Sport und Freizeit

Wolf, Steffen
~: Kinderfilm in Europa

Wolfart, Ursula
Simon, Elisabeth: Bibliothekswesen in den USA

Wolff, Johannes
Minerva-Fachserie Medizin. ~: Die funktionelle ...

Wolff, Ludwig
Almanach für Freunde der Schauspielkunst

Wolffheim, Elsbeth
Gesellschaft für Exilforschung ...

Wolfrum, Rüdiger
Handbuch Vereinte Nationen

Woll, Helmut
Minerva-Fachserie Wirtschafts- und Sozialwissen-
schaften. ~: Die Untauglichkeit ...
- ~: Monopol ...

Wolschke-Bulmahn, Joachim
Arbeiten zur sozialwissenschaftlich orientierten
Freiraumplanung
- Bd. 11

Wolter, John A.
IFLA Publications Vol. 31

Woods, Elaine W.
UBCIM Publication Series. Woods, Elaine W.: The
MINISIS

Woodworth, David
Years Work in Serials 1985

Woolls, Blanche
IFLA Publications Vol. 55
- Bd. 66/67

Wormell, Irene
Libri

Woronitzin, Sergej
Bibliographie der Sozialforschung in der
Sowjetunion 1960 - 1970

Worters, Garance
American Biographical Archive

Wright, Derek J.
Health Information for All ...

Wronkow, Ludwig
Dortmunder Beiträge zur Zeitungsforschung Bd. 46

Wüst, Dorothea
Verhandlungen des Deutschen Bundestages ... 11.
Wahlperiode. Sachregister ...

Wüst, Ernst
~: Lexicon Aristophaneum

Wüst, Karl
Wüst, Ernst: Lexicon Aristophaneum

Wüster, Eugen
Infoterm Series Vol. 1
- Vol. 2
~: The road to Infoterm

Wulf, Bernd R.
Beiträge zur empirischen Kriminologie 5

Wulf, Joseph
~: Das Dritte Reich und seine Vollstrecker
Poliakov, Leon / ~: Das Dritte Reich und die
Juden
- Das Dritte Reich und seine Denker
- Das dritte Reich und seine Diener

Wulff, Hans J.
Bibliographie der Filmbibliographien

Wunsch, G.
Gluschkow, Viktor Michaijlovič: Einführung in
die technische ...

X

Xiao Hong
Berliner Chinastudien Bd. 3

Y

Young, George
~: Accomodation Services in Britain ...

Yü-Dembski, Dagmar
Berliner China-Studien Bd. 18
Publizistikwissenschaftlicher Referatedienst (prd)

Z

Zacharias, Claus O.
Wirtschaftsinformatik und Quantitative ... 9

Zachert, Ursula
Verzeichnis medizinischer und
naturwissenschaftlicher Drucke ... Reihe A
- Reihe B
- Reihe C

Zänker, Uwe
Qualifizierung und wissenschaftlich-technischer
Fortschritt ... Bd. 3
Texte - Dokumente - Berichte Bd. 11
- Bd. 14

Zahlan, Anne R.
International Documents on Palestine 1971

Zahn, Peter
Beiträge zur Bibliothekstheorie und Bibliotheksgeschichte

Zahn-Harnack, Agnes
Die Frauenfrage in Deutschland. Bd. 1

Zaiss, Konrad
Minerva-Fachserie Pädagogik. ~:
Entwicklungsfaktoren ...

Zapf, Georg W.
~: Augsburgs Buchdruckergeschichte ...

Zapf-Schramm, Thomas
Beiträge zur Kommunalwissenschaft. 33

Zarusky, Jürgen
Widerstand als „Hochverrat" 1933-1945

Zesen, Philipp von
Bibliographien zur deutschen Barockliteratur. 1

Zeidler, Ursel
Verzeichnis medizinischer und
naturwissenschaftlicher Drucke ... Reihe A

Zeilinger, Heidi
Internationale Bibliographie zur Deutschen
Klassik 1750-1850. Flg. 39

Register der Institutionen / Index of Institutions

[Aufgeführt werden die herausgebenden Institutionen,
die in den bibliographischen Einträgen genannt werden /
Listing institutions named as editor in the bibliographic entries]

A

Africa Bibliographic Centre, Dar es Salaam
Africa Index to Continental Periodical ...
Akademie der Künste zu Berlin
Analytische Bibliographien deutschsprachiger
literarischer Zeitschriften
Der Bestand Preussische Akademie der Künste
Nachlässe und Sammlungen zur deutschen Kunst ...
**Akademie der Architekenkammer Nordrhein-
Westfalen, Köln**
Wohnen in der Stadt. Neue Wohnformen
**Akademie der Künste, Tanzarchiv der DDR,
Leipzig**
Petermann, Kurt: Tanzbibliographie
**Akademie der Pädagogischen Wissenschaften der
UdSSR**
Allgemeine Grundlagen der marxistischen
Pädagogik
American Academy in Rome
Ancient Roman Architecture
American Antiquarian Society, Worcester
Bibliography of American Imprints to 1901
Amerika Gedenkbibliothek, Berlin
Berliner Adressbuch
25 Jahre Amerika-Gedenkbibliothek
**Arbeitsgemeinschaft ausseruniversitärer
historischer Forschungseinrichtungen der
Bundesrepublik Deutschland, München**
Jahrbuch der historischen Forschung in ...
**Arbeitsgemeinschaft der Archivrestauratoren,
Bibliotheksrestauratoren und Graphikrestaura-
toren, Freiburg im Breisgau**
Internationaler Restauratorentag
**Arbeitsgemeinschaft der Bibliotheken und Doku-
mentationsstellen der Osteuropa-, Südosteuropa-
und DDR-Forschung in der Bundesrepublik
Deutschland und West-Berlin, Berlin**
Topographie der Osteuropa- Südosteuropa- ...
**Arbeitsgemeinschaft der Kunstbibliotheken,
München**
Deutsche Kunstbibliotheken
**Arbeitsgemeinschaft der Parlaments- und
Behördenbibliotheken, Bonn**
Bibliotheksarbeit für Parlamente und Behörden
**Arbeitsgemeinschaft der Spezialbibliotheken e.V.
(ASpB), Jülich**
NTWZ. Verzeichnis Neuer Technisch-
Wissenschaftlicher ...
**Arbeitsgemeinschaft Katholisch-Theologischer
Bibliotheken, Trier**
Handbuch der katholisch-theologischen
Bibliotheken

**Arbeitsgruppe Bedarfsbemessung wissenschaft-
licher Hochschulen im Finanzministerium Baden-
Württemberg**
Beiträge zur Bedarfsmessung an
wissenschaftlichen Hochschulen. 4
**Arbeitsgruppe Bibliotheksplan Baden-
Württemberg, Freiburg**
Gesamtplan für das wissenschaftliche
Bibliothekswesen
**Arbeitsgruppe FOPRODA der Hochschul-
Informations-System GmbH, Hannover**
HIS-Brief 40
**Arbeitskreis für die Dokumentation Sozial-
wissenschaftlicher Forschung; Bonn**
Forschungsarbeiten in den Sozialwissenschaften.
1970
**Arbeitskreis „Kinder- und Jugendbücherarbeit"
bei d. Arbeitsstelle für d. Bibliothekswesen, Berlin**
Systematik für Kinder- und Jugendbüchereien
Archiv Bibliographia Judaica e.V., Frankfurt a. M.
Dokumentation zur jüdischen Kultur in
Deutschland 1840-1940
Konfrontation und Koexistenz
Lexikon deutsch-jüdischer Autoren
**Archiv der sozialen Demokratie der Friedrich-
Ebert-Stiftung, Bonn**
Inventar zu den Nachlässen der ...
Paul, Hans-Holger: Inventar zu den Nachlässen
der ...
Archives générales du Royaume, Brüssel
Guides to the Sources for the History ... 3rd Series.
Vol. 1
Archives Nationales du Canada, Ottawa
Conservation in Archives ...
Documents that move and speak ...
Toward International Descriptive Standards for
Archives
**The Arthur and Elizabeth Schlesinger Library on
the History of Women in America, Radcliffe
College, Cambridge**
The Black Women Oral History Project
L'Association Marionette et Thérapie, Salzburg
Leleu-Rouvray, Geneviève / Gladys Langevin:
Bibliographie Internationale de la Marionette
**Association of College and Research Libraries,
Boston**
New Horizons for Academic Libraries
**Ausschuß für Wirtschaftliche Verwaltung in
Wirtschaft u. Öffentlicher Hand e.V. (AWV),
Frankfurt a. M.**
Gekeler, Otto / Klaus-Dieter Herdt; Walter
Oberender: Warenkatalogisierung und
Kommunikation ...

**Ausstellungs- und Messe-GmbH des
Börsenvereins des Deutschen Buchhandels,
Frankfurt a. M.**
Printed for children
Who's Who at the Frankfurt Book Fair

B

Barber Institute of Fine Arts, Birmingham
Fenlon, Iain: Catalogue of the printed music ...
Bauakademie Berlin
Bibliographie Bauwesen - Architektur - Städtebau
**Bayerische Akademie der Wissenschaften,
München**
Bibliographia Academica Germaniae
Die Bundesrepublik Deutschland und Frankreich:
Dokumente 1949-1963
**Bayerische Landeszentrale für neue Medien
(BLM), München**
Communication Research and Broadcasting. 11
Bayerische Staatsbibliothek, München
Die Bayerische Staatsbibliothek in historischen ...
Bayerische Staatsbibliothek. Alphabetischer
Katalog
Bayerische Staatsbibliothek. Katalog der
Geschichtszeitschriften
Bayerische Staatsbibliothek. Katalog der
Musikdrucke
Bayerische Staatsbibliothek. Katalog der
Musikzeitschriften
Bayerische Staatsbibliothek. Katalog der
Notendrucke
Bayerische Staatsbibliothek. Katalog der
Osteuropazeitschriften
Bayerische Staatsbibliothek. Katalog 1501-1840
Körperschaftsnamendatei Index of corporate
bodies
New Contents Slavistics
**Bayerisches Landesamt für Denkmalpflege,
München**
Baudenkmäler in Bayern
**Bayerisches Staatsministerium für Unterricht,
Kultus, Wissenschaft und Kunst, München**
Neue Informations- und
Kommunikationstechnologien für
wissenschaftliche ...
Berliner Zentralbibliothek, Berlin
25 Jahre Amerika-Gedenkbibliothek
Künstler der jungen Generation
Systematik der Amerika-Gedenkbibliothek ...
Bibliotheca Bodmeriana, Cologny-Geneve
Papyri Bodmer XXX-XXXVII
Bibliotheca Vaticana, Rome
Bibliotheca Palatina
Bibliothèque Municipale de Grenoble
Catalogue Général Auteurs des Livres ...

Bibliothèque Nationale, Paris
Catalogue collectif des ouvrages en langue arabe ...
Bibliothèque Sainte Geneviève, Paris
Catalogue des Ouvrages Imprimés au ...
Bildarchiv Foto Marburg - Deutsches Dokumentationszentrum für Kunstgeschichte der Philipps-Universität Marburg
Ägypten-Index Bilddokumentation zur Kunst ...
Benelux-Kunst-Index
Digitales Informationssystem für Kunst und Sozialgeschichte ...
Index photographique de l'art en France
Italien-Index
Literatur und Archiv. 4
Marburger Index. Heusinger, Lutz: DV-Anleitung
 - Heusinger, Lutz: Handbuch
 - Heusinger, Lutz: Marburger Informations-, ...
Marburger Index. Inventar der Kunst in Deutschland
Marburger Index. Österreich-Index
Marburger Index. Schweiz-Index
Marburger Index. Wegweiser zur Kunst ...
Spanien- und Portugal-Index
Bildungswerk des Verbands Deutscher Tonmeister, Nürnberg
Bericht der Tonmeistertagung
Bodleian Library, Oxford
Der Kirchenkampf
Börsenverein der Deutschen Buchhändler zu Leipzig
Börsenblatt für den Deutschen Buchhandel ...
Kapp, Friedrich / Johann Goldfriedrich: Geschichte des Deutschen ...
The British Library, London
British Catalogue of Music 1957-1985
British Library General Catalogue of Printed Books 1976-1982
The British Library General Catalogue of Printed Books to 1975
The British Library General Catalogue of Printed Books 1982-1985
The British Library General Catalogue of Printed Books 1986-1987
The British Library General Catalogue of Printed Books 1988-1989
The British Library General Catalogue of Printed Books 1990-1992
The British Library General Catalogue of Printed Books 1993-1994
The British Library General Catalogue of Printed Books 1995-1996
The British Library General Catalogue of Printed Books 1997-1998
The British Library General Subject Catalogue 1975-1985
The British Library General Subject Catalogue 1986-1990
Catalogue of Cartographic Materials in the ...
Catalogue of Dated and Datable Manuscripts
The Catalogue of Printed Music in the ...
Index of Conference Proceedings Received ...
International Bibliography of Biography 1970-1987
Büchereizentrale Schleswig-Holstein, Flensburg
SfB-Systematik für Bibliotheken
Bundesarchiv Koblenz
Lageberichte (1920-1929) und Meldungen ...
Die Tagebücher von Joseph Goebbels. Sämtliche Fragmente

Bundesinstitut für Berufsbildungsforschung, Berlin
Literaturinformationen zur Berufsbildungsforschung
Bundesministerium der Verteidigung, Bonn
Verteidigungs-Dokumentation
Bundesministerium für Wissenschaft, Forschung und Kunst, Wien
ÖVK-WB Österreichischer Verbund-Katalog ...

C

Caritasverband d. Erzdiözese München u. Freising, München
Soziokulturelle Herausforderungen ... 5
Catholic University of America, Washington D.C
Continuing Education: Issues and Challenges
Center for Documentation and Studies, Abu Dhabi
United Nations Resolutions on Palestine 1947-1972
Centre for Educational Research and Innovation (CERI), Genf
Lernsysteme
Centre international des publications oecuméniques des liturgies (CIPOL), Paris
Corpus des Liturgies Chrétiennes sur Microfiches
Centre National de la Recherche Scientifique, Paris
Répertoire International des Médiévistes
Centro di Ricerca, Roma
Kecskeméti, Károly: La Hongrie et le Reformisme ...
Cinémathèque Municipale de Luxembourg
Filmkultur zur Zeit der Weimarer Republik
Clark University, Worcester
Filmkultur zur Zeit der Weimarer Republik
Comité International d'Histoire de l'Art, Paris
Glossarium Artis
Comité International des Sciences Historiques, Paris
International Bibliography of Historical Sciences
Commission Française du Guide des Sources de l'Histoire des Nations, Paris
Guides to the Sources for the History ... / 3rd Series. Vol. 2
 - 3rd Series. Vol. 5
Commission of the European Commuities →
Kommission der Europäischen Gemeinschaften
Committee on Business Archives of the International Council Archives, Paris
Business Archives
Committee on Conservation, International Council on Archives, Paris
ICA Handbooks Series. 4
Conference of Directors of National Libraries, London
IFLA Publications 40/41
UBCIM Publication Series. International Transfers ...
Conference of Social Science Councils and Analogous Bodies (CNSSC), Washington D.C.
International directory of social science ...
Conseil International des Archives, Paris
Actes de la conférence internationale ...
Archivum
The Capitals of Europe
Guides to the Sources for the History ...
ICA Handbooks Series
International Archival Bibliography
International Journal of Archives
Politique de Préservation du Patrimoine Archivistique

Continuing Library Education Network and Exchange (CLENE), Washington D.C.
Directory of continuing education opportunities ...
Who's who in continuing education
Council of Europe, Strasbourg
Computer-based aids to parliamentary work
EUDISED. European Educational Research Yearbook
EUDISED R & D Bulletin

D

Danish National Archives, Kopenhagen
Guides to the Sources for the History ... 3rd Series. Vol. 3. Part 1
Department Computer & Letteren d. Rijksuniversiteit in Utrecht
Digitales Informationssystem für Kunst und Sozialgeschichte ...
Deutsch-Französisches Institut, Ludwigsburg
France - Allemagne
Deutsch-Französisches Jugendwerk, Ludwigsburg
France - Allemagne
Deutsch-Italienische Vereinigung e.V., Frankfurt a. M.
Sanna, Salvatore A.: Sardinien - Bibliographie
Deutsche Afrika-Gesellschaft, Bonn
Commercial Radio in Africa
Deutsche Akademie der Wissenschaften Berlin
Abhandlungen und Sitzungsberichte der Deutschen ...
Deutsche Bibliothek Frankfurt am Main
Berthold, Werner / Brita Eckert; Frank Wende: Deutsche Intellektuelle ...
Inventar zu den Nachlässen emigrierter ...
RAK-Anwendung in der Deutschen Bibliothek
 - 2. neubearb. Aufl.
Richard A. Bermann alias Arnold Höllriegel
Deutsche Bibliothekskonferenz, Berlin
Deutsches Bibliotheksadressbuch
Deutsche Blindenstudienanstalt, Marburg
Bestandskatalog Archiv- und internationale Dokumentationsstelle ...
Deutsche Bücherei Leipzig
Bibliographie der versteckten Bibliographien
Deutsche Nationalbibliographie
 - Ergänzung 2
Sachkatalog der Deutschen Bücherei Leipzig ...
Deutsche Forschungsgemeinschaft, Bonn
Internationale Bibliographie zur deutschsprachigen Presse ...
Studies on the organizational structure and ...
Deutsche Gesellschaft für Dokumentation e.V. (DGD), Frankfurt am Main
Deutscher Dokumentartag
DGD-Schriftenreihe
International Symposium Patent Information
Internationales Symposium über Patentinformation und ...
Nachrichten für Dokumentation (NfD)
Vickery, Brian Campell: Zur Theorie von Dokumentationssystemen
Deutsche Gesellschaft für Dokumentation, Berlin
Die Dokumentation und ihre Probleme
Deutsche Gesellschaft für Kartographie, München
Bibliographia Cartographica
Deutsche Gesellschaft für Photographie e.V., Berlin
Schriftenreihe der Deutschen Gesellschaft für Photographie e.V.
 - Bd. 3

Deutsche Gesellschaft für Wirtschaftsgeschichte, Köln
German Yearbook on Business History
Deutsche Presseforschung, Bremen
Studien zur Publizistik
Deutsche Staatsbibliothek Berlin
Winter, Ursula: Die europäischen Handschriften der ...
Studien zum Buch- und Bibliothekswesen
Theoretische Literatur zum Kinder- und Jugendbuch
Von der Wirkung des Buches
Deutsche UNESCO-Kommission, Bonn
Art Nouveau
Die deutschsprachige UNESCO-Literatur
Kulturaustausch zwischen Orient und Okzident
Kulturförderung und Kulturpflege in der Bundesrepublik Deutschland
Ohne Seil und Haken
Seminarberichte der Deutschen UNESCO-Kommission
UNESCO-Konferenzberichte
Der vergeudete Reichtum
Wasted Wealth
Wohnen in der Stadt. Neue Wohnformen
Deutscher Akademikerinnenbund e.V., Bonn
Die Frauenfrage in Deutschland
Die Frauenfrage in Deutschland. Neue Folge
Deutscher Bundesrat, Bonn
Protokolle des Vermittlungsausschusses des Deutschen ...
Thesaurus für Parlamentsmaterialien
Verhandlungen des Deutschen Bundestages ...
Deutscher Bundestag, Bonn
Protokolle des Vermittlungsausschusses des Deutschen ...
Thesaurus für Parlamentsmaterialien
Verhandlungen des Deutschen Bundestages ...
Deutscher Bundestag, Gruppe Datenverarbeitung, Bonn
Struktur des Thesaurus für Parlamentsmaterialien
PARTHES
Deutscher Musikverlegerverband e.V., Bonn
Bonner Katalog. Verzeichnis ...
Bonner Katalog. CD-ROM-Edition
Deutscher Werkbund Rheinland-Pfalz e.V., Mainz
... in Szene, gesetzt
Deutsches Archäologisches Institut, Rom
Index der Antiken Kunst und Architektur
Deutsches Bibliotheksinstitut, Berlin
Bibliotheksautomatisierung, Benutzererwartungen und Serviceleistungen
Der Einsatz von Kleincomputern in Bibliotheken ...
Internationale Bibliographie der Reprints. Bd. 2
Deutsches Bucharchiv München
Buchwissenschaftliche Beiträge
Deutsches Exilarchiv der Deutschen Bibliothek, Frankfurt a. M.
Berthold, Werner / Brita Eckert; Frank Wende: Deutsche Intellektuelle ...
Inventar zu den Nachlässen emigrierter ...
Deutsches Forschungsinstitut, Kyoto
Bibliographischer Alt-Japan-Katalog 1542-1853
Deutsches Historisches Institut, London
Akten der Britischen Militärregierung in ...
Deutsches Historisches Institut, Paris
Documentation et Recherches
Dokumentation Westeuropa
Deutsches Historisches Museum, Berlin
Digitales Informationssystem für Kunst und Sozialgeschichte ... Tl. 003: D. pol. Plakat d. DDR

- Tl. 008: Plakate ...
- Tl. 010: Politische Abzeichen ...
- Tl. 012: 1848 - Politik, Propaganda ...
- Tl. 014: Plakate der SBZ ...
Deutsches Institut für Filmkunde, Wiesbaden
Wandlungen im Film Junge deutsche Produktion
Deutsches Komitee der AIESEC, Köln
Management technologischer Innovationen
Deutsches Komitee für Reprographie, Frankfurt
Wörterbuch der Reprographie
Deutsches Lackinstitut GmbH (DLI), Frankfurt a. M.
Minerva-Fachserie Technik. Arbeitsschutz und ...
Deutsches Literaturarchiv Marbach am Neckar
Deutsche literarische Zeitschriften 1880-1945
Deutsche literarische Zeitschriften 1945-1970
Deutsches Museum, München
Das Bamberger Blockbuch
Bibliothek des Deutschen Museums
Deutsches Museum - Bildarchiv
Wissenschaftsgeschichte. Vogel, Kurt: Beiträge zur ...
Deutsches Musikarchiv, Berlin
Bonner Katalog. Verzeichnis ...
Bonner Katalog. CD-ROM-Edition
Deutsches Patentamt (DPA), München
Bibliothek des Deutschen Patentamtes München
International Symposium Patent Information
Internationales Symposium über Patentinformation und ...
Deutsches Rundfunkarchiv, Frankfurt
Bild- und Tonträger-Verzeichnisse
Rundfunkstudien Bd. 3
Deutsches Übersee-Institut Hamburg
Länderkatalog Afrika der Übersee-Dokumentation ...
Direction des Archives de France, Paris
Actes de la conférence internationale ... (10e) Copenhague 1967
Document Supply Centre of the British Library, Boston Spa
Index of Conference Proceedings Received ...
Dokumentations- und Kulturzentrum der deutschen Sinti und Roma, Frankfurt a. M
Memorial Book
Dokumentationsring Pädagogik (DOPAED), Berlin
Bibliographie Pädagogik
- Jg. 17 - 22
Mayer, Rudolf A. / Diethelm Prauss: Fachinformation Bildung
Sträter, Hans H.: Thesaurus Pädagogik
Thesaurus Pädagogik
Dr. Johannes-Lepsius-Archiv an der Martin-Luther-Universität, Halle-Wittenberg
Deutschland, Armenien und die Türkei 1895-1925

E

EREW Institut, Viersen
Berichte z. Erziehungstherapie u. Eingliederungshilfe
European Association for Health Information and Libraries (EAHIL), Brüssel
Health Information for All ...
European Association of Information Services (EUSIDIC), London
Information Trade Directory
European Centre for Higher Education (CEPES), Bukarest
Multilingual Lexicon of Higher Education

European Consortium for Political Research, University of Essex
Directory of European Political Scientists
Political Manifestos in the Post ...
European Foundation for Library Cooperation, Brüssel
Knowledge for Europe: Librarians and ...
European Foundation for Library Cooperation/Groupe de Lausanne
Library Automation and Networking ...
Evangelisches Zentralarchiv, Berlin
Kirchlicher Zentralkatalog beim Evangelischen Zentralarchiv ...

F

Fachbereich Politische Wissenschaft d. Freien Universität Berlin
Pressearchiv zur Geschichte Deutschlands sowie ...
Fachgruppe Presse-, Rundfunk- und Filmarchivare im Verein Deutscher Archivare, Bonn
Presse-, Rundfunk- und Filmarchive ...
Fédération Internationale d'Information et de Documentation (FID), Den Haag
Function and Organization of a National Documentation ...
Market Economy and Planned Economy
Fédération Internationale des Archives du Film (FIAF), Brüssel
Film, Television, Sound Archives Series
International Directory of Cinematographers ...
Federation Internationale des Associations des Bibliothecaires et d'Institutions, Den Haag
UBCIM Publication Series. Manuel Unimarc Version française
UBCIM Publications - New Series. 16
Forschungsgemeinschaft 20. Juli e.V., Berlin
Bibliographie 'Widerstand'
Forschungsgruppe Hochschulkapazität, Universität Mannheim
HIS-Brief 36
Forschungsinstitut „Brenner Archiv" (Innsbruck)"
Literatur und Archiv. 1
Forschungsinstitut der Friedrich-Ebert-Stiftung, Bonn
Schriftenreihe der Georg-von-Vollmar-Akademie. 2
Forschungsinstitut des Deutschen Museums für die Geschichte der Naturwissenschaften und Technik, München
Das Bamberger Blockbuch
Wissenschaftsgeschichte. Vogel, Kurt: Beiträge zur ...
Forschungsinstitut für Wirtschaftspolitik an der Universität Mainz, Mainz
Studien des Forschungsinstitutes für Wirtschaftspolitik ...
Forschungsstelle der Deutschen Gesellschaft für die Vereinten Nationen, Bonn
Handbuch Vereinte Nationen
Hüfner, Klaus / Jens Naumann: The United Nations System
Forschungsstelle für vergleichende Erziehungswissenschaft der Philipps-Universität Marburg
Texte - Dokumente - Berichte
Fototeca Unione, Rom
Ancient Roman Architecture
Franklin Institut, München
Gillespie, Paul D. / Paul Katzenberger; John Page: Problems of document delivery ...

Abkürzungen / Abbreviations

Sachbegriffe / Keywords

a.	und / and
Abb.	Abbildungen / illustrations
Abh.	Abhandlung / treatise
Abt.	Abteilung / department
amerik.	Amerikanisch / American
Anh.	Anhang / appendix
approx.	ungefähr / approximately
arab.	arabisch / Arabic
Arb.	Arbeit / work
Assist.	Unterstützung / assistance
Aufl.	Auflage / print run
Auftr.	Auftrag / order
Ausg.	Ausgabe / edition
Auslfg.	Auslieferung / distribution
AVM	Audiovisuelle Materialien / audio-visual material
Bd.	Band / volume
bearb.	bearbeitet / edited
Begleitbr.	Begleitbroschüre / accompanying leaflet
Begleith.	Begleitheft / accompanying booklet
Beitr.	Beitrag / contribution
Bibliogr.	Bibliographie / bibliography
Bild.	Bildung / education
Bildtl.	Bildteil / illustrated section
Br	Broschiert / paperback
ca.	ungefähr / approximately
CD	Compact Disc
CD-ROM	Compact Disc Read-Only-Memory
collab.	Zusammenarbeit / collaboration
COM	Computer Outprint on Microfiche
comp.	zusammengestellt / compiled
Coprod.	Koproduktion / co-production
corr.	corrigée: korrigiert / corrected
cplt.	komplett / complete
dt.	deutsch / German
Dok.	Dokumentation / documentation
Ebr	englische Broschur / English Brochure
Ed.	Ausgabe / Edition
ed.	herausgegeben / edited
Einl.	Einleitung / introduction
eingel.	eingeleitet / introduced
einz.	einzeln / separate
empir.	empirisch / empirical
engl.	englisch / English
enl.	erweitert / enlarged
erg.	ergänzt / supplemented
ersch.	erschienen / published

erw.	erweitert / expanded
et al.	et aliter
Ethnol.	Ethnologie / ethnology
exp.	erweitert / expanded
Facs.	Faksimile / Facsimile
farb.	farbig / colour
Flg.	Folge / issue
Forsch.	Forschung / research
Frwd.	Vorwort / Foreword
frz.	französisch / French
Gb	Gebunden / bound (hardcover)
Ges.	Gesellschaft / society
Gesamtw.	Gesamtwerk / complete work
Gh	Klammerheftung / saddle stich
Graf.	Grafiken / illustrations
Handb.	Handbuch / guide
Hard	Gebunden / hardcover
Hrsg.	Herausgeber / editor
i.Kass.	in Kassette / boxed
Illu.	Illustration / illustration
in Zus.-Arb. m.	in Zusammenarbeit mit / in collaboration with
Inform.	Information
Intern.	International
Introd.	Einleitung / Introduction
In Verb.	In Verbindung / in collaboration
ISBN	International Standard Book Number
i.Sch.	im Schuber / boxed
ISSN	International Standard Serial Number
ital.	italienisch / Italian
IuD	Information und Dokumentation / Information and Documentation
i.Vb.	In Vorbereitung / in preparation
Jg.	Jahrgang / year
jährl.	jährlich / annually
Kte.	Karte / map
Konf.	Konferenz / conference
Koprod.	Koproduktion / co-production
kplt.	komplett / complete
Kst.	Kunststoff / plastic
Kt.	Kartoniert / softbound
Ktn.	Karten / maps
Lfg.	Lieferung / instalment
Lin	Linson: Gebunden / hardcover
Ln	Leinen / cloth
m.	mit / with
Mill.	Millionen / million
Mitarb.	Mitarbeit / collaboration

Mitgl.	Mitglied / member
Mitw.	Mitwirkung / collaboration
Nachdr.	Nachdruck / reprint
Nachtr.	Nachtrag / addition
Nachw.	Nachwort / appendix
Nr.	Nummer / number
Orig.	Original / original
Org.	Organisation / organization
p.	Seite / page
Päd.	Pädagogik / teaching
Pb	Paperback
privil.	privilegiert / privileged
Prof.	Professor
publ.	veröffentlicht / published
RAK	Regeln der alphabetischen Katalogisierung / alphabetical cataloguing rules
Rechtswiss.	Rechtswissenschaft / legal science
Red.	Redaktion / editor
Reg.	Register / index
rev.	überarbeitet / revised
Rundf.	Rundfunk / radio
S.	Seite / page
Schriftenr.	Schriftenreihe / series
Sem.-Ber.	Seminar-Bericht / seminar report
Soft	Broschur / softcover
Sozialökon.	Sozialökonomie / social economy
Sozialwiss.	Sozialwissenschaft / social science
span.	spanisch / Spanish
Staatswiss.	Staatswissenschaft / international relations
Stud.	Studien / studies
Suppl.	Ergänzungsband / Supplement
SWI	Schlagwortindex / keyword index
Tab.	Tabelle / table
Taf.	Tafeln / charts
techn.	technisch / technical
Tl.	Teil / part
Tlbd	Teilband / part volume
transl.	übersetzt / translated
u.	und / and
u.d.T.	unter dem Titel / former title
überarb.	überarbeitet / revised
übers.	übersetzt / translated
umgearb.	umgearbeitet / rearranged
unveränd.	unverändert / unchanged
Univ.	Universität / university

Verb.	Verbindung / connection
verb.	verbessert / improved
vergl.	vergleichen / compare
verm.	vermehrt / increased
Vlg.	Verlag / publisher
Vlg.-Nr.	Verlagsnummer / publisher's number
vol.	Band / volume
vollst.	vollständig / complete
Vorw.	Vorwort / foreword
VR	Volksrepublik / People's Republic
w.	mit / with
Wirtsch.	Wirtschaft / economy
zahlr.	zahlreich / numerous
zus.	zusammen / all together
Zus.-Arb.	Zusammenarbeit / collaboration
zsgst.	zusammengestellt / compiled
ZK	Zentralkomitee / central committee

Institutionen / Institutions

AGB	Amerika-Gedenkbibliothek, Berlin
AiDOS	Archiv- und internationale Dokumentationsstelle für das Blinden- und Sehbehindertenwesen, Marburg/Lahn
AIESEC	Association Internationale des Etudiants en Sciences Economiques et Commerciales / Internationale Vereinigung der Studenten der Wirtschaftswissenschaften
ASpB	Arbeitsgemeinschaft der Spezialbibliotheken e.V., Jülich
AWV	Ausschuß für Wirtschaftliche Verwaltung in Wirtschaft und Öffentlicher Hand e.V.
BLM	Bayerische Landeszentrale für neue Medien, München
BSB	Bayerische Staatsbibliothek, München
CEPES	Centre Européen pour l'Enseignement Supérieur
CERI	Centre for Educational Research and Innovation, Genf
CIDOC	Comité International de l'ICOM pour la Documentation

CIPOL	Centre International des publications oecuméniques des liturgies
CLENE	Continuing Library Education Network and Exchange, Washington D.C.
CNRS	Centre National de la Recherche Scientifique, Paris
CNSSC	Conference of Social Science Councils and Analogous Bodies, Washington D.C.
DAF	Deutsche Arbeitsfront, Berlin
DFG	Deutsche Forschungsgemeinschaft, Bonn
DGD	Deutsche Gesellschaft für Dokumentation e.V.
DLI	Deutsches Lackinstitut GmbH, Frankfurt a. M.
DOPAED	Dokumentationsring Pädagogik, Berlin
DPA	Deutsches Patentamt, München
EAHIL	European Association for Health Information and Libraries, Brüssel
EUSIDIC	European Association of Information Services, London
FIAF	Fédération Internationale des Archives du Film / International Federation of Film Archives
FID	Fédération Internationale d'Information et de Documentation, Den Haag
FOPRODA	Forschungsprojektdatei (Arbeitsgruppe FOPRODA der Hochschul-Informations-System GmbH, Hannover)
GFD	Gesellschaft für sozial- und erziehungswissenschaftliche Forschung und Dokumentation
GID	Gesellschaft für Information und Dokumentation, Frankfurt/M.
GMD	Gesellschaft für Mathematik und Datenverarbeitung, St Augustin
GTF	Gesellschaft für Technologiefolgenforschung e. V., Berlin
HIS	Hochschul-Informations-System GmbH, Hannover
IAALD	International Association of Agricultur Librarians and Dokumentalists, Brüssel
IALL	International Association of Law Libraries, Marburg
ICA	International Council on Archives, Paris
ICOM	International Council of Museums, Budapest

IERD	International Exposition of Rural Development
IFG	Institut Frau und Gesellschaft, Bonn
IFLA	International Federation of Library Associations, Den Haag
infoterm	International Information Centre for Terminology, Wien
IRA	International Reading Association, Newark
IRSCL	International Research Society on Children's Literature
ISI	Institut für Systemtechnik und Innovationsforschung der Fraunhofer Gesellschaft, Karlsruhe
IZ	Informationszentrum Sozialwissenschaften, Bonn
IZC	Internationales Institut für Zeichentrick- und Comicforschung, Köln
IZI	Internationales Zentralinstitut für das Jugend und Bildungsfernsehen, München
OKR	Stiftung Ostdeutscher Kulturrat, Bonn
RLG	Research Libraries Group Ltd., Mountain View, Ca.
RISM	Répertoire International des Sources Musicales, Frankfurt a. M.
SCOLMA	Standing Conference on Library Materials on Africa, London
SVD	Schweizerische Vereinigung für Dokumentation
UIA	Union of International Associations, Brüssel
UNESCO	United Nations Educational, Scientific and Cultural Organization, Paris
VBB	Verein der Bibliothekare an Öffentlichen Bibliotheken e.V., Reutlingen
VDB	Verein Deutscher Bibliothekare, Hannover
VDI	Verein Deutscher Ingenieure, Düsseldorf
WFSF	World Futures Studies Federation, Bacalod City
WIPO	World Intellectual Property Organization, Genf
ZI 6	Zentralinstitut für sozialwissenschaftliche Forschung der freien Universität Berlin
ZMD	Zentralstelle für Maschinelle Dokumentation, Frankfurt a. M.

50 years
K.G. Saur Publishing

Karl-Josef Kutsch and Leo Riemens
Großes Sängerlexikon
(Biographical Dictionary of Singers)
New, enlarged complete edition
Jubilee edition May 1999. 5 vols. 4,024 pp. PB.
DM 498.00
ISBN 3-598-11419-2

Bibliothek der Deutschen Literatur
(Library of German literature)
Microfiche edition based on the *Taschengoedeke*
Ed. by Kulturstiftung der Länder
1990-1998. C. 10 Mio pages on 20,681 Diazo fiches.
Incl. Bibliography and Index. Reader factor 42x.
DM 9,800.00
ISBN 3-598-50000-9

Albin Lesky
Geschichte der griechischen Literatur
(History of Greek Literature)
3rd rev. edition
Jubilee edition May 1999. 1,023 pp. HB.
DM 98.00
ISBN 3-598-11423-0

Georg Witkowski
Geschichte des literarischen Lebens in Leipzig
(History of the Literary Life in Leipzig)
Reprint
1994. 520 pp. HB.
DM 48.00
ISBN 3-598-11195-9

August Bebel
Ausgewählte Reden und Schriften
(August Bebel. Selected Speeches and Papers)
Ed. with the support of the International Institute for Social
History, Amsterdam.
Vols. 1,2,6 Reprint of the original edition by Dietz Verlag,
Berlin, Vols. 3-5, 7ff comp. by A. Beske and E. Müller
1995-1997. 10 vols. in 14 part vols. HB.
DM 880.00
ISBN 3-598-11372-2

Digital Information System for Art and Social History
(DISKUS)
CD-ROM edition
Ed. by Bildarchiv Foto Marburg – Deutsches Dokumentations-
zentrum für Kunstgeschichte der Philipps-Universität Marburg
and the Department Computer & Letteren
Jubilee edition. 5 Discs. Set **DM 298.00.**
ISBN 3-598-40320-8
Printed Portraits 1500-1618 from the Prints
Collection of the Germanisches Nationalmuseum
1995. Ed. by the Germanischen Nationalmuseum, Nuremberg
Italian Drawings of the 14th to 18th century
in the Berlin Kupferstichkabinett
1995. Ed. by the Kupferstichkabinett, Berlin
Political Allegories and Satires from the Prints
Collection of the Germanisches Nationalmuseum
1996. Ed. by Germanisches Nationalmuseum, Nuremberg
1848 – Politics, Propaganda, Information and
Entertainment from the Printing Press
1998. Ed. by the Deutsches Historisches Museum, Berlin
Picture Gallery Berlin
1998. Ed. by Gemäldegalerie, Staatliche Museen zu Berlin –
Preußischer Kulturbesitz

Gesamtverzeichnis des deutschsprachigen
Schrifttums 1700-1910 (GV)
(Bibliography of German-Language Publications 1700-1910)
Microfiche edition
1986. 160 vols. and supplementary volume on
795 Diazo fiches. Reader factor 24x.
DM 980.00
ISBN 3-598-30590-7

Gesamtverzeichnis des deutschsprachigen
Schrifttums 1911-1965 (GV)
(Bibliography of German-Language Publications 1911-1965)
Microfiche edition
1984. 150 vols. on 400 Diazo fiches. Reader factor 24x.
DM 980.00
ISBN 3-598-30455-2

Special jubilee prices are valid till 31.12.1999

K·G·Saur Verlag
Postfach 70 16 20 · 81316 München · Germany · Tel. +49 (0)89 7 69 02-232
Fax +49 (0)89 7 69 02-250 · e-mail: CustomerService_Saur@csi.com · http://www.saur.de